T0231567

Improving Food Quality with Novel Food Processing Technologies

Improving Food Quality with Novel Food Processing Technologies

EDITED BY

Özlem Tokuşoğlu • Barry G. Swanson

CRC Press
Taylor & Francis Group
Boca Raton London New York

CRC Press is an imprint of the
Taylor & Francis Group, an **informa** business

CRC Press
Taylor & Francis Group
6000 Broken Sound Parkway NW, Suite 300
Boca Raton, FL 33487-2742

First issued in paperback 2017

© 2015 by Taylor & Francis Group, LLC
CRC Press is an imprint of Taylor & Francis Group, an Informa business

No claim to original U.S. Government works

ISBN-13: 978-1-4665-0724-1 (hbk)
ISBN-13: 978-1-138-19988-0 (pbk)

This book contains information obtained from authentic and highly regarded sources. Reasonable efforts have been made to publish reliable data and information, but the author and publisher cannot assume responsibility for the validity of all materials or the consequences of their use. The authors and publishers have attempted to trace the copyright holders of all material reproduced in this publication and apologize to copyright holders if permission to publish in this form has not been obtained. If any copyright material has not been acknowledged please write and let us know so we may rectify in any future reprint.

Except as permitted under U.S. Copyright Law, no part of this book may be reprinted, reproduced, transmitted, or utilized in any form by any electronic, mechanical, or other means, now known or hereafter invented, including photocopying, microfilming, and recording, or in any information storage or retrieval system, without written permission from the publishers.

For permission to photocopy or use material electronically from this work, please access www.copyright.com (http://www.copyright.com/) or contact the Copyright Clearance Center, Inc. (CCC), 222 Rosewood Drive, Danvers, MA 01923, 978-750-8400. CCC is a not-for-profit organization that provides licenses and registration for a variety of users. For organizations that have been granted a photocopy license by the CCC, a separate system of payment has been arranged.

Trademark Notice: Product or corporate names may be trademarks or registered trademarks, and are used only for identification and explanation without intent to infringe.

Library of Congress Cataloging-in-Publication Data

Improving food quality with novel food processing technologies / edited by Özlem Tokuşoğlu, Barry G. Swanson.
    pages cm
    Summary: "Improving food quality, specifically properties such as rheological, physicochemical, and sensorial aspects, is always a goal of food and beverage manufacturers. During the past decade, novel processing technologies including high hydrostatic pressure (HHP), ultrasound, pulse electric field (PEF), and advanced heating technologies containing microwave, ohmic heating, and radio frequency have frequently been applied in the processing of foods and beverages. This book addresses maintaining and improving food quality through the use of these novel food processing technologies"-- Provided by publisher.
    Includes bibliographical references and index.
    ISBN 978-1-4665-0724-1 (hardback)
    1. Food industry and trade--Sanitation. 2. Processed foods. 3. Food--Preservation. I. Tokuşoğlu, Özlem, editor. II. Swanson, Barry Grant, editor.

TP373.6.I524 2014
338.4'7664--dc23
                           2014022855

Visit the Taylor & Francis Web site at
http://www.taylorandfrancis.com

and the CRC Press Web site at
http://www.crcpress.com

To my mother, Özden Tokuşoğlu, a retired teacher and to my father, Armağan Tokuşoğlu, a retired senior colonel, for their great emotional support and cordial encouragements.

Özlem Tokuşoğlu

Contents

PART III *Improving Food Quality with Pulsed Electric Field Technologies*

Editors

Özlem Tokuşoğlu, PhD, is an associate professor and faculty member of the Department of Food Engineering at Celal Bayar University, Manisa, Turkey. She earned a bachelor's degree (1992) and master's degree (1996) from Ege University, Izmir, Turkey in the Department of Chemistry and a PhD from Ege University in the Department of Food Engineering (2001). She worked as a researcher at Ege University from 1993 to 2001, and was a research associate in the Food Science and Nutrition Department at the University of Florida–Gainesville from 1999–2000. Dr. Tokuşoğlu was an assistant professor at Celal Bayar University, Manisa, Turkey, during 2002–2007.

Dr. Tokuşoğlu was a visiting professor in the Food Science and Nutrition Department at the University of Florida, Gainesville during 1999–2000 and at the School of Food Science, Washington State University, Pullman during April–May 2010. Her study focuses on food quality control, food chemistry, food safety and toxicology, shelf-life of foods, and innovative food processing technologies in foods, beverages, and functional products. Her specific study areas are phenolics, phytochemicals, bioactive antioxidative components, food additives, bioactive lipids, their determination by instrumental techniques, their effects on the quality of food and beverages, oil-fats and functional food technologies, and the novel food processing effects on food bioactives, food toxicants, and shelf-life of foods and beverages.

Tokuşoğlu has conducted academic research studies, delivered keynote addresses, and made academic presentations at Geneva, Switzerland in 1997; Gainesville, Florida, in 1999; Anaheim–Los Angeles, California, in 2002; Sarawak, Malaysia in 2002; Chicago, Illinois, in 2003; Szczyrk, Katowice, Poland in 2005; Ghent, Belgium in 2005; Madrid, Spain in 2006; New Orleans, Louisiana, in 2008; Athens, Greece in 2008; Anaheim–Los Angeles, California, in 2009; Skopje/Üsküp, the Republic of Macedonia in 2009, Chicago, Illinois, in 2010; Munich, Germany in 2010; Jamshoro, Sindh-Hyderabad, Pakistan in 2011; New Orleans, Louisana, in 2011; Boston, Massachusetts, in 2011; Natick, Massachusetts, in 2011; Damghan, Tehran, Iran in 2011; Osnabrück, Germany in 2011, Otsu, Kyoto, Japan in 2012; Chicago, Illinois, in 2013; Philadelphia, Pennsylvania, in 2013; Las Vegas, Nevada, in 2014; and San Francisco-Albany, California, in 2014. She has professional affiliations with the Institute of Food Technologists (IFT) and American Oil Chemists' Society (AOCS) in the United States and has a professional responsibility with the Turkey National Olive and Olive Oil Council (UZZK) as a research and consultative board member.

As conference chair, Professor Tokuşoğlu organized and directed the International Congress titled *ANPFT2012* (Advanced Nonthermal Processing in Food Technology: Effects on Quality and Shelf-Life of Food and Beverages) in May 7–10, 2012 at Kusadasi-Aegean, Turkey. She served as an organizing committee member at the 2nd International Conference and Exhibition on Nutritional Science & Therapy Conference in July 2013 in Philadelphia, USA and cochair at the *Food*

Technology 2014 conference (3rd International Conference and Exhibition on Food Processing and Technology) in July 2014 in Las Vegas, USA. She is currently chair of the Food Technology 2015 conference in August 10–12 in London, UK. Dr. Tokuşoğlu is an editorial board member of the *International Journal of Food Science and Technology (IJFST)* and the *Polish Journal of Food and Nutrition Sciences (PJFNS)*. Dr. Tokuşoğlu has several editorial and reviewer assignments in the Science Citation Index (SCI) and international index covered journals. She has published a scientific edited book titled *Fruit and Cereal Bioactives: Chemistry, Sources and Applications* by CRC Press, Taylor & Francis Group, and another book titled *Food By-Product Based Functional Food Powders* by CRC Press, Taylor & Francis Group, is in progress. She has published many research papers in peer-reviewed international journals, international book chapters, and international presentations (as oral and posters) presented at international congresses and other organizations. She was the principal administrator and advisor for the theses of four master's students, and currently one doctorate student and one master's student are under her supervision.

Barry G. Swanson, PhD, is Emeritus Regents Professor of the School of Food Science at Washington State University and the University of Idaho. Dr. Swanson's research interests range from studies of legume protein digestibility and storage quality in collaboration with the Institute for Nutrition in Central America and Panama (INCAP) supported by the USAID Collaborative Research Support Program (CRSP), to initial studies with sucrose fatty acid polyesters, syntheses of fat substitutes, alternative fat replacers and methods to improve the quality of reduced fat cheeses. More recent research interests focused on the implementation of ultra-high pressure to improve cheese yield and the hydrophobic functional properties of whey proteins. Dr. Swanson has coauthored more than 200 research manuscripts and 35 book chapters. He takes pride in having mentored 47 MS and 24 PhD students who now are successfully pursuing professional careers across the United States and around the world. Dr. Swanson received a College of Agricultural, Human, and Natural Resource Sciences (CAHNRS) Faculty Excellence in Research Award in 2001 and was invited to Michigan State University as a prestigious G. Malcolm Trout Visiting Scholar in 2004. In July 2005, he was recognized as one of ISI Thomson Citation Index's Most Highly Cited Researchers and is ranked 22nd among international authors in agricultural sciences, 1996–2006, by *Science Watch* 17(4), Thomson Scientific.

Dr. Swanson was elected a Fellow of IFT (Institute of Food Technologists) in 2002, and a Fellow of IUFoST (International Union of Food Science and Technology) in 2006. He is a retired editor of the *Journal of Food Processing and Preservation*. Dr. Swanson served for 6 years as executive secretary to the Washington State University (WSU) Faculty Senate, and served as interim director of the merged WSU and University of Idaho (UI) School of Food Science. He was promoted to the prestigious rank of Regents Professor at WSU and elected to the IFT Board of Directors in 2009. The professor retired in May 2011 and is currently serving on the IFT Education Advisory Panel and 2013 AMFE Food Chemistry Program Sub-Panel.

Contributors

Malek Amiali
Department of Bioresource
  Engineering
McGill University
Montreal, Quebec, Canada

Gustavo V. Barbosa-Cánovas
Biological Systems Engineering
Washington State University
Pullman, Washington

Faruk T. Bozoğlu
Department of Food Engineering
Middle East Technical University
  (METU)
Ankara, Turkey

Sencer Buzrul
TAPDK
Ankara, Turkey

Sónia Marília Castro
Department of Chemistry
University of Averio
Aveiro, Portugal

Om Prakash Chauhan
Defence Food Research Laboratory
Mysore, India

Ana Isabel Loureiro Correia
Department of Chemistry
University of Averio
Aveiro, Portugal

Ivonne Delgadillo
Department of Chemistry
University of Aveiro
Aveiro, Portugal

Michael A. Kempkes
Diversified Technologies, Inc.
Bedford, Massachusetts

Olga Martín-Belloso
Department of Food Technology
UTPV-CeRTA
University of Lleida
Lleida, Spain

Michael O. Ngadi
Department of Bioresource
  Engineering
McGill University
Montreal, Quebec, Canada

Alexandre Nunes
Department of Chemistry
University of Aveiro
Aveiro, Portugal

Isabella Odriozola-Serrano
Department of Food Technology
UTPV-CeRTA
University of Lleida
Lleida, Spain

Jorge Alexandre Saraiva
Department of Chemistry
University of Averio
Aveiro, Portugal

Kambiz Shamsi
School of Biosciences
Taylor's University
Selangor, Malaysia

Toru Shigematsu
Faculty of Applied Life Sciences
Niigata University of Pharmacy and
    Applied Life Sciences (NUPALS)
Niigata, Japan

Barry G. Swanson
School of Food Science
Washington State University
Pullman, Washington

Stefan Toepfl
Deutsches Institut für
    Lebensmitteltechnik DIL e.V.
Quakenbrueck, Germany

Özlem Tokuşoğlu
Department of Food Engineering
Celal Bayar University
Manisa, Turkey

Shigeaki Ueno
Faculty of Education
Saitama University
Saitama, Japan

L.E. Unni
Defence Food Research Laboratory
Mysore, India

Halil Vural
Department of Food Engineering
Hacettepe University
Ankara, Turkey

L. Juan Yu
Department of Bioresource
    Engineering
McGill University
Montreal, Quebec, Canada

Howard Q. Zhang
Western Regional Research Center
    (WRRC)
Albany, California

Part I

Introduction

1 Introduction to Improving Food Quality by Novel Food Processing

Özlem Tokuşoğlu and Barry G. Swanson

CONTENTS

1.1 INTRODUCTION

Consumers around the world are better educated and more demanding in their identification and purchase of quality health-promoting foods. The food industry and regulatory agencies are searching for innovative technologies to provide safe and stable foods for their clientele. Thermal pasteurization and commercial sterilization of foods provide safe and nutritious foods that, unfortunately, are often heated beyond a safety factor that results in unacceptable quality and nutrient retention. Nonthermal processing technologies offer unprecedented opportunities and challenges for the food industry to market safe, high-quality health-promoting foods. The development of nonthermal processing technologies for food processing is providing an excellent balance between safety and minimal processing, between acceptable economic constraints and superior quality, and between unique approaches and traditional processing resources (Zhang et al., 2011). Nonthermal food processing is often perceived as an alternative to thermal food processing; yet, there are many nonthermal preparatory unit operations as well as food processing and preservation opportunities and challenges that require further investigation by the food industry. Nonthermal technologies are useful not only for inactivation of microorganisms and enzymes, but also to improve yield and development of ingredients and marketable foods with novel quality and nutritional characteristics (Bermudez-Aguirre and Barbosa-Canovas, 2011).

Nonthermal processing is effectively combined with thermal processing to provide improved food safety and quality. Nonthermal processing facilitates the development of innovative food products not previously envisioned. Niche markets for food products and processes will receive greater attention in future years.

3

Nonthermal technologies successfully decontaminate, pasteurize, and potentially pursue commercial sterilization of selected foods while retaining fresh-like quality and excellent nutrient retention. The quest for technologies to meet consumer expectations with optimum quality-safe processed foods is the most important priority for future food science research. Zhang et al. (2011) listed the relevant factors to consider when conducting research into novel nonthermal and thermal technologies such as: (1) target microorganisms to provide safety; (2) target enzymes to extend quality shelf life; (3) maximization of potential synergistic effects; (4) alteration of quality attributes; (5) engineering aspects; (6) conservation of energy and water; (7) potential for convenient scale-up of pilot-scale processes; (8) reliability and economics of technologies; and (9) consumer perception of the technologies. "The search for new approaches to processing foods should be driven, above all, to maximize safety, quality, convenience, costs, and consumer wellness" (Zhang et al., 2011).

Morris et al. (2007) conclude that nonthermal unit operations in food processing interest food scientists, manufacturers, and consumers because the technologies expose fresh foods to minimal impact on nutritional and sensory qualities, yet presumably provide safe shelf-stable foods by inactivating pathogenic microorganisms and spoilage enzymes. The presumption that nonthermal processing is energy efficient and environmentally friendly adds to contemporary popularity. Additional benefits to the food industry include the provision of food safety, value-added heat-labile foods, and new market opportunities.

Nonthermal food-processing technologies are extensive with high hydrostatic pressure (HHP), pulsed electric fields (PEFs), ultrasonics, ultraviolet light, ionizing irradiation (electron beams), and hurdle technologies leading the way. In addition, pulsed x-rays, pulsed high-intensity light, high-voltage arc discharge, magnetic fields, dense-phase carbon dioxide, plasma, ozone, chlorine dioxide, and electrolyzed water are receiving attention individually and as a hurdle in minimal processing protocols (Morris et al., 2007; Sun, 2005; Tokuşoğlu, 2012).

The authors in this book devote attention to improving food functionality with HHP and PEFs. The focus on improving the quality and retaining bioactive constituents of fruits and vegetables and improving the quality of dairy, egg, meat, and seafood products with HHP is evident in many chapters. The inclusion of modeling reviews and simulations of HHP inactivation of microorganisms and the relative effects of HHP processing on food allergies and intolerances broaden the scope of the information provided. Improving food functionality with PEF processes is focused on dairy and egg products, fruit juices, and wine. A chapter attending to industrial applications of HHP and PEF systems and potential commercial quality and shelf life of food products concludes this discussion.

HHP, ultra-high pressure (UHP), and ultra-high-pressure processing (HPP) are different names and acronyms for equivalent nonthermal processes employing pressures in the range of 200–1000 MPa with only small increases in processing temperature. The UHPs inactivate microbial cells by disrupting membrane systems, retaining the biological activity of quality, sensory, and nutrient cell constituents, thus extending the shelf lives of foods. High pressures inactivate enzymes by altering the secondary and tertiary structures of proteins, changing functional integrity, biological activity, and susceptibility to proteolysis. HHP processing of dairy proteins reduces the size of casein micelles, denatures whey proteins, increases calcium solubility, and induces color changes (Morris et al., 2007). The use of HHP to increase the yield of cheese curd from milk and accelerate the proteolytic ripening of Cheddar cheeses are promising improvements to the economics for the dairy food industry. The most widely available commercial applications of HHP include pasteurization of guacamole, tomato salsas, oysters, deli-sliced meats, and yogurts. The provision of HHP processing to provide a preservation method for thermally labile tropical fruits is very promising. It is stated that HHP provides pathogen inactivation, shelf-life extension, unwanted enzyme inactivation, gives innovative fresh products, reduced sodium products and clean-labelling (Figure 1.1a).

PEF processing exposes fluid foods to microsecond bursts of high-intensity electric fields, 10–100 kV/cm, inactivating selected microorganisms by electroporation, a disruption of cell membranes. PEF processing reliably results in five-log reduction in selected pathogenic microorganisms, resulting in minimal detrimental alterations in

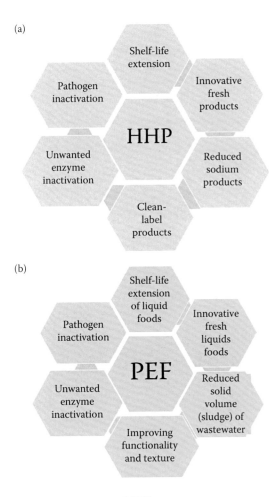

FIGURE 1.1 The usage area of HHP and PEF.

physical and sensory properties of the fluid foods. PEF adequately pasteurizes acid (pH < 4.5) fruit juices and research is continuing on uniform adequate pasteurization of milk and liquid eggs. The commercial application of PEF to improve the extraction yield of fruit juices and bioactive components of plant materials is in progress. PEF inactivation of enzymes is inconsistent and nonuniform, resulting in plant products subject to short shelf lives at ambient temperatures. It is expressed that PEF provides pathogen inactivation, shelf-life extension of liquid foods, unwanted enzyme inactivation, improves functionality and texture of foods, gives innovative fresh liquid foods and reduced solid volume (sludge) of wastewater (Figure 1.1b). Although PEF is identified as a nonthermal process, temperature increases during PEF processing result in fluid foods at 35–50°C, requiring cooling prior to packaging. The presence of particulates or bubbles in fluid foods subjected to PEF will result in dielectric breakdown, arcing, and scorching of the food. Homogenization and vacuum degassing are necessary to minimize the hazards associated with PEF processing of fluid foods. Technical issues that must be addressed to commercialize PEF for approval as an adequate food pasteurization technology include: (1) consistent and uniform generation of high-intensity electric fields; (2) identification of critical electric field intensities for uniform microbial inactivation; (3) identification of homogenization and vacuum-degassing techniques to assure the absence of particulates and air cells that promote arcing; and (4) identification of flow rates, temperature control, cooling, and aseptic packaging parameters to obtain processing uniformity and safe handling practices (Morris et al., 2007).

HHP and PEFs processing of foods continues with a focus on heat-labile acid fruits, vegetables, and dairy foods that meet consumer expectations for a minimally process, safety, fresh-like quality, and convenience. Nonthermal preservation extends shelf life without the addition of preservatives while retaining expected fresh-like appearance, sensory, and nutrient quality. It will be necessary to combine nonthermal and thermal preservation technologies to inactivate heat-resistant spores, potentially contaminating low-acid foods. Commercial nonthermal processing success stories such as pasteurized guacamole, oysters, salsa, yogurt, refrigerated meats, and improved yields of fruit juices, and bioactive compounds from herbs and other plant materials will demonstrate the efficacy and economic success of the technologies in niche markets. Successful research and identification of economic benefits, including energy and water conservation as well as demonstrated safety and fresh-like quality attributes will improve consumer perception of nonthermal technologies and result in further development by the food industry around the world.

REFERENCES

Bermudez-Aguirre D., Barbosa-Canovas G.V. 2011. *Introduction to Nonthermal Processing Technologies for Food*, Zhang H.Q., Barbosa-Canovas G.V., Balasubramaniam V.M., Dunne C.P., Farkas D.F., Yuan J.T.C., eds. John Wiley & Sons, Inc., Ames, IA.

Morris C., Brody A.L., Wicker L. 2007. Nonthermal food processing/preservation technologies: A review with packaging implications. *Packag. Technol. Sci.* 20:275–286.

Sun D.W. 2005. *Emerging Technologies for Food Processing. Food Science and Technology, International Series*. Elsevier Academic Press, London, UK. ISBN 0-12-676757-2.

Tokuşoğlu Ö. 2012. *ANPFT2012 Proceeding Book. International Congress. Advanced Nonthermal Processing in Food Technology: Effects of Quality and Shelf Life of Food and Beverages*, Tokuşoğlu O., ed. May 07–10, 2012, Kuşadasi-Turkey. 321pp. Celal Bayar University Publishing, Turkey. ISBN: 978-975-8628-33-9.

Zhang H.Q., Barbosa-Canovas G.V., Balasubramaniam V.M., Dunne C.P., Farkas D.F., Yuan J.T.C. 2011. Preface to *Nonthermal Processing Technologies for Food*, Zhang H.Q., Barbosa-Canovas G.V., Balasubramaniam V.M., Dunne C.P., Farkas D.F., Yuan J.T.C., eds. John Wiley &Sons, Inc., Ames, IA.

Part II

Improving Food Quality with High-Pressure Processing

2 High-Pressure Processing of Bioactive Components of Foods

Özlem Tokuşoğlu

CONTENTS

2.1 INTRODUCTION

Phenolic compounds are naturally derived bioactive substances that have health-promoting, and/or nutraceutical and medicinal properties. Phenolics occur as plant secondary metabolites that are widely distributed in the plant kingdom and represent an abundant antioxidant component of the human diet.

Recently, there is a great demand for high-quality and convenient products with natural flavor and taste, and a greater appreciation for the fresh appearance of minimally processed food. High-pressure processing (HPP) is a nonthermal processing method that holds promise for retaining wholesomeness and freshness of the processed food products. HPP is an emerging technology that can be used instead of thermal process for pasteurization and sterilization. Recent work provides studies to illustrate the ability of this nonthermal food preservation technology, regarding the preservation of the phenolic bioactives of plant foods and health-related compounds.

2.2 PHENOLICS AS BIOACTIVE COMPOUNDS

Phenolic compounds occur as plant secondary metabolites. Their ubiquitous presence in plants and plant foods, favors animal consumption and accumulation in tissues. Polyphenols are widely distributed in the plant kingdom and represent an abundant antioxidant component of the human diet. Interest in the possible health benefits of polyphenols has increased due to the corresponding antioxidant capacities.

Recent evidence shows that there is a great interest to anticarcinogenic effects of polyphenolic compounds, as well as the potential to prevent cardiovascular and cerebrovascular diseases. As the name suggests, phytochemicals working together with chemical nutrients found in fruits, cereals, and nuts may help slow the aging process and reduce the risk of many diseases, including cancer, heart disease, stroke, high blood pressure, cataracts, osteoporosis, and urinary tract infections (Cheynier, 2005; Meskin et al., 2003; Tokuşoğlu and Hall, 2011).

Polyphenols divide into several subgroups including flavonoids, hydroxybenzoic and hydroxycinnamic acids, lignans, stilbenes, tannins, and coumarins that have specific physiological and biological effects (Andersen and Markham, 2006; Meskin et al., 2003; Tokuşoğlu, 2001; Figure 2.1).

Flavonoids are a major group of polyphenols (Figure 2.2) that include flavan-3-ols, flavonols, flavones, flavanones, isoflavones, anthocyanidins, anthocyanins, flavononols, and chalcons as subgroups (Figure 2.3), which are distributed in plants and foods of plant origin (Crozier et al., 2006; Tokuşoğlu and Hall, 2011).

Phenolic compounds including flavonoids play some important roles in fruits such as visual appearance, taste, and aroma. In addition to these, phenolic compounds have health-promoting benefits (Thomas-Barberan and Espin, 2001). These bioactive compounds have been found to be important in the quality of plant-derived foods (Thomas-Barberan and Espin, 2001). Anthocyanins are a type of phenolic compounds classified under flavonoids group of phenolic compounds, which are water-soluble glycosides of anthocyanidins (Kong et al., 2003; Tokuşoğlu and Yldrm, 2012).

Phenolic contents of the fruits obviously vary from fruit to fruit. This difference may depend on the methods used both for the extraction of the phenolic compounds and methods for analysis. Also, the phenolic compound composition in fruits is

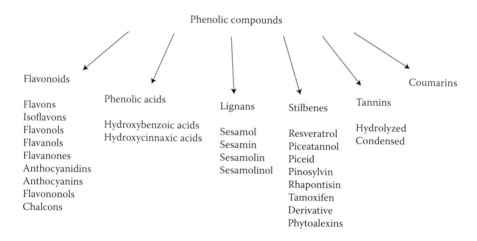

FIGURE 2.1 Family of phenolic compounds. (Adapted from Andersen, Q. M., and Markham, K. R. 2006. *Flavonoids. Chemistry, Biochemistry, and Applications,* CRC Press, Taylor & Francis, Boca Raton, FL; Tokuşoğlu, O., and Hall, C. 2011. *Fruit and Cereal Bioactives, Chemistry, Sources and Applications,* CRC Press, Taylor & Francis, Boca Raton, FL.)

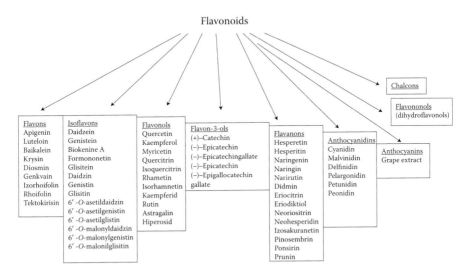

FIGURE 2.2 General chemical structure and numbering pattern for common food flavonoids. Attached R groups: H or OH substituents.

affected by some intrinsic factors, such as using different genus, species, or cultivars, and extrinsic factors, such as the time of the collection of fruits, location, environmental factors, and storage. In addition to these intrinsic and extrinsic factors, some food-processing technologies can also affect the composition of plant phenolics (Tokuşoğlu, 2001).

Flavonoids

Chalcons

Flavononols
(dihydroflavonols)

Flavons	Isoflavons	Flavonols	Flavon-3-ols	Flavanons	Anthocyanidins	Anthocyanins
Apigenin	Daidzein	Quercetin	(+)–Catechin	Hesperetin	Cyanidin	Grape extract
Luteloin	Genistein	Kaempferol	(−)–Epicatechin	Hesperitin	Malvinidin	
Baikalein	Biokenine A	Myricetin	(−)–Epicatechingallate	Naringenin	Delfinidin	
Krysin	Formononetin	Quercitrin	(−)–Epicatechin	Naringin	Pelargonidin	
Diosmin	Glisitein	Isoquercitrin	(−)–Epigallocatechin	Narirutin	Petunidin	
Genkvain	Daidzin	Rhametin	gallate	Didmin	Peonidin	
Izorhoifolin	Genistin	Isorhamnetin		Eriocitrin		
Rhoifolin	Glisitin	Kaempferid		Eriodiktiol		
Tektokirisin	6′ -O-asetildaidzin	Rutin		Neoriositrin		
	6′ -O-asetilgenistin	Astragalin		Neohesperidin		
	6′ -O-asetilglistin	Hiperosid		Izosakuranetin		
	6′ -O-malonyldaidzin			Pinosembrin		
	6′ -O-malonylgenistin			Ponsirin		
	6′ -O-malonilglisitin			Prunin		

FIGURE 2.3 Flavonoid family in food plants. (Adapted from Tokuşoğlu, Ö. 2001. *The Determination of the Major Phenolic Compounds (Flavanols, Flavonols, Tannins and Aroma Properties of Black Teas*. PhD thesis. Department of Food Engineering, Bornova, Izmir, Turkey: Ege University; Merken, H. M., and Beecher, G. R. 2000. *Journal of Agriculture Food Chemistry*, 48(3), 579–595; Tokuşoğlu, Ö., and Hall, C. 2011. *Fruit and Cereal Bioactives: Sources, Chemistry and Applications*. Boca Raton, FL, USA: CRC Press, Taylor & Francis Group, 459pp. ISBN: 9781439806654; ISBN-10:1439806659.)

2.3   HIGH-PRESSURE PROCESSING AND ITS PREFERENCES AND ADVANTAGES

Phenolic compounds in fruits and vegetables decrease by conventional and traditional heat-treatment processes. These thermal treatments are the most used methods to extend the shelf life of foods by the microorganism and enzyme inactivation, while heat causes irreversible losses of nutritional compounds, undesirable alterations in physicochemical properties, and changes of their antioxidant properties (Plaza et al., 2006; Wang and Xu, 2007).

Many factors including temperature, pH, oxygen, enzymes in the presence of copigments, metallic ions, ascorbic acid (AA), sulfur dioxide, as well as sugars may affect the stability of the anthocyanins. During pasteurization and storage, several red-fruit derivatives lose their bright-red colors and become dull-red colors. Similarly, the polyphenol content decreases in several liquid, semisolid, or solid foodstuffs by heat treatments (Ferrari et al., 2011). Many food manufacturers have investigated alternative techniques to thermal pasteurization to facilitate the preservation of unstable nutrients and bioactives in foods and beverages.

Nonthermal technologies have been reported to be a good option for obtaining food and beverages with a fresh-like appearance while preserving their nutritional quality (Odriozola-Serrano et al., 2009; Zabetakis et al., 2000). At that point, the potential use of these emerging technologies, such as "High Hydrostatic Pressure (HHP)" or "Pulsed Electrical Fields (PEF)," are important because they inactivate microorganisms and undesirable enzymes to a certain extent and can avoid the negative effects of heat pasteurization (Toepfl et al., 2006).

Recently, there has been an increasing interest for nonthermal technologies as HPP to preserve fruits, vegetables, daily foods, and beverages (Barbosa-Cánovas et al., 1998). Great technological and research efforts have been made to obtain foods and beverages by HPP without the quality and nutritional damage caused by heat treatments. HHP or ultra-HPP or HPP is one technology that has begun to fulfill its potential to satisfy both consumer and scientific requirements, and is a leading alternative in replacing thermal processing in some food applications in the drive to meet increasing consumer demand for foods featuring improved organoleptic qualities and higher acceptance (Patterson et al., 2008; Bevilacqua et al., 2010; Tokuşoğlu and Doona, 2011a).

HPP can be used to obtain a high-quality food/beverage and increases its shelf life while maintaining its physicochemical, nutritional characteristics, and bioactive profiles (Tokuşoğlu, 2011; Tokuşoğlu, 2012a,b; Tokuşoğlu and Doona, 2011a,b; Tokuşoğlu et al., 2010).

The technology is especially beneficial for heat-sensitive products (Barbosa-Cánovas et al., 2005; Tokusoglu and Doona, 2011). HPP can be conducted at ambient or moderate temperatures, thereby eliminating thermally induced cooked off-flavors. Compared to thermal processing, the HPP of foods results in products with a fresher taste, better appearance, and texture. Foods are processed in batch (for solid products) or continuous and semicontinuous systems (for liquid products) in a pressure range of 50–1000 MPa; process temperature during pressure treatment can be from below 0°C to above 100°C, while exposure time usually ranges from seconds to 20 min

(Bevilacqua et al., 2010; Corbo et al., 2009; Patterson et al., 2008). HPP technology has been successfully applied in several industrial sectors such as meat, seafood, dairy food, fruit juices, fruit, and vegetable products (Figure 2.4). HPP has been found to inactivate several microorganisms and enzymes. However, it has less effect on low-molecular-weight food components such as vitamins, pigments, flavoring agents, and other nutritional compounds. HPP conditions in the range of 300–700 MPa at moderate initial temperatures (around ambient) are generally sufficient to inactivate vegetative pathogens for pasteurization processes, some enzymes, or spoilage organisms to extend shelf life. HPP can also increase the extraction capacity of phenolic constituents, and higher levels of bioactive compounds and phytochemicals are preserved in HPP-treated samples (Oms-Oliu et.al., 2012b; Tokusoglu and Doona, 2011).

Consumer perception of food quality depends not only on microbial quality but also on other food factors such as biochemical and enzymatic reactions and structural changes. In this context, HPP can have an effect on food yield and on sensory qualities such as food color and texture. High pressures (HPs) can also be used to enhance extraction of compounds from foods. Recent studies have shown that high-pressure extraction (HPE) can shorten processing times, and provide higher extraction yields while having less negative effects on the structure and antioxidant activity of bioactive constituents. The use of HPE enhances mass transfer rates, increases cell permeability, and increases diffusion of secondary metabolites (Cheftel, 1995; Dornenburg and Knoor, 1993; Tokusoglu and Doona, 2011).

HHP increases the dissolution rate of the bioactives. A rapid permeation is observed under HPE owing to the large differential pressure between the cell interior and the exterior of cell membranes (Zhang et al., 2005). This situation increases the solvent penetration through the broken membranes into cells or increases the mass transfer rate due to increased permeability (Shouqin et al., 2005). This means the higher the hydrostatic pressure is, the more solvent can enter into the cell. More compounds can permeate the cell membrane that could cause the higher yield of extraction (Shouqin et al., 2005; Zhang et al., 2005).

In other words, the extraction capacity of phenolic constituents has been increased by HHP and HPP-treated samples that retain higher levels of bioactive compounds (Tokusoglu et al., 2010; Tokusoglu and Doona, 2011; Zhang et al., 2005). Studies on HPP effects on total phenolics determined that these compounds were either

FIGURE 2.4 HHP. (Centre for Nonthermal Processing of Food (CNPF), Washington State University (WSU), Pullman, WA, USA.) (Photo: Tokuşoğlu by Frank Younce, 2010.)

unaffected or actually increased in concentration and/or extractability, following treatment with HP.

2.4 HHP APPLICATIONS ON PHENOLIC AND ANTIOXIDANT BIOACTIVES OF VEGETABLES

In one study given by Vázquez-Gutiérrez et al. (2013), HHP (100–600 MPa/1–3 min/25°C) affected the microstructure and antioxidant properties of onions (cv. Doux). Owing to the fact that onions have antioxidant properties and are an important source of bioactive compounds such as phenols, HHP also affected the extractability of potential health-related compounds of studied onions (Vázquez-Gutiérrez et al., 2013). In this study, it is shown that vitamin C (AA) did not show significant alterations, while the extracted phenolic content and antioxidant activity increased at pressures of 300 or 600 MPa of HHP. Vázquez-Gutiérrez et al. (2013) concluded that HHP produced changes in membrane permeability and disruption of cell walls, favoring the phenolic compounds releasing from the tissue and, in consequence, improving their extractability (Vázquez-Gutiérrez et al., 2013).

Jung et al. (2013) stated the potential effectiveness of HHP on the alterations in quality-related properties of carrot and spinach. In the study described by Jung et al. (2013), better retention of AA and carotenoids was observed as the carrots and spinaches were treated at 100, 300, and 500 MPa for 20 min compared to the thermal processing (Jung et al., 2013). It was shown that the flavonoid amounts were increased with increasing pressure levels, leading to the enhanced antioxidant activity and also, it was determined that the residual polyphenoloxidase (PPO) activities were decreased in carrot and spinach as 6.9–15.1% and 21.3–31.1%, respectively. Jung et al. (2013) reported that HHP could be used as an alternative technology for improving quality of vegetables (Jung et al., 2013).

HHP (400 MPa/10 min, 500 MPa/5 min, and 600 MPa/2.5 min) and high-temperature short time (HTST) (110°C/8.6 s) processing of purple sweet potato nectar was reported by Wang et al. (2012). The quality-related aspects including the microorganism level, total phenolics, anthocyanins, antioxidant capacity, color, and shelf-life prediction during 12 weeks of storage at 4°C and 25°C were determined. It was reported that the purple sweet potato nectar samples stored at 4°C showed better quality and longer shelf life when compared with those stored at 25°C and longer shelf life was observed in HHP-treated samples compared to HTST-treated samples (Wang et al., 2012). The shelf life, estimated in accordance with the zero-order reaction, was 29.256, 35.862, 32.821, and 32.499 weeks for HTST, 400 MPa/10 min, 500 MPa/5 min, and 600 MPa/2.5 min treated purple sweet potato nectar at 4°C, respectively. By comparison, it was 6.343, 7.256, 8.466, and 7.951 weeks for HTST, 400 MPa/10 min, 500 MPa/5 min, and 600 MPa/2.5 min treated purple sweet potato nectar at 25°C, respectively, by Wang et al. (2012).

Figure 2.5 shows the changes of antioxidant capacity (DPPH) in purple sweet potato nectar during 12 weeks of storage at 4°C (a) and 25°C (b). It was determined that the DPPH antioxidant capacity decreased by 23.76–26.97% and 28.27–41.62% in purple sweet potato nectar at 4°C and 25°C after the 12-week storage.

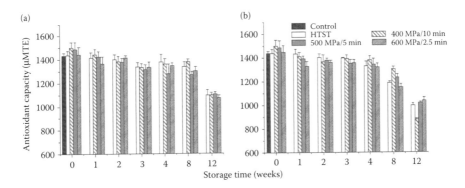

FIGURE 2.5 The antioxidant capacity (DPPH) changes in purple sweet potato nectar during 12 weeks of storage at 4°C (a) and 25°C (b). (Adapted from Wang, Y. et al. 2012. *Innovative Food Science and Emerging Technologies,* 16, 326–334.)

The DPPH antioxidant capacity in HTST-treated samples was higher than that in 400 MPa/10 min treated samples, while lower than 600 MPa/2.5 min, and 500 MPa/5 min treated samples at 25°C (Wang et al., 2012). It was also shown that sweet potato nectar samples stored at 4°C had higher DPPH antioxidant capacity than those stored at 25°C by Wang et al. (2012).

Low-pressure treatments (100–200 MPa for 10–20 min) on green peppers caused a decrease of 10–15% of the initial vitamin C, while in red peppers, these treatments resulted in a 10–15% increase in vitamin C (Barrett and Lloyd, 2011).

Van Eylen et al. (2007) studied the HP (600–800 MPa) and temperature (30–60°C) stabilities of sulforaphane and phenylethyl isothiocyanate in broccoli juice. It was concluded that isothiocyanates are relatively thermolabile and pressure stable. Van Eylen et al. (2007) also stated that myrosinase activity was stabilized by using mild pressure treatments, and thus leading to products with increased isothiocyanate content.

2.5 HHP APPLICATIONS ON PHENOLIC AND ANTIOXIDANT BIOACTIVES OF FRUITS

It has been reported that the anthocyanins of different liquid foods (red-fruit juices) are stable to HHP treatment at moderate temperatures. The nutraceutical and sensorial properties are strictly related to the anthocyanin and polyphenol content in pomegranate juice at room temperature. It was reported that the stability or preservation of bioactive compounds of red-fruit juices is contradictory. The concentration of red-fruit-based bioactives decreases with the intensity of the treatment in terms of pressure level and processing time (Ferrari et al., 2010).

Ferrari et al. (2010) stated the effects of HHP on the polyphenol contents and anthocyanin levels of several red-fruit-based products (strawberry and wild strawberry mousses, pomegranate juice). 500 MPa/50°C/10 min and 400 MPa/25°C/5 min of HHP conditions were applied for mousse samples (strawberry and wild

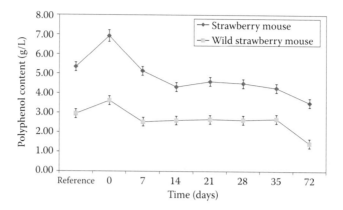

FIGURE 2.6 The polyphenol level of HP-treated strawberry mousse and wild strawberry mousse evaluated at fixed storage times under refrigerated conditions (4°C). (Adapted from Ferrari, G., Maresca, P., and Ciccarone, R. 2010. *Journal of Food Engineering*, 100 (2), 245–253.)

strawberry mousses) and pomegranate juice samples, respectively. It was found that HPP treatment at moderate temperatures promoted the extractability of colored pigments and increased the polyphenol levels of fruits (Ferrari et al., 2010). Figure 2.6 shows that polyphenol level of HP-treated strawberry mousse and wild strawberry mousse evaluated at fixed storage times under refrigerated conditions (4°C) whereas Figure 2.7 shows the polyphenol content of HP-treated samples of pomegranate juice evaluated at fixed storage times under refrigerated conditions (4°C) (Ferrari et al., 2010). The results by Ferrari et al. (2010) showed the potentiality of

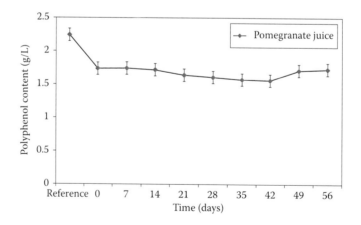

FIGURE 2.7 The polyphenol content of HP-treated samples of pomegranate juice evaluated at fixed storage times under refrigerated conditions (4°C). (Adapted from Ferrari, G., Maresca, P., and Ciccarone, R. 2010. *Journal of Food Engineering*, 100 (2), 245–253.)

the HP processing for the treatment of products rich in thermolabile nutraceutical constituents.

Barba et al. (2012) reported quality changes of blueberry juice during 56 days of refrigerated storage at 4°C after HPP and PEF processing. The study as described by Barba et al. (2012), shows blueberry juice was processed by HP (600 MPa/42°C/5 min). It was determined that it was decreasing lower than 5% in AA content compared with the untreated blueberry juices. At the end of refrigerated storage, unprocessed blueberry juices and similarly PEF-treated juices showed 50% of AA losses whereas they showed 31% losses of AA for HPP-treated blueberry juices (Barba et al., 2012).

Figure 2.8 shows AA remaining levels in untreated, HP- and PEF-treated blueberry juices stored in refrigeration at 4°C. It was found that the AA of HPP-treated blueberry juices maintained more stability during storage time and HPP preserved antioxidant activity (21% losses) more than unprocessed (30%) juices and PEF-treated (48%) juices after 56 days at 4°C (Barba et al., 2012). It was concluded that the second conservation treatment such as refrigerated storage must join to nonthermal technologies and HPP can be a potentially useful unit operation for preserving bioactive compounds in blueberry juices during refrigerated storage (Barba et al., 2012).

Varela-Santos et al. (2012) stated the HHP processing (350–550 MPa for 30, 90, and 150 s) effects on microbial, physicochemical quality, and bioactives of pomegranate juices during 35 days of storage at 4°C. The applied HHP treatment at or over 350 MPa for 150 s resulted in about 4.0 log-cycles reduction of microbial load and these treatments were able to extend the microbiological shelf life of pomegranate juice stored in the above-mentioned conditions. In the study reported by

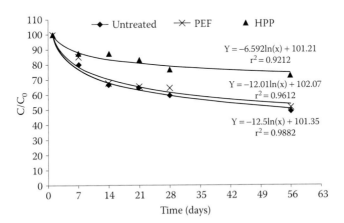

FIGURE 2.8 AA remaining levels in untreated, HP-, and PEF-processed blueberry juices stored in refrigeration at 4°C. The lines interpolating the experimental data points show the fit of a 1.4th-order reaction model. (Adapted from Barba, F. J. et al. 2012. *Innovative Food Science and Emerging Technologies*, 14, 18–24.)

Varela-Santos et al. (2012), the phenolic levels of pomegranate juices showed an increase in the first 3 days and started to decrease after 5 days. It was shown that the phenolic of pomegranate juices reached a steady-state level after 10 days, remaining constant for those treated samples stored at 4°C, until the end of the study (Varela-Santos et al., 2012). It was revealed that HHP has a remarkable effect on the antioxidant activities, IC50, with much lower values when pressure increases; therefore, a smaller IC50 value was obtained at 500 MPa that corresponds to a higher antioxidant activity. The DPPH scavenging activities of the pomegranate juice expressed as an IC50 value were 11– 20 mg/mL at starting point. At 450 and 550 MPa, it exhibited the strongest antioxidant capacity (11–13 mg/mL), followed by the control sample (14 mg/mL). Pérez-Vicente et al. (2004) put forward that the increasing antioxidant activity in pomegranate juices could be due to the extraction of some of the hydrolyzable tannins, present in the fruit rind, and/or related to the increase in ellagic acid, ellagic structures polymerized into ellagitannins, and/or anthocyanin polymers formed during the storage period of fruit (Pérez-Vicente et al., 2004).

Vázquez-Gutiérrez et al. (2011) showed the effects of HHP (for 200/400 MPa treatment during 1/3/6 min) on the microstructure of persimmon fruit cv. "Rojo Brillante" during two different ripening stages including with and without deastringency treatment (95–98% CO_2), and on some bioactive compounds levels. Vázquez-Gutiérrez et al. (2011) reported that HHP treatment produced a significant effect on the persimmon structure by affecting the integrity of cell walls and membranes. Vázquez-Gutiérrez et al. (2011) stated that much of the soluble tannins spread outside vacuoles, carotenoid substances were released from the chromoplasts, cell walls were degraded, and extractability was affected. It was put forward that the HHP application has been induced with the precipitation of soluble tannins in "*Rojo Brillante*" persimmons, which could be related to the loss of astringency (Vázquez-Gutiérrez et al., 2011).

Food matrix and processing parameters are effective on retaining of phenolic compounds. The combination with other emerging methods (ultrasound, γ-irradiation, carbon dioxide, and antimicrobial agents) can also help to retain nutritional and health-related characteristics of these compounds. HHP process condition parameters (pressure, temperature, and time) are important for phenolic quality and quantity (Tokusoglu and Doona, 2011; Tokusoglu et al., 2010).

Qiu et al. (2006) revealed that the highest stability of lycopene (Figure 2.9) in tomato purée was found when pressurized at 500 MPa and stored at $4 \pm 1°C$

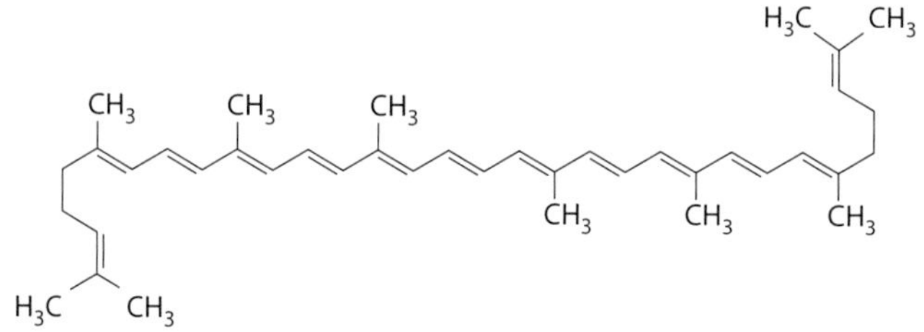

FIGURE 2.9   Lycopene.

in the study, which retained most of the total lycopene content in tomato purée (6.25 ± 0.23 mg/100 g) (Qiu et al., 2006). Table 2.1, parts (a) and (b) show total lycopene losses in lycopene standard (as percentage) and total lycopene content in tomato purée (as mg/100 g), respectively, as a function of storage time at 4 ± 1°C, at six different HHP conditions (Qiu et al., 2006).

It was found that 500 and 600 MPa of pressure led to the highest reduction of lycopene, while 400 MPa could retain the maximal stability of lycopene (Qiu et al., 2006). The highest stability of lycopene in tomato purée was found when pressurized at 500 MPa and stored at 4 ± 1°C. It was established that HHP is an alternative preservation method for producing ambient-stable tomato products in terms of lycopene conservation (Qiu et al., 2006).

The caffeic acid increasing in tomato juices after 28 days of storage could be directly associated with residual hydroxylase activities, which convert coumaric acid into caffeic acid. It was stated that total phenolics in tomato-based beverages and tomato purées appeared to be relatively resistant to the effect of HP (Patras et al., 2009; Barba et al. 2010).

After HPP treatment of tomato purée for 400 MPa/25°C/15 min, the AA and total AA contents decreased as 40% and 30%, respectively (Sánchez-Moreno et al., 2006). Individual carotenoids including β-carotene, β-cryptoxanthin, zeaxanthin, and lutein with antioxidant activity in tomato-based soup appeared to be resistant to a HPP treatment of 400 MPa at 40°C for 1 min, thus resulting into a better preservation of the antioxidant activity in comparison with the thermally pasteurized activity (Sánchez-Moreno et al., 2005).

It was reported that key antioxidants (cyanidin-3-glycoside, pelargonidin-3-glucoside, and AA) in strawberry and blackberry purées (Figure 2.10) and the antioxidant activity of these purées were quantified after various HPP treatments (400, 500, and 600 MPa/15 min/10–30°C) and thermal treatments (70°C/2 min) (Patras et al., 2009).

400, 500, and 600 MPa/15 min/10–30°C and thermal treatment (70°C/2 min) applications of strawberry and blackberry purées were performed by Patras et al. (2009) (Table 2.2). Table 2.2 shows the antioxidant indices of HPP-treated and thermally processed strawberry and blackberry purées. It was found that the three different pressure treatments did not cause any significant changes in AA levels. Following thermal processing ($P_{70} \geq 2$ min), the AA content degraded by 21% compared to the unprocessed purée (Patras et al., 2009). Similarly, no significant alterations in anthocyanin compounds were observed in HPP-treated and HPP-unprocessed purées, while conventional thermal treatments significantly reduced the anthocyanin levels. Antioxidant activity of HPP-treated strawberry and blackberry purées was significantly higher than in thermally processed purées (Patras et al., 2009)

It was stated that dry-weight content of vitamin C in strawberry and blackberry purées was significantly higher in HPP-treated samples (Barrett and Lloyd, 2011). It was reported that the level of retention of AA in guava purée proceeded according to the following decreasing order: (400 MPa for 15 min) > (88–90°C for 24 s) > (600 MPa for 15 min) (Yen and Lin, 1999).

After HPP processing (400 MPa, 40°C, and 1 min), orange juice presented a significant increase on the extractability of each individual flavanone with regard to untreated juice and hence on total flavanone content (15.46%) (Plaza et al.,

TABLE 2.1

Total Lycopene Losses in Lycopene Standard (a) (as Percentage) and Total Lycopene Content in Tomato Purée (b) (as mg/100 g) as a Function of Storage Time at 4 ± 1°C, at Six Different HHP Conditions

a. Lycopene

Storage Time (Days)	Untreated (0 MPa)	Pressure Applied (MPa)					
		100	200	300	400	500	600
0	2.10 ± 0.02	2.10 ± 0.02	2.11 ± 0.02	2.11 ± 0.02	2.13 ± 0.02	20.8 ± 1.12	56.3 ± 3.02
2	3.05 ± 0.23	2.10 ± 0.02	2.11 ± 0.02	2.11 ± 0.02	2.13 ± 0.02	20.8 ± 1.12	56.3 ± 3.02
4	5.22 ± 0.34	2.40 ± 0.05	2.52 ± 0.09	2.34 ± 0.07	2.29 ± 0.09	21.7 ± 1.19	57.4 ± 3.34
8	6.13 ± 0.40	2.49 ± 0.07	2.63 ± 0.09	2.45 ± 0.09	2.39 ± 0.11	22.7 ± 1.21	57.4 ± 3.34
16	7.89 ± 0.44	4.21 ± 0.23	3.29 ± 0.28	3.78 ± 0.22	2.70 ± 0.28	25.7 ± 1.41	60.4 ± 3.76

b. Tomato Purée

Storage Time (Days)	Untreated (0 MPa)	Pressure Applied (MPa)					
		100	200	300	400	500	600
0	5.16 ± 0.12	5.33 ± 0.13	5.39 ± 0.11	5.48 ± 0.12	5.55 ± 0.12	6.25 ± 0.23	5.10 ± 0.10
2	5.18 ± 0.13	5.39 ± 0.12	5.42 ± 0.12	5.50 ± 0.13	5.50 ± 0.13	6.20 ± 0.21	5.11 ± 0.11
4	5.18 ± 0.13	5.37 ± 0.12	5.43 ± 0.12	5.51 ± 0.13	5.50 ± 0.13	6.21 ± 0.20	5.10 ± 0.12
8	5.17 ± 0.13	5.37 ± 0.13	5.40 ± 0.15	5.51 ± 0.13	5.48 ± 0.14	6.19 ± 0.22	5.08 ± 0.10
16	4.37 ± 0.10	5.17 ± 0.12	5.22 ± 0.16	5.26 ± 0.12	5.18 ± 0.13	6.11 ± 0.23	4.88 ± 0.12

Source: Adapted from Qiu, W. et al. 2006. *Food Chemistry*, 97, 516–523.

Note: Values reported are means of triplicate determinations ($n = 3$) ± SD.

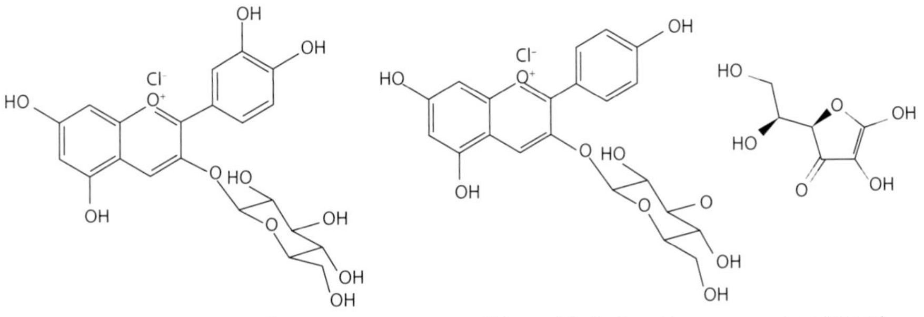

Cyanidin-3-glycoside   Pelargonidin-3-glycoside   Ascorbic acid

FIGURE 2.10   Key antioxidants of strawberry and blackberry purées.

2011). Regarding the main flavanones identified in orange juice, HP treatments (400 MPa/40°C/1 min) increased the content of naringenin by 20% and by 40% the content of hesperetin in comparison with an untreated orange juice (Oms-Oliu et al., 2012). These data are in accordance with those obtained by other authors showing higher extraction of phenolic compounds due to HPP and the levels of phenols increased significantly in HP-treated (600 MPa, 20°C, and 15 min) strawberry and blackberry purées (9.8% and 5.0%, respectively).

The litchi is the sole member of the genus *Litchi* in the soapberry family Sapindaceae and it is a tropical fruit tree. The litchi (*Litchi chinensis* Sonn.) is a fragranced fruit with a sweet taste. After 30 min of HPE of litchi fruit pericarp (LFP), the extract yield, total phenolic level, 1,1-diphenyl-2-picrylhydrazyl radical scavenging activity (DPPH), and superoxide anion scavenging ability were carried out (Prasad et al., 2009a). The extraction yield by treatments of 400 MPa HPE for 30 min was 30%, while that by conventional extraction (CE, control) was 1.83%. There was no significant difference in the total phenolic content (as mg/g DW) among the two extraction methods (HPE and CE). It was found that the DPPH radical scavenging activity obtained by HPE (400 MPa) was the highest (74%) level (Prasad et al., 2009a) (Table 2.3).

Epicatechin (EC) and epicatechingallate (ECG) were identified and quantified as the major flavonoids of litchi, while catechin (C) and procyanidin B$_2$ (Pro B$_2$) (Figure 2.11) were identified as the minor litchi phenolic compounds (Prasad et al., 2009a). The total flavonoid content of litchi was 0.65, 0.75, 0.29, and 0.07 mg/g dry weight by HPE at 200 and 400 MPs, ultraextraction (UE), and CE, respectively. It was reported that the yield of flavonoid extraction increased 2.6 times in comparison with UE, and up to 10 times compared with CE (Prasad et al., 2009a) (Table 2.3).

The longan fruit ("dragon eyes") (*Dimpcarpus longan* Lour.) is edible, extremely sweet, juicy, and succulent in superior agricultural varieties, and apart from ingested fresh, is also often used in East Asian soups, snacks, desserts, and sweet-and-sour foods, either fresh or dried, sometimes canned with syrup in supermarkets. Prasad et al. (2009b) indicated the extraction of longan fruit pericarp by various pressures of HPP (200 – 500 MPa/2.5 – 30 min of duration at 30 – 70°C) and by different solvent concentration (25 – 100%, v/v) and solid-to-liquid ratio (1:25–1:100, w/w) (Prasad et al., 2009b).

TABLE 2.2
Antioxidant Indices of HPP-Treated and Thermally Processed Strawberry and Blackberry Purées

Treatment	Antiradical Power (g/L)$^{-1}$		Total Phenols (mg GAE/100 g DW)[e]		Anthocyanin (mg/100 g DW)		Ascorbic Acid (mg/100 g DW)	
	Strawberry	Blackberry	Strawberry	Blackberry	Strawberry	Blackberry	Strawberry	Blackberry
Unprocessed	1.55 ± 0.07[a]	2.86 ± 0.23[a]	855.02 ± 6.52[a]	1694.19 ± 3.0[a]	202.27 ± 0.50[a]	1004.90 ± 8.60[a]	633.10 ± 9.31[a]	nd
Thermal	1.16 ± 0.01[b]	2.78 ± 0.26[a]	817.01 ± 5.26[b]	1633.62 ± 8.4[a]	145.82 ± 6.40[b]	975.28 ± 7.90[b]	496.11 ± 0.04[b]	nd
HPP (400 MPa)	1.25 ± 0.05[b]	3.87 ± 1.11[a]	859.03 ± 6.56[a]	1546.26 ± 8.0[a]	173.34 ± 6.51[ab]	1039.21 ± 4.51[a]	574.30 ± 3.93[c]	nd
HPP (500 MPa)	1.30 ± 0.02[ab]	3.70 ± 0.57[a]	926.00 ± 5.93[a]	1724.65 ± 0.7[b]	202.53 ± 5.40[a]	1014.21 ± 0.10[a]	577.10 ± 6.52[c]	nd
HPP (600 MPa)	1.33 ± 0.02[a]	4.80 ± 1.79[b]	939.01 ± 0.99[c]	1778.44 ± 6.0[b]	204.30 ± 1.60[a]	1014.47 ± 1.00[a]	599.11 ± 0.60[c]	nd

Source: Adapted from Patras, A. et al. 2009. Effect of thermal and high pressure processing on antioxidant activity and instrumental colour of tomato and carrot purées. *Innovative Food Science Emerging Technology*, 10, 16–22.

Note: Values are mean ± standard deviation, $n = 3$, and mean values in a column with different letters are significantly different at $p < 0.05$; nd = not detected.

[a] Dry weight.
[b] Expressed as mg/100 g DW pelargonidin-3-glucoside.
[c] Expressed as mg/100 g DW cyanidin-3-glucoside.

TABLE 2.3

HPP, UE, and CE Effects on Litchi Phenolics

			Extraction Methods	
Flavonoids (mg/g DW)[a]	CE	UE	HPE at 200 MPa	HPE at 400 MPa
Epicatechin	0.0414 ± 0.001	0.16 ± 0.04	0.32 ± 0.002	0.348 ± 0.06
Epicatechin gallate	0.0121 ± 0.003	0.06 ± 0.01	0.019 ± 0.04	0.2527 ± 0.04
Catechin	0.0002 ± 0.0	0.0020 ± 0.0005	0.0016 ± 0.001	0.0160 ± 0.07
Procyanidin B2	0.0175 ± 0.0003	0.0731 ± 0.0011	0.14 ± 0.03	0.1346 ± 0.03

Source: Adapted from Prasad, K. N. et al. 2009a. *Journal of Food Process Engineering*, 32(6), 828–843.

Note: Values reported are means of triplicate determinations ($n = 3$) ± SD; DW[a], dry weight; CE, conventional extraction; UE, ultrasonic extraction; HPE, high-pressure extraction.

The extraction yield, total phenolics, and scavenging activities of superoxide anion radical and DPPH radical by HPE were determined and compared with those from a CE for longan fruit pericarp. The HPE provided a higher extraction yield and required a shorter extraction time compared to CE. In addition, the total phenolics and the antioxidant activities of HPE were higher than those produced by CE (Table 2.4) (Prasad et al., 2009b).

Corrales et al. (2008) examined the extraction capacity of anthocyanins that could be used as natural antioxidants or colorants from grape by-products by HPP and other emerging techniques (Figure 2.12). The HPP at 600 MPa showed feasibility and selectivity for extraction purposes. The heat-treatment effect at 70°C combined with the effect of different emerging novel technologies such as ultrasonics (35 KHz), HHP (600 MPa), and PEF (3 kV cm^{-1}) showed a great feasibility and selectivity for extraction purposes. By 1 h of extraction, the total phenolic levels of grape by-products subjected to HHP technology were 50% higher than in the control samples (Figure 2.12). Using novel technology applications, the antioxidant activity of the pomace extracts increased with PEF as fourfold, with HHP

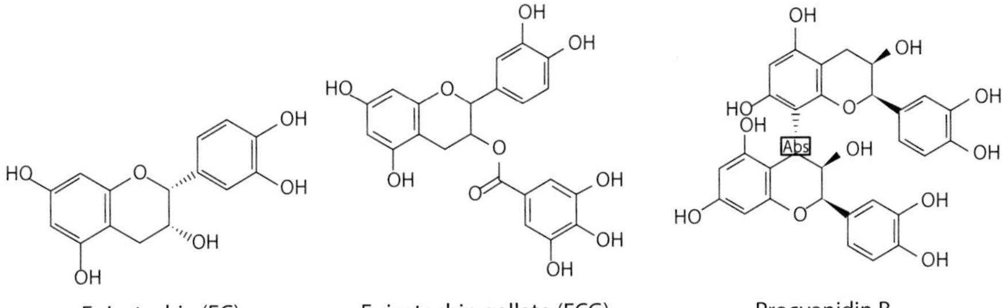

Epicatechin (EC)   Epicatechin gallate (ECG)   Procyanidin B$_2$

FIGURE 2.11 Major phenolics in litchi (*Litchi chinensis* Sonn.) fruit, EC, ECG, and Pro B$_2$.

TABLE 2.4
DPPH Scavenging Activity for Longan Fruit

	DPPH Scavenging Activity (%)	
	50 µg/mL	100 µg/mL
CE extract	50.1 ± 0.2c	76.6 ± 0.5b
HPE extract	75 ± 0.2b	77.7 ± 0.2a
Ascorbic acid	80 ± 2a	80.4 ± 0.6a
BHT™	77.4 ± 2a	80 ± 1a

Source: Adapted from Prasad, K. N. et al. 2009b. *Innovative Food Science and Emerging Technologies*, 10, 155–159.

Note: Conventional extraction (CE): 50% ethanol, 1:50 (w/v) solid/liquid ratio, and 12 h of extraction time at 30°C and high-pressure extraction (HPE): 50% ethanol, 1:50 (w/v) solid/liquid ratio, 500 MPa pressure, and 2.5 min of extraction time at 30°C. For each treatment means within a column followed by different letters were significantly different at the 5% level.

as threefold, and with ultrasonics as twofold higher than the control extraction (Corrales et al., 2008).

Anthocyanins have been reported to be stable to HP treatments in different fruit juices such as strawberry juice, blackcurrant juice, and raspberry juice (Oms-Oliu et al., 2012). Combined pressure and temperature application of blueberry-pasteurized juice led to a slightly faster degradation of total anthocyanins during storage

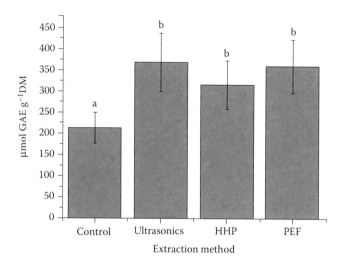

FIGURE 2.12   Total phenolic content (µmol GAE g^{-1} DM) from grape by-products extracted by ultrasonics, HHP, and PEF. (Adapted from Corrales, M. et al. 2008. *Innovative Food Science Emerging Technology*, 9, 85–91.)

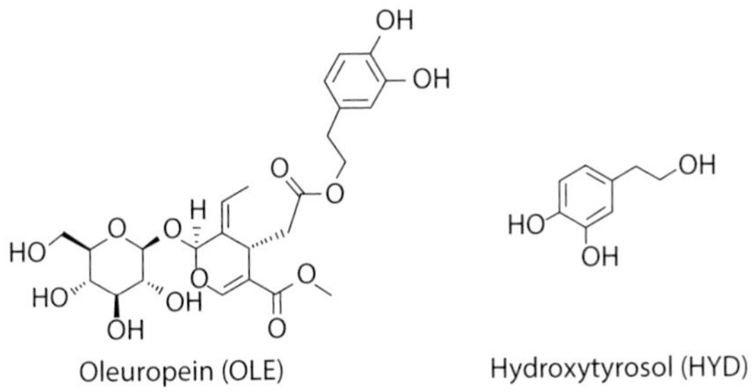

FIGURE 2.13 The major phenolics of table olives.

compared to conventional heat treatments (Buckow et al., 2010). Pressure seems to accelerate anthocyanin degradation at elevated temperatures. This can be related to condensation reactions, involving covalent association of anthocyanins with other flavanols present in fruit juices.

Tokuşoğlu et al. (2010) reported that the total phenolics of table olives increased (2.1–2.5)-fold after HPP (as mg gallic acid equivalent/100 g). Phenolic hydroxytyrosol (Figure 2.13) in olives increased as average (0.8–2.0)-fold, whereas phenolic oleuropein (Figure 2.13) decreased as average (1–1.2)-fold after HPP (as mg/kg DW). Antioxidant activity values varied from 17.238 to 29.344 mmol Fe^{2+}/100 g for control samples, and 18.579 to 32.998 mmol Fe^{2+}/100 g for HPP-treated samples (Tokuşoğlu et al., 2010).

HP causes a significant reduction in the activity of the enzymes although apparent enzymatic activation of PPO in some HP-treated strawberry samples (300 MPa/60°C/30 s) may be caused by the release of membrane- bound enzymes due to pressurization (Terefe et al., 2009). The vitamin A content of persimmon purée increased by 45% with HP processing. It was found that the total carotenoid content was significantly higher in all carrot purées treated with HP. Following the 600 MPa/20°C/15 min treatment, total carotenoids increased by 58% as compared to raw carrots. de Ancos etal. (2000) stated that pressure treatments at 50 and 300 MPa/15 min/25°C for Spain originated Rojo Brillante persimmon fruit purees and at 50 and 400 MPa/15 min/25°C for Sharon persimmon fruit purees increased the amount of extractable carotenoids (9–27%), which are related with the increase of vitamin A value (75–87 RE/100 g) (de Ancos et.al., 2000). Butz et al. (2002) studied the effects of both HP (600 MPa/25°C) and thermal processing (118°C/20 min) and found that neither preservation method resulted in a significant change in total carotenoids in fruit and vegetable juices, or pieces of apple, peach, and tomato.

HPP is an excellent food-processing technology that has the potential to retain the bioactive constituents with health properties in plant foods. HPP-treated foods retain more of their fresh-like features and can be marketed at a premium over their thermally processed counterparts.

REFERENCES

Andersen, Q. M., and Markham, K. R. 2006. *Flavonoids. Chemistry, Biochemistry, and Applications.* Boca Raton, FL: CRC Press, Taylor & Francis.

Barba, F. J., Esteve, M. J., and Frigola, A. 2010. Ascorbic acid is the only bioactive that is better preserved by hydrostatic pressure than by thermal treatment of a vegetable beverage. *Journal of Agricultural and Food Chemistry*, 58, 10070–10075.

Barba, F. J., Jäger, H., Meneses, N., Esteve, M. J., Frígola, A., and Knorr, D. 2012. Evaluation of quality changes of blueberry juice during refrigerated storage after high-pressure and pulsed electric fields processing. *Innovative Food Science and Emerging Technologies*, 14, 18–24.

Barbosa-Cánovas, G. V., Pothakamury, U. R., Palou, E., and Swanson, B. 1998. Emerging technologies in food preservation. In *Nonthermal Preservation of Foods*. New York: Marcel Dekker, 1–9.

Barbosa-Cánovas, G. V., Tapia María, S., and Pilar Cano, M. 2005. *Novel Food Processing Technologies*. New York: Marcel Dekker.

Barrett Diane, M., and Lloyd, B. 2011. Advanced preservation methods and nutrient retention in fruits and vegetables. *Journal of Science Food Agriculture*, 2012; 92, 7–22.

Bevilacqua, A., Campaniello, D., and Sinigaglia, M. 2010. Chapter 8: Use of high pressure processing for food preservation. In *Application of Alternative Food-Preservation Techniques to Enhance Food Safety and Stability*. A. Bevilacqua, M. R. Corbo, and M. Sinigaglia, eds. Bentham Science Publishers Ltd., Sharjah, 114–142.

Buckow, R., Kastell, A., Shiferaw Terefe, N., and Versteeg, C. 2010. Pressure and temperature effects on degradation kinetics and storage stability of total anthocyanins in blueberry juice. *Journal of Agricultural and Food Chemistry*, 58, 10076e–10084e.

Butz, P., Edenharder, R., Garcia, A. F., Fister, H., Merkel, C., and Tauscher, B. 2002. Changes in functional properties of vegetables induced by high pressure treatment. *Food Research International*, 35, 25–300.

Cheftel, J. C. 1995. Review: High pressure, microbial inactivation and food preservation. *Food Science Technology International*, 1, 75–90.

Cheynier, V. 2005. Polyphenols in foods are more complex than often thought. *The American Journal of Clinical Nutrition*, 81 (Suppl), 223–229.

Corbo, M. R., Bevilacqua, A., Campaniello, D., D'Amato, D., Speranza, B., and Sinigaglia, M. 2009. Prolonging microbial shelf life of foods through the use of natural compounds and non-thermal approaches. *International Journal of Food Science and Technology*, 44, 223–241.

Corrales, M., Toepfl, S., Butz, P., Knorr, D., and Tauscher, B. 2008. Extraction of anthocyanins from grape by-products assisted by ultrasonic, high hydrostatic pressure or pulsed electric fields: A comparison. *Innovative Food Science Emerging Technology*, 9, 85–91.

Crozier, A., Jaganath, I. B., and Clifford, M. N. 2006. Phenols, polyphenols and tannins: An overview. In *Plant Secondary Metabolites*. A. Crozier, M. N. Clifford, and H. Ashihara, eds. Oxford: Blackwell Publishing, Ltd., 1–24.

de Ancos, B., Gonzalez, E., and Cano, M. P. 2000. Effect of high-pressure treatment on the carotenoid composition and the radical scavenging activity of persimmon fruit purees. *Journal of Agricultural Food Chemistry*, 48, 542–3548.

Dornenburg, H., and Knoor, D. 1993. Cellular permeabilization of cultured plant tissues by high electric field pulses or ultra high pressure for the recovery of secondary metabolites. *Food Biotechnology*, 7, 35–48.

Ferrari, G., Maresca, P., and Ciccarone, R. 2011. The effects of high hydrostatic pressure on the polyphenols and anthocyanins in red fruit products. *Procedia Food Science*, 1, 847–853.

Ferrari, G., Maresca, P., and Ciccarone, R. 2010. The application of high hydrostatic pressure for the stabilization of functional foods: Pomegranate juice. *Journal of Food Engineering*, 100 (2), 245–253.

Jung, L. S., Lee, S. H., Kim, S., and Ahn, J. 2013. Effect of high hydrostatic pressure on the quality-related properties of carrot and spinach. *Food Science and Biotechnology*, 22(1), 189–195.

Kong, J., Chia, L., Goh, N., Chia, T., and Brouillard, R. 2003. Analysis and biological activities of anthocyanins. *Phytochemistry*, 64, 923–933.

Merken, H. M., and Beecher, G. R. 2000. Measurement of food flavonoids by high performance liquid chromatography: A review. *Journal of Agricultural Food Chemistry*, 48(3), 579–595.

Meskin, M. S., Bidlack, W. R., Davies, A. J., Lewis, D. S., and Randolph, R. K. 2003. *Phytochemicals: Mechanisms of Action*. Boca Raton, FL: CRC Press.

Odriozola-Serrano, I., Soliva-Fortuny, R., and Martin-Belloso, O. 2009. Impact of high intensity pulsed electric fields variables on vitamin C, anthocyanins and antioxidant capacity of strawberry juice. *LWT—Food Science and Technology*, 42, 93–100.

Oms-Oliu, G., Isabel Odriozola-Serrano, I., Robert Soliva-Fortuny, R., Pedro Elez Martinez, P., and Olga Martin-Belloso, O. 2012a. Stability of health related compounds in plant foods through the application of non thermal processes. *Trends in Food Science and Technology*, 23, 111–123.

Oms-Oliu, G., Odriozola-Serrano, I., and Martín-Belloso, O. 2012b. The effects of non-thermal technologies on phytochemicals. Chapter 5. In *Phytochemicals—A Global Perspective of Their Role in Nutrition and Health*. V. Rao, ed., Hard cover, InTech, Rijeka, Croatia, 538pp, ISBN: 978-953-51-0296-0, InTech.

Patras, A., Brunton, N., Da Pieve, S., Butler, F., and Downey, G. 2009. Effect of thermal and high pressure processing on antioxidant activity and instrumental colour of tomato and carrot purees. *Innovative Food Science Emerging Technology*, 10, 16–22.

Patterson, M. F., Linton, M., and Doona, C. J. 2008. Introduction to high pressure processing of foods. Chapter 1. In *High Pressure Processing of Foods*. C. J. Doona, and E. Florence Feehery, eds. USA: Wiley-Blackwell Publishing, 29–42.

Plaza, L., Sánchez-Moreno, C., De Ancos, B., Elez-Martinez, P., Martin-Belloso, O., and Cano, M. P. 2011. Carotenoid and flavanone content during refrigerated storage of orange juice processed by high-pressure, pulsed electric fields and low pasteurization. *LWT—Food Science and Technology*, 44, 834–839.

Plaza, L., Sánchez-Moreno, C., Elez-Martínez, P., de Ancos, B., Martín-Belloso, O., and Cano, M. P. 2006. Effect of refrigerated storage on vitamin C and antioxidant activity of orange juice processed by high-pressure or pulsed electric fields with regard to low pasteurization. *European Food Research and Technology*, 223, 487–493.

Prasad, K. N., Yang, B., Ruenroengklin, N., and Jiang, Y. 2009a. Application of ultrasonication or high-pressure extraction of flavonoids from litchi fruit pericarp. *Journal of Food Process Engineering*, 32(6), 828–843.

Prasad, K. N., Yang, E., Yi, C., Zhao, M., and Jiang, Y. 2009b. Effects of high pressure extraction on the extraction yield, total phenolic content and antioxidant activity of longan fruit pericarp. *Innovative Food Science and Emerging Technologies*, 10, 155–159.

Qiu, W., Jiang, H., Wang, H., and Gao, Y. 2006. Effect of high hydrostatic pressure on lycopene stability. *Food Chemistry*, 97, 516–523.

Sánchez-Moreno, C., Cano, M. P., De Ancos, B., Plaza, L., Olmedilla, B., Granado, F., Elez-Martínez, P., Martín-Belloso, O., and Martín, A. 2005. Intake of Mediterranean vegetable soup treated by pulsed electric fields affects plasma vitamin C and antioxidant biomarkers in humans. *International Journal of Food Sciences and Nutrition*, 56, 115–124.

Sánchez-Moreno, C., Plaza, L., De Ancos, B., and Cano, M. P. 2006. Impact of high-pressure and traditional thermal processing of tomato purée on carotenoids, vitamin C and antioxidant activity. *Journal of the Science of Food and Agriculture*, 86, 171–179.

Shouqin, Z., Jun, X., and Changzheng, W. 2005. High hydrostatic pressure extraction of flavonoids from propolis. *Journal of Chemical Technology and Biotechnology*, 80(1), 50–54.

Terefe, N. S., Matthies, K., Simons, L., and Versteeg, C. 2009. Combined high pressure—mild temperature processing for optimal retention of physical and nutritional quality of strawberries (*Fragaria* x *ananassa*). *Innovative Food Science and Emerging Technologies*, 10, 297–307.

Toepfl, S., Mathys, A., Heinz, V., and Knorr, D. 2006. Review: Potential of high hydrostatic pressure and pulsed electric fields for energy efficient and environmentally friendly food processing. *Food Reviews International*, 22, 405–423.

Tokuşoğlu, Ö. 2012a. *ANPFT2012 International Congress-Proceeding Book. Advanced Nonthermal Processing in Food Technology: Effects of Quality and Shelf Life of Food and Beverages*. Ö. Tokuşoğlu, ed. May 07–10, 2012, Kuşadasi-Turkey, 321pp. Turkey: Celal Bayar University Publishing, ISBN: 978-975-8628-33-9.

Tokuşoğlu, Ö. 2012b. High pressure processing (HPP) strategies on polyphenolic bioactives and shelf life stability in foods and beverages. In *7th International Conference on High Pressure Bioscience and Biotechnology*. L1–14, Session 2, 1600–1645. *Book of Abstracts*. p. 40. October 29–November 2, 2012, Otsu, Kyoto, Japan. Keynote Lecture.

Tokuşoğlu, Ö., and Hall, C. 2011. *Introduction to Bioactives in Fruits and Cereals* (Chapter 1—Part I. Introduction). In *Fruit and Cereal Bioactives: Sources, Chemistry and Applications*. Ö. Tokuşoğlu and C. Hall, eds. Boca Raton, Florida, USA: CRC Press, Taylor & Francis Group, 459pp. ISBN: 9781439806654; ISBN-10: 1439806659.

Tokuşoğlu, Ö., and Doona, C. 2011a. *High Pressure Processing Technology on Bioactives in Fruits and Cereals* (Chapter 21—Part IV. *Functionality, Processing, Characterization and Applications of Fruit and Cereal Bioactives*). In *Fruit and Cereal Bioactives: Sources, Chemistry and Applications*. Ö Tokuşoğlu and C. Hall, eds. Boca Raton, Florida, USA: CRC Press, Taylor & Francis Group, 459pp. ISBN: 9781439806654; ISBN-10: 1439806659.

Tokuşoğlu, Ö., and Doona, C. J. 2011b. High pressure processing (HHP) strategies on food functionality, quality and bioactives: Biochemical and microbiological aspects. *Nonthermal Processing Division Workshop 2011*. October 12–14, Osnabrück, Germany. Oral Presentation.

Tokuşoğlu, Ö., and Yıldırım, Z. 2012. Effects of cooking methods on the anthocyanin levels and antioxidant activity of a local Turkish sweetpotato [*Ipomoea batatas* (L.) Lam.] cultivar Hatay Kırmızı: Boiling, steaming and frying effects. *Turkish Journal of Field Crops*, 17(2), 87–90.

Tokuşoğlu, Ö. 2001. *The Determination of the Major Phenolic Compounds (Flavanols, Flavonols, Tannins and Aroma Properties of Black Teas*. PhD thesis. Department of Food Engineering, Bornova, Izmir, Turkey: Ege University.

Tokuşoğlu, Ö. 2011. *Effects of High Pressure and Ultrasound on Food Phytochemicals and Quality Improving. Invited Research Conference*. U.S. Army Natick Soldier Research, Development & Engineering Center, Natick, Massachusetts, MA, USA. June 17, 2011, Friday 13:00 p.m., 15:00 p.m. Oral Presentation.

Tokuşoğlu, Ö., Alpas, H., and Bozoğlu, F. T. 2010. High hydrostatic pressure effects on mold flora, citrinin mycotoxin, hydroxytyrosol, oleuropein phenolics and antioxidant activity of black table olives. *Innovative Food Science and Emerging Technologies*, 11(2), 250–258.

Tokuşoğlu, Ö., Bozoğlu, F., and Doona, C. J. 2010. High pressure processing strategies on phytochemicals, bioactives and antioxidant activity in foods. Technical Research Paper. 2010 IFT *Annual Meeting + Food Expo. Book of Abstracts* p.115. Division: Non-thermal processing; Session: 100-3/Novel bioactives: Approaches for the search, evaluation and processing for nutraceuticals, July 19, 2010, 9:15 a.m.–9:35 a.m.; Room N426b, Mc Cormick Place, Chicago/Illinois, USA. Oral Presentation.

Tomas-Barberan, F. and Espin, J. C. 2001. Phenolic compounds and related enzymes as determinants of quality of fruits and vegetables. *Journal of the Science of Food and Agriculture*, 81, 853–876.

Varela-Santos, E., Ochoa-Martinez, A., Tabilo-Munizaga, G., Reyes, J.E., Pérez-Won, M., Briones-Labarca, V., and Morales-Castro, J. 2012. Effect of high hydrostatic pressure (HHP) processing on physicochemical properties, bioactive compounds and shelf-life of pomegranate juice. *Innovative Food Science and Emerging Technologies*, 13, 13–22.

Van Eylen, D., Oey, I., Hendrickx, M., and Van Loey, A. 2007. Kinetics of the stability of broccoli (*Brassica oleracea* cv. Italica) myrosinase and isothiocyanates in broccoli juice during pressure/temperature treatments. *Journal of Agricultural and Food Chemistry*, 55, 2163–2170.

Vázquez-Gutiérrez, J. L., Plaza, L., Hernando, I., Sanchez-Moreno, C., Quiles, A., de Ancos, B., and Cano, M. P. 2013. Changes in the structure and antioxidant properties of onions by high pressure treatment. *Food and Function*, 4(4), 586–591.

Vázquez-Gutiérrez, J. L., Quiles, A., Hernando, I., and Pérez-Munuera, I. 2011. Changes in the microstructure and location of some bioactive compounds in persimmons treated by high hydrostatic pressure. *Postharvest Biology and Technology*, 61, 137–144.

Wang, W., and Xu, S. 2007. Degradation kinetics of anthocyanins in blackberry juice and concentrate. *Journal of Food Engineering*, 82, 271–275.

Wang, Y., Liu, F., Cao, X., Chen, F., Hu, X., and Liao, X. 2012. Comparison of high hydrostatic pressure and high temperature short time processing on quality of purple sweet potato nectar. *Innovative Food Science and Emerging Technologies*, 16, 326–334.

Yen, G., and Lin, H. T. 1999. Changes in volatile flavor components of guava juice with high-pressure treatment and heat processing and during storage. *Journal of Agricultural and Food Chemistry*, 47, 2082–2087.

Zabetakis, I., Leclerc, D., and Kajda, P. 2000. The effect of high hydrostatic pressure on the strawberry anthocyanins. *Journal of Agricultural and Food Chemistry*, 48, 2749–2754.

Zhang, S., Xi, J., and Wang, C. 2005. High hydrostatic pressure extraction of flavonoids from propolis. *Journal of Chemistry Technology Biotechnology*, 80, 50–54.

3 High-Pressure Processing for Improved Dairy Food Quality

Özlem Tokuşoğlu, Barry G. Swanson, and Gustavo V. Barbosa-Cánovas

CONTENTS

3.1 INTRODUCTION

Milk has been used by humans since the beginning of recorded time to provide both fresh and storable nutritious food. In some countries in the world, almost half the milk produced is consumed as fresh-pasteurized whole, low-fat, or skim milk. Most milk is manufactured into more stable dairy products of worldwide commerce, such as butter, cheese, yogurt, ice cream, condensed milk and dried milk, and milk powder (Bandler, 2013).

High-pressure processing (HPP) technology provides advantages to the dairy foods such as killing bacteria, extending shelf life, maintaining the texture, enhancing desired attributes, and it has been produced as value-added dairy food products by high hydrostatic pressure (HHP). HPP is a cold-processing technique that has important applications in processed dairy products with improved characteristics. HPP influences the physicochemical and technological properties of milk and dairy products whereas it homogenizes milk and inactivates the microorganisms in milk, milk-based products, and premixes of dairy products (Bermúdez-Aguirre and Barbosa-Cánovas, 2012; Chawla et al., 2011; Kadam et al., 2012; Rastogi et al., 2007; San Martín et al., 2002; Taşkın and Tokuşoğlu, 2012).

3.2 ABOUT HIGH-PRESSURE PROCESSING

Among the modern technologies in the food industry, the most important technologies are those involving nonthermal treatment of the product. To overcome the negative effects on heat-labile nutrients in the food, several nonthermal processing or "cold-processing" techniques including HHP technology have been developed. The application of HPP as a method for microbial inactivation has been acknowledged in the food industry. The effectiveness of HPP on microbial inactivation has to be studied in great detail to ensure the food quality and safety (Barbosa-Cánovas, 1998, 2005). HPP (100–1000 MPa) is one of the most promising methods for food treatment and preservation at room temperature (Cheftel, 1992), and is of great concern because of its potential to achieve interesting functional effects (Bermúdez-Aguirre and Barbosa-Cánovas, 2012; Rastogi et al., 2007; San Martín et al., 2002; Toepfl, 2006; Trujillo et al., 2002).

High-pressure technology is increasingly being used to produce value-added food products. It is a leading alternative in replacing thermal processing in some food applications in order to meet increasing consumer demand for foods with improved organoleptic qualities and higher acceptance. HPP is a technology that potentially addresses many, if not all, of the most recent challenges faced by the food industry. It can facilitate the production of food products that have the quality of fresh foods, but the convenience and profitability associated with shelf-life extension (Chawla et al., 2011; McClements et al., 2001; San Martín et al., 2002; Tokuşoğlu, 2011, 2012; Tokuşoğlu and Doona, 2011; Tokuşoğlu et al., 2010).

In the HPP method, the pressure vessel is filled with a food product and pressure is applied for a desired time and then it is depressurized (Chawla et al., 2011). Vessels with a volume of several thousand liters are in use, with typical operating pressures in the range of 100–500 MPa and holding times of about 5–10 min. Laboratory-scale high-pressure (HP) equipment capable of reaching pressures up to 1000 MPa is also available (Huppertz et al., 2002; Myllymaki, 1996). This technology can be conducted at ambient or moderate temperatures, hence eliminating thermally induced cooked off-flavors. Compared to thermal processing, HPP-treated foods result in products with a fresher taste, better appearance, and texture (Bermudez-Aguirre and Barbosa-Cánovas, 2011; Patterson et al., 2008).

Recently, HPP applications on milk and dairy products have greatly increased (Garcia-Amezquita et al., 2013; Gaucheron et al., 1997; Johnston, 1992; Liu et al., 2005a,b; Martinez-Rodriguez et al., 2012; Masson et al., 2011; Penna et al., 2007; Pereda et al., 2007, 2008; Taskin and Tokuşoğlu, 2012). In this chapter, HPP effects on milk and dairy foods quality, physicochemical characteristics, functionality, and shelf life are addressed. Table 3.1 shows the current utilization of HHP on milk and dairy industry.

3.3 EFFECTS OF HIGH-PRESSURE PROCESSING ON MICROORGANISMS AND SHELF LIFE OF MILK AND DAIRY FOODS

Milk, being a functional perishable foodstuff, is usually heat treated with a specific time–temperature combination, to obtain safe, acceptable, and appropriate shelf life.

TABLE 3.1

Recent Utilization of HHP Processing in Milk and Dairy Industry

Utilization	HHP Pressure Range (MPa)	Milk and Dairy Products
Microbial inactivation	200–800	Raw milk, milk-based mixes, predairy mixes, and dairy products
Cheese making and ripening of cheese	200–700	M and C cheeses
Fat globule size reduction	200	Milk, cheese, and yogurt
Viscosity improvement	200–400	Yogurt and ice cream
Reducing of rennet coagulation time (curdling time reduction)	200–670	M and C cheeses

Utilizing heat treatment extends the shelf life of milk, but simultaneously destroys its natural nutrients and bioactives. Raw milk has often been a vehicle in outbreaks of staphylococcal food poisoning (Chawla et al., 2011; De Buyser et al., 2001).

High-temperature treatments may result in undesirable effects in milk, such as off-flavor, nonenzymatic browning, and denaturation of certain protein and vitamins (Brinez et al., 2007; Diels et al., 2005; Kheadr et al., 2002). Over a century ago, HPP was demonstrated to be an effective method to extend the shelf life of milk and other food products, and the first commercial HP-treated product appeared in the market in 1991 in Japan (Patterson and Kilpatrick, 1998; Smelt, 1998; Trujillo, 2002).

Most studies report that HHP causes a number of alterations in morphology, cell wall, thermotropic phase in cell membrane lipids, dissociation of ribosomes, biochemical reactions, and loss of genetic functions of the microorganisms (Cheftel, 1995; Ritz et al., 2000; Wouters et al., 1998).

HHP inactivates the pathogen microorganisms of milk and extends the shelf life of the milk and dairy products. Heat treatment extends the shelf life, but at the same time, damages its nutrients. Hence, to preserve its nutritional value, HHP treatment can be considered as an alternative preservation (Chawla et al., 2011; O'Reilly et al., 2001). Current studies have proven that HHP applications at 400–600 MPa give raw milk the same quality as that of pasteurized milk.

HHP is effective for destroying pathogenic and spoilage microorganisms, but owing to the presence of pressure-resistant spores, to achieve a shelf life of 10 days at a storage temperature of 10°C a pressure treatment of 400 MPa for 15 min or 600 MPa for 3 min at 20°C is necessary (Rademacher and Kessler, 1997). It was reported that HHP treatment at 400 MPa applied to cheese at room temperature can easily inactivate yeast and mold spores. 6-Log cycle reductions of *Penicillium roqueforti* spores were found in cheddar (C) cheese slurry by applied HHP conditions (400 MPa/20 min/20°C) (O'Reilly et al., 2000).

It was put forward that vegetative forms of yeasts and molds are the most pressure sensitive compared to their spores (Smelt, 1998) and result in 99.99% of reduced activity of psychotropics, lactococci, and total bacterial count (Pereda et al., 2007) along with a reduction in yeast and mold growth (Daryaei et al., 2006).

Pressures between 300 and 600 MPa are shown to be an effective method to inactivate microorganisms, including most infectious food-borne pathogens (O'Reilly et al., 2001; Trujillo et al., 2002). It is proven from many studies that HHP treatment at 400–600 MPa gives raw milk the same quality as that of pasteurized milk. However, HHP does not give the quality of sterilized milk, owing to the presence of pressure-resistant spores. The physicochemical environment can adversely change the resistance of a bacterial species to pressure. Gram-positive microorganisms are more resistant to pressure than Gram-negative microorganisms; for example, Gram-positive organisms need an application of 500–600 MPa at 25°C for 10 min to achieve inactivation while Gram-negative organisms can be inactivated with 300–400 MPa with the same time–temperature combination. Factors such as the water activity and pH also influence the extent to which foods need to be treated to eliminate pathogenic microorganisms (Chawla et al., 2011).

It is revealed that dynamic HHP treatment inactivates three major food pathogens including *Listeria monocytogenes* LSD 105-1, *Escherichia coli* O157:H7 ATCC 35150, and *Salmonella enteritidis* ATCC 13047 present in milk by Vachon et al. (2002).

Microbial inactivation in cheese using HP depends on the species, starter cultures, and characteristics of cheese such as composition and acidity (O'Reilly et al., 2001). Prior to the cheese-making process, pressurizing of milk resulted in a positive inactivation of pathogenic *L. monocytogenes* without any negative changes in Camembert cheese (Linton et al., 2008). Similar observations were recorded in soft cheeses with *Staphylococcus aureus* (Lopez-Pedemonte et al., 2007). Moreover, during storage, growth of yeasts, molds, and mesophilic bacteria is considerably reduced compared to unpressurized cheese samples, extending the shelf life of the product (O'Reilly et al., 2001).

Salmonella, E. coli, Shigella, and *S. aureus* are significant food-borne pathogens and are commonly found in raw milk. Interest in the HHP application to milk pasteurization has recently increased (Trujillo, 2002). Using HP, not only the pathogens can be inactivated, but also the quality characteristics such as flavor, taste, vitamins, and nutrients can be improved (Trujillo, 2002).

HP treatment applied in the range of 300–500 MPa/15 min resulted in lactococci inability to acidify milk, although the peptidolytic activity remained constant even after treatment (Krasowska et al., 2005).

However, *S. aureus* appears to be highly resistant to pressure, and the level of *S. aureus* in a complex medium was decreased below the detection limit (8–9 log-units reduction) using only 900 MPa/5 min of HP treatment without heating by Jofré et al. (2010). It was asserted that lower pressures required either longer treatments or combination with temperature (Fioretto et al., 2005; Gervilla et al., 1999).

Yang et al. (2012) reported the inactivation of food-borne pathogens (*Salmonella, E. coli, Shigella,* and *S. aureus*) in raw milk using HHP to determine the optimal inactivation conditions and further understand the mechanisms of HHP on pathogens inactivation in food (Yang et al., 2012) (Figure 3.1). It was concluded that 300 MPa treatment/30 min duration/25°C was the optimum condition for inactivation of *Salmonella, E. coli, Shigella,* and *S. aureus* (Yang et al., 2012) (Figure 3.1). It was determined that the injured cells could not be recovered, and the growth rate of survivors was much lower than that of the untreated cells based on the transmission electron microscope (TEM) micrographs. Moreover, pulsed-field gel electrophoresis (PFGE) showed neither corresponding deoxyribonucleic acid (DNA) bands with same molecular weight nor DNA bands with same brightness could be found in the lanes between HHP-treated pathogens and HHP-untreated pathogens (Yang et al., 2012). It was also indicated that HHP processing can be applied to inactivate pathogens in food, and the inactivation is mainly due to cell membrane damage, cell wall rupture, and chromosome DNA degradation by Yang et al. (2012).

The microorganism resistance to HHP is highly variable, based on the microorganism type, physiological state of the microorganism, and the food matrix. Fat, protein, mineral, and sugars in the food content can serve as protectors for microorganisms and can increase microbial resistance to pressure (Black et al., 2007).

It was reported that microbial inactivation in cheese using HP depends on the species, starter cultures, and characteristics of cheese such as composition and acidity (O'Reilly et al., 2001).

Using antimicrobial substances such as bacteriocins, lysozyme, or essential oils can improve the efficacy of HHP treatments (Corbo et al., 2009; Evrendilek and Balasubramaniam, 2011; Vurma et al., 2006). The inactivation of *S. aureus* strains was performed by HHP treatment or combined HHP applications with natural antimicrobials (nisin, enterocin AS-48, cinnamon oil, and clove oil) in rice pudding (Pérez Pulido et al., 2012). It was detected that treatments at 600 MPa for 10 min reduced initial populations of staphylococci (7.9 log CFU/g) below detectable levels of 1 log CFU/g in the puddings. With applications at 500 MPa for 5 min, viable counts were decreased 2.9 log singly or in combination with nisin (200 and 500 IU/g), enterocin AS-48 (25 and 50 mg/g), cinnamon oil (0.2% vol/wt), or clove oil (0.25% vol/wt) (Pérez Pulido et al., 2012).

As it is known, bacteriophages are regarded as natural antibacterial agents because they are able to specifically infect and lyze undesirable bacteria (García et al., 2010). Effectiveness of bacteriophages was proven in milk preservation and

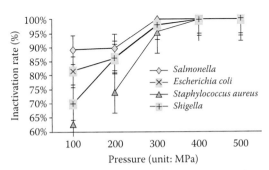

Notes: The abscissa axis indicates different pressure, and the ordinate axis indicates inactivation rate under HHP treatment. MPa: million pascal.

FIGURE 3.1 Pressure effect on inactivation rates of four pathogens at 25°C with 30 min duration. (Adapted from Yang B. et al. 2012. *Food Control* 28, 273–278.)

were confirmed the advantages of bacteriophages for *S. aureus* infecting dairy products. These phages can be used as biocontrol agents (García et al., 2009). It was reported that a bacteriophage preparation effective against *L. monocytogenes* titled Listex™ P100 (EBI Food Safety, www.ebifoodsafety.com) received the U.S. Food and Drug Administration (FDA) approval for use on cheese (U.S. FDA/CFSAN, 2006) and on all food products (U.S. FDA/CFSAN, 2007).

In the study described by Tabla et al. (2012), the effect of a cocktail of two bacteriophages (IPLA35 and IPLA88) and HHP against *S. aureus* in milk was performed. Bacteriophages can be regarded as valuable hurdles on minimally processed foods (Tabla et al., 2012).

It was found that 400 MPa was found to be the most appropriate pressure used in combination with bacteriophages (IPLA35 and IPLA88). The initial *S. aureus* contamination was reduced below the detection limit (<10 CFU/mL) by using the combined treatment, compared to each single treatment. It was also determined that bacteriophage performance in pressurized milk against *S. aureus* provided a milder opportunity for hydrostatic pressure treatment (Tabla et al., 2012).

3.4   HPP EFFECTS ON WHEY PROTEIN FUNCTIONALITY AND QUALITY OF MILK AND DAIRY PRODUCTS

HP technology improves the functional properties of whey protein as well as in other milk components and native constituents. Pressurization of milk causes conformational changes in milk proteins.

With HPP treatment, the size and number of casein micelles increases as spherical particles change to form chains or clusters of submicelles. HHP causes casein micelles disintegration into smaller particles, with a decrease in milk turbidity and lightness, and an increase in viscosity of the milk. This application improves rennet or acid coagulation of milk without detrimental effects on the important quality characteristics, such as taste, flavor, vitamins, and nutrients, and also reduces the rennet coagulation time (Huppertz et al., 2006; Sivanandan et al., 2008; Trujillo et al., 2002).

HP (up to 400 MPa) has been applied to milk processing, successfully. It was stated that HPP resulted in alterations in its manufacturing and final characteristics of milk products such as yogurt, cheese, and cream (Bermudez-Aguirre and Barbosa-Cánovas, 2011). The improved hardness, surface hydrophobicity, solubility, gelation, and emulsifying properties were determined in HHP-treated whey protein (Lee et al., 2006a).

It has been known that whey and whey products have been used successfully in the food sector for years. Whey proteins are very important food ingredients because of their functional properties. Whey proteins increase milk nonfat solids, are highly soluble, and improve the nutritional value, foaming, and emulsification properties of ice cream and frozen desserts (Jayaprakasha and Brueckner, 1999; Morr and Ha, 1993). The undesirable changes due to heat treatments affect nutritional attributes as well as protein denaturation, which decreases protein solubility and foaming properties of whey proteins (Martin et al., 2002) and also severe thermal treatments (above 70°C) result in protein denaturation, accompanied by loss of aqueous solubility and foaming properties (Kester and Richardson, 1984). In food systems, especially in foams and emulsions, the functional behavior of processed whey may not meet consumer expectations if heated at temperatures >70°C, which is traditionally used to concentrate dry whey protein concentrate (WPC) powder (Richert et al., 1974). The previous research stated that WPC heating at moderate temperatures (40–50°C) improves its foaming properties, while heat treatment at >70°C results in impairment of foaming properties by irreversible whey protein denaturation and aggregation (deWit and Klarenbeek, 1984; Richert et al., 1974). Besides, spray-drying process tends to result in considerable decreases in WPC functionality due to the high-temperature necessity. Improvements in functional properties may be achieved by modifying the protein structure using physical treatments instead of heat (Kato et al., 1983).

The functional properties of WPC can be modified by using HHP processing, involving elevated pressures ranging from 100 to 1000 MPa, with or without the addition of heat (Rodiles-Lopez et al., 2008).

Lim et al. (2008) reported that the maintenance of protein solubility after HHP indicates that HHP-treated WPC might be appropriate for applications to food systems. Untreated WPC exhibited the smallest overrun (OR) percentage, whereas the largest percentage for OR and foam stability was obtained for WPC treated at 300 MPa for 15 min. It was found that HHP–WPC treated at 300 MPa for 15 min acquired larger OR than commercial WPC 35. WPC treated at 300 or 400 MPa for 15 min and 600 MPa for 0 min significantly exhibited greater foam stability than commercial WPC 35. HHP treatment was beneficial to enhance OR and foam stability of WPC, showing promise for ice cream and whipping cream applications (Lim et al., 2008).

Figures 3.2 and 3.3 indicated the whipping time and foam stability for commercial WPC 35, untreated, and HHP-treated Washington State University (WSU)-WPC in different HHP conditions. It is shown that the HHP treatment conditions that promoted functionality of proteins, including large OR and foam stability, were observed after 300 MPa for 15 min. The HHP treatment conditions that promoted functionality of proteins, including large OR and foam stability, were observed after 300 MPa for 15 min in the study described by Lim et al. (2008). In that study, it is stated that the particle size reduction occurring during HHP treatments of whey proteins partially explains improved foaming properties of WSU-WPC owing to HHP treatment increased OR and foam stability of fluid WSU-WPC. Lim et al. (2008) determined that the improvements of OR and foam stability may potentially enhance body and texture of whipping cream or ice cream.

It was reported that HHP modified WPC functionally (Lim et al., 2008). Lim et al. (2008) stated the HHP modification of WPC for improved body and texture of low-fat ice cream. Commercial WPC 35 powder was reconstituted to equivalent total solids as WSU-WPC (8.23%). Three batches of low-fat ice-cream mix were produced to contain WSU-WPC without HHP, WSU-WPC with HHP (300 MPa for 15 min), and WPC 35 without HHP and all low-fat ice-cream mixes contained 10% WSU-WPC or WPC 35. The OR and foam stability of ice-cream mixes were determined after whipping for 15 min. Ice creams were manufactured using standard ice cream ingredients and processing (Lim et al., 2008). It was stated that the ice-cream mix containing HHP-treated WSU-WPC exhibited the greatest OR and foam stability, confirming the effect of HHP on foaming properties of whey proteins in a complex system. Figure 3.4 shows whipping time and OR of low-fat ice-cream mix containing

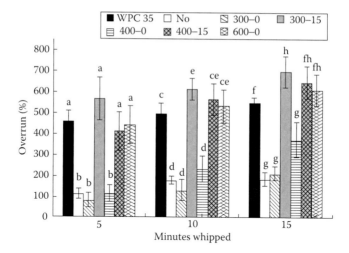

FIGURE 3.2 Whipping time and OR for commercial WPC 35 (no HHP) or WSU-WPC (No = no HHP; 300–0 = 300 MPa for 0 min; 300–15 = 300 MPa for 15 min; 400–0 = 400 MPa for 0 min; 400–15 = 400 MPa for 15 min; and 600–0:600 MPa for 0 min) treated with selected HHP–time combinations. [a–h]Different letters denote significant differences for individual whipping times ($P \leq 0.05$). Vertical lines correspond to standard deviation. (Adapted from Lim S.Y., Swanson B.G., and Clark S. 2008. *Journal of Dairy Science* 91, 1299–1307.)

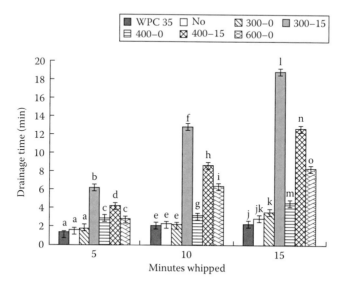

FIGURE 3.3 Whipping time and foam stability for commercial WPC 35 (no HHP) or WSU-WPC (No = no HHP; 300–0 = 300 MPa for 0 min; 300–15 = 300 MPa for 15 min; 400–0 = 400 MPa for 0 min; 400–15 = 400 MPa for 15 min; and 600–0 = 600 MPa for 0 min) treated with selected HHP–time combinations. [a–o]Different letters denote significant differences for individual whipping times ($P \leq 0.05$). Vertical lines correspond to standard deviation. (Adapted from Lim S.Y., Swanson B.G., and Clark S. 2008. *Journal of Dairy Science* 91, 1299–1307.)

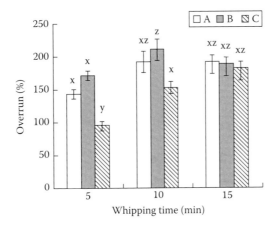

FIGURE 3.4 Whipping time and OR of low-fat ice-cream mix (A = untreated WSU-WPC; B = HHP-treated WSU-WPC; and C = untreated WPC 35). [x-z]Different letters indicate that significant differences exist within whipping times; $p \leq 0.05$; vertical lines indicate SD. (Adapted from Lim S.Y., Swanson B.G., and Clark S. 2008. *Journal of Dairy Science* 91, 1299–1307.)

untreated WPCs and HHP-treated WSU-WPC. It was also reported that ice cream containing HHP-treated WSU-WPC significantly indicated a greater hardness than ice cream produced with untreated WSU-WPC or WPC 35 (Lim et al., 2008).

It is known that ice-cream mix viscosity is an important parameter in determining flow behavior (Goff et al., 1994). Instrumental viscosity for low-fat ice-cream mix containing HHP-treated WSU-WPC was significantly greater than the viscosity for low-fat ice-cream mix containing untreated WSU-WPC or WPC 35 (Figure 3.5) (Lim et al., 2008). Figures 3.5 and 3.6 indicate that the apparent viscosity of ice-cream

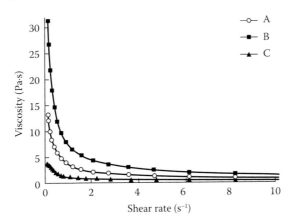

FIGURE 3.5 Apparent viscosity at 10°C on shear rate (0.1–10 s⁻¹, upward rate) of low-fat ice-cream mix formulation (A = untreated Washington WSU-WPC; B = HHP-treated WSU-WPC; and C = untreated WPC 35). (Adapted from Lim S.Y., Swanson B.G., and Clark S. 2008. *Journal of Dairy Science* 91, 1299–1307.)

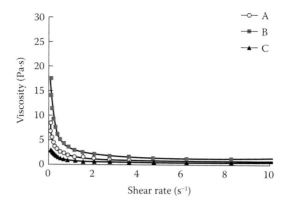

FIGURE 3.6 Apparent viscosity at 10°C on shear rate (10–0.1 s⁻¹ downward rate) of low-fat ice-cream mix formulation (A = WSU-WPC; B = HHP-treated WSU-WPC; and C = untreated WPC 35). (Adapted from Lim S.Y., Swanson B.G., and Clark S. 2008. *Journal of Dairy Science* 91, 1299–1307.)

mixes was decreased by an increase in shear rate, although experimental observations were made at small shear rates (between 0.1 and 10 s⁻¹) (Lim et al., 2008).

HHP treatment of whey proteins indicated a positive effect on low-fat ice-cream mix OR and foam stability and low-fat ice-cream hardness, demonstrating that HHP affects the low-fat ice-cream acceptability even when HHP-treated whey proteins are used at a concentration as low as 0.82% TS and 0.3% protein in a formulation of low-fat ice cream. It was shown that HHP treatment of fresh WPC may provide benefits for applications requiring foaming properties in the dairy food industry (Lim et al., 2008).

Lim et al. (2008) stated the HHP modification of WPC for improved functional properties.

Liu et al. (2005b) studied the effect of HHP on hydrophobicity of WPC and recorded that HHP treatment of WPC yields an increase in the number of binding sites that leads to certain modifications in proteins that enhance hydrophobicity and shows promising results for improving functional properties of foods. Similar observations for improved hardness, surface hydrophobicity, solubility, gelation, and emulsifying properties were recorded in whey proteins functionality (Lee et al., 2006a).

As it is known, ANS (8-anilino-1-naphthalenesulfonic acid-anionic, aromatic probe) and CPA (*cis*-parinaric acid-anionic, aliphatic probe) probes are widely used to assay hydrophobicity of food proteins (Yang et al., 2001). The hydrophobic probe-binding behavior of WPC is affected by the holding time of pressurization. In the study stated by Liu et al. (2000b), the ANS extrinsic fluorescence of HHP-treated WPC exhibited three distinct regions (Figure 3.7). HHP treatment up to 2.5 min resulted in a 1.4-fold increase in the ANS extrinsic fluorescence intensity. HHP treatment for 30 min resulted in a twofold increase in the ANS fluorescence intensity (Liu et al., 2005b). It was reported that HHP resulted in an increase in intrinsic tryptophan fluorescence intensity and a 4 nm red shift

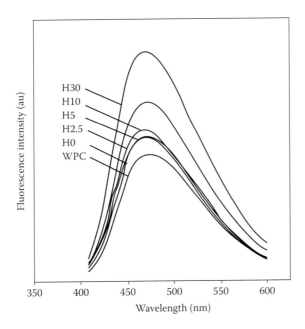

FIGURE 3.7 Mean of three independent extrinsic ANS emission spectra of WPC solutions affected by HHP at 600 MPa and 50°C for various holding times from 0 to 30 min (H0–H30). (Adapted from Liu X. et al. 2005b. *Innovative Food Science and Emerging Technologies* 6, 310–317.)

after 30 min of treatment, which indicate alterations in the polarity of tryptophan residues microenvironment of whey proteins from a less polar to a more polar environment (Liu et al., 2005b). During HHP treatments, conformational changes of whey proteins and aggregation affect the hydrophobicity and binding properties of WPC. Liu et al. (2005b) put forward that HHP-treated WPC may display improved functionality and provide opportunities for increasing WPC utilization in the food industry.

Padiernos et al. (2009) reported that the HHP modification of WPC for use in low-fat whipping cream improves foaming properties.

It is known that good-quality whipping cream must exhibit a short whipping time, a large OR, and stable foams (Jakubczyk and Niranjan, 2006). It was exhibited that low-fat whipping cream containing no WPC had the least foam stability and low-fat whipping cream containing HHP-treated WPC had the greatest foam stability. Table 3.2 shows that mean foam stability of low-fat whipped creams expressed as the amount of drainage. In the study described by Padiernos et al. (2009), it was revealed that low-fat whipping cream containing HHP-treated WPC had significantly greater foam stability than low-fat whipping cream containing untreated WPC (Table 3.1; $p \leq 0.05$). It was stated that improvements in the foaming properties of low-fat whipping creams with the incorporation of WPC may positively affect the dairy sector by increasing the incorporation of HHP-treated WPC in reduced-fat dairy products.

TABLE 3.2
Foam Stability of Low-Fat Whipped Creams as the Amount of Drainage

Low-Fat Whipped Cream	Amount of Drainage[a] (%)
Without WPC	12.13[b]
Untreated WPC	6.19[c]
HHP-treated WPC	3.66[d]

Source: Adapted from Padiernos C.A. et al. 2009. *Journal of Dairy Science* 92, 3049–3056.

[a] Weight of drainage obtained after 12 h divided by the weight of low-fat whipping cream at the beginning.

[b-d] Different superscripts in a column indicate that a significant difference exists between means.

HP treatment also enhances pepsin hydrolysis of β-lactoglobulin (β-LG) at 400 MPa, reduction in antigenicity, and immunoglobulin E (IgE) binding of β-LG, which further opens the possibility of obtaining hypoallergenic hydrolyzates of β-LG (Chicón et.al., 2008). A pressure treatment of 500 MPa at 25°C denatures lactoglobulin. Denaturation of immunoglobulins and lactalbumins occurs only at the highest pressures, particularly at temperature above 50°C, which gives an idea of preservation of colostrum immunoglobulins that otherwise gets damaged during heat treatment (Felipe et al., 1997).

3.5 HPP EFFECTS ON MILK FAT AND MILK SUGAR

Gervilla et al. (2001) stated that hydrostatic pressure up to 500 MPa catalyzes modifications in size and distribution of milk fat globules of ewes' milk. HHP treatments at 25°C and 50°C showed a tendency to increase the number of small globules (1–2 μm) whereas at 4°C, the tendency gets reversed (Gervilla et al., 2001). No damage was found in the milk fat globule membrane and it was stated that these modifications in distribution of milk fat globules could be due to the aggregation and disintegration of fat globule membrane (Gervilla et al., 2001). Besides, HHP-treated milk showed the advantage of stability against creaming off, when done at 25°C and 50°C, but at 4°C, a reverse phenomenon was observed (Gervilla et al., 2001).

In ewes' milk, free fatty acids (FFAs) content by lipolysis of milk fat showed that HHP treatments of 100–500 MPa at 4°C, 25°C, and 50°C did not increase FFA levels (Gervilla et al., 2001). Even some treatments at 50°C gave lower FFA content than that of fresh raw milk. It was reported that this phenomenon is of great interest to avoid production of off-flavors originated from lipolytic rancidity in milk (Gervilla et al., 2001).

Buchheim and Frede (1996) reported that HHP treatment induced the fat crystallization, shortened the time required to achieve a desirable solid fat content, and thereby reduced the aging time of ice-cream mix. It was put forward that 600 MPa/2 min pressurization treatment improved the whipping ability of cream and is possibly owing to better crystallization properties of milk fat (Eberhard et al., 1999).

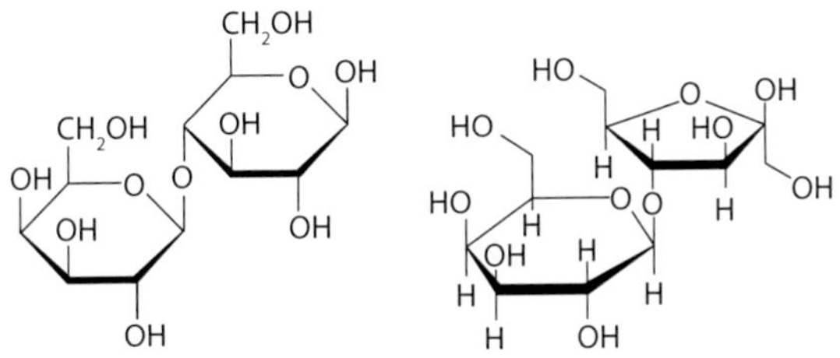

FIGURE 3.8 Lactose (main sugar in milk) (left) and lactulose (right).

Messens et al. (1999) revealed that cheeses made from HHP-treated milk indicated a similar level of lipolysis as in cheeses made from raw milk, whereas the lipolysis level in cheese made from pasteurized milk was lower. It was explained that these situations can be originated from heat-sensitive/partial pressure-resistant properties of the indigenous milk lipase (Messens et al., 1999).

Needs et al. (2000) reported the lower values of fracture stress in set yogurts made from HHP-treated milk (60 MPa for 15 min) compared to heat-treated milk (Needs et al., 2000). Serra et al. (2008) stated that yogurt prepared from milk that was ultra-HP homogenized at 200 and 300 MPa at 30°C and 40°C, considered modifications induced in the fat fraction that could delay the lipid oxidation and lower the degree of lipolysis.

Lactose ($C_{12}H_{22}O_{11}$) (Figure 3.8, left) is the main carbohydrate of milk and it is known as milk sugar. Lactose is a disaccharide sugar derived from galactose and glucose that is found in milk. Lactose makes up around 2 ~ 8% of milk (by weight), although the amount varies among species and individuals (Linko, 1982). It was stated that milk products may isomerize to lactulose (Figure 3.8, right) by heating and then degrade to form acids and other sugars. It was determined that no alterations in these compounds have been observed after HHP (100–400 MPa for 10–60 min at 25°C). It was also identified that no Maillard or lactose isomerization reaction occurs in milk after HHP treatment (Lopez-Fandino et al., 1996).

3.6 EFFECTS OF HHP ON PHYSICOCHEMICAL PROPERTIES, RHEOLOGY, OVERALL QUALITY, AND BIOACTIVES OF DAIRY PRODUCTS (CHEESE, ICE CREAM, ETC.)

In HP-treated milk, whey proteins are found to be associated with casein micelles as well as in the serum phase of milk (Anema, 2008). HP treatment also affects casein micelles profoundly. The average micelle size in HP-treated milk may be increased or decreased, depending on the treatment pressure, temperature, and time (Anema et al., 2005). It was reported that the hydration of casein micelles increased by HHP, as a result of increasing viscosity of milk (Huppertz et al., 2004). It was reported that the extent of micellar disruption under HP decreased with increasing concentration of micellar casein (Huppertz and De Kruif, 2006).

Escobar et al. (2011) stated HP homogenization as an alternative processing in the manufacturing of Queso fresco (QF) cheese. On the basis of the study given by Escobar et al. (2011), the highest cheese yield, moisture content, and crumbliness were

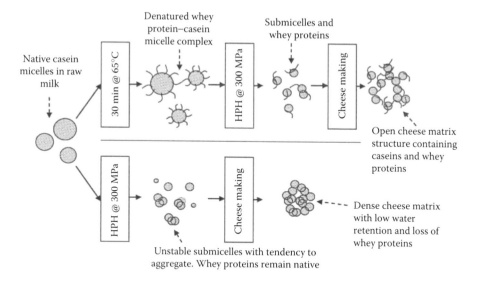

FIGURE 3.9 Proposed structural changes in cheese made from whole milk processed using thermal processing and HP homogenization. (Adapted from Escobar D. et al. 2011. *Journal of Dairy Science* 94(3), 1201–1210.)

obtained for thermally processed milk subjected to HP homogenization at 300 MPa. It was reported that HP homogenization has a strong potential for the manufacturing of QF with excellent yield and textural structure. Figure 3.9 shows the proposed structural alteration in cheese made from thermal-processed and HP-homogenizated whole milk (Escobar et al., 2011). It was stated that the combination of high-pressure homogenization (HPH) and thermal processing of milk resulted in cheeses with increased yield and moisture content. It was suggested that the combination of thermal processing and HP homogenization promoted thermally induced denaturation of whey protein, together with homogenization-induced dissociation of casein micelles. It was also found that the combined effect resulted in QF containing a thin casein–whey matrix that is able to retain sweet whey better (Escobar et al., 2011).

It was indicated the less micellar disruption in HP-treated milk with increasing solids content of the milk (Anema, 2008). HP-induced gelation of milk was postulated to require milk solids concentrations as >25% and pressures >400 MPa (Velez-Ruiz and Barbosa-Canovas, 1998) and disruption of casein micelles under pressure (Fertsch et al., 2003).

Capellas et al. (2001) stated that *L** (lightness) and *a** (redness) values decreased, while *b** increased at HHP-treated (500 MPa/5, 10, and 15 min/10°C) mat'o cheeses compared to untreated mat'o cheese samples. Sandra et al. (2004) found the cheeses to be more yellowish after 1 d post-HHP treatment (400 MPa/20 min/20°C) than control QF cheeses, but not after 8 d.

Lower lightness and higher chroma values were determined in garrotxa cheeses at HHP treated under 400 MPa/5 min/14°C conditions than those of control garrotxa samples by Saldo et al. (2002).

Rynne et al. (2008) reported that increasing pressure intensity and holding time did not affect $L*$ values. It was determined that HHP- (400 MPa/10 min) treated cheeses had altered color values and were discriminated from the control cheese by descriptive sensory analysis. The $a*$ (redness) decreased and $b*$ (yellowness) increased compared to control C cheeses. It was found that HP-treated cheese also had significantly higher fracture strain, fracture stress levels, and significantly lower flowability than those of control C cheese along with ripening (Rynne et al., 2008).

Recent studies showed that HHP treatments did not change total solid, ash, fat, protein, moisture, and nutrient levels in cheese (Koca et al., 2011; Rynne et al., 2008; Sheehan et al., 2005).

In the study given by Koca et al. (2011), the 200 and 400 MPa/22°C/5 and 15 min of applications resulted in significantly softer, less springy, less gummy, and less chewy cheese and these applied HHP conditions did not affect moisture, protein, and fat contents of white-brined cheeses. It was put forward that the cheese became more greenish and yellowish with the increase in pressure level (Koca et al., 2011).

It was found that pressures up to 300 MPa (5 min/25°C) applied to 1-month-old cheese had no significant effect and at 800 MPa (5 min/25°C), cheese had similar fracture stress and Young's modulus as control cheese whereas pressure applied to 4-month-old cheese increased fracture work (Wick et al., 2004).

In the study given by O'Reilly et al. (2002a), low-moisture Mozzarella cheese (LMMC) was exposed to HP treated at different stages (Figure 3.10). It was stated that HP treatment resulted in an increase in the flowability and a reduction in the melt time on heating at 280°C, especially at storage times ≤15 days. It was reported that reduced time was required to attain satisfactory cooking performance (by 15 d) (O'Reilly et al., 2002) (Figure 3.10).

In the research stated by Lee et al. (2006b), experimental low-fat-processed cheese foods were prepared from cheese base and C cheese with the addition of 5% (w/w) untreated whey protein (LWP) or 5% (w/w) whey protein treated at 690 MPa for 5 min (LHP). Lee et al. (2006b) stated that low-fat-processed cheese foods prepared from ultra-HP-treated whey protein resulted in acceptable firmness and meltability. However, the texture was undesirable due to sandy or grainy texture.

Ice cream is a smooth and soft frozen mixture of a combination of components of milk, cream, sweeteners, stabilizers, emulsifiers, flavoring, and possibly other ingredients such as egg products and coloring (Goff and Hartel, 2004). As common ice cream stabilizers, gelatin (E441), guar gum (E412), sodium carboxymethyl cellulose (CMC) (E466), locust bean gum (carob bean gum) (LBG) (E410), carrageenan (Irish moss) (E407), xanthan (E415), alginates, microcrystalline cellulose (cellulose gel) (MCC) (E460), and local hydrocolloids (salep, a type of glycomannan, etc.) have been used (Goff and Sahagian, 1996). Stabilizers have very beneficial functions in ice cream, but their excessive use may create problems; these limitations include undesirable melting characteristics, excessive mix viscosity, and contribution to a heavy, soggy body, also impart off-flavors (if not kept in a dry and cool environment). It is stated that ice-cream mixes exhibit non-Newtonian pseudoplastic behavior, meaning that there is a nonlinear relationship between shear stress and shear rate, with the apparent viscosity decreasing with increasing shear rate (Bahramparvar and Tehrani,

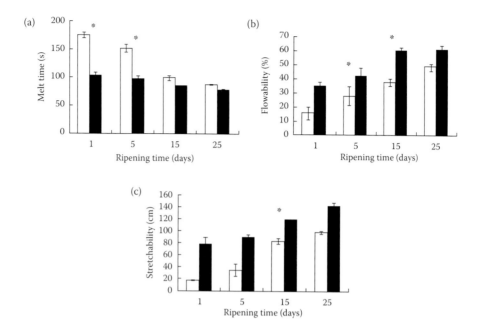

FIGURE 3.10 (See color insert.) Comparison of the heat-induced functional parameters: (a) melt time, (b) flowability, and (c) stretchability, in control LMMC stored for 1, 5, 15, and 25 days at 4°C (□) or cheeses that were HP treated after these storage times at 400 MPa for 20 min at 25°C (■). Error bars represent standard deviations and* denotes significant difference between control and treated samples, $p < 0.05$. (Adapted from O'Reilly C.E. et al. 2002a. *Innovative Food Science and Emerging Technologies* 3, 3–9.)

2011). It is known that ice cream viscosity is influenced by mix composition (mainly stabilizer and protein), by type and quality of the ingredients, by processing and by handling of the mix, by concentration (total solid content), and by temperature. As the viscosity increases, the resistance to melting and the smoothness of texture increases, but the rate of whipping decreases (Goff and Hartel, 2013; Marshall et al., 2003).

Controlling the viscosity of ice-cream mixes at pasteurization temperatures was more important. Recently, the effects of HP treatment on ice-cream mix and ice cream prepared therefrom have been studied.

HP provided opportunities to structure the serum phase of ice-cream mixes, the effects of HP treatment on the physicochemical properties of ice-cream mixes, and manufactured ice cream was revealed (Huppertz et al., 2011). It was shown that HHP treatment had little effect on the size of the milk fat globules, but increased the viscosity of the ice- cream mix, considerably. It was stated that the viscosity of HP-treated ice-cream mix increased with increasing pressure and treatment time and with increasing fat, milk solids nonfat, and sucrose content of the mix (Huppertz et al., 2011; Goof et al., 1994).

It was stated the HHP in various treatment pressure (300, 400, and 500 MPa) and holding conditions (10/30 min at 25 ± 1°C) and holding time on the rheology, overall quality, and antioxidant activity (AA) of high-processed strawberry, huckleberry,

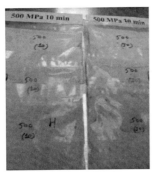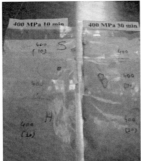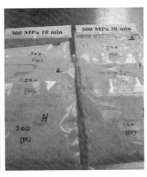

FIGURE 3.11   (**See color insert.**) HHP-processed ice-cream mixes at 300/400/500 MPa, 10/30 min. (Photo: Tokuşoğlu, 2010, WSU.)

and blackberry ice-cream mixes (BIMs) (Tokuşoğlu et al., 2011) (Figure 3.11). It also performed the preliminary studies on storage stability and shelf-life stability of HHP-processed berry ice creams via quality parameters detection.

It was found that the viscosity of HP-treated ice-cream mix increased with increasing pressure and treatment time and with increasing phenolic content ($p \leq 0.005$). For huckleberry ice creams (HIMs), and strawberry ice creams (SIMs), 400 MPa/10 min HHP processed samples showed the highest viscosity whereas the viscosity of blackberry ice creams increased with increasing pressure and treatment time (Figure 3.12).

In the third study given by Tokuşoğlu et al. (2011), the general form of the Weibull distribution function, which is a two-parameter function, for viscosity is given by

$$f(v) = \left(\frac{k}{c}\right)\left(\frac{v}{c}\right)^{k-1} \exp\left[-\left(\frac{v}{c}\right)^k\right]$$

where $f(v)$ is the probability of observing viscosity, k is the dimensionless Weibull shape parameter (or factor), and c is shear rate, as the Weibull-scale parameter. Table 3.3 shows the viscosity, shear rate, and shear stress correlations of blackberry ice cream processed at 300 MPa/10 min by using the Weibull model. The Weibull model explained the viscosity–shear rate–shear stress relations of three types of berry ice creams ($r = 0.997$). For blackberry ice cream, 300 MPa/10 min of HHP treatment, estimated viscosity was found as the equation shown below (Tokuşoğlu et al., 2011):

Estimated viscosity = 3.271 + 20.912 × (1/shear rate) ($R^2 = 0.99$)

The diagrams of Weibull model applications on HHP-processed (300–400–500 MPa/10/30 min) ice creams were exhibited in Figure 3.13 (Tokuşoğlu et al., 2011).

It was stated that K (consistent coefficient, Pa.s) values for HPP-treated ice-cream mixes increased significantly ($P < 0.05$) whereas there was little increasing in K and

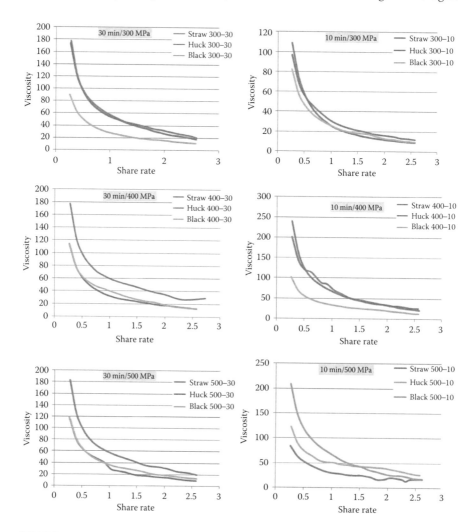

FIGURE 3.12 (**See color insert.**) HP treatment and duration time on the viscosity/shear rate results of strawberry, huckleberry, and BIMs. (Adapted from Tokuşoğlu Ö. et al. 2011. Rheological, microbial quality characteristics and antioxidant activity of high hydrostatic pressure (HHP) processed berry ice-cream mixes: Weibull modeling and response surface methodology approach for treatment pressure and holding time effects. *Nonthermal Processing Division Workshop 2011—Innovation Food Conference.* October 12–14, Osnabrück, Germany. Oral Presentation.)

little decreasing in n for the control (untreated) ice creams. By using HHP-processing treatments, the viscosity of the serum phases increased, and lower OR values in three types of ice creams while melting rates of studied ice creams decreased (Tokuşoğlu et al., 2011). For blackberry, huckleberry, and SIMs, OR values did not change significantly in 300 MPa/10 min treatment ($p \leq 0.05$) whereas the highest reducing OR was found for HIM treated at 400 MPa/10 min.

TABLE 3.3
The Correlations of Viscosity, Shear Rate, and Shear Stress of Blackberry Ice Cream Processed at 300 MPa/10 min

Fruit	Pressure	Interval			Correlations				
					Interval Shear Rate	Shear Rate	Shear Stress	Viscosity	Interval Viscosity
Black	300	10	Interval shear rate	Pearson correlation	1	-1817	-1699	1997	-1781
				Sig. (two tailed)		1000	1001	1000	1000
				N	19	19	19	19	19
			Shear rate	Pearson correlation	-1817	1	1410	-1854	1992
				Sig. (2tailed)	1000		1081	1000	1000
				N	19	19	19	19	19
			Shear stress	Pearson correlation	-1699	1410	1	-1665	1301
				Sig. (2tailed)	1001	1081		1002	1211
				N	19	19	19	19	19
			Viscosity	Pearson correlation	1997	-1854	-1665	1	-1823
				Sig. (2tailed)	1000	1000	1002		1000
				N	19	19	19	19	19
			Interval viscosity	Pearson correlation	-1781	1992	1301	-1823	1
				Sig. (2tailed)	1000	1000	1211	1000	
				N	19	19	19	19	19
		30	Interval shear rate	Pearson correlation	1	-1818	-1939	1998	-1812
				Sig. (two tailed)		1000	1000	1000	1000
				N	19	19	19	19	19
			Shear rate	Pearson correlation	-1818	1	1840	-1850	1998
				Sig. (2-tailed)	1000		1000	1000	1000
				N	19	19	19	19	19
			Shear stress	Pearson correlation	-1939	1840	1	-1945	1818
				Sig. (2-tailed)	1000	1000		1000	1000
				N	19	19	19	19	19

Source: From Tokuşoğlu Ö. et al. 2011. Rheological, microbial quality characteristics and antioxidant activity of high hydrostatic pressure (HHP) processed berry ice-cream mixes; Weibull modeling and response surface methodology approach for treatment pressure and holding time effects. *Nonthermal Processing Division Workshop 2011—Innovation Food Conference.* October 12–14, Osnabrück, Germany. Oral presentation.

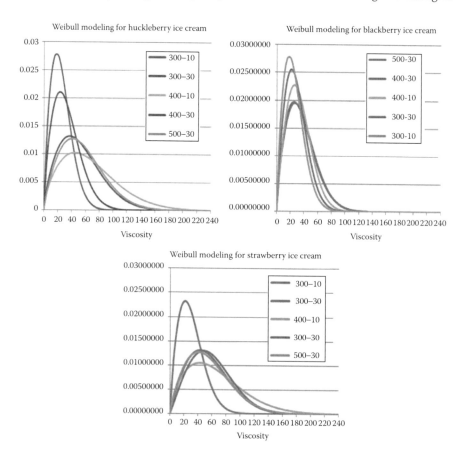

FIGURE 3.13 The explanation of the viscosity–shear rate–shear stress relations by Weibull model applications on huckleberry, blackberry, and SIMs. (Adapted from Tokuşoğlu Ö. et al. 2011. Rheological, microbial quality characteristics and antioxidant activity of high hydrostatic pressure (HHP) processed berry ice-cream mixes: Weibull modeling and response surface methodology approach for treatment pressure and holding time effects. *Nonthermal Processing Division Workshop 2011—Innovation Food Conference.* October 12–14, Osnabrück, Germany. Oral Presentation.)

It was stated that HHP was a cold-processing technique for improving the quality of berry ice-cream mixes and it was found that the improved rheological characteristics were obtained. The development of the Weibull model approach for HPP can be very useful for evaluating of the industrial ice-cream quality (Tokuşoğlu et al., 2011).

It was reported that total anthocyanin (TA) bioactives increased about twofold in BIMs, about 1.95-fold in HIMs whereas they increased to about 2.09-fold in SIMs after 300 MPa/10 min of HHP processing. It has been determined that higher levels of bioactives retained in HPP-treated samples (Tokuşoğlu et al., 2011) (Figure 3.14).

In HIMs, 1,1-diphenyl-2-picrylhydrazyl (DPPH) AA increased as 2.48-fold at 300 MPa of HHP processing, and as 3.53-fold at 400 MPa of HHP processing (Figure 3.15).

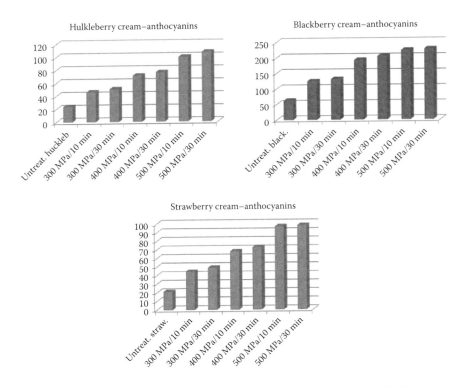

FIGURE 3.14 The anthocyanin levels in HHP-processed (300–400–500 MPa/10–30 min) berry ice-cream mixes. (From Tokuşoğlu Ö. et al. 2011. Rheological, microbial quality characteristics and antioxidant activity of high hydrostatic pressure (HHP) processed berry ice-cream mixes: Weibull modeling and response surface methodology approach for treatment pressure and holding time effects. *Nonthermal Processing Division Workshop 2011—Innovation Food Conference*. October 12–14, Osnabrück, Germany. Oral Presentation.)

Figure 3.15 shows the AA of HIMs by HHP processing (300–400–500 MPa; 10/30 min) and compares with strong antioxidants (butylated hydroxyanisole [BHA] and ascorbic acid) (Tokuşoğlu et al., 2011). The correlation on total phenolics and AA in HIMs was $R^2 = 0.978$, $y = 0.223x–52.456$ and $R^2 = 0.989$, and $y = 0.268x–40.566$ for 300 and 400 MPa of treatments, respectively (Tokuşoğlu et al., 2011) (Figure 3.15).

It is known that cheese ripening is a slow and expensive process owing to high storage costs. For that matter, reducing aging time without significantly affecting other quality properties would provide noteworthy savings to cheese manufacturers (El Soda and Awad, 2011).

It has been stated that during cheese ripening, biochemical and microbiological alterations occur that result in the development of the flavor and texture characteristic of the variety (McSweeney, 2004). Biochemical changes in cheese during ripening may be grouped into primary (lipolysis, proteolysis, and metabolism of residual lactose and of lactate and citrate) or secondary (metabolism of fatty acids and of amino acids) events (McSweeney, 2004). It has also been stated that lipolysis in cheese is

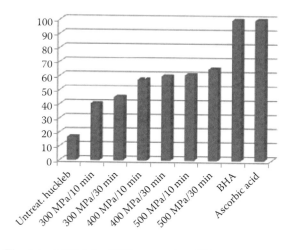

FIGURE 3.15   The AA of HIMs by HHP processing (300–400–500 MPa;10/30 min) and comparing with BHA and ascorbic acid. (Adapted from Tokuşoğlu Ö. et al. 2011. Rheological, microbial quality characteristics and antioxidant activity of high hydrostatic pressure (HHP) processed berry ice-cream mixes: Weibull modeling and response surface methodology approach for treatment pressure and holding time effects. *Nonthermal Processing Division Workshop 2011—Innovation Food Conference*. October 12–14, Osnabrück, Germany. Oral Presentation.).

catalyzed by lipases from various sources, particularly the milk and cheese microflora in varieties where this coagulant is used, by enzymes from rennet paste (McSweeney, 2004). Numerous research groups have assessed the application of HHP treatments to accelerate the ripening of cheese (Martínez-Rodríguez et al., 2012).

Rynne et al. (2008) stated the HP-treated (400 MPa for 10 min at room temperature) full-fat C cheese results. On the basis of the study given by Rynne et al. (2008), HP treatment had little or no effect on primary proteolysis in C cheese and, consistent with measured plasmin and chymosin activities, it was unaffected by the HP application. In this study, total lactate and D(−)- and L(+)-lactate concentrations were significantly reduced in applicated HP-treated conditions in comparison with the control C cheese during ripening (Rynne et al., 2008).

It was also determined that HHP enhanced the physical ripening of cream for butter making (Buchheim and Frede, 1996).

O'Reilly et al. (2002a) reported that dynamic measurement of the viscoelastic alterations on heating the mozzarella (M) cheese from 20°C to 82°C showed that the HP application resulted in an increase in the fluidity of the heated cheese, as measured by the phase angle, especially in 1-d-old cheese. So, that accelerated ripening of LMMC was induced by HP treatment (at 400 MPa for 20 min), which may lead to the development of an industrial process if cost-effective commercial HP equipment was available.

It is known that plasmin is the main indigenous proteinase in milk responsible for hydrolyzing αs2- and β-caseins at the same rate and αs1-casein at a slower rate (Nielsen, 2002). It was reported that 800 MPa treatment for 60 min at 8°C did not inactivate plasmin in 14-d-old C cheese, while at 20°C, its activity was reduced by 15% compared to controls, and at 30°C, its activity was reduced up to 50%

(Huppertz et al., 2004). In a specific study on cheese stated by Rynne et al. (2008), it was shown that pressure applied at 400 MPa/room temperature/10 min (1 day after manufacture) did not affect the plasmin and chymosin activities in the cheese; even the sensorial evaluation test was unfavorable, as there were changes in cheese color (Rynne et al., 2008).

Malone et al. (2002) observed that *Lactococcus lactis* ssp. *cremoris* MG1363 cell suspensions lyzed more rapidly at 300 MPa treatment than those treated at 100–200/5 min and 600–800 MPa/5 min applications. In the study given by O'Reilly et al. (2002b), in contrast, cell lysis of *L. lactis* strains 303, 223, 227, and AM2 did not occur in C cheese when subjected to HHP at 100–400 MPa/25°C/20 min treatmant (O'Reilly et al., 2002a).

Meanwhile, it was accessed that the published studies regarding the pressure effects on the activity of proteolytic enzymes of adjunct secondary cultures (for *P. roqueforti* and *Penicillium camemberti*), have an important role in smear and mold-ripened cheese.

O'Reilly et al. (2001) stated that some starter cultures such as *P. roqueforti* were studied in buffer and the cheese matrix, showing an easier inactivation in buffer (O'Reilly et al., 2001).

Bermúdez-Aguirre and Barbosa-Cánovas (2012) stated the nonthermal approaches on the ω-3 fortified cheeses including QF, C, and M. In the study reported by Bermúdez-Aguirre and Barbosa-Cánovas (2012), HHP, pulsed electric fields (PEFs), and ultrasound (US) were used to increase the retention of ω-3. In C and M, HHP was found to be the best method (5.49 mg/g and 6.64 mg/g) for ω-3 retention. It was stated that the spoilage (4°C) was fast during storage of QF and M, even though HHP was able to delay microbial growth in M (Bermúdez-Aguirre and Barbosa-Cánovas, 2012).

REFERENCES

Anema S.G., Lowe E.K., and Stockman G. 2005. Particle size changes and casein solubilisation in high-pressure-treated skim milk. *Food Hydrocolloids* 19, 257–267.

Anema S.G. 2008. On heating milk, the dissociation of kappa-casein from the casein micelles can precede interactions with the denatured whey proteins. *Journal of Dairy Research* 75(4), 415–421.

Bahramparvar M., and Tehrani M.M. 2011. Application and functions of stabilizers in ice cream. *Food Reviews International* 27(4), 389–407.

Bandler D.K. 2013. *Milk Flavor and Quality: From Cow to Consumer. Part: Dairy Foods.* Encyclopaedia Britannica, Inc., Chicago, IL, USA.

Barbosa-Cánovas G.V., Tapia María S., and Pilar Cano M. 2005. *Novel Food Processing Technologies.* New York: Marcel Dekker.

Barbosa-Cánovas G.V., Pothakamury U.R., Palou E., and Swanson B. 1998. Emerging technologies in food preservation. In *Nonthermal Preservation of Foods*, 1–9p. New York: Marcel Dekker.

Bermúdez-Aguirre D., and Barbosa-Cánovas G.V. 2012. Fortification of Queso fresco, Cheddar and Mozzarella cheese using selected sources of omega-3 and some nonthermal approaches. *Food Chemistry* 133(3), 787–797.

Bermudez-Aguirre D., and Barbosa-Cánovas G.V. 2011. An update on high hydrostatic pressure, from the laboratory to industrial applications. *Food Engineering Review* 3, 44–61.

Black E.P., Huppertz T.H.M., Kelly A.L., and Fitzgerald G.F. 2007. Baroprotection of vegetative bacteria by milk constituents: A study of *Listeria innocua*. *International Dairy Journal* 17, 104–110.

Brinez W.J., Roig-Sagues A.X., Herrero M.M.H., and Lopez B.G. 2007. Inactivation of *Staphylococcus* spp. strains in whole milk and orange juice using ultra high pressure homogenisation at inlet temperatures of 6 and 20 degrees C. *Food Control* 18, 1282–1288.

Buchheim W., and Frede E. 1996. Use of high pressure treatment to influence the crystallization of emulsified fats. *DMZ Lebensmind Milchwirtsch* 117, 228–237.

Capellas M., Mor-Mur M., Sendra E., and Guamis B. 2001. Effect of high pressure processing on physico-chemical characteristics of fresh goat's milk cheese (Mató). *International Dairy Journal* 11, 165–173.

Chawla R., Patil G.R., and Singh A.K. 2011. High hydrostatic pressure technology in dairy processing: A review. *Journal of Food Science and Technology* 48(3), 260–268.

Cheftel J.C. 1995. Review: High pressure, microbial inactivation and food preservation. *Food Science Technology International* 1, 75–90.

Cheftel J.C. 1992. Effects of high hydrostatic pressure on food constituents: An overview. In *High Pressure and Biotechnology*, eds., Balny C., Hayashi R., Heremans K., and Masson P. Elsevier Science, Amsterdam, The Netherlands, pp. 195–209.

Chicón R., López-Fandiño R., Alonso E., and Belloque J. 2008. Proteolytic pattern, antigenicity, and serum immunoglobulin E binding of β-lactoglobulin hydrolysates obtained by pepsin and high-pressure treatments. *Journal of Dairy Science* 91, 928–938.

Corbo M.R., Bevilacqua A., Campaniello D., D'Amato D., Speranza B., and Sinigaglia M. 2009. Prolonging microbial shelf life of foods through the use of natural compounds and nonthermal approaches: A review. *International Journal of Food Science and Technology* 44, 223–241.

Daryaei H., Coventry M.J., Versteeg C., and Sherkat F. 2006. Effects of high-pressure treatment on shelf-life and quality of fresh lactic curd cheese. *Australian Journal of Dairy Technology* 61, 186–188.

De Buyser M.L., Dufour B., Maire M., and Lafarge V. 2001. Implication of milk and milk products in food-borne diseases in France and in different industrialised countries. *International Journal of Food Microbiology* 67, 1–17.

deWit J.N., and Klarenbeek G. 1984. Effect of various heat treatments on structure and solubility of whey proteins. *Journal of Dairy Science* 67, 2701–2710.

Diels A.M.J., Callewaert L., Wuytack E.Y., Masschalck B., and Michiels C.W. 2005. Inactivation of *Escherichia coli* by high-pressure homogenisation is influenced by fluid viscosity but not by water activity and product composition. *International Journal of Food Microbiology* 101, 281–291.

Eberhard P., Strahm W., and Eyer H. 1999. High pressure treatment of whipped cream. *Agrarforschung* 6, 352–354.

El Soda M., and Awad S. 2011. Accelerated cheese ripening. In *Encyclopedia of Dairy Sciences*. Vol. 1. Eds., Fuquay J.W., Fox P.F., and McSweeney P.H.L., pp. 795–798. Oxford: Elsevier Science.

Escobar D., Clark S., Ganesan V., Repiso L., Waller J., and Harte F. 2011. High-pressure homogenization of raw and pasteurized milk modifies the yield, composition, and texture of Queso fresco cheese. *Journal of Dairy Science* 94(3), 1201–1210.

Evrendilek G.A., and Balasubramaniam V.M. 2011. Inactivation of *Listeria monocytogenes* and *Listeria innocua* in yogurt drink applying combination of high pressure processing and mint essential oils. *Food Control* 22, 1435–1441.

Felipe X., Capellas M., and Lawm A.J.R. 1997. Comparison of the effect of high pressure treatment and heat pasteurization on the whey proteins in goat's milk. *Journal of Agricultural Food Chemistry* 45, 627–631.

Fertsch, B., Müller, M., and Hinrichs, J. 2003. Firmness of pressure-induced casein and whey protein gels modulated by holding time and rate of pressure release. *Innovative Food Science and Emerging Technologies* 4(2), 143–150.

Fioretto F., Cruz C., Largeteau A., Sarli T.A., Demazeau G., and El Moueffak A. 2005. Inactivation of *Staphylococcus aureus* and *Salmonella enteritidis* in tryptic soy broth and caviar samples by high pressure processing. *Brazilian Journal of Medical and Biological Research* 38, 1259–1265.

Garcia-Amezquita L.E., Primo-Mora Á.R., Guerrero-Beltran J.A., Barbosa-Cánovas G.V., and Sepulveda D.R. 2013. Rennetability of cheese-making milk processes by non-thermal technologies. *Journal of Food Process Engineering* 36, 247–253.

García P., Rodríguez L., Rodríguez A., and Martínez B. 2010. Food biopreservation: Promising strategies using bacteriocins, bacteriophages and endolysins. *Trends in Food Science and Technology* 21, 373–382.

García P., Madera C., Martínez B., Rodríguez A., and Suárez J.E. 2009. Prevalence of bacteriophages infecting *Staphylococcus aureus* in dairy samples and their potential as biocontrol agents. *Journal of Dairy Science* 92, 3019–3026.

Gaucheron F., Famelart M.H., Mariette F., Raulot K., Michel F., and Le Graet Y. 1997. Combined effects of temperature and high pressure treatment on physiological characteristics of skim milk. *Food Chemistry* 59, 439–447.

Gervilla R., Ferragut V., and Guamis B. 2001. High hydrostatic pressure effects on colour and milk-fat globule of ewes' milk. *Journal of Food Science* 66, 880–885.

Gervilla R., Sendra E., Ferragut V., and Guamis B. 1999. Sensitivity of *Staphylococcus aureus* and *Lactobacillus helveticus* in ovine milk subjected to high hydrostatic pressure. *Journal of Dairy Science* 82, 1099–1107.

Goff H.D., and Hartel R.W. 2013. *Ice Cream*, 7th edn. New York: Springer.

Goff H.D., Davidson V.J., and Cappi E. 1994. Viscosity of ice cream mix at pasteurization temperatures. *Journal of Dairy Science* 77, 2207–2213.

Goff H.D., and Sahagian M.E. 1996. Freezing of dairy products. In *Freezing Effects on Food Quality*, Jeremiah L.E., ed., pp. 299–335. New York: Marcel Dekker, Inc.

Huppertz T., Kelly A.L., and Fox P.F. 2002. Effects of high pressure on constituents and properties of milk. *International Dairy Journal* 12, 561–572.

Huppertz T., Fox P.F., and Kelly A.L. 2004. Susceptibility of plasmin and chymosin in Cheddar cheese to inactivation by high pressure. *Journal of Dairy Research* 71, 496–499.

Huppertz T., Kelly A.L., and Fox P.F. 2006. High pressure induced changes in ovine milk: Effects on casein micelles and whey proteins. *Milchwissenschaft* 61, 394–397.

Huppertz T., Fox P.F., de Kruit K.G., and Kelly A.L. 2006. High pressure induced changes in bovine milk proteins: A review. *Biochimica et Biophysica Acta (BBA)—Protein and Proteomics* 1764(3), 593–598.

Huppertz T., and de Kruif C.G. 2006. Disruption and reassociation of casein micelles during high pressure treatment: Influence of whey proteins. http://www.aseanfood.info/Articles/11023879.pdf

Huppertz T., Smiddy M.A., Goff H.D., and Kelly A.L. 2011. Effect of high pressure treatment of mix on ice cream manufacture. *International Dairy Journal* 21, 718–726.

Jakubczyk E., and Niranjan K. 2006. Transient development of whipped cream properties. *Journal of Food Engineering* 77, 79–83.

Jayaprakasha H.M., and Brueckner H. 1999. Whey protein concentrate: A potential functional ingredient for food industry. *Journal of Food Science Technology* 36, 189–204.

Jofré A., Aymerich T., Bover-Cid S., and Garriga M. 2010. Inactivation and recovery of *Listeria monocytogenes*, *Salmonella enterica* and *Staphylococcus aureus* after high hydrostatic pressure treatments up to 900 MPa. *International Microbiology* 13, 105–112.c

Johnston D.E., Austin B.A., and Murphy R.J. 1992. Effects of high hydrostatic pressure on milk. *Milchwissenschaft* 47(12), 760–763.

Kadam P.S., Jadhav B.A., Salve R.V., and Machewad G.M. 2012. Review on the high pressure technology (HPT) for food preservation. *Journal of Food Process Technology* 3, 135. doi:10.4172/2157-7110.1000135.

Kato A., Osako Y., Matsudomi N., and Kobayashi K. 1983. Changes in the emulsifying and foaming properties of proteins during heat denaturation. *Agricultural Biological Chemistry* 47, 33–37.

Kester J.J., and Richardson T. 1984. Modification of whey protein to improve functionality. *Journal of Dairy Science* 67, 2757–2774.

Kheadr E.E., Vachon J.F., Paquin P., and Fliss I. 2002. Effect of dynamic high pressure on microbiological, rheological and microstructural quality of Cheddar cheese. *International Dairy Journal* 12, 435–446.

Koca N., Balasubramanian V.M., and Harper J.W. 2011. High-pressure effects on the microstructure, texture, and color of white-brined cheese. *Journal of Food Science* 76, 399–404.

Krasowska M., Reps A., and Jankowska A. 2005. Effect of high pressures on the activity of selected strains of lactic acid bacteria. *Milchwissenschaft* 60, 382–385.

Lee W., Clark S., and Swanson B.G. 2006a. Functional properties of high hydrostatic pressure-treated whey protein. *Journal of Food Process Preservation* 30, 488–501.

Lee W., Clark S., and Swanson B.G. 2006b. Low-fat process cheese food containing ultra-high presure-treated whey protein. *Journal of Food Process Preservation* 2006 (30), 164–179.

Lim S.Y., Swanson B.G., Ross C.F., and Clark S. 2008. High hydrostatic pressure modification of whey protein concentrate for improved body and texture of lowfat ice cream. *Journal of Dairy Science* 91, 1308–1316.

Lim S.Y., Swanson B.G., and Clark S. 2008. High hydrostatic pressure modification of whey protein concentrate for improved functional properties. *Journal of Dairy Science* 91, 1299–1307.

Linko P. 1982. Lactose and lactitol. In *Natural Sweeteners*, Birch G.G., and Parker K.J., eds., pp. 109–132.London and New Jersey: Applied Science Publishers, ISBN 0-85334-997-5.

Linton M., Mackle A.B., and Upadhyay V.K., Kelly A.L., and Patterson M.F. 2008. The fate of *Listeria monocytogenes* during the manufacture of Camembert type cheese: A comparison between raw milk and milk treated with high hydrostatic pressure. *Innovative Food Science Emergency* 9, 423.

Liu X., Powers J.R., Swanson B.G., Hill H.H., and Clark S. 2005a. High hydrostatic pressure affects flavor-binding properties of whey protein concentrate. *Journal of Food Science* 70, C581–C584.

Liu X., Powers J.R., Swanson B.G., Hill H.H., and Clark S. 2005b. Modification of whey protein concentrate hydrophobicity by high hydrostatic pressure. *Innovative Food Science and Emerging Technologies* 6, 310–317.

Lopez-Fandino R., Carrascosa A.V., and Olano A. 1996. The effects of high pressure on whey protein denaturation and cheese-making properties of raw milk. *Journal of Dairy Science* 79, 929–936.

Lopez-Pedemonte T., Roig-Sagues A.X., de Lamo S., Hernandez-Herrero M., and Guamis B. 2007. Reduction of counts of *Listeria monocytogenes* in cheese by means of high hydrostatic pressure. *Journal of Food Microbiology* 24, 59–66.

Malone A.S., Shellhammer T.H., and Courtney P.D. 2002. Effects of high pressure on the viability, morphology, lysis and cell wall activity of *Lactococcus lactis* subsp. *cremoris*. *Applied Environment Microbiology* 68, 4357–4363.

Marshall R.T., Goff H.D., and Hartel R.W. 2003. *Ice Cream*, 6th edn. New York: Kluwer Academic, ISBN 0-306-47700-9, 366pp.

Martínez-Rodríguez Y., Acosta-Muñiz C., Olivas G.I., Guerrero-Beltrán J., Rodrigo-Aliaga D., and Sepúlveda D.R. 2012. High hydrostatic pressure processing of cheese. *Comprehensive Reviews in Food Science and Food Safety* 11, 399–416.

Martin M.F.S., Barbosa-Cánovas G.V., and Swanson B.G. 2002. Food processing by high hydrostatic pressure. *CRC Critical Review Food Science Nutrition* 33, 431–476.

Masson L.M.P., Rosenthal A., Calado V.M.A., Deliza R., and Tashima L. 2011. Effect of ultra-high pressure homogenization on viscosity and shear stress of fermented dairy beverage. *LWT—Food Science and Technology* 44, 495–501.

McClements J.M.J., Patterson, M.F., and Linton, M. 2001. The effect of growth stage and growth temperature on high hydrostatic pressure inactivation of some psychrotrophic bacteria in milk. *Journal of Food Protection* 64 (4), 514–522.

McSweeney Paul L.H. 2004. Biochemistry of cheese ripening. *International Journal of Dairy Technology* 57(2/3), 126–144.

Messens W., Arevalo J., Dewettinck K., and Huyghebaert A. 1999. Proteolysis and visco-elastic properties of high pressure treated Gouda cheese. In *Advances in High Pressure Bioscience and Biotechnology*, Ludwig H., ed., *Proceedings of the International Conference of High Pressure Bioscience and Biotechnology*, Heidelberg, August 30–September 3, 1998. Springer-Verlag, Berlin, pp. 445–448.

Morr C.V., and Ha E.Y.W. 1993. Whey protein concentrates and isolates: Processing and functional properties. *CRC Critical Review of Food Science Nutrition* 33, 431–476.

Myllymaki O. 1996. High pressure food processors. In *High Pressure Processing of Food and Food Components—A Literature Survey and Bibliography*, Ohlsson T., ed., pp. 29–46. Goteborg, Kompendiet.

Needs E.C., Capellas M., Bland P., Manoj P., MacDougal D.B., and Gopal P. 2000. Comparison of heat and pressure treatments of skimmed milk, fortified with whey protein concentrate, for set yoghurt preparation: Effects on milk proteins and gel structure. *Journal od Dairy Research* 67, 329–348.

Nielsen S.S. 2002. Plasmin system and microbial proteases in milk: Characteristics, roles, and relationship. *Journal of Agricultural Food Chemistry* 50, 6628–6634.

O'Reilly C.E., Murphy P.M., Kelly A.L., Guinee T.P., Auty M.A.E., and Beresford T.P. 2002a. The effect of high pressure treatment on the functional and rheological properties of Mozzarella cheese. *Innovative Food Science and Emerging Technologies* 3, 3–9.

O'Reilly C.E., O'Connor P.M., Murphy P.M., Kelly A.L., and Beresford T.P. 2002b. Effects of high-pressure treatment on viability and autolysis of starter bacteria and proteolysis in Cheddar cheese. *International Dairy Journal* 12, 915–922.

O'Reilly C.E., Kelly A.L., Murphy P.M., and Beresford T.P. 2001. High pressure treatment: Applications in cheese and manufacture and ripening. *Trends in Food Science Technology* 12, 51.

O'Reilly C.E., O'Connor P.M., Kelly A.L., Beresford T.P., and Murphy P.M. 2000. Use of hydrostatic pressure for inactivation of microbial contaminants in cheese. *Applied Environment Microbiology* 66, 4890–4896.

Padiernos C.A., Lim S.Y., Swanson B.G., Ross C.F., and Clark S. 2009. High hydrostatic pressure modification of whey protein concentrate for use in low-fat whipping cream improves foaming properties. *Journal of Dairy Science* 92, 3049–3056.

Patterson M.F., Linton M., and Doona C.J. 2008. Introduction to high pressure processing of foods. Chapter 1. In *High Pressure Processing of Foods*, Doona C.J., and Florence E.F., eds., pp. 29–42. USA: Wiley-Blackwell Publishing.

Patterson M.F., and Kilpatrick D.J. 1998. The combined effect of high hydrostatic pressure and mild heat on inactivation of pathogens in milk and poultry. *Journal of Food Protection* 61, 432–436.

Penna A.L.B., Subba Rao-Gurram, and Barbosa-Cánovas G.V. 2007. High hydrostatic pressure processing on microstructure of probiotic low-fat yoğurt. *Food Research International* 40, 510–519.

Pereda J., Ferragut V., Buffa M., Guamis B., and Trujillo A.J. 2007. Proteolysis of ultrahigh pressure homogenized treated milk during refrigerated storage. *Food Chemistry* 111, 696.

Pereda J., Ferragut V., Quevedo J.M., Guamis B., and Trujillo A.J. 2007. Effects of ultra-high pressure homogenization on microbial and physicochemical shelf-life of milk. *Journal of Dairy Science* 90, 1081–1089.

Pérez Pulido R., Toledo del Árbol J., José Grande Burgos M., and Gálvez A. 2012. Bactericidal effects of high hydrostatic pressure treatment singly or in combination with natural anti-microbials on *Staphylococcus aureus* in rice pudding. *Food Control* 28, 19–24.

Rademacher B., and Kessler H.G. 1997. High pressure inactivation of microorganisms and enzymes in milk and milk products. In *High Pressure Bioscience and Biotechnology*, Heremans K., ed., pp. 291–293. Leuven University Press, Leuven.

Rastogi N.K., Raghavarao K.S.M.S., Balasubramaniam V.M., Niranjan K., and Knorr D. 2007. Opportunities and challenges in high pressure processing of foods. *Critical Reviews in Food Science and Nutrition* 47, 69–112.

Richert S.H., Morr C.V., and Cooney C.M. 1974. Effect of heat and other factors upon foaming properties of whey protein concentrates. *Journal of Food Science* 39, 42–48.

Ritz M., Freulet M., Orange N., and Federighi M. 2000. Effects of high hydrostatic pressure on membrane proteins of *Salmonella typhimurium*. *International Journal of Food Microbiology* 55, 115–119.

Rodiles-Lopez J.O., Jaramillo-Flores M.E., Gutierrez-Lopez G.F., Hernandez-Arana A., Fosado-Quiroz R.E., Barbosa-Canovas G.V., and Hernandez-Sanchez H. 2008. Effect of high hydrostatic pressure on bovine α-lactalbumin functional properties. *Journal of Food Engineering* 87, 363–370.

Rynne N.M., Beresford T.P., Guinee T.P., Sheehan E., Delahunty C.M., and Kelly A.L. 2008. Effect of high-pressure treatment of 1 day-old full-fat Cheddar cheese on subsequent quality and ripening. *Innovative Food Science and Emerging Technologies* 9, 429–440.

Saldo J., Sendra E., and Guamis B. 2002. Colour changes during ripening high-pressure-treated hard caprine cheese. *High Pressure Research* 22, 659–63.

Sandra S., Standford M.A., and Meunier Goddik L. 2004. The use of high-pressure processing in the production of queso fresco cheese. *Journal of Food Science* 69, 153–58.

San Martín M.F., Barbosa-Cánovas G.V., and Swanson B.G. 2002. Food processing by high hydrostatic pressure. *Critical Reviews in Food Science and Nutrition* 42(6), 627–645.

Serra M., Trujillo A.J., Pereda J., Gumais B., and Ferragut V. 2008. Quantification of lipolysis and lipid oxidation during cold storage of yoghurts produced from milk treated by ultra-high pressure homogenization. *Journal of Food Engineering* 89, 99–104.

Sheehan J.J., Huppertz T., Hayes M.G., Kelly A.L., Beresford T.P., and Guinee T.P. 2005. High pressure treatment of reduced-fat Mozzarella cheese: Effects on functional and rheological properties. *Innovative Food Science and Emerging Technologies* 6, 73–81.

Sivanandan L., Toledo R.T., and Singh R.K. 2008. Effect of continuous flow high-pressure throttling on rheological and ultrastructural properties of soymilk. *Journal of Food Science* 73, 288–295.

Smelt, J. P. P. M. 1998. Recent advances in the microbiology of high pressure processing. *Trends in Food Science and Technology* 9, 152–158.

Tabla R., Martínez B., Rebollo J.E., González J., Ramírez M.R., Roa I., Rodríguez A., and García P. 2012. Bacteriophage performance against *Staphylococcus aureus* in milk is improved by high hydrostatic pressure treatments. *International Journal of Food Microbiology* 156, 209–213.

Taskin B., and Tokuşoğlu Ö. 2012. High pressure processing on dairy foods quality, functionality, and shelf life. In *ANPFT2012 Proceeding Book. Advanced Nonthermal Processing in Food Technology: Effects of Quality and Shelf Life of Food and Beverages.* May 07–10, 2012, Kuşadasi-Turkey. Oral Presentation. pp. 149–154. Celal Bayar University Publishing, Turkey. Oral Presentation.

Toepfl S., Mathys A., Heinz V., and Knorr D. 2006. Review: Potential of high hydrostatic pressure and pulsed electric fields for energy efficient and environmentally friendly food processing. *Food Reviews International* 22, 405–423.

Tokuşoğlu Ö. 2012. *ANPFT2012 International Congress-Proceeding Book. Advanced Nonthermal Processing in Food Technology: Effects of Quality and Shelf Life of Food and Beverages.* Edited by associate professor Dr. Özlem Tokuşoğlu. May 07–10, 2012, Kuşadasi-Turkey. 321pp. Celal Bayar University Publishing, Turkey. ISBN: 978-975-8628-33-9.

Tokuşoğlu Ö., and Doona C. 2011. *High Pressure Processing Technology on Bioactives in Fruits and Cereals* (Chapter 21—Part IV. *Functionality, Processing, Characterization and Applications of Fruit and Cereal Bioactives*). In *Fruit and Cereal Bioactives: Sources, Chemistry and Applications Book,* eds., Özlem T., and Hall C., 459pp. Boca Raton, FL, USA: CRC Press, Taylor & Francis Group.

Tokuşoğlu Ö. 2011. Effects of high pressure and ultrasound on food phytochemicals and quality improving. *Invited Research Conference.* U.S. Army Natick Soldier Research, Development & Engineering Center, Natick, MA, USA. June 17, 2011, Friday 13:00 p.m., 15:00 p.m. Oral Presentation.

Tokuşoğlu Ö., Swanson B.G., Younce F., and Barbosa-Cánovas G.V. 2011. Rheological, microbial quality characteristics and antioxidant activity of high hydrostatic pressure (HHP) processed berry ice-cream mixes: Weibull modeling and response surface methodology approach for treatment pressure and holding time effects. *Nonthermal Processing Division Workshop 2011—Innovation Food Conference.* October 12–14, Osnabrück, Germany. Oral Presentation.

Tokuşoğlu Ö., and Doona C.J. 2011. High pressure processing (HHP) strategies on food functionality, quality and bioactives: Biochemical and microbiological aspects. *Nonthermal Processing Division Workshop 2011.* October 12–14, Osnabrück, Germany. Oral Presentation.

Tokuşoğlu Ö., Bozoğlu F., and Doona C.J. 2010. High pressure processing strategies on phytochemicals, bioactives and antioxidant activity in foods. Technical Research Paper.2010IFT *Annual Meeting + Food Expo. Book of Abstracts.* p.115. Division: Nonthermal processing; Session: 100-3/Novel bioactives: Approaches for the search, evaluation and processing for nutraceuticals), July 19, 2010, 9:15 a.m.– 9:35 a.m.; Room N426b, Mc Cormick Place, Chicago/IL, USA. Oral Presentation.

Trujillo A.J., Capellas M., Saldo J., Gervilla R., and Guamis B. 2002. Applications of high-hydrostatic pressure on milk and dairy products: A review. *Innovative Food Science and Emerging Technologies* 3(4), 295–307.

Trujillo A.J. 2002. Applications of high-hydrostatic pressure on milk and dairy products. *High Pressure Research* 22, 619–626.

U.S. FDA/CFSAN, 2006. Agency response letter: GRAS Notice No. GRN 000198. http://www.accessdata.fda.gov/scripts/fcn/gras_notices/612853A.PDF 2006.

U.S. FDA/CFSAN, 2007. Agency response letter: GRAS Notice No. GRN 000218. http://www.accessdata.fda.gov/scripts/fcn/gras_notices/701456A.PDF2007.

Vachon J.F., Kheadr E.E., Giasson J., Paquin P., and Fliss I. 2002. Inactivation of foodborne pathogens in milk using dynamic high pressure. *Journal of Food Protection* 65, 345–352.

Velez-Ruiz J.F., and Barbosa-Canovas G.V. 1998. Rheological properties of concentrated milk as a function of concentration, temperature and storage time. *Journal of Food Engineering* 35, 177–190.

Vurma M., Chung Y.-K., Shellhammer T.H., Turek E.J., and Yousef A.E. 2006. Use of phenolic compounds for sensitizing *Listeria monocytogenes* to high-pressure processing. *International Journal of Food Microbiology* 106, 263–269.

Wick C., Nienaber U., Anggraeni O., Shellhammer T.H., and Courtney P.D. 2004. Texture, proteolysis and viable lactic acid bacteria in commercial Cheddar cheeses treated with high pressure. *Journal of Dairy Research* 71, 107–115.

Wouters P.C., Glaasker E., and Smelt J.P.P.M. 1998. Effects of high pressure on inactivation kinetics and events related to proton efflux in *Lactobacillus plantarum. Applied and Environmental Microbiology* 64, 509–514.

Yang B., Shi Y., Xia X., Xi M., Wang X., Ji B., and Meng J. 2012. Inactivation of foodborne pathogens in raw milk using high hydrostatic pressure. *Food Control* 28, 273–278.

Yang J., Dunker A.K., Powers J.R., Clark S., and Swanson B.G. 2001. β-lactoglobulin molten globule induced by high pressure. *Journal of Agricultural and Food Chemistry* 49, 3236–3243.

4 Improving Quality of Agrofood Products by High-Pressure Processing

Shigeaki Ueno

CONTENTS

4.1 INTRODUCTION

High hydrostatic pressure processing (HPP) is known as one of nonthermal processing for fresh food in the food industry (Eshtiaghi et al., 1994, Oey et al., 2008). HPP has been applied to food processing for controlling food quality, such as the hardness of vegetables (Araya et al., 2007), pressure-shift freezing of seafood (Chevalier et al., 2000), gelatinization of starch (Kawai et al., 2007, 2012), and inactivation of microorganisms (Hasegawa et al., 2012, Koseki et al., 2008, Shigematsu et al., 2010a, Ueno et al., 2011). Food treated by HPP has been shown to keep its original freshness, flavor, taste, and color changes are minimal (Dede et al., 2007). While the structure of high-molecular-weight molecules such as proteins and carbohydrates can be altered by HPP, smaller molecules such as volatile compounds, pigments, vitamins, and other compounds connected with the sensory, nutritional, and health promoting are unaffected (Oey et al., 2008, Patras et al., 2009a).

HPP is based on two fundamental principles: the Le Chatelier principle, which proposes that pressure favors all structural reactions and changes that involve a decrease in volume and the isostatic principle, which proposes that the distribution of pressure is proportional in all parts of a foodstuff irrespective of its shape, size, and food composition, yielding highly homogeneous products (Barba et al., 2012b, Valdez-Fragoso et al., 2011). Butz et al. reported that high-pressure treatment did not induce loss of vitamins, antioxidants, and antimutagens in tomatoes (Butz et al., 2002). Matser et al. investigated the color, texture, and activity of the browning enzyme polyphenoloxidase in mushrooms treated to high pressure, and discussed the possibility of the treatment as an alternative to conventional blanching of mushrooms (Master et al., 2000). The polyphenoloxidase activity and color of mushrooms pressurized at 950 MPa were comparable to those of mushrooms blanched for 10 min in boiling water. High-pressure treatment produced a slightly firmer product than blanching under these conditions. Boynton et al. investigated the color, texture, and microbial levels of precut mangos treated by high pressure, and showed that high-pressure treatment was effective in reducing microbial levels of precut mangos during storage (Boynton et al., 2002). Krebbers et al. evaluated the effect of high-pressure treatment on microbial flora, texture, color, ascorbic acid content, and peroxidase activity of whole green beans and pointed out that high-pressure treatment had potential as a substitute for conventional preservation techniques (Krebbers et al., 2002).

4.1.1   MASS TRANSFER PROMOTION

Conventional food processing such as heat and freezing have been used to preserve agrofood products. However, these techniques deteriorate the quality of agrofood products. Sterilization is by extensive thermal treatment caused by slow heat penetration into the core of the products and subsequent slow cooling, inducing off-flavor formation, texture softening, and destruction of colors and vitamins (Krebbers et al., 2002). Cell breakage, cell separation, and impaired texture are frequently observed in freeze-thawed plant tissue. HPP also deteriorates the quality of agrofood products due to the cell damage it induces. Cellular disruption due to HPP may cause loss of mango-flavor compounds (Boynton et al., 2002). Although HPP of green beans lowered peroxidase activity, ascorbic acid content decreased during storage after the treatment due to chemical breakdown caused by damaged cell membranes and subsequent increased diffusion and reaction of substrates (Krebbers et al., 2002). As damage to cells due to the treatments influences the quality of agrofood products after processing and during preservation, estimating the damage to cells is important.

Cell disruption due to both heat and freeze treatments increases the drying rate of agrofood products, such as banana (Dandamrongrak et al., 2002). HPP damages the cell membrane (Kato et al., 2002, Rastogi and Niranjan, 1998) and facilitates mass transfer in plant materials (Ahromrit et al., 2006, Nunez-Mancilla, 2011). Under high-pressure conditions, approximately above 100 MPa, lipid-phase transitions occur. HPP above 100 MPa could, thus, result in destruction of the internal structure and membrane structure of foodstuffs, then accelerate mass transfer, changes in food texture, and physicochemical properties (Dörnenburg and Knorr, 1993, Islam et al., 2003, Kato et al., 1999, Tangwongchai et al., 2000). Regarding the relationship between protein denaturation and pressure, many enzymes are deactivated at pressures >600 MPa (Jacobo-Velazquez and Hernandez-Brenes, 2010). However, certain enzymes are reportedly still active even at a high pressure of 600 MPa (Knorr et al., 2006). HPP with approximately between 100 and 600 MPa can induce a transformation of cellular foodstuffs into an alternative form, where membrane systems are destroyed but certain enzymes are still active. In the alternative form of foodstuffs, mass transfer inside can be accelerated and certain biochemical reactions can proceed.

The damage to cells induced by HPP was estimated by comparing the relative drying rate of samples (RDR) after high-pressure pretreatment with the RDR after chloroform vapor, heat, and freeze–thaw pretreatments (Ueno et al., 2009c). Figure 4.1

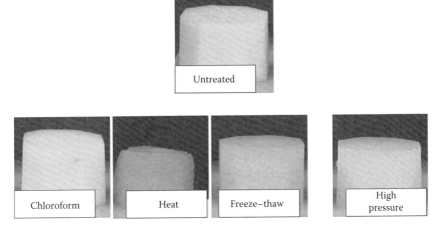

FIGURE 4.1 Photographs of pressurized Japanese radish disks. Chloroform, chloroform vapor treatment for 12 h; heat, heat treatment at 120°C for 20 min; freeze–thaw, frozen at –20°C for 12 h, and thawed in water; and high pressure, HPP at 400 MPa for 5 min at 25°C.

shows the drying characteristic curves (drying rate against moisture content) of radish disks. The drying rate was estimated from the slopes of drying curves of Figure 4.2. In all pretreatments, the drying rate of pretreated radish disks was greater than for untreated radish disks at any moisture content during the drying test. The increase in the drying rate was marked for radish disks both heated for 20 min and freeze

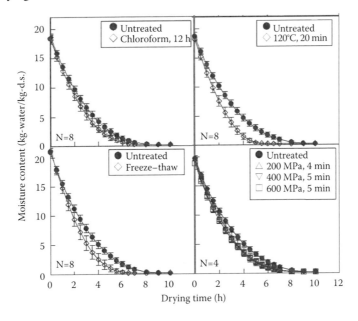

FIGURE 4.2 The effect of pretreatment on the drying curves of radish disks during air ventilation drying. Means and standard deviations are shown (N = 4 or 8).

thawed. The radish disks pretreated by high pressure showed a higher drying rate than untreated radish disks, but it was the same among radish disks treated at different pressures. High-pressure pretreatment (100–700 MPa) increases mass transfer rates during osmotic dehydration of pineapples, and the diffusivity value of water reaches a plateau after an increase in pressure (Rastogi and Niranjan, 1998). In this study, the extent of the increase in the drying rate of radish disks after chloroform vapor pretreatment for 12 h was almost the same as after high-pressure pretreatment.

To investigate in detail the influence of pretreatments on the drying rate of the Japanese radish, the RDR at the moisture content of 0.5, 5, 10, and 15 kg-water/kg-d.s. was calculated (Figure 4.3). The RDR greater than unity means that pretreatment increased the drying rate of Japanese radishes. All RDRs were greater than unity at all moisture contents. Therefore, all pretreatments tested in our study increased the drying rate of radish disks.

By comparing RDRs, the pretreatments may be divided into two groups: (1) heat and freeze–thaw pretreatments that induced a greater increase in the drying rate; (2) high-pressure and chloroform vapor pretreatments that induced a moderate increase in the drying rate. During the freeze–thaw pretreatment, both the cell membrane and cell wall were often damaged by the formation of ice crystals. Heat pretreatment also damaged both the cell membrane and cell wall. The RDR after heat pretreatment was similar to the freeze–thaw pretreatment, indicating a corresponding degree of damage to cells by the two pretreatments. In the living root of plants, water moves through the apoplast, transmembrane, and symplast pathways. The damage

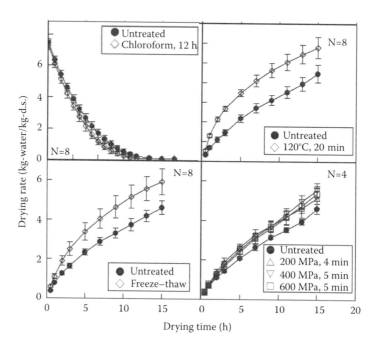

FIGURE 4.3   The effects of pretreatment on the drying rates of radish disks during air ventilation drying. Means and standard deviations are shown (N = 4 or 8).

to the cell membrane alone reduced the resistance to water movement through the cell membrane. Therefore, sample disks in which the cell membrane was damaged showed a lower drying rate than sample disks in which both the cell membrane and cell wall were damaged. The cell membrane might be selectively damaged by chloroform vapor pretreatment, because the RDR after chloroform vapor pretreatment was moderate. In addition to these physical treatments, chloroform vapor treatment may also facilitate mass transfer in biological materials. Toh-e et al. showed that *Saccharomyces cerevisiae* cell membranes became permeable to a substrate of intracellular alkaline phosphatase with chloroform vapor without inactivating the enzyme (Toh-e et al., 1976). Thus, cell membranes composed of phospholipids would be selectively damaged by chloroform vapor treatment. Not only cell membranes, but also cell walls, are damaged by heat and freeze treatments, but chloroform vapor treatment does not damage cell walls severely. A severer damage to the plant tissue would cause a higher mass transfer rate, such as the drying rate. Therefore, the drying rate could be used to estimate cell damage due to each treatment.

The RDR after high-pressure pretreatment was similar to the RDR after chloroform vapor pretreatment and tended to be lower than the RDR after heat and freeze–thaw pretreatments (Ueno et al., 2009c). Thus, high-pressure treatment was thought to damage the cell membrane, and not to damage the cell wall severely. Knockaert et al. reported that concerning the structural properties of carrots, HPP can be a good alternative to thermal processing, that is, the high-intensity pasteurization by high pressure (600 MPa/20 min/45°C) resulted in carrots that were 2.6× harder as compared to the thermal equivalent. β-eliminative depolymerization of pectin, which results in pectin solubilization, is known as the main cause for thermal softening of carrots (Knockaert et al., 2011). However, this reaction is retarded under high pressure that results in a better hardness and microstructure of pressurized carrots. As a result of the reduced pectin solubilization, less cell wall separation occurred in the pressurized carrots compared to the thermally treated carrots and less pectin leached out during HPP compared to thermal processing.

4.1.2 Effect of HPP on Antioxidant Activity

Fresh vegetables and fruits are rich sources of various essential nutrients and biologically active phytochemicals such as carotenoids and polyphenols. Thermal processing has been shown in many studies to affect the antioxidant activities of vegetables (Dewanto et al., 2002, Zhang and Hamauzu, 2004). Thermal-processed vegetables result in a decrease in antioxidant activities; however, Patras et al. have also reported an increase in antioxidant activities (Patras et al., 2009b). The effects of HPP on the functional properties and components in vegetables and fruits (or fruit beverages) have also been reported (Barba et al., 2012a, Chauhan et al., 2011, Keenan et al., 2010, Wang et al., 2012). Goznalez-Cebrino et al. reported that neither thermal treatment nor HPP (400 or 600 MPa/7 min) was completely effective for the inactivation of "Songold" plum polyphenol oxidase. Polyphenols were relatively resistant to the effect of both HPP and thermal treatment (Gonzalez-Cebrino et al., 2012). Tokusoglu et al. reported that the total phenolics of black table olives were increased to 2.1–2.5-fold (as mg of garlic acid/100 g) after HPP of 250 MPa for 5 min at 35°C (Tokusoglu et al., 2010). They also reported that antioxidant activity values varied from 17.24 to 29.34 mmol $Fe^{2+}/100$ g for control samples whereas 18.58–33.00 mmol $Fe^{2+}/100$ g for pressurized samples. According to Mcnerney et al., HPP had differential effects on water-soluble antioxidant activity, depending on the vegetable type (Mcnerney et al., 2007). They investigated that, for broccoli, HPP had no effect on antioxidant activity; for carrots, there was a modest reduction in antioxidant activity at 400 MPa for 2 min at ambient temperature; and for green beans, water-soluble antioxidant activity was increased by HPP at both 400 and 600 MPa for 2 min (Mcnerney et al., 2007).

We pressurized onions (*Allum cepa*) at the pressure range from 50 to 400 MPa for 5 min at 25°C, and analyzed the components and antioxidant activity during storage at 25°C for 4 days with respect to the high-pressure-induced damage of the cells (Ueno et al., 2009a). We analyzed the dielectric properties of pressurized onions to evaluate the damage level. The pressurized onions at 50–400 MPa were partially damaged; the radius of the Cole–Cole arc decreased with increasing the pressure level (Figure 4.4). Additionally, the Cole–Cole arc disappeared over 400 MPa. The existence of the arc implied the turgor pressure between the inside and outside of the membrane structures. Thus, the disappearance of the arc led to the loss of turgor pressure. The quantitative analysis had also investigated the partially destroyed onions. Tangwongchai et al. observed the high-pressure-treated tomato by scanning electron microscopy and the cell structure was partially destroyed at 200 MPa; the degree of destruction over 300 MPa was similar to that of 200 MPa (Tangwongchai et al., 2000). Ohnishi et al. reported that Cole–Cole arcs disappeared after freezing–thawing or chloroform vapor treatment (Ohnishi et al., 2003).

Onions were partially destroyed by HPP; thus, the antioxidant activities of pressurized onions were analyzed by diphenyl-2-picrylhydrazyl (DPPH) radical scavenging method during storage at 25°C. The antioxidant activities of pressurized onions increased with increasing the pressure level over 200 MPa for 5 min, and the pressurized onions at 400 MPa showed the maximum antioxidant activities during

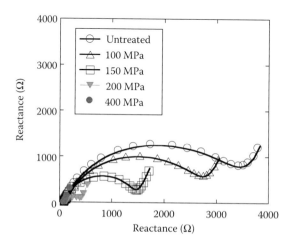

FIGURE 4.4 Typical Cole–Cole arcs of pressurized onions at 100, 150, 200, and 400 MPa for 5 min at 25°C.

storage (Figure 4.5). However, the antioxidant activities of HPP-treated onions at 400 MPa reached the maximum value 3 days after HPP, and that of 200 and 400 MPa reached the maximum value 4 days after HPP reached a plateau or decreased. The antioxidant activities of untreated onions showed 0.7 µmol-Trolox/g-sample before HPP; however, those of pressurized onions during storage were 2.5–4.0 µmol-Trolox/g-sample. As for untreated onions, the antioxidant activities ranged from 0.24 to 1.56 µmol-Trolox/g-sample (Oki et al., 2005, Sellapan and Akoh, 2002). The HPP and subsequent storage enabled to increase the antioxidant activity on onions.

To investigate the component that increased the antioxidant activity, we applied HPLC (high-performance liquid chromatography) to analyze the pressurized onions. If certain enzyme reactions have occurred after HPP, the distribution of the compounds in pressurized onions should be different from that of untreated onions. We analyze the concentrations of the compounds in pressurized onions compared with the original untreated onions by using the HPLC after HCl hydrolysis of glycoside. The HPLC chromatograms of the pressurized samples after 3 days of storage had several peaks at different retention times similar to those of untreated onions (Figure 4.6). The peaks at the retention time of 27.3 min increased by HPP, while that at 23.8 min decreased by HPP. To distinguish the HPP effect on the component, the peak areas of HPLC chromatograms were applied to the scatter plot. The peak areas of untreated and HPP-treated onions were located at the horizontal and vertical axis. Thus, the plot having the quantitative values of the untreated and pressurized onion was plotted to investigate the effect of HPP on the concentrations of pressure-induced changes. If the slope of a line fitted to the data points is 1.0, it means

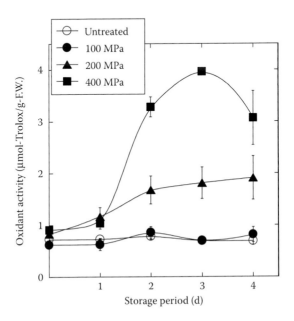

FIGURE 4.5 Antioxidant activity of pressurized onions during storage. Means and standard deviations are shown (N = 3).

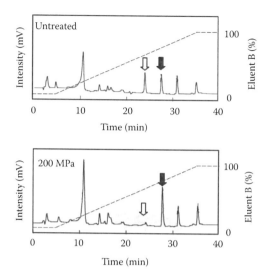

FIGURE 4.6 HPLC chromatograms of the methanol extract from pressurized onions at 200 MPa after 4 days of storage at 25°C. The white and black arrows indicate the peak of 23.8 and 27.3 min, respectively.

the concentrations of certain components are the same in both types of the sample. In contrast, if the slope of a line fitted to the data points is >1.0, it means HPP caused an increase in the concentrations of the certain component.

The data sets of the points with the retention times of 14.3, 15.9, 16.7, 23.8, 27.3, 35.1, and 40.7 min were investigated (Figure 4.7). The scatter plot showed that the component with a retention time of 27.3 min increased by HPP with greater than a slope of

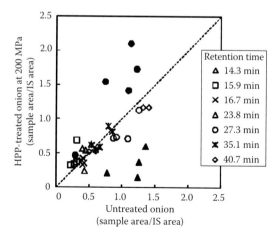

FIGURE 4.7 Scatter plot of the peak area ratio between HPP-treated at 200 MPa and HPP-untreated onions after 4 days of storage at 25°C (N = 5). The values of pressurized samples were significantly different from untreated samples with the retention time of 23.8 ($p < 0.01$) and 27.8 min ($p < 0.05$).

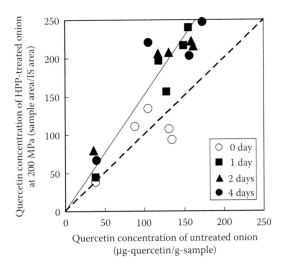

FIGURE 4.8 Scatter plot of quercetin concentrations between HPP-treated at 200 MPa and HPP-untreated onions after 4 days of storage at 25°C (N = 5).

1.0, while that of 23.8 min decreased. The component with a retention time of 27.3 min was investigated to be a quercetin, which was the most abundant flavonoid in onions.

To investigate the effect of the storage period on concentrations of quercetin, the concentrations of the quercetin in HPP-treated onions were analyzed during storage for 4 days (Figure 4.8). The concentrations of quercetin in HPP-treated onions (200 MPa, 5 min) immediately after HPP (day 0) showed that data plots having both concentrations of untreated and HPP-treated onion were around the unity. Therefore, HPP of 5 min at 200 MPa did not result in increasing the concentration of quercetin. In contrast, the concentrations of quercetin during storage showed the slope of 1.5, which resulted in the increase in the concentrations of quercetin by HPP and subsequent storage at 25°C. These results of HPP-induced changes in quercetin concentrations would lead to an increase in the antioxidant activities of HPP-treated onions.

Roldan-Marin et al. reported that processing onions with treatments that combine low temperature (5°C) with HPP of 100 and 400 MPa at a constant time (5 min) significantly increased the extractability of quercetin-4′-glucoside, total quercetin, and quercetin-3,4′-diglucoside, yielding an increase in their contents of 33%, 26%, and 17%, respectively, compared with untreated onions (Roldan-Marin et al., 2009). Roldan-Marin concluded that the increase in antioxidant activity was caused by promoting the extractability by HPP. Additionally, we suggest that an increase in antioxidant activity during storage was caused not only by promoting the extractability, but also by the endogenous enzymatic reaction of quercetin generation.

4.1.3 NEW COMPOSITION PRODUCTION

As mentioned above, HPP enables to increase functional compounds in agrofood products. In this section, the new composition production in agrofood products by HPP is introduced.

Brassica vegetables are rich in nutrients, which decrease the risk of some chronic diseases (Romani et al., 2006). *Brassica rapa* is comparatively abundant in diastase and peroxidase, which are frequently used as a source of pharmaceutical products. Thus, we selected *B. rapa* root (turnip) as a model of cellular foodstuff and investigated the HPP effect on a biochemical change of *B. rapa* root during storage after pressure treatment.

According to the dielectric analysis by a Cole–Cole plot, the untreated samples without HPP, which were expected to have intact membrane structures, showed a regular Cole–Cole arc and those with HPP were smaller in diameter than the untreated samples (Figure 4.9). The Cole–Cole arc completely disappeared at 400 MPa. In this study, cellular membrane structures in *B. rapa* root would be partially destroyed at 200 MPa, and completely destroyed over 400 MPa. The pressurized turnip roots were stored at 4°C or 25°C for 7 days. The high-pressure-treated *B. rapa* roots with 400 and 600 MPa seemed to form a green–blue compound during storage at 4°C in the polyethylene pouch (Figure 4.10). The color of the sample turned darker according to the storage time. Samples with high-pressure treatment at 200 MPa slightly turned green–blue and this color formation did not always emerge. High-pressure-treated samples at 100 and 150 MPa showed no change in color during storage, being white soon after pressure release. The typical values of pH of the untreated samples were stable, during 7-d storage, being approximately 6.3. In contrast, those of high-pressure-treated samples decreased approximately from 5.9 to 5.3. In addition, untreated samples without HPP also showed no change in color during storage. Pressurized samples stored at 25°C were partially digested into small fluid particles and the color turned light pink during storage.

To evaluate the influence of the oxygen transmission rate (OTR) of pouches used on the formation of green–blue compounds, low OTR polyethylene–polyamide nylon pouches (Magic Vac, Flaem Nuova, San Martino della Battaglia, Italy) were used instead of polyethylene pouches. The OTR, provided by the manufacturer, for the polyethylene pouches and the polyethylene–polyamide nylon pouches were 2000 and 70 mL m^{-2} d^{-1} atm^{-1}, respectively. The high- pressure-treated samples in

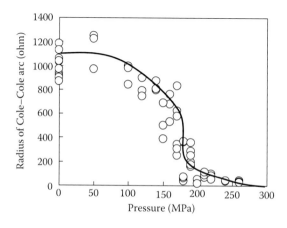

FIGURE 4.9 Radius of the Cole–Cole arc of pressurized turnip roots.

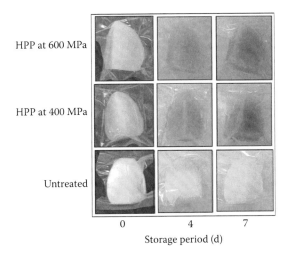

HPP at 600 MPa

HPP at 400 MPa

Untreated

0   4   7

Storage period (d)

FIGURE 4.10   *B. rapa* roots greening with HPP at 400 and 600 MPa in polyethylene pouches.

low OTR pouches remained translucent white during storage at all pressure levels tested; the green–blue compounds never formed. Although the two types of pouches apparently seemed to keep vacuum conditions during 7-d storage, the different color formation obtained would be caused by the different OTR of the pouches. The green–blue compound formation, thus, was suggested to be related to oxygen. The boiled and HPP-treated turnip root sealed in both bags never changed color. These results suggested that the internal enzymes related to the oxygen caused the green–blue formation in pressurized turnip. The antioxidant activities of the green–blue turnip were analyzed by the oxygen radical absorbance capacity (ORAC) method. The antioxidant activities of untreated turnip roots were slightly increased during 7-d storage (Figure 4.11). In contrast, the antioxidant activities of HPP-treated turnip roots, which turned a green–blue color, significantly increased during storage. Thus, the green–blue turnip roots showed the higher antioxidant activity. These results suggest that the reaction for formation of this unique green–blue color pigment was promoted by accelerations of mass transfer with the destruction of membrane systems caused by HPP, which resulted in abundant antioxidant substances in turnip roots. This color formation reaction was also suggested to contain O_2-dependent steps by the two types of pouches used.

Duarte-Vazquez et al. reported that peroxidase was relatively abundant in *B. rapa* and played important physiological roles in protective mechanisms in physically damaged or infected tissues (Duarte-Vazquez et al., 2003). Certain enzymes, such as peroxidase, might be involved in the O_2-dependent formation of green–blue compounds in the sample root, although further studies to isolate and identify the green–blue color pigment and to characterize its synthesis pathways and related enzymes are still needed.

There have been several studies concerning changes in the color of plant tissues and cellular foodstuffs, promoted by environmental conditions or food processing.

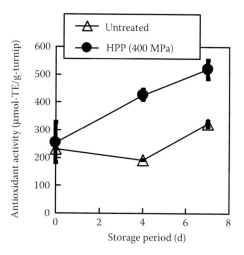

FIGURE 4.11   Antioxidant activity of pressurized turnip roots at 400 MPa for 5 min during storage at 25°C.

Fukuoka and Enomoto reported that a high soil temperature caused internal browning in radish root and this heat stress affected the enzyme activity of polyphenol biosynthesis and the ascorbate–glutathione cycle (Fukuoka and Enomoto, 2002). The production of brown substances in plant tissue was mainly caused by enzymatic oxidation of phenolic compounds. Some cellular foodstuffs such as sweet potato and burdock turned marked green during some types of food processing (Namiki et al., 2001). We described the color formation of green–blue compounds in *B. rapa* root promoted by HPP. The mechanism of this color formation would be based on a biochemical pathway for a unique green–blue pigment synthesis, which contains O_2-dependent steps and possibly enzymatic reactions that contain the cell damage-induced promotion of mass transfer and pH shift to acidic environment. Pressure-induced changes in the apparent catalytic rate may be due to changes in the enzyme–substrate interaction, changes in the reaction mechanisms, or the effect of a particular rate-limiting step on the overall catalytic rate. Provided that the cellular membrane or the membrane of intercellular organelles is altered, intercellular enzymes may be released from extracellular fluids or cell cytoplasm, hereby facilitating enzyme–substrate interactions.

4.1.4 ENZYMATIC REACTION PROMOTION

The effects of HPP on enzymes could be related to reversible or irreversible changes in protein structure (Hernandez and Cano, 1998). However, loss of catalytic activity can differ depending on the enzyme, the nature of the substrates, and the temperature and length of processing (Buckow et al., 2009, 2010, Cano et al., 1997). We have shown that 200-MPa HPP treatment applied to onions can increase the level of the functional compound, quercetin, resulting in increased antioxidant activity (Ueno et al., 2009a). We have also found that HPP-treated turnip roots (*B. rapa*) turn a green–blue color during 7 days of storage (Ueno et al., 2009b). Since these alterations are caused by HPP-induced enzymatic reactions in the agricultural products, the ability to control such HPP-induced enzymatic reactions for certain components is desirable. To investigate the HPP effect on enzymatic reactions, we chose soybean for the sample as follows.

The soybean is a highly nutritious plant food that has commercial value as a raw food material for processing into various soy products. One of the attractive components of soybeans is γ-aminobutyric acid (GABA), which is produced primarily by the α-decarboxylation of L-glutamic acid (Glu) and is catalyzed by glutamate decarboxylase (GAD) (Oh and Choi, 2001). GABA functions in the lowering of high blood pressure and the alleviation of pain and anxiety (Waagepetersen et al., 1999). Therefore, GABA possesses a great potential as a bioactive component in foods, and the development of a high level of GABA in foodstuffs has been investigated (Saikusa et al., 1994a,b, Snedden et al., 1995). However, the accumulation of GABA through the use of HPP has hardly been reported (Sasagawa et al., 2006).

Environmental stresses such as salt stress, lower temperature, and drought stress, also induced changes in amino acid distribution of soybean without addition of protease (Wallace et al., 1984). In these reports, GABA, a nonprotein amino acid, also increased. Salt stress on soybean roots strongly promoted the activity of diamine oxidases to stimulate polyamines degradation (Xing et al., 2007). After a longer stress period, soybean plants were able to adjust their amino acid metabolism to the unfavorable conditions (Simon-Sarkadi et al., 2006). Some researchers investigated that GABA accumulation was caused by environmental stress-induced proteolysis, but these reports were mainly about soybean plants and not soybean cotyledon. The effect of germination and breeding on the chemical composition of soybean cotyledon was reported (Clarke et al., 2000, Matsuyama et al., 2009). GABA content increased in the early germination stage of soybean. Soybean cotyledon without germination, which is enriched in nutritional ingredients, is desired as a raw material for value-added soybean food. Sasagawa et al. reported that the concentration of GABA in pressurized brown rice increased during storage at moderate temperature (Sasagawa et al., 2006). This finding could be well explained by the alternation of agrofood products induced by HPP described above. In the noble use of HPP enabling addition of values on foodstuffs, it is important to analyze biochemical changes of agrofood products during storage after HPP.

High pressure is a kind of stress on plant tissues. Therefore, application of HPP on soybeans is expected to lead to proteolysis, which will result in a novel technique for increasing amino acid content. In this study, we examine the high-pressure-induced changes in free amino acid distribution in pressurized soybeans, and investigate the effect of HPP on the reaction related to amino acid metabolism.

The precursors of target components have been enriched to increase the speed of bioproduction of certain target components (Gueven and Knorr, 2011, Oller et al., 2009), and precursor feeding is an effective tool for improving the accumulation of target compounds. Although the HPP-induced changes in compound levels result from random enzymatic reactions, the control of enzymatic reactions to induce a large accumulation of target compounds has hardly been reported for pressurized agrofood products (Shigematsu et al., 2010b). The generation of free amino acids in water-soaked and HPP-treated soybean seeds has been investigated (Ueno et al., 2010) and, as compared with untreated, Glu-soaked soybeans, the HPP-treated, Glu-soaked (0.05 g/mL Glu) soybeans have higher GABA concentrations, suggesting that both proteolysis and specific Glu metabolism are accelerated by HPP. However, not only the effect of the substrate concentration on the enzymatic reaction activity but also the generation–behavior of GABA in HPP-treated soybeans remains unclear. In this study, we focused on the GAD enzymatic reaction. We treated water-soaked and Glu-soaked soybean seeds with HPP and investigated the effects of the HPP on the bioconversion of Glu into GABA with respect to the substrate concentration dependence on the GABA generation. Kinetic analysis has also been investigated in the HPP-treated soybean seeds with precursor soaking (Glu concentrations of 0, 0.01, 0.03, 0.05, 0.08, 0.1, or 0.5 g/mL) using the reaction kinetics of the GABA generation.

The time course of GABA concentration in water-soaked or Glu-soaked soybeans (0.01 and 0.05 g/mL) is shown in Figure 4.12. The concentrations of GABA produced in the untreated, water-soaked soybeans varied from 0.89 to 1.7 μmol/g, and the GABA concentrations in the HPP-treated, water-soaked soybeans varied from 1.1 to 2.3 μmol/g. The Glu concentrations in the untreated and HPP-treated soybeans at 0 d increased with the increasing Glu concentrations in the soaking solution: the Glu concentrations of 0.01–0.5 g/mL in the soaking solution led to soybean Glu concentrations of 3.0–120 μmol/g in the untreated soybeans and 2.3–93 μmol/g in the HPP-treated soybeans. HPP caused a slight increase in GABA concentration in water-soaked samples, but would be predicted to cause a greater increase in Glu-soaked samples, since Glu is a precursor of GABA. The highest GABA concentration was 4.2 μmol/g (12 μmol/g dry basis, d.b.) in Glu-soaked (Glu = 0.05 g/mL) pressurized soybean at 2 d, while in water-soaked pressurized soybean at 0 d, it was 0.90 μmol/g. GABA concentration eventually increased five-fold during storage.

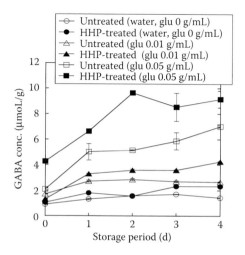

FIGURE 4.12   GABA concentrations in presoaked in Glu solution and pressurized soybeans.

The Glu concentrations in both samples decreased during the storage period. In contrast, the concentrations of GABA increased under all test conditions (0.01–0.1 g/mL) during storage ($p < 0.05$), with the exception of the 0.5 g/mL concentration. The maximum GABA concentration in this study was 6.8 μmol/g in the HPP-treated, Glu-soaked (Glu = 0.08 g/mL) soybeans, and the maximum GABA concentration in the untreated, Glu-soaked (Glu = 0.5 g/mL) soybeans was 5.5 μmol/g.

Typically, GABA levels in plant tissue are low, ranging from 0.03 to 2.0 μmol/g (Shelp et al., 1999). Martinez-Villaluenga and others reported that the GABA concentration in dry soybean seeds was 2.4 μmol/g d.b., while after 3-d germination, it was 8.0 μmol/g d.b (Martinez-Villaluenga et al., 2006). GABA concentrations in mechanically or cold-stimulated soybean leaves rise to 2.0 μmol/g (Wallace et al., 1984). In our study, GABA concentration reached up to 12 μmol/g d.b. in Glu-soaked pressurized soybean at 2 d. Higher accumulation of GABA in soybean cotyledon was performed by the combination of precursor soaking and HPP. To investigate the effect of HPP on GAD activities, we extracted the crude GAD from HPP. Crude GAD was obtained from HPP-treated soybean at 200 MPa after the homogenate, ammonium sulfate saturation, dialysis, and separation by diethylaminoethyl (DEAE) column. The fraction with the maximum GAD activity was applied for the sodium dodecyl sulfate polyacrylamide gel electrophoresis (SDS-PAGE) (Figure 4.13). The crude GAD from HPP-treated soybean had the same band pattern as that of the untreated soybean. Using this crude GAD, the secondary structural analysis by circular dichroism (CD) spectrum had shown that the crude GAD from HPP-treated soybean almost maintained the conformational properties after HPP (Table 4.1). Additionally, the GAD activities of the crude GAD from HPP-treated soybean still showed the similar activities compared with those of the untreated soybeans.

The high GABA production by HPP could be caused by the combination of several phenomena. The first reason for the high GABA production rate in HPP-treated soybean could be caused by the HPP effect on the specific membrane

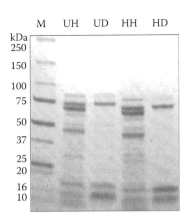

FIGURE 4.13 SDS-PAGE of the extracted GAD fraction from pressurized soybean after HPP of 200 MPa for 5 min at 25°C. M: Marker, UH: Homogenate from untreated soybean, UD: GAD of DEAE-fraction from untreated soybean, HH: Homogenate from HPP-treated soybean (200 MPa), HD: GAD of DEAE-fraction from HPP-treated soybean (200 MPa).

TABLE 4.1

Secondary Structure Analysis of Purified GAD from Pressurized Soybean (200 MPa)

Sample	α-Helix (%)	β-Sheet (%)	β-Turn (%)	Other (%)
Untreated	17	31	14	17
HPP of 200 MPa	16	30	11	13

proteins. HPP-treated soybean showed similar values of kinetic parameters with the extracted GAD. It was because the intercellular structure of the HPP-treated soybean was partially damaged; thus, the GAD in HPP-treated soybean behaved as the GAD itself without regulating the biosynthesis. However, the untreated soybean maintained the intact structures such as the membrane transporters that transported Glu, which would lead to the different kinetics from the HPP-treated soybean. The GAD from soybean was proven to have a calmodulin-binding domain at C-terminus, and is stimulated by Ca^{2+}/calmodulin (Snedden et al., 1995). HPP could result in the structural changes of Ca^{2+} channel proteins and/or the Glu transporters. The second reason was caused by the HPP effect on the diffusion of Glu as the precursor of GABA. If the GAD was localized in specific organelles such as vacuole for storage protein, the HPP-induced structural destruction would lead to the GAD diffusion inside the soybean cotyledons, which would accelerate the encounter between Glu and GAD in the HPP-treated soybeans. However, GAD is a cytosolic enzyme (Breitkreuz and Shelp, 1995), and it has been suggested that stress-induced GABA synthesis is the result of cytosolic acidosis and the consequent stimulation of GAD (Shelp et al., 1999). From the viewpoint of the enzymatic reaction, the Glu diffusion inside soybean cells would be promoted. Therefore, one possible explanation for the higher GABA production rate in the HPP-treated soybeans with the Glu feeding is that a Glu concentration gradient may not exist around GAD because of the HPP-related damage on the membrane systems, which may lead to the diffusion of Glu between the internal and external regions. Thus, the encounter rate between GAD and Glu may be accelerated, and the GABA production may be increased despite the weak affinity between GAD and Glu. Another reason was caused by the HPP-induced pH changes in cellular materials. The intercellular structure such as vacuole showed the low pH values. If these intercellular structures were ruptured by HPP, the intercellular pH could lead to lower pH values. Even if the intercellular pH was maintained by the intact biosynthesis system, the HPP-treated and damaged cells could lose the intact biosynthesis system. Therefore, HPP-induced decrease in pH could result in improving the enzymatic activity by shifting to the optimum pH level.

Since the GABA production rate in the HPP-treated soybeans showed a dependence on the initial Glu concentration, HPP-treated food products with precursor feeding have a great potential to increase the concentrations of target components.

Angiotensin-I-converting enzyme (ACE) inhibitory activities of the pressurized soybean were investigated. The ACE inhibitory activities of the HPP-treated soybean at 600 MPa increased while increasing the pressurizing time; however, the ACE

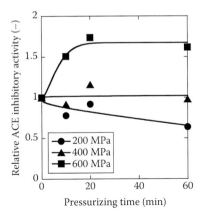

FIGURE 4.14   Relative ACE inhibitory activities of pressurized soybean at 200 (●), 400 (▲), and 600 MPa (■) for 10, 20, or 60 min. Relative ACE inhibitory activity was estimated from the ratio of ACE inhibitory of pressurized soybean/ACE inhibitory activity of untreated soybean.

activity reached a plateau when the pressurizing time exceeded 20 min (Figure 4.14). HPP on soybean resulted in ACE inhibitory activity; thus, HPP might generate the ACE inhibitory peptides.

HPP and subsequent storage resulted in improving the quality of agrofood products, such as functional properties and the distribution of functional compounds. Additionally, the combination of HHP and precursor feeding is a novel technique for the high accumulation of functional compounds. These innovative applications of HPP may prove useful for developing healthier agrofood products for the coming future.

REFERENCES

Ahromrit A., Ledward D., and Niranjan K., High pressure induced water uptake characteristics of Thai glutinous rice, *J. Food Eng.*, 72, 225–233, 2006.

Araya X. T., Hendrickx M., Verlinden B., Buggenhout S., Smale N., Stewart C., and Mawson A., Understanding texture changes of high pressure processed fresh carrots: A microstructural and biochemical approach, *J. Food Eng.*, 80, 873–884, 2007.

Barba F. J., Cortes C., Esteve M. J., and Frigola A., Study on antioxidant capacity and quality parameters in an orange juice-milk beverage after high-pressure processing treatment, *Food Bioprocess Technol.*, 5, 2222–2232, 2012a.

Barba F. J., Zesteve M. J., and Frigola A., High pressure treatment effect on physicochemical and nutritional properties of fluid foods during storage, *Compr. Rev. Food Sci. Food Saf.*, 11, 307–322, 2012b.

Boynton B. B., Sims C. A., Sargent S., Balaban M. O., and Marshall M. R., Quality and stability of precut mangos and carambolas subjected to high-pressure processing. *J. Food Sci.*, 67, 409–415, 2002.

Breitkreuz K. E., and Shelp B. J., Subcellular compartment of the 4-aminobutyrate shunt in protoplasts from developing soybean cotyledons, *Plant Physiol.*, 108, 99–103, 1995.

Buckow R., Weiss U., and Knorr D., Inactivation kinetics of apple polyphenol oxidase in different pressure–temperature domains, *Innov. Food Sci. Emerg. Technol.*, 10, 441–448, 2009.

Buckow R., Kastell A., Terefe N. S., and Versteeg C., Pressure and temperature effects on degradation kinetics and storage stability of total anthocyanins in blueberry juice, *J. Agric. Food Chem.*, 58, 10076–10084, 2010.

Butz P., Edenharder R., Fernández García A., Fister H., Merkel C., and Tauscher B., Changes in functional properties of vegetables induced by high pressure treatment, *Food Res. Int.*, 35, 295–300, 2002.

Cano M. P., Hernandez A., and De Ancos B., High pressure and temperature effects on enzyme inactivation in strawberry and orange products, *J. Food Sci.*, 62, 85–88, 1997.

Chauhan O. P., Raju P. S., Ravi N., Roopa N., and Bawa A. S., Studies on retention of antioxidant activity, phenolics and flavonoids in high pressure processed black grape juice and their modeling, *Int. J. Food Sci. Technol.*, 46, 2562–2568, 2011.

Chevalier D., Sentissi M., Havet M., and Le Bail A., Comparison of air-blast and pressure shift freezing on Norway lobster quality, *J. Food Sci.*, 65, 329–333, 2000.

Clarke E. J., and Wiseman J., Development in plant breeding for improved nutritional quality of soya bean 1. Protein and amino acid content, *J. Agric. Sci.*, 134, 111–124, 2000.

Dandamrongrak R., Young G., and Mason R., Evaluation of various pre-treatments for the dehydration of banana and selection of suitable drying models, *J. Food Eng.*, 55, 139–146, 2002.

Dede S., Alpas H., and Bayindirli A., High hydrostatic pressure treatment and storage of carrots and juice: Antioxidant activity and microbial safety, *J. Sci. Food Agric.*, 87, 773–872, 2007.

Dewanto X. W., Adrom K., and Liu R., Thermal processing enhances the nutritional value of tomatoes by increasing the total antioxidant activity, *J. Agric. Food Chem.*, 50, 3010–3014, 2002.

Dörnenburg H., and Knorr D., Cellular permeabilization of cultured plant tissues by high electric field pulses or ultra high pressure for the recovery of secondary metabolites, *Food Biotechnol.*, 7, 35–48, 1993.

Duarte-Vazquez M., Whitaker J., Rojo-Dominguez A., Garcia-Almendarez E., and Regalado C., Isolation and thermal characterization of an acidic isoperoxidase from turnip roots, *J. Agric. Food Chem.*, 51, 5096–5102, 2003.

Eshtiaghi M., Stute R., and Knorr D., High-pressure and freezing pretreatment effects on drying, rehydration, texture and color of green beans, carrots and potatoes, *J. Food Sci.*, 59, 1168–1170, 1994.

Fukuoka N., and Enomoto T., Enzyme activity changes in relation to internal browning of *Raphanus* roots sown early and late, *J. Hort. Sci. Biotechnol.*, 77, 456–460, 2002.

Gueven A., and Knorr D., Isoflavonoid production by soy plant callus suspension culture, *J. Food Eng.*, 103, 237–243, 2011.

Gonzalez-Cebrino F., Garcia-Parra J., Contador R., Tabla R., and Ramirez R., Effect of high-pressure processing and thermal treatment on quality attributes and nutritional compounds of "Songold" plum puree, *J. Food Sci.*, 77, C866–C873, 2012.

Hasegawa T., Hayashi M., Nomura K., Hayashi M., Kido M., Ohmori T., Fukuda M., Iguchi A., Ueno S., Shigematsu T., Hirayama M., and Fujii T., High-throughput method for a kinetics analysis of the high-pressure inactivation of microorganisms using microplate, *J. Biosci. Bioeng.*, 113, 788–791, 2012.

Hernandez A., and Cano M. P., High-pressure and temperature effects on enzyme inactivation in tomato puree, *J. Agric. Food Chem.*, 46, 266–270, 1998.

Islam S. M., Igura N., Shimoda M., and Hayakawa I., Effects of low hydrostatic pressure and moderate heat on texture, pectic substances and color of carrot, *Eur. Food Res. Technol.*, 217, 34–38, 2003.

Jacobo-Velazquez D. A., and Hernandez-Brenes C., Biochemical changes during the storage of high hydrostatic pressure processed avocado paste, *J. Food Sci.*, 75, S264–S270, 2010.

Kato M., and Hayashi R., Effects of high pressure on lipids and biomembranes for understanding high-pressure-induced biological phenomena, *Biosci. Biotechnol. Biochem.*, 63, 1321–1328, 1999.

Kato M., Hayashi R., Tsuda T., and Taniguchi K., High pressure-induced changes of biological membrane—Study on the membrane bound Na^+/K^+–ATPase as a model system. *Eur. J. Biochem.*, 269, 110–118, 2002.

Kawai K., Fukami K., and Yamamoto K., State diagram of potato starch–water mixtures treated with high hydrostatic pressure, *Carbohydr. Polym.*, 67, 530–535, 2007.

Kawai K., Fukami K., and Yamamoto K., Effect of temperature on gelatinization and retrogradation in high hydrostatic pressure treatment of potato starch–water mixture, *Carbohydr. Polym.*, 87, 314–321, 2012.

Keenan D. F., Brunton N. P., Gormley T. R., Butler F., Tiwari B. K., and Patras A., Effect of thermal and high hydrostatic pressure processing on antioxidant activity and colour of fruits smoothies, *Innov. Food Sci. Emerg. Technol.*, 11, 551–556, 2010.

Knockaert G., Roeck A. D., Lemmens L., Buggenhout S.V., Hendricks M., and Loey R. A., Effect of thermal and high pressure processes on structure and health-related properties of carrots (*Daucus carota*), *Food Chemi.*, 125, 903–912, 2011.

Knorr D., Heinz V., and Buckow R., High pressure application for food biopolymers, *Biochim. Biophys. Acta*, 1764, 619–631, 2006.

Koseki S., Mizuno Y., and Yamamoto T., Use of mild-heat treatment following high-pressure processing to prevent of pressure-injured *Listeria monocytogenes* in milk, *Food Microbiol.*, 25, 288–293, 2008.

Krebbers B., Matser A. M., Koets M., and Van den Berg R. W., Quality and storage-stability of high-pressure preserved green beans, *J. Food Eng.*, 54, 27–33, 2002.

Martinez-Villaluenga C., Kuo Y., Lambein F., Frias J., and Vidal-Valverde C., Kinetics of free amino acids, free non-protein amino acids, and trigonelline in soybean (*Glycine max* L.) and lupin (*Lupinus angustifolius* L.) sprouts, *Eur. Food Res. Technol.*, 224, 177–186, 2006.

Matser A. M., Knott E. R., Teunissen P. G. M., and Bartels P. V., Effect of high isostatic pressure on mushrooms. *J. Food Eng.*, 45, 11–16, 2000.

Matsuyama, A., Yoshimura K., Shimizu, C., Murano Y., Takeuchi H., and Ishimoto M., Characterization of glutamate decarboxylase mediating gamma-aminobutyric acid increase in the early germination stage of soybean, *J. Biosci. Bioeng.*, 107, 538–543, 2009.

Mclnerney J. K., Seccafien C. A., Stewart C. M., and Bird A. R., Effects of high pressure processing on antioxidant activity, and total carotenoid content and availability, in vegetables, *Innov. Food Sci. Emerg. Technol.*, 8, 543–548, 2007.

Namiki M., Yabuta G., Koizumi Y., and Yano M., Development of free radical products during the greening reaction of caffeic acid esters (or chlorogenic acid) and a primary amino compound, *Biosci. Biotechnol. Biochem.*, 65, 2131–2136, 2001.

Nunez-Mancilla Y., Perez-Won M., Vega-Galvez A., Arias V., Tabilo-Munizaga G., Briones-Labarca V., Lemus-Mondaca R., and Scala K. D., Modeling mass transfer during osmotic dehydration of strawberries under high hydrostatic pressure conditions, *Innov. Food Sci. Emerg. Technol.*, 12, 338–343, 2011.

Oey I., Plancken I., Loey A., and Hendrickx M., Does high pressure processing influence nutritional aspects of plant based food system? *Trends Food Sci. Technol.*, 19, 300–308, 2008.

Oh S.-H., and Choi W.-G., Changes in the levels of gamma-aminobutyric acid and glutamate decarboxylase in developing soybean seedlings, *J. Plant Res.*, 114, 309–313, 2001.

Ohnishi S., Fujii T., and Miyawaki O., Freezing injury and rheological properties of agricultural products, *Food Sci. Technol. Res.*, 9, 367–371, 2003.

Oki T., Abe H., Masuda M., Nakagawa T., and Suda I., Determination of quercetins in onion extracted with pressurized liquid, *Nippon Shokuhin Kogaku Kogaku Kaishi*, 52, 424–428, 2005.

Oller A. L. W., Agostini E., Milrad S. R., and Medina M. I., *In situ* and de novo biosynthesis of vitamin C in wild type and transgenic tomato hairy roots: A precursor feeding study, *Plant Sci.*, 177, 28–34, 2009.

Patras A., Brunton N., Pieve S. D., and Butler F., Impact of high pressure processing on total antioxidant activity, phenolic, ascorbic acid, anthocyanin content and colour of strawberry and blackberry purees, *Innov. Food Sci. Emerg. Technol.*, 10, 308–313, 2009a.

Patras A., Brunton N., Butler F., and Gerard D., Effect of thermal and high pressure processing on antioxidant activity and instrumental colour of tomato and carrot puree, *Innov. Food Sci. Emerg. Technol.*, 10, 16–22, 2009b.

Rastogi N. K., and Niranjan K., Enhanced mass transfer during osmotic dehydration of high pressure treated pineapple, *J. Food Sci.*, 63, 508–511, 1998.

Roldan-Marin E., Sanchez-Moreno C., Llooria R., Ancos B. D., and Cano M. P., Onion high-pressure processing: Flavonol content and antioxidant activity, *LWT—Food Sci. Technol.*, 43, 835–841, 2009.

Romani A., Vignolini P., Isolani L., Ieri F., and Heimler D., HPLC–DAD/MS characterization of flavonoids and hydroxycinnamic derivatives in turnip tops (*Brassica rapa L.* subsp *sylvestris L.*), *J. Agric. Food Chem.*, 54, 1342–1346, 2006.

Saikusa T., Horino T., and Mori Y., Accumulation of gamma-aminobutyric acid (GABA) in the rice germ during water soaking, *Biosci. Biotechnol. Biochem.*, 58, 2291–2292, 1994a.

Saikusa T., Horino T., and Mori Y., Distribution of free amino-acids in the rice kernel and kernel fractions and the effect of water soaking on the distribution, *J. Agric. Food Chem.*, 42, 1122–1125, 1994b.

Sasagawa A., Naiki Y., Nagashima S., Yamakura M., Yamazaki A., and Yamada A., Process for producing brown rice with increased accumulation of GABA using high-pressure treatment and properties of GABA-increased brown rice, *J. Appl. Glycosci.*, 53, 27–33, 2006.

Sellapan S., and Akoh C., Flavonoids and antioxidant capacity of Georgia-grown Vidalia onions, *J. Agric. Food Chem.*, 50, 5338–5342, 2002.

Shelp B., Bown A., and Mclean M., Metabolism and functions of gamma-aminobutyric acid, *Trends Plant Sci.*, 4, 446–452, 1999.

Shigematsu T., Nasuhara Y., Nagai G., Nomura K., Ikarashi K., Hirayama M., Hayashi M., Hayashi M., Ueno S., and Fujii T., Isolation and characterization of beroprotective mutants of *Saccharomyces cerevisiae* obtained by UV mutagenesis, *J. Food Sci.*, 75, M509–M514, 2010a.

Shigematsu T., Murakami M., Nakajima K., Uno Y., Sakano A., Narahara Y., Hayashi M., Ueno S., and Fujii T., Bioconversion of glutamic acid to γ-aminobutyric acid (GABA) in rice grain induced by high pressure treatment, *Jpn. J. Food Eng.*, 11(4), 189–199, 2010b.

Simon-Sarkadi L., Kocsy G., Varhegyi A., Galiba G., and Ronde J., Stress-induced changes in the free amino acid composition in transgenic soybean plants having increased proline content. *Biol. Plantarum.*, 50, 793–796, 2006.

Snedden W. A., Arazi T., Fromm H., and Shelp B. J., Calcium–calmodulin activation of soybean glutamate–decarboxylase, *Plant Physiol.*, 108, 543–549, 1995.

Tangwongchai R., Ledward D., and Ames J., Effect of high pressure treatment on the texture of cherry tomato, *J. Agric. Food Chem.*, 48, 1434–1441, 2000.

Toh-e A., Nakamura H., and Oshima Y., A gene controlling the synthesis of non specific alkaline phosphatase in *Saccaromyces cerevisiae*, *Biochim. Biophys. Acta*, 428, 182–192, 1976.

Tokusoglu O., Alpas H., and Bozoglu F., High hydrostatic pressure effects on mold flora, citrinin mycotoxin, hydroxytyrosol, oleuropein phenolics and antioxidant activity of black table olives, *Innov. Food Sci. Emerg. Technol.*, 11, 250–258, 2010.

Ueno S., Shigematsu T., Kuga K., Saito M., Hayashi M., and Fujii T., High-pressure induced transformation of onion, *Japan J. Food Eng.*, 10, 37–43, 2009a.

Ueno S., Hayashi M., Shigematsu T., and Fujii T., Formation of green–blue compounds in *Brassica rapa* root by high pressure processing and subsequent storage, *Biosci. Biotechnol. Biochem.*, 73, 943–945, 2009b.

Ueno S., Izumi T., and Fujii T., Estimation of damage to cells of Japanese radish induced by high pressure with drying rate as index, *Biosci. Biotechnol. Biochem.*, 73, 1699–1703, 2009c.

Ueno S., Watanabe T., Nakajima K., Murakami M., Hayashi M., Shigematsu T., and Fujii T., Generation of free amino acids and γ-aminobutyric acid in water-soaked soybean by high-hydrostatic pressure processing, *J. Agric. Food Chem.*, 58, 1208–1213, 2010.

Ueno S., Shigematsu T., Hasegawa T., Higashi J., Anzai M., Hayashi M., and Fujii T., Kinetic analysis of *E. coli* inactivation by high hydrostatic pressure with salts, *J. Food Sci.*, 76, M47–M53, 2011.

Valdez-Fragoso A., Mujica-Paz H., Welti-Chanes J., and Torres J. A., Reaction kinetics at high pressure and temperature: Effects on milk flavor volatiles and on chemical compounds with nutritional and safety importance in several foods, *Food Bioprocess Technol.*, 4, 986–995, 2011.

Waagepetersen H. S., Sonnewald U., and Schouboe A., Synthesis of vesicular GABA from glutamine involves TCA cycle metabolism in neocortical neurons, *J. Neurochem.*, 73, 1335–1342, 1999.

Wallace W., Secor J., and Schrader L., Rapid accumulation of γ-aminobutyric acid and alanine in soybean leaves in response to an abrupt transfer to lower temperature, darkness, or mechanical manipulation, *Plant Physiol.*, 75, 170–175, 1984.

Wang R., Wang T., Zheng Q., Hu X., Zhang Y., and Liao X., Effects of high hydrostatic pressure on color of spinach puree and related properties, *J. Sci. Food Agric.*, 92, 1417–1423, 2012.

Xing S., Jun Y., Hau Z., and Liang L., Higher accumulation of g-aminobutyric acid induced by salt stress through stimulating the activity of diamine oxidases in *Glycine max* (L.) Merr. roots. *Plant Physiol. Biochem.*, 45, 560–566, 2007.

Zhang D., and Hamauzu Y., Phenolics, ascorbic acid, carotenoids and antioxidant activity of broccoli and their changes during conventional and microwave cooking, *Food Chem.*, 88, 503–509, 2004.

5 High-Pressure Processing for Freshness, Shelf-Life Quality of Meat Products and Value-Added Meat Products

Özlem Tokuşoğlu and Halil Vural

CONTENTS

5.1 INTRODUCTION TO HIGH PRESSURE FOR MEAT AND MEAT PRODUCTS

Consumers demand products that are safe, nutritious, convenient, varied, attractive (in appearance, texture, and flavor), and also innovative. High-pressure processing (HPP) can be used as an alternative process to heat pasteurization and for shelf-life extension of a wide range of products.

Emerging non-thermal processing technologies are becoming widespread in the food industry, mainly as post-packaging interventions for food safety assurance and stabilization of all-natural, preservative-free propositions. Techniques such as HPP prove to be effective and economically feasible, showing now consistent double-digit growths. High pressure (HP) has become an innovative preservation technology with great potential for heat-sensitive food. HP is one such non-thermal processing technology that involves the application of hydrostatic pressure to inactivate microorganisms and extend the shelf life of meat and meat products with minimal effects on nutritional and sensory quality (Aydoğdu and Tokuşoğlu, 2012; Cheftel, 1995; Hugas et al., 2002; Rubio et al., 2007; Simonin et al., 2012; Tokuşoğlu, 2012; Vural and Tokuşoğlu, 2012).

HP in comparison to sterilization can give an energy saving of around 20% (Toepfl et al., 2006) and can be applied in packaged foods, eliminating possible recontamination after the treatment (Toepfl et al., 2006). Various studies have shown that HP treatment can alter meat structure, color, and lipid oxidation levels (Cheftel and Culioli, 1997; McArdle et al., 2010).

It was concluded that the application of HP offers some interesting opportunities for the processing of muscle-based food products. At present, almost half of the 200,000 tons annually of HP-treated foods are meat, fish, or seafood products (Heinz and Buckow, 2010).

The diverse nutrient composition of meat makes it an ideal environment for the growth and propagation of meat spoilage microorganisms and common food-borne pathogens. It is essential that adequate preservation technologies are applied to maintain the safety and quality of meat and meat products. HP is an emerging processing technique, used mainly for its preservative effects. Pressures in the range of 100–800 MPa are applied on meat products (Cheftel and Culioli, 1997; Vural and Tokuşoğlu, 2012) although commercial pressure vessels have a limit at <700 MPa (Torres and Velazquez, 2005). It is stated that pressures above 300 MPa help to inactivate microorganisms, making the product microbiologically safer. It can be explained the high level of microbial inactivation reached at pressures of 300 MPa or higher (Pedras et al., 2012).

Besides, HP affects the processes of meat protein denaturation, solubilization, aggregation, and gelation (Cheftel and Culioli, 1997; Iwasaki et al., 2006; Jimenez Colmenero, 2002; Sikes et al., 2009).

HHP induces changes on meat properties (e.g., texture and structure) and can be used for the development of new products or to improve the functional properties of meat (Tewari et al., 1999). It is known that meat texture is modified by HPP (Cheftel and Culioli, 1997). It was shown that pressure-induced gels provide generally smoother, more glossy, less firm, and more elastic gels with improved water-holding capacity (WHC), compared to thermally-induced gels (Cheftel and Culioli, 1997; Jimenez Colmenero, 2002). HP also preserves micro-nutrients better than thermal treatment (Aymerich et al., 2008).

Since pressure can affect the texture and gel-forming properties of myofibrillar proteins, it has been suggested as a physical and additive-free process to tenderize and soften meat and meat products (Cheftel and Culioli, 1997; Jung et al., 2000; Macfarlane, 1985; Sun and Holley, 2010;).

The effectiveness of HHP for microbial control depends on factors such as the process parameters, pressure level, temperature, and exposure time, as well as intrinsic factors of the food itself, such as pH, strain, and growth stage of microorganisms, and food matrix (Hugas et al., 2002).

HPP is generally applied at the post-packaging stage so that it will avoid further contamination during later food processing. At ambient temperatures, vegetative microorganisms and enzymes can be inactivated by applying a pressure of 400–600 MPa (Claude, 1995). HPP at 400–600 MPa was effective in controlling most major foodborne pathogens (*Escherichia coli* O157:H7, *Listeria monocytogenes*, *Salmonella* spp. *Salmonella aureus*, and so on) present in various meat products such as vacuum-packaged ground beef, cooked ham, and dry-cured ham (Black et al., 2010;

Jofré et al., 2009). Industrial HPP on meat applications are performed to extend the shelf life at low and medium temperatures, where inactivation of vegetative pathogen microorganisms at pressures above 400 MPa is possible (Buckow and Heinz, 2008). Table 5.1 shows some HP applications in meat products (Aymerich et al., 2008).

It was shown that HPP as an alternative in order to reduce the salt (NaCl) level in meat and meat products, has been established with respect to the processing of beef sausage (Sikes et al., 2009), pork (Iwasaki et al., 2006), and frankfurters (Crehan et al., 2000). It was concluded that HP below 200 MPa increased the binding properties of meat batters at a low salt concentration (Mandava et al., 1994) by increasing the meat protein solubilization (Sikes et al., 2009). It was shown the polyphosphates are utilized in association with NaCl to improve the binding properties and solubilization of myofibrils in meat products (Desmond, 2006; Xiong et al., 2000), and it was recorded that using polyphosphates induced a reduction in the salt content of meat products by enhancing the saltiness and improving the water-binding capacity (Weiss et al., 2010).

HPP is a very promising technology for the preservation of sliced cooked and cured meat products, and it shows a big potential for the innovative development of new products with a relatively low energy consumption. HPP uses an isostatic pressure at room temperature and between 100 and 600 MPa. The pressure chamber is loaded and closed, degassed, and the pressure is transmitted by the pumps through a liquid, generally water. HPP at low or moderate temperature causes destruction of microbial vegetative cells and enzyme-inactivation, without altering the sensory characteristics of the meat product and leaving the vitamins intact.

HPP for the meat industry has a place as 31% in overall food sector (Figure 5.1). HPP allows the decontamination of foods with minimal impact on their nutritional and sensory properties. The use of HPP to reduce microbial loads is important for

TABLE 5.1
Some HP Applications in Meat Products

Microorganism	Product	Initial Counts	Reduction	Process	Reference
Salmonella enteritidis strains	Chicken breast fillets	7	4.8	400 MPa, 15 min, 12°C	Morales et al. (2009)
C. freundii *P. florescens* *L. innocua*	Minced beef muscle	7	>5 after treatment	300 MPa, 10 min, 20°C 200 MPa, 20 min, 20°C 400 MPa, 20 min, 20°C	Carlez et al. (1993)
Total microflora	Minced beef muscle	~6.8	>4 after 10 days (3°C)	450 MPa, 20 min, 20°C	Carlez et al. (1994)
E. coli O157:H	Raw minced meat	5.9	5 after treatment	700 MPa, 1 min, 15°C	Gola et al. (2000)
Aerobic total count	Marinated beef loin	6.5	>4.5 after 120 days (4°C)	600 MPa, 6 min, 31°C	Garriga et al. (2004)

Source: Adapted from Aymerich T., Picouet P.A., and Monfort J.M. 2008. *Meat Science*, 78, 114–129.

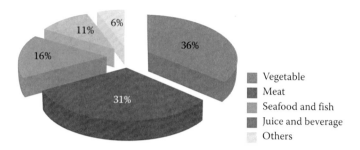

FIGURE 5.1 HPP in food technology and high pressure application ratio for meat sector (Adapted from Anonymous, 2008. NC Hyperbaric, Spain.)

red meat, poultry, and seafood. HPP not only reduces the bacterial content of meat products but also give many beneficial effects on meat product quality, freshness, and shelf life. HPP improves the WHC and decreased cooking losses of low fat burgers, reduces the negative impact of reducing fat via using the conjunction with meat extenders, and reduces the need for fat, salt, and phosphate in meat products without significantly changing their quality. Also, HPP-processed breakfast sausages with reduced fat, low salt, and low phosphate have been accepted by consumers.

It is stated that HP levels can promote the lipid oxidation, but mid-range levels have no impact on fatty acid (FA) composition; however, mild pressure treatments minimally influence meat quality while improving meat hygiene. HHP increases precipitation of sarcoplasmic proteins onto myofibrils as pressure level increases. In this context, for meat quality and shelf life; microorganism inactivation and meat safety, meat protein quality, meat color quality, textural quality, meat lipid quality, and sensory quality are most important (Figure 5.2).

The effects of HP on meat structure, including globular proteins, myofibrillar proteins, myofibrillar protein gelation, intramuscular connective tissue, and filamentous structure, were shown. Also stated were the HP effects on the meat enzymes

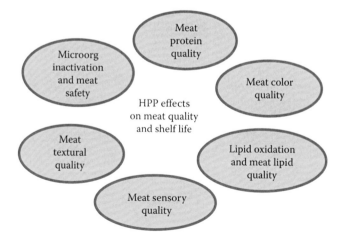

FIGURE 5.2 HPP effects on meat quality and shelf life. (Compiled by Tokuşoğlu, Ö.)

including meat endogenous proteases, lysosomal enzymes (cathepsins) and meat proteases (calpain), the meat color stability, texture quality, and lipid oxidation in meats (Buckow et al., 2013).

It was stated that the pressure treatment at 300–500 MPa denatured the tissue proteins and increased their proteolytic susceptibility, so, autolytic activity induced by endogenous acid proteases was increased by the pressure applied. It was also stated that in neutral and alkaline conditions, autolytic activity was reduced by pressurization at more than 400 MPa, at which level there was slight inactivation of neutral and alkaline proteases (Ohmori et al., 1992; 1991).

Ohmori et al. (1992, 1991) reported that lysosomal enzymes such as cathepsins were released from fractionated lysosomes after pressurization (100–500 MPa/10 min at 25°C). The increased activity of cathepsins B, D, and L in bovine muscle pressurized at 100–500 MPa/2–20°C/5 min (Homma et al., 1994) was also reported. It was determined that lysosomal enzymes – cathepsins B, D, and L activity – decreased gradually in a crude extract with increasing pressure, with a significant decrease of cathepsin D activity at 500 MPa (Dufour et al., 1995). It is known that calpain is a calcium-dependent intracellular cysteine protease and it participates in a number of cell processes. The necrosis that occurrs due to the level of calpain is considerably reduced during meat ageing (Sentandreu et al., 2002). Homma et al. (1995) reported a HP sensitivity of the calpain inhibitor calpastatin, which might contribute to a higher calpain activity in the case of pressure treatments below 200 MPa.

Sentandreu et al. (2002) reported the role of muscle endopeptidases and their inhibitors in meat tenderness. It is known that the myofibrillar component is greatly influenced by post-slaughter processing procedures, and post-mortem meat tenderization is generally assumed to result from softening of myofibrillar components by endopeptidases including cathepsins and calpains (Kemp et al., 2010). It was shown that the HPP positively affected the meat tenderness (Homma et al., 1995; Ichinoseki et al., 2007).

It was shown that the lightness (L^*) values of red meat increased significantly after pressure treatments above 250 MPa, while the redness (a^*) values decreased at 400–500 MPa, resulting in a gray-brown meat with a cooked aspect (Jung et al., 2003; Tintchev et al., 2010). It was reported that vacuum-packaged, HP-treated (400, 500, or 600 MPa) ham during 60 days of storage at 8°C gave stable red color (de Alba et al., 2012b). It was stated that the increased lightness (L* values, CIELAB) was observed during treatment at 400 MPa, and this pressure level also modified the color intensity (Clariana et al., 2012). It was reported that 600 MPa processing improved red color stability for minced, cured, restructured ham dried to 20% weight loss by Quick-Dry-Slice® (QDS) handling (Bak et al., 2013).

5.2   HPP EFFECTS ON THE QUALITY OF BEEF, FRANKFURTERS, AND LAMBS

HPP has been shown to reduce juiciness but improve tenderness in fresh beef and beef-based foods in comparison to control samples (Ludikhuyze and Hendrickx, 2001). HHP-treated beef samples at 200, 300, and 400 MPa at two different temperatures 20°C and 40°C showed that high pressure had no effect on polyunsaturated/

saturated fatty acid (PUFA/SFA) or omega 6/omega 3 (n6/n3) ratio. The temperature at which HPP was applied had a significant effect on the sum of SFA, monounsaturated (MONO) fatty acid, and PUFA (McArdle et al., 2010). It is shown that HPP at 40°C showed higher SFA and PUFA and lower MONO fatty acid compared to HPP processing at 20°C. In this context, it is stated that HP at low or moderate temperatures improves the microbiological quality of the meat with minimal effects on the meat quality (McArdle et al., 2010).

It is shown that the mild pressure treatments (200 MPa) would minimally affect the meat quality parameters, while improving meat hygiene (Mullen, 2012). HPP at higher pressure levels would promote lipid oxidation. It was found that HPP in the range 200–400 MPa did not alter the fatty acid profile of beef and HPP had a less effect than cooking over a 30-day storage period (Mullen, 2012).

Kang et al. (2013) stated the effect of HPP on fatty acid composition in Korean native black goat (KNBG) meat, and reported that FA content in KNBG meat was not significantly ($p > 0.05$) different among the control goats and those subjected to HPP.

Mullen (2012) stated that for lamb meat, the highest pressure level (600 MPa) had a detrimental effect on texture quality, oxidation, and water-binding properties. While all pressure treatments altered meat quality to some extent in some cases, this effect was not as pronounced as conventional cooking. It was shown that HPP may have the potential to be used as a pretreatment of meat to reduce the cooking time (Mullen, 2012).

Tintchev et al. (2013) stated the impact of HP/temperature treatment on structure modification and functional sensory properties of frankfurter batter. Frankfurters belong to the group of classical raw-cooked meat products and enjoy worldwide popularity. Frankfurter sausage batter can be defined as poly-dispersed systems consisting of a liquid continuous phase (water and soluble proteins, ions), a dispersed liquid phase (fat droplets), and a dispersed solid phase (nonsolvated muscle fiber particles, connective tissue, and spices) (Tintchev et al., 2013). In the study described by Tintchev et al. (2013), the degree of solubilization of meat proteins, particularly of myosin, was identified as a key process with significant effect on the batter's structural properties.

It was found that the maximal solubilization level was at 200 MPa/40°C IT for all formulations that were found to be treatment time-dependent (Tintchev et al., 2013). It was stated that the major role in the solubilization, aggregation, and gelation processes occurring in the aqueous phase was due to the myosin S-1 and S-2, N-terminal, C-terminal, the myosin light chains (MLC), and actin during the high HP/temperature treatment (Tintchev et al., 2013).

It was stated that the WHC was characterized by the drip loss parameter for the pressure levels of 300 and 600 MPa and initial temperature up to 40°C (Figure 5.3) (Tintchev et al., 2013). Figure 5.3 shows the influence of reduced NaCl on WHC expressed as drip loss by combined HP–T treatment, plotted as a function of temperature after pressurizing.

It was found that a decrease of drip loss with increasing pressure and temperature was detected, with a minimum observed at 600 MPa and 40°C IT. After HP treatment, a positive effect on drip loss reduction at low pressurization gradient

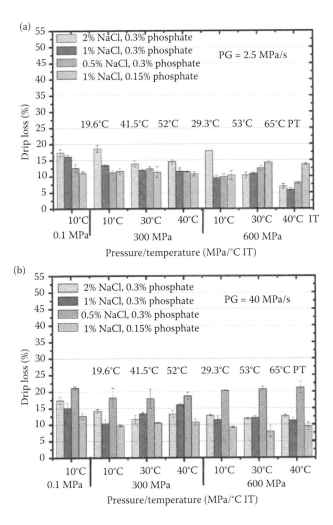

FIGURE 5.3 Influence of reduced NaCl on WHC expressed as drip loss by combined HP–T treatment, plotted as a function of temperature after pressurizing at (a) PG = 2.5 MPa/s and (b) PG = 40 MPa/s to desired set pressure for a 240 s holding time. (Adapted from Tintchev F. et al. 2013. *Meat Science*, 94, 376–387.)

(PG) was seen for all formulations. It was stated that the improvement of the WHC at low PG was more marked for the reduced NaCl formulation (0.5% NaCl) because of some structural alterations provoked through low PG (Figure 5.3) (Tintchev et al., 2013).

Figure 5.4 shows the effect of HPP and salts (NaCl and PP-polyphosphate) on the mechanism of water binding of meat proteins. The level of molecule opening is assumed to be pressure-dependent, as shown in Figure 5.4. (Tintchev et al., 2013). It was stated that opening the protein molecule provoked by the repulsing effect of phosphates was proposed to be promoted through penetration, by adding more water

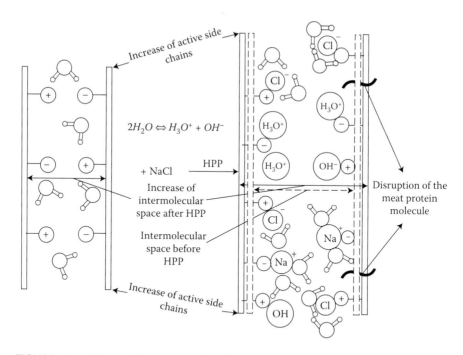

FIGURE 5.4 Effect of HP and salts (NaCl and PP-polyphosphate) on the mechanism of water binding of meat proteins (Adapted from Tintchev F. et al. 2013. *Meat Science*, 94, 376–387.)

($-\Delta V$) into the same volume during pressurization (Tintchev et al., 2013). Tintchev et al. (2013) stated that the process of ionic strength increase, caused by water dissociation during HPP, could additionally enhance the solubilization effect on the meat proteins. So, hydrogen (H_3O^+) and hydroxide (OH^-) ions can be attracted to the protein-charged groups (Tintchev et al., 2013).

It was found that the pressure treatment of about 100 MPa for 2.5 min at 40–60°C post-rigor muscle caused a significant increase in tenderness (Bouton et al., 1977). In the study reported by Ma and Ledward (2004), the texture profile analysis of post-rigor beef clearly showed that treatment at moderate pressure (200 MPa) and at 60–70°C temperature led to a very significant decrease in hardness (Figure 5.5). The effect was more marked at 70°C compared to 60°C. It was stated that the treatment at this pressure level at lower temperatures led to the expected increase at 20°C whereas little alteration at 40°C (Ma and Ledward, 2004).

5.3 HPP EFFECTS ON THE QUALITY OF HAM PRODUCTS

The use of HHP extended the shelf life of commercial-sliced, vacuum-packed cooked or dry-cured ham. Previous studies reported small changes in the proximate composition after pressurization of both dry-cured and cooked ham (Hugas et al., 2002). It is stated that the treatment at 400 MPa increased the instrumental color

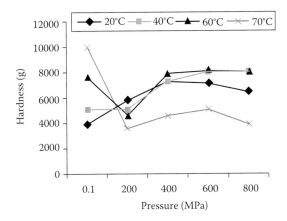

FIGURE 5.5   The effect of different pressure/temperature regimes on the hardness of a post-rigor beef muscle. (Adapted from Ma H.J. and Ledward D.A. 2004. *Meat Science*, 68(3), 347–355.)

measurement of lightness (L* values, CIELAB). This level of pressure also modified the hardness, chewiness, saltiness, and color intensity (Clariana et al., 2012).

HP pressurization of sliced skin vacuum-packed dry-cured ham at 400 and 900 MPa did not cause alterations in the main physicochemical parameters studied (fat, moisture, collagen). Only protein percentage was increased with HPP at 900 MPa. Pressurization at 400 MPa produced a slight increase on SOD activity (Clariana et al., 2012)

HPP may influence the enzyme activities in meat, increasing or decreasing their activity. The accumulation of purines and pyrimidines or its conversion into oxypurines in meat products is ruled by enzymes such as nucleotidases, adenosine deaminase, nucleoside phosphorylase, or xanthine oxidase, which can be affected by HHP technology (Eisenmenger and Reyes-De-Corcuera, 2009); for example, an increase of inosine monophosphate, and consequently the "umami" flavor, has been reported in rabbit muscle treated with HPP at 450 MPa (Mori et al., 2007). The effect of high pressure (at 600 and 900 MPa) on the levels of purines and pyrimidines was evaluated in dry-cured and cooked ham. Pressurization of dry-cured ham did not modify purines and pyrimidines contents. On the contrary, treatment at 600 and 900 MPa caused a decrease in guanosine and an increase in adenosine, respectively (Clariana et al., 2011a).

Villamonte et al. (2013) stated the effect of HP treatment (350 MPa, 6 min, 20°C) combined with sodium chloride (1.5–3.0%) and phosphates (0.25–0.5%) on the texture, water retention, color, and thermal properties in pork meat batters. In the study described by Villamonte et al. (2013), a clear hardening effect of HP on meat batters was reported and the hardness of the cooked pork meat batters (CPMB) was negatively correlated with the total denaturation enthalpy of the pork meat proteins ($p < 0.05$). It was stated that the HP denaturation of proteins was the cause of the hardening effect and the highest hardness was obtained after pressure treatment in the absence of salt and phosphates, which led to 50% of the total protein denaturation.

It was found that salt induced 29% of the total protein denaturation at high concentration (3% salt) and had no effect on the texture of the CPMB (Villanmonte et al., 2013).

Villanmonte et al. (2013) reported that the synergetic effect of salt and polyphosphates on water-binding properties was enhanced by HPP. It was suggested that HP could allow the production of meat products with less salt and without polyphosphates and with optimal technological properties.

Ham products are convenient to *L. monocytogenes* growth (Glass and Doyle, 1989; Lianou et al., 2007). *L. monocytogenes* has the unique ability among pathogens to grow at refrigerated temperatures, and it is problem for ready-to-eat (RTE) foods consumed without recooking (Kathariou, 2002). It is found that *L. monocytogenes* contamination occurs mostly during post-lethality processes including slicing, packaging, or other handling operations (Lin et al., 2006; Vorst et al., 2006).

It was shown that the HHP application of meats after packaging reduced the number of *L. monocytogenes* in hams (Aymerich et al., 2005; Jofré et al., 2009; Marcos et al., 2008; Pietrzak et al., 2007).

Recently, Myers et al. (2013) stated the effect of HP, sodium nitrite, and salt level on the growth of *L. monocytogenes* on RTE ham. It was stated that the growth of *L. monocytogenes* was evaluated for up to 182 days after inoculation on RTE-sliced ham formulated with sodium nitrite (0 or 200 ppm), sodium chloride (1.8 or 2.4%), and treated (no treatment or 600 MPa) with HHP. It was found that HP at 600 MPa/3 min resulted in a 4.07–4.35 log CFU/g reduction in *L. monocytogenes*. Table 5.2 shows the $\log_{10}$ reduction in plate count of *L. monocytogenes* after inoculation with a 5-strain cocktail of *L. monocytogenes* at a level of 10^5 CFU/g followed by HHP treatment at 600 MPa/3 min and measurement ca. 2 h after HHP (day 0) (Table 5.2) (Myers et al., 2013).

In the study reported by Myers et al. (2013), HP utilization greatly reduced the counts of *L. monocytogenes* by more than 3 log CFU/g, but when the inoculation level was ca. >3.4 log CFU/g, some survivors were able to repair and grow at >120 days after HHP treatment when stored at 4.4°F. It was found that HHP is effective for reducing

---

TABLE 5.2

For Ham, Log_{10} Reduction in Plate Count of *L. monocytogenes* after Inoculation with a 5-strain Cocktail of *L. monocytogenes* at a Level of 10^5 CFU/g Followed by HHP Treatment at 600 MPa for 3 min and Measurement ca 2 h after HHP (day 0) (S.E. = 0.39)

Treatments	$\log_{10}$ Reduction (CFU/g)
Ham-uncured low salt	4.35[a]
Ham-uncured high salt	4.29[a]
Ham-cured low salt	4.15[a]
Ham-cured high salt	4.07[a]

Source: Adapted from Myers K. et al. 2013. *Meat Science*, 93, 263–268.

[a] Means with the same superscript show no differences among means in a column ($p > 0.05$).

the number of *L. monocytogenes* on the RTE-processed meat products than that of sodium nitrite concentration, salt concentration, and meat species (Myers et al., 2013).

de Alba et al. (2012a) revealed the combined effect of sodium nitrite with HP treatments on the inactivation of *E. coli* BW25113 and *L. monocytogenes* NCTC 11994. It was stated that the bactericidal effect of acidified sodium nitrite alone or when combined with HHP treatment was examined with *E. coli* BW25113 and *L. monocytogenes* NCTC 11994. The powerful synergistic effect of HHP plus nitrite were found at pH 4.0, but not at higher pH values (de Alba et al., 2012a). Figure 5.6 indicates the effect of HHP and nitrite separately or in combination with *Escherichia coli*. Log$_{10}$ cycles of inactivation at pH 4 immediately after 300 MPa/8 min treatment (0 h) and after 24 h storage at 4°C was performed (Figure 5.6). de Alba et al. (2012a) stated that the low concentrations of nitrite (17–34 ppm) were sufficient to increase the bactericidal effect of HHP on *L. monocytogenes*.

Rubio et al. (2013) reported the use of safe enterococci as bioprotective cultures in low-acid fermented sausages combined with HHP. The sausage mixture was prepared with lean pork/pork back fat, (1:1) ingredients, and additives including NaCl, NaNO$_2$, KNO$_3$, sodium ascorbate, dextrose, and black pepper. Then the meat batter was inoculated by adding a cocktail containing three strains of *L. monocytogenes* and three strains of *S. aureus* to achieve a final concentration of ca. 4×10^3 CFU/g per sausage for each species (Rubio et al., 2013). In the study described by Rubio et al. (2013), three bacteriocinogenic, non-aminogenic and non-virulent Enterococcus strains (*Enterococcus faecium* CTC8005, *Enterococcus devriesei* CTC8006 and *Enterococcus casseliflavus* CTC8003) were used as starter cultures in low-acid fermented sausages to obtain their competitiveness and their bioprotective potential against *L. monocytogenes* and *S. aureus*. It was found that the HHP application (600 MPa/5 min) at the end of ripening

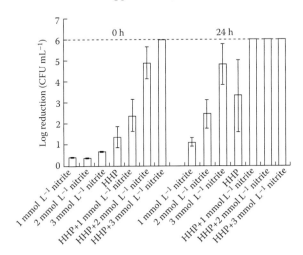

FIGURE 5.6 Effect of HHP and nitrite separately or in combination with *Escherichia coli*. Log$_{10}$ cycles of inactivation at pH 4.0 immediately after 300 MPa/8 min treatment (0 h) and after 24 h storage at 4°C. Results are means of three observations ± standard deviations (error bars) and dashed line indicates maximum detectable log reduction. (Adapted from de Alba M. et al. 2012a. *Letters in Applied Microbiology* 56, 155–160.)

(day 21) gave an immediate reduction in the counts of Enterobacteriaceae to levels <1 log CFU/g and promoted a decrease of 1 log unit in the counts of *S. aureus* (Rubio et al., 2013). It was detected that the counts of *L. monocytogenes* ca. reduced by *E. faecium* CTC8005. 2 log CFU/g immediately after stuffing and, in combination with HP promoted a further reduction of 1 log CFU/g in the pathogen counts (Rubio et al., 2013). Figure 5.7 shows the behavior of Enterococcus on *L. monocytogenes* and *S. aureus* during ripening of sausage without bacteriocinogenic Enterococcus culture and inoculated with bacteriocinogenic Enterococcus cultures (Rubio et al., 2013). Rubio et al. (2013) revealed that the combination of *E. faecium* CTC8005 and HHP was the most efficient for antilisterial approach.

It was found that lightness (L*) increased whereas redness (a*) and yellowness (b*) decreased for the 20% QDS-subjected hams, but no changes occurred in color for the 50% QDS hams (Bak et al., 2012). Bak et al. (2013) stated the effect of HP (600 MPa, 13°C, 5 min) and residual oxygen (0.02–0.30%) on the color stability of minced-cured restructured ham at different levels of drying (20%, 50% weight loss), pH (low, normal, high), and NaCl (15, 30 g/kg). 50% QDS-subjected hams to HP treatment had varying effects on Δa* ((a*(dayx) − a*(day1)) (Figure 5.8), but overall, the red color was more stable during storage up to 20 days (Bak et al., 2013).

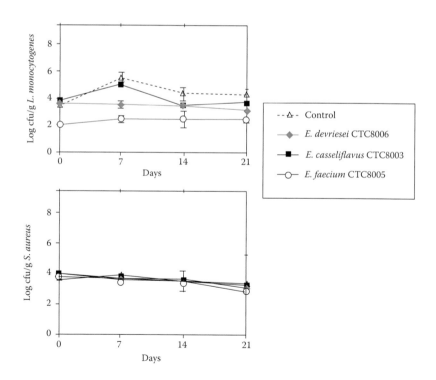

FIGURE 5.7 Behavior of enterococcus on *L. monocytogenes* and *S. aureus* during ripening of sausage without bacteriocinogenic Enterococcus culture and inoculated with bacteriocinogenic Enterococcus cultures. Values are the means of triplicates and error bars indicate the standard deviation. (Adapted from Rubio R. et al. 2013. *Food Microbiology* 33, 158–165.)

It was concluded that HP at 600 MPa improved color stability measured as Δa^* of minced, cured, restructured ham dried to 20% weight loss by QDS with an initial color within the normal range during 20 days of chill storage under applied illumination conditions with determined O_2 levels (0.02–0.30% O_2). In the study reported by Bak et al. (2013), it was also stated that the HP effect on the color stability of minced, cured, restructured ham dried to 50% weight loss by QDS during the 20-day storage period was much smaller, yet still with a slightly better color stability for HP-treated than for non-HP-treated hams.

de Alba et al. (2012b) reported a stable red color of vacuum-packaged, HP-treated (400, 500, or 600 MPa) Serrano ham during 60 days of storage at 8°C. Andrés et al. (2006) performed the HP treatment effect (200 or 400 MPa) on the color stability of dry-cured Iberian ham packaged in modified atmosphere packaged (MAP) containing either 0% O_2 or 5% O_2. It was also shown that hams stored without oxygen had higher a* values than hams stored with oxygen (Andrés et al., 2006). Even though, the a* values decreased within the first three days for both non-HP-treated and HP-treated hams, they then stabilized with time and had the same values on day 39 as on day 1 (Andrés et al., 2006).

It has been reported that in RTE meat treated with 600 MPa at 20°C for 180 s no significant deterioration in sensory quality was perceived (Zhou et al., 2010), while Clariana et al. (2011b) reported that HHP at 600 MPa at 15°C for 6 min modified the color of commercial dry-cured ham and the sensory attributes.

5.4   HPP EFFECTS ON THE QUALITY OF POULTRY MEAT PRODUCTS

Fresh chicken meat is a highly perishable product due to its biological composition. Many interrelated factors influence the shelf life and freshness of meat such as holding temperature, atmospheric oxygen (O_2), endogenous enzymes, moisture, light and most importantly, microorganisms. With the increased demand for high quality, convenience, safety, fresh appearance, and an extended shelf life in fresh meat products, alternative nonthermal preservation technologies, such as HHP, superchilling, active packaging, and natural biopreservation technologies, were used to investigate their effects on microbial population, meat quality, and sensory characteristics of chicken breast fillets (Antunes et al., 2003).

HHP, a nonthermal technology, is of primary interest because it can inactivate product-spoiling microorganisms and enzymes at low temperatures without altering the sensory or nutritional characteristics of the chicken meat. Pressure processing is usually carried out in a steel cylinder containing a liquid pressure-transmitting medium such as water, with the sample being protected from direct contact by using sealed flexible packaging. Maintaining the sample under pressure for an extended period of time does not require any additional energy apart from that required to maintain the chosen temperature (Cheftel and Culioli, 1997). HPP is used for the preservation of chicken, owing to its positive effects on protease activity, textural properties, taste, and flavor. The shelf life of chicken meat can be prolonged by HP processing.

In the particular case of chicken meat, the prevalence of pathogens such as *Salmonella* may be high (Antunes et al., 2003; Naugle et al., 2006). HHP is a

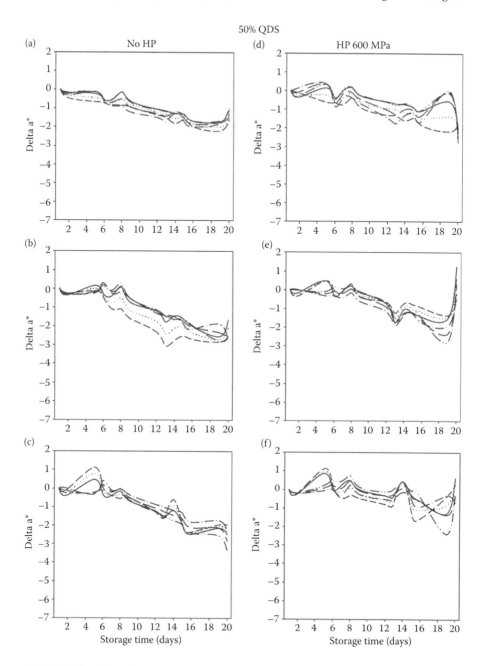

FIGURE 5.8   The redness (Δa*) (a*(day$x$) − a*(day1)) as a function of storage time (days) for 50% QDS cured, restructured ham of either low or high salt and either 0.05%, 0.15%, or 0.25% O_2 for each of the combinations. (a) Low pH_{24} and no HP, (b) Normal pH_{24} and no HP, (c) High pH_{24} and no HP, (d) Low pH_{24} and HP, (e) Normal pH_{24} and HP, and (f) High pH_{24} and HP. (Adapted from Bak K.H. et al. 2013. *Meat Science*, 95, 433–443.)

powerful tool to control risks associated with *Salmonella* spp. and *L. monocytogenes* in raw or marinated chicken meats (Hugas et *al.*, 2002).

HPP of chicken breast fillets inoculated with *Salmonella* achieved 4.0 and 4.6 log reductions after one 10 min-cycle and three 3 min-cycles at 400 MPa, respectively, with no recovery of injured cells during later refrigerated storage (Morales et al., 2009). Also, the shelf life of chicken meat can be prolonged by HP processing (O'Brien and Marshall, 1996). Pressures of 450 and 600 MPa almost completely eliminated all the three major pathogens *S. typhimurium* (KCTC 1925), *E. coli* (KCTC 1682), and *L. monocytogenes* (KCTC 3569) and therefore improved the safety of chicken breast fillets.

It is stated that the color and texture of chicken breast fillets were significantly affected by HPP at 400 MPa. Alterations brought about by HP treatments resembled those caused by cooking to an end-point temperature of 72°C in a water bath (Del Olmo et al., 2010). It was concluded that the use of HP extends the shelf life of vacuum-packaged chicken patties by up to 3 weeks, based on the condition of storage during refrigeration. After this period, the number of mesophilic, psychrotrophic, and lactic acid bacteria in the pressure-treated patties has been found 5–6 logarithmic cycles lower than that in the control products. It has been proposed that meat companies should apply HHP processing (400–600 MPa) for the shelf-life extension of processed, sliced meats. 600 MPa was enough to inactivate most microorganisms with no detection during storage life (4°C, 120 days).

HHP renders the chicken meat more stable due to its ability to reduce the number of spoilage and pathogenic microorganisms, and to inactivate certain meat enzymes (Patterson, 2005). It was stated that untreated raw ground chicken with an initial microbial load of 5.25 log CFU/g showed about 7.5 log CFU/g after 7 days at 4°C, whereas they remained below 7 log CFU/g after 70 days at 4°C in raw ground chicken pressurized at 408 MPa for 10 min (Patterson, 2005). Similarly, bacterial counts in minced chicken pressurized at 500 MPa for 15 min at 40°C did not increase significantly after storage for 6 months at 3°C, whereas they reached 7 log CFU/g after 8 days at 3°C (Linton et al., 2004).

Kruk et al. (2009) stated that HHP (300, 450, and 600 MPa) was used to investigate its effect on microbial population and sensory characteristics of chicken breast fillets. In the study reported by Kruk et al. (2009), commercially available chicken breast samples were completely sterilized by irradiation and then inoculated with either *Escherichia coli*, *L. monocytogenes* or *Salmonella typhimurium* for pathogen resistance experiments. Besides, another group of samples was only pressurized, grilled, and served to semi-trained sensory panels for sensory assessment (Kruk et al., 2009).

Table 5.3 shows the effects of HPP on microbial populations of chicken breast fillet (Kruk et al., 2009). It was determined that the increased pressure of 450 and 600 MPa almost completely inactivated all three strains of pathogens and improved safety of chicken breast fillets (Kruk et al., 2009). It was found that the 600 MPa HHP treatment reduced bacteria count by 6–8 log CFU/g for 7–14 days while the 450 MPa treatment reduced bacteria count by 4–8 log (CFU/g) for 3–14, based on the microorganism type (Kruk et al., 2009).

Figure 5.9 shows the effect of pressure on flavor, juiciness, and aroma strength of a chicken breast fillet. It was detected that the increased pressure impacted on flavor,

TABLE 5.3

Effects of HPP on Microbial Populations of Chicken Breast Fillet (as log (CFU/g))

Pathogen	High Pressure (MPa)	Storage (day) 0				
		0	3	7	14	SEM[1]
E. coli KCTC 1682	0	8.45^w	7.98%	7.84%	8.01%	0.163
	300	6.76^{a}%	5.39^{by}	5.97^{b}%	6.88^{a}%	0.173
	450	nd^{bz}	nd^{bz}	1.30^{aby}	3.62^{ay}	0.717
	600	nd^{bz}	nd^{bz}	nd^{by}	1.95^{az}	0.125
	SEM[2]	0.149	0.110	0.670	0.327	
S. typhimurium KCTC 1925	0	6.17%	6 74%	6.69%	6.84%	0.348
	300	5.53%	5.26^y	5.06%	5.38%	0.255
	450	2.82^y	nd^z	1.48^y	1.00^y	0.738
	600	nd^{bz}	nd^{bz}	nd^{by}	1.00^{by}	0.500
	SEM[2]	0.230	0.224	0.763	0.543	
L. monocytogenes KCTC 3569	0	7.35^{a}%	6.08^{b}%	5.63^{b}%	6.92^{a}%	0.115
	300	4.13^y	4.38^y	4.40^{xy}	4.89^y	0.314
	450	nd^z	nd^z	nd^z	nd^z	–
	600	nd^z	nd^z	nd^z	nd^z	–
	SEM[2]	0.097	0.063	0.103	0.297	

Source: Adapted from Kruk Z. et al. 2009. The effect of high pressure on microbial population and sensory characteristics of chicken. *Proceedings of the 55th International Congress of Meat Science and Technology (ICoMST)*, PE6.02. Copenhagen, Denmark, 16–21 August 2009.

Note: Values with different letters (a–c) within the same row differ significantly ($P < 0.05$). Values with different letters (w–z) within the same column differ significantly ($P < 0.05$). SEM[1] = standard errors of the mean ($n = 16$). SEM[2] = standard emus of the mean ($n = 16$). nd = not detected (<2.0 log CFU/g).

aroma strength, and juiciness; the 300 MPa pressure significantly reduced flavor, pleasantness, and juiciness, and 450 MPa produced breast fillets with the weakest aroma (Figure 5.9) (Kruk et al., 2009). In an another detailed study described by Kruk et al. (2011), it was also shown that increasing pressure increased cooking loss and color by increasing L*, a* and b* values and elevated pressure increased hardness, cohesiveness, gumminess and chewiness, as well as improved freshness of meat by reducing VBN. It was also found that the pressure of 450 MPa and higher pressures induced the lipid oxidation (Kruk et al., 2011).

It is stated that the HPP treatment is an effective technology in reducing bacterial spoilage and extending the shelf life of chicken breast fillets, however, it may have a negative impact on some quality and sensory characteristics based on the applied pressure; further researches are needed on chicken fillets and HHP (Kruk et al., 2011).

It was shown that the pink color of fresh turkey breast muscle quickly altered at pressures higher than 300 MPa at 10°C (Tintchev et al., 2010), and it was not as

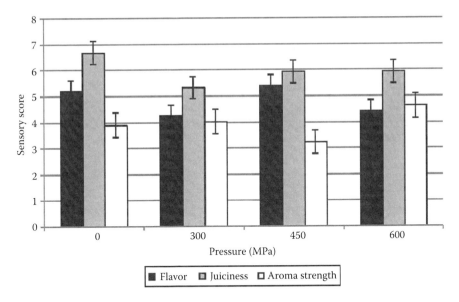

FIGURE 5.9 The effect of pressure on flavor, juiciness, and aroma strength of chicken breast fillet. (Adapted from Kruk Z. et al. 2009. The effect of high pressure on microbial population and sensory characteristics of chicken. *Proceedings of the 55th International Congress of Meat Science and Technology (ICoMST)*, PE6.02. Copenhagen, Denmark, 16–21 August 2009.)

sensitive to HPP as the color of chicken breast. On the contrary, it was reported that chicken meat only slightly paled after pressurization at 500 MPa/10 min in the study described by Yoshioka et al. (1992).

Improvement of functional and rheological properties of turkey breast meat proteins with different ultimate pHs at 24 h post-mortem (pH_{24}) was attempted using HPP (up to 200 MPa for 5 min at 4°C). Pressures of 50 and 100 MPa were found to increase the WHC of low pH meat (Chan et al., 2011).

Myers et al. (2013) found that HP at 600 MPa/3 min resulted in a 3.85–4.29 log CFU/g reduction in *L. monocytogenes*. Table 5.4 shows the log_{10} reduction in plate count of *L. monocytogenes* after inoculation with a 5-strain cocktail of *L. monocytogenes* at a level of 10^5 CFU/g followed by HHP treatment at 600 MPa/3 min and measurement ca. 2 h after HHP (day 0) (Myers et al., 2013).

5.5 HPP EFFECTS ON THE QUALITY OF VALUE-ADDED MEAT PRODUCTS

Technologies have been developed to improve the functionality of low-value beef (or chicken) and to incorporate it into value-added products, including restructured, coated, and emulsion-type meat products. Also, in recent years, there has been a growing demand among consumers for RTE meat products, which are microbiologically safe and possess superior sensory attributes and nutritional quality with an accompanying shelf-life extension. Consumers require such products that are safe, nutritious, convenient, varied, attractive (in appearance, texture, odor, and taste) and innovative.

TABLE 5.4

For Turkey, $\log_{10}$ Reduction in Plate Count of *L. monocytogenes* after Inoculation with a 5-strain Cocktail of *L. monocytogenes* at a Level of 10^5 CFU/g Followed by HHP Treatment at 600 MPa for 3 min and Measurement ca 2 h after HHP (day 0) (S.E. = 0.39)

Turkey-uncured low salt	4.10[a]
Turkey-uncured high salt	4.06[a]
Turkey-cured low salt	4.29[a]
Turkey-cured high salt	3.85[a]

Source: Adapted from Myers K. et al. 2013. *Meat Science*, 93, 263–268.

[a] Means with the same superscript show no differences among means in a column ($p > 0.05$)

It is stated that HPP effects meat structure including globular proteins, myofibrillar proteins, gelation, intramuscular connective tissue, and filamentous structure, it effects meat enzymes including meat endogenous proteases, lysosomal enzymes (cathepsins) and meat proteases (calpain). HHP, also effects meat color stability, meat texture quality and lipid oxidation level (Scheme 5.1).

Value-added meat products are further processed meat products with increasing convenience to consumer through decreasing preparation time, minimizing preparations steps, allowing use of specific parts, taking risks out of kitchen, and increasing value of product.

Value-added meat products (VAMPs) may be divided into six different types, as shown in the above mentioned Scheme 5.2. These are emulsion-based value-added meat products (VAMPs), nugget type VAMPs, combination type VAMPs, restructured type VAMPs, enrobed type VAMPs, and fruit /vegetable incorporated type VAMPs.

HPP effects on meat and meat products
- **For meat structure**
  - *Globular proteins,*
  - *Myofibrillar proteins,*
  - *Myofibrillar protein, gelation,*
  - *Intramuscular connective tissue,*
  - *Filamentous structure*
- **For Meat Enzymes**
  - *Meat endogenous proteases,*
  - *Lysosomal enzymes (cathepsins),*
  - *Meat proteases (calpain)*
- **For meat color stability**
- **For texture quality**
- **For lipid oxidation in meats**

SCHEME 5.1   HPP effects on meat and meat product quality. (Compiled by Tokuşoğlu, 2013).

It is known that typical commercial beefburgers contain 23–25% fat, which is too high in view of consumers' demands for healthier meat products. Since the fat level reduction in meat products is at the expense of their organoleptic properties, texture, and appeal, technological ways are needed to compensate for fat reduction. Troy et al. (2001) stated the HPP application to raw ground beef and the ability of HPP to compensate for fat reduction in beefburgers.

In the study reported by Troy et al. (2001), the acceptability of beefburgers containing 15% fat with and without the application of 300 MPa pressure to the ground beef component of the beefburgers was examined, and tapiocaline as a functional ingredient for low-fat beefburgers was also examined. Table 5.3 shows the formulation for reduced-fat beefburgers with added ingredients and HPP (Troy et al., 2001). It was stated that where fat was reduced to 15%, it was replaced by water or a combination of water and tapiocaline in the study given by Troy et al. (2001).

Troy et al. (2001) reported that the highest cook yields were recorded in beefburgers with 15% fat and 300 MPa pressure and/or a combination of HP and added tapiocaline (Table 5.5). It was found that the percentage WHC was similar for all of the products and the 15% fat beefburgers with tapiocaline and 300 MPa pressure were acceptable in this regard (Troy et al., 2001). It was reported that the reduced fat (15%) beefburgers showed optimum sensory scores with or without HP when tapiocaline was used and also the overall flavor, overall texture, and juiciness scores were enhanced. Besides, the combination of pressure treatment and tapiocaline addition gave less reduction in diameter of the beefburgers on cooking, and it was concluded that HPP and tapiocaline addition improved the quality characteristics of reduced-fat beefburgers (Troy et al., 2001).

It was found that HPP was able to control the pathogenic and spoilage microorganisms of the value-added meats in the study described by Aydoğdu & Tokuşoğlu (2012). The applied conditions at 450 MPa with short processing times of 5 min at 25°C inactivated about 4 log units (CFU) of the vegetative pathogenic and spoilage microorganism and increased the shelf life stability of manufactured value-added meats.

Scheme 5.3 shows the HHP-processed chicken nugget product including grape pomace incorporation. It was stated that untreated raw chicken nugget with an initial microbial load of 7.6 log CFU/g showed about 9.3 log CFU/g after one week at 4°C,

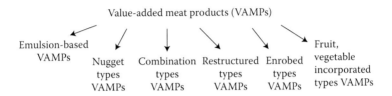

SCHEME 5.2   The consumed value-added meat products (VAMP). (Adapted from Aydoğdu T. and Tokuşoğlu Ö. 2012. High pressure processing (HPP) on freshness, shelf life of value-added meat (VAM) products. In *7th International Conference on High Pressure Bioscience and Biotechnology.* Poster Presentation. P2–17. Book of Abstracts. p. 105. October 29–November 2, 2012, Otsu, Kyoto, Japan.)

TABLE 5.5

Formulation for Reduced-Fat Beefburgers with Added Ingredients and HPP)

	Beef (kg)	Fat (kg)	Spice (kg)	Water (kg)	Tapiocaline (kg)
20% fat control	2.6	0.75	0.15	0.04	-
20% fat + HP	2.6	0.75	0.15	0.04	-
15% fat control	2.6	0.61	0.15	0.14	-
15% fat + HP	2.6	0.61	0.15	0.14	-
15% fat control + tapiocaline	2.6	0.61	0.15	0.105	0.035
15% fat + HP + tapiocaline	2.6	0.61	0.15	0.105	0.035

Source: Adapted from Troy D.J., Crehan C., Mullen A.M., and Desmond E. 2001. Development of value-added beef products, *Teagasc Report*, The National Food Centre, Dunsinea, Castleknock, Dublin. ISBN 1841702617.

while HHP-treated nuggets at 450 MPa/5 min/25°C remained about 3.7 log CFU/g after two months at 4°C.

It was reported that the applied HP did not alter the FA profiles of value-added nuggets ($p > 0.05$) and the FA composition was similar in both control and HPP-treated samples. The determined HPP conditions in the preliminary studies were able

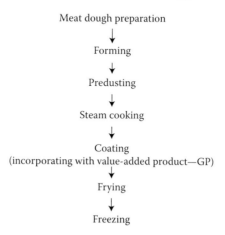

Commercial production flow chart of value-added Turkish chicken nugget

Meat dough preparation
↓
Forming
↓
Predusting
↓
Steam cooking
↓
Coating
(incorporating with value-added product—GP)
↓
Frying
↓
Freezing

SCHEME 5.3   The commercial production flowchart of value-added Turkish chicken nugget (Lezita Abalioğlu CBU project-USI-2012-01. (Adapted from Aydoğdu T. and Tokuşoğlu Ö. 2012. High pressure processing (HPP) on freshness, shelf life of value-added meat (VAM) products. In *7th International Conference on High Pressure Bioscience and Biotechnology.* Poster Presentation. P2–17. Book of Abstracts. p. 105. October 29–November 2, 2012, Otsu, Kyoto, Japan.)

to improve freshness protection and the overall quality of the Turkish value-added meat products (Aydoğdu and Tokuşoğlu, 2012).

Ma and Ledward (2013) stated that, subjected to HP at ambient temperatures, the shelf life of fresh meat is increased while the meat takes on a cooked appearance as the actomyosin denatures at about 200 MPa and the myoglobin denatures and/or co-precipitates with other proteins at about 400 MPa. Besides, the unsaturated lipids in the meat become more susceptible to oxidation at 400 MPa HHP and higher pressures. Owing to the iron released from complexes (haemosiderin and ferritin) present in meat and/or alterations in the lipid membrane, this oxidation phenomenon may occur. Ma and Ledward (2013) also reported that pre-rigor meat is subjected to pressures of about 100–150 MPa, below those necessary to cause color alterations. It becomes significantly more tender than its untreated counterpart, and this may also be a commercially attractive process for the food industry (Ma and Ledward, 2013).

HPP is currently being used by the meat industry as a post-processing technology to extend shelf life and improve the safety of various RTE meat products. HPP, a nonthermal processing technology, is of primary interest since it can inactivate product-spoiling microorganisms and undesired enzymes at low temperatures without altering the nutritional and sensory characteristics of the meat products. HPP can also be used to improve the functional properties of muscle proteins and the tenderness of meat, and to improve the texture of meat and meat color quality. The effectiveness of HHP for microbial quality and functional improvement depends on factors including the process parameters, level of pressure, process temperature, exposure time, and intrinsic factors (pH, strain, and growth stage of microorganisms) of meat including pH, strain, and growth stage of microorganisms.

REFERENCES

Andrés A.I., Adamsen C.E., Møller J.K.S., Ruiz J., and Skibsted L.H. 2006. High-pressure treatment of dry-cured Iberian ham. Effect on colour and oxidative stability during chill storage packed in modified atmosphere. *European Food Research and Technology*, 222, 486–491.

Anonymous. 2008. *NC Hyperbaric*, Spain

Antunes P., Réu C., Sousa J.C., Peixe L., and Pestana N. 2003. Incidence of Salmonella from poultry products and their susceptibility to antimicrobial agents. *International Journal of Food Microbiology*, 82, 97–103.

Aydoğdu T. and Tokuşoğlu Ö. 2012. High pressure processing (HPP) on freshness, shelf life of value-added meat (VAM) products. In: *Seventh International Conference on High Pressure Bioscience and Biotechnology*. Poster Presentation. P2–17. Book of Abstracts. p. 105. October 29–November 2, 2012, Otsu, Kyoto, Japan.

Aymerich T., Picouet P.A., and Monfort J.M. 2008. Decontamination technologies for meat products. *Meat Science*, 78, 114–129.

Aymerich T., Jofre A., Garriga M., and Hugas M. 2005. Inhibition of *Listeria monocytogenes* and *Salmonella* by natural antimicrobials and high hydrostatic pressure in sliced cooked ham. *Journal of Food Protection*, 68(1), 173–177.

Bak K.H., Lindahl G., Karlsson A.H., Lloret E., Gou P., Arnau J., and Orlien V. 2013. The effect of high pressure and residual oxygen on the color stability of minced cured restructured ham at different levels of drying, pH, and NaCl. *Meat Science*, 95, 433–443.

Bak K.H., Lindahl G., Karlsson A.H., Lloret E., Ferrini G., Arnau J., and Orlien V. 2012. High pressure effect on the color of minced cured restructured ham at different levels of drying, pH, and NaCl. *Meat Science*, 90, 690–696.

Black E.P., Hirneisen K.A., Hoover D.G., and Kniel K.E. 2010. Fate of *Escherichia coli* O157:H7 in ground beef following high-pressure processing and freezing. *Journal of Applied Microbiology*, 108:1352–1360.

Bouton P.E., Harris P.V., Macfarlane J.J., and O'Shea J.M. 1977. Effect of pressure treatments on the mechanical properties of pre- and post-rigor meat. *Meat Science*, 1(4), 307–318.

Buckow R. and Heinz V. 2008. High pressure processing—a data-base of kinetic information. *Chemie Ingenieur Technik*, 80(8):1081–1095.

Buckow R., Sikes A., and Tume R. 2013. Effect of high pressure on physicochemical properties of meat. *Critical Reviews in Food Science and Nutrition*, 53, 770–786.

Carlez A., Rosec J.P., Richard N., and Cheftel J.C. 1993. High pressure inactivation of *Citrobacter freundii*, *Pseudomonas fluorescens* and *Listeria innocua* in inoculated minced beef muscle. *Lebensmittel-Wissenschaft und-Technologie*, 26, 357–363.

Carlez A., Rosec J.P., Richard N., and Cheftel J.C. 1994. Bacterial growth during chilled storage of pressure-treated minced meat. *Lebensmittel-Wissenschaft und-Technologie*, 27, 48–54.

Chan Jacky T.Y., Omana D.A., and Betti M. 2011. Application of high pressure processing to improve the functional properties of pale, soft, and exudative (PSE)-like turkey meat. *Innovative Food Science and Emerging Technologies*, 12, 216–225.

Cheftel J.C. 1995. Review: High-pressure, microbial inactivation and food preservation. *Food Science and Technology International*, 1(2-3), 75–90.

Cheftel J.C. and Culioli J. 1997. Effect of high pressure on meat: A review. *Meat Science*, 46(3), 211–236.

Clariana M., Hortós M., García-Regueiro J.A., and Castellari M. 2011a. Effect of high pressure processing on the level of some purines and pyrimidines nucleosides and bases in dry cured and cooked ham. *Meat Science*, 89, 533–535.

Clariana M., Guerrero L., Sárraga C., Díaz I., Valero A., and García-Regueiro J.A. 2011b. Influence of high pressure application on the nutritional, sensory and microbiological characteristics of sliced skin vacuum packed dry-cured ham: Effects along the storage period. *Innovative Food Science Emerging Technologies*, 12, 456–465.

Clariana M., Guerrero L., Sárraga C., and Garcia-Regueiro J.A. 2012. Effects of high pressure application (400 and 900 MPa) and refrigerated storage time on the oxidative stability of sliced skin vacuum packed dry-cured ham. *Meat Science*, 90, 323–329.

Crehan C.M., Troy D.J., and Buckley D.J. 2000. Effects of salt level and high hydrostatic pressure processing on frankfurters formulated with 1.5 and 2.5% salt. *Meat Science*, 55, 123–130.

de Alba M., Bravo D., Medina M., Park S.F., and Mackey B.M. 2012a. Combined effect of sodium nitrite with high pressure treatments on the inactivation of *Escherichia coli* BW25113 and *Listeria monocytogenes* NCTC 11994. *Letters in Applied Microbiology*, 56, 155–160.

de Alba M., Montiel R., Bravo D., Gaya P., and Medina M. 2012b. High pressure treatments on the inactivation of *Salmonella enteritidis* and the physicochemical, rheological and color characteristics of sliced vacuum packaged dry cured ham. *Meat Science*, 91, 173–178.

Del Olmo A., Morales P., Ávila M., Calzada J., and Nuñez M. 2010. Effect of single-cycle and multiple-cycle high-pressure treatments on the colour and texture of chicken breast fillets. *Innovative Food Science and Emerging Technologies* 11, 441–444.

Desmond E. 2006. Reducing salt: A challenge for the meat industry. *Meat Science*, 74, 188–196.

Dufour E., Dalgalarrondo M., Herve G., Goutefongea R., and Haertle T. 1995. Proteolysis of type III collagen by collagenase and cathepsin B under high hydrostatic pressure. *Meat Science*, 42, 261–269.

Eisenmenger M.J. and Reyes-De-Corcuera J.I. 2009. High pressure enhancement of enzymes: A review. *Enzyme and Microbial Technology*, 45(5), 331–347.

Garriga M., Grebol N., Aymerich M.T., Monfort J.M., and Hugas M. 2004. Microbial inactivation after high-pressure processing at 600 MPa in commercial meat products over its shelf life. *Innovative Food Science and Emerging Technologies*, 5, 451–457.

Glass K.A. and Doyle M.P. 1989. Fate of *Listeria monocytogenes* in processed meat products during refrigerated storage. *Applied and Environmental Microbiology*, 55(6), 1565–1569.

Gola S., Mutti P., Manganelli E., Dazzi M., Squarcina N., Ghidini M., et al. 2000. Behaviour of pathogenic *E. coli* in a model system and in raw minced meat treated by HP: Microbiological and technological aspects. *Industria Conserve*, 75, 13–25.

Heinz V. and Buckow R. 2010. Food preservation by high pressure. *Journal of Consumer Protection and Food Safety*, 5(1), 73–81.

Homma N., Ikeuchi Y., and Suzuki A. 1994. Effects of high pressure treatment on the proteolytic enzymes in meat. *Meat Science*, 38, 219–228.

Homma N., Ikeuchi Y., and Suzuki A. 1995. Levels of calpain and calpastatin in meat subjected to high pressure. *Meat Science*, 41, 251–260.

Hugas M., Garriga M., and Monfort J.M. 2002. New mild technologies in meat processing: high pressure as a model technology. *Meat Science*, 62, 359–371.

Ichinoseki S., Nishiumi T., and Suzuki A. 2007. Effect of high hydrostatic pressure on collagen fibril formation. *Proceedings of Fourth International Conference on High Pressure Bioscience and Biotechnology*, Tsukuba, Japan, pp. 225–231.

Iwasaki T., Noshiroya K., Saitoh N., Okano K., and Yamamoto K. 2006. Studies on the effect of hydrostatic pressure pretreatment on thermal gelation of chicken myofibrils and pork meat patty. *Food Chemistry*, 95(3), 474–483.

Jimenez Colmenero F. 2002. Muscle protein gelation by combined use of high pressure/temperature. *Trends in Food Science and Technology*, 13, 22–30.

Jofr´e A., Aymerich T., Grebol N., and Garriga M. 2009. Efficiency of high hydrostatic pressure at 600 MPa against food-borne microorganisms by challenge tests on convenience meat products. *LWT – Food Science and Technology*, 42(5), 924–928.

Jung S., de Lamballerie-Anton M., and Ghoul M. 2000. Modifications of ultrastructure and myofibrillar proteins of post-rigor beef treated by high pressure. *Food Science Technology* 33, 313–319.

Jung S., Ghoul M., and de Lamballerie-Anton M. 2003. Influence of high pressure on the color and microbial quality of beef meat. *Food Science Technology* 36, 625–631.

Kang G., Cho S., Seong P., Park B., Kim S., KimD., Kim Y., Kang S., and Park K. 2013. Effects of high pressure processing on fatty acid composition and volatile compounds in Korean native black goat meat. *Meat Science*, 94, 495–499.

Kathariou S. 2002. *Listeria monocytogenes* virulence and pathogenicity, a food safety perspective. *Journal of Food Protection*, 65(11), 1811–1829.

Kemp C.M., Sensky P.L., Bardsley R.G., Buttery P.J., and Parr T. 2010. Tenderness- an enzymatic view. *Meat Science*, 84, 248–256.

Kruk Z., Yun H., Rutley D., Lee E.J., Kim Y.J., and Jo C. 2009. The effect of high pressure on microbial population and sensory characteristics of chicken. *Proceedings of the 55th International Congress of Meat Science and Technology (ICoMST)*, PE6.02. Copenhagen, Denmark, 16–21 August 2009.

Kruk Zbigniew A., Yun H., Rutley D.L., Lee E.J., Kim J.Y., and Jo C. 2011. The effect of high pressure on microbial population, meat quality and sensory characteristics of chicken breast fillet. *Food Control*, 22, 6–12.

Lianou A., Geornaras I., Kendall P.A., Belk K.E., Scanga J.A., Smith G.C. et al. 2007. Fate of *Listeria monocytogenes* in commercial ham, formulated with or without antimicrobials, under conditions simulating contamination in the processing or retail environment and during home storage. *Journal of Food Protection*, 70, 378–385.

Lin C.M., Takeuchi K., Zhang L., Dohm C.B., Meyer J.D., Hall P.A. et al. 2006. Cross-contamination between processing equipment and deli meats by *Listeria monocytogenes*. *Journal of Food Protection*, 69(1), 71–79.

Linton M., McClements J.M.J., and Patterson M.F. 2004. Changes in the microbiological quality of vacuum packaged, minced chicken treated with high hydrostatic pressure. *Innovative Food Science and Emerging Technologies*, 5, 151–159.

Ludikhuyze L. and Hendrickx M. 2001. Effects of high pressure on chemical reactions related to food quality. In M. Hendrikx, and D. Knorr (eds.), *Ultra High Pressure Treatment of Foods*. New York: Kluwer Academic.

Ma H. and Ledward D.A. 2013. High pressure processing of fresh meat—Is it worth it? *Meat Science*, 95, 897–903.

Ma H.J. and Ledward D.A. 2004. High pressure/thermal treatment effects on the texture of beef muscle. *Meat Science*, 68(3), 347–355.

Mandava R., Fernández I., and Juillerat M.A. 1994. Effect of high hydrostatic pressure on sausage batters. *Proceedings 40th International Congress on Meat Science and Technology*, S-VIB, 11, (The Hague, Holland). Zeist, The Netherlands: TNO Nutrition and Food Research.

Marcos B., Aymerich T., Monfort J.M., and Garriga M. 2008. High pressure processing and antimicrobial biodegradable packaging to control *Listeria monocytogenes* during storage of cooked ham. *Food Microbiology*, 25(1), 177–182.

Macfarlane J.J. 1985. High pressure technology and meat quality. In: *Developments in Meat Science*, pp. 155–184. Lawrie R.A., ed. Elsevier Applied Science Publishers: New York, NY.

Morales P., Calzada J., Rodríguez B., De Paz M., and Nuñez M. 2009. Inactivation of *Salmonella enteritidis* in chicken breast fillets by single-cycle and multiple-cycle high pressure treatments. *Foodborne Pathogens and Disease*, 6, 577–581.

Mori S., Yokoyama A., Iguchi R., Yamamoto S., Suzuki A., Mizunoya W., et al. 2007. Effect of high pressure treatment on cytoplasmic 5′-nucleotidase from rabbit skeletal muscle. *Journal of Food Biochemistry*, 31(3), 314–327.

McArdle R., Marcos B., Kerry J.P., and Mullen A. 2010. Monitoring the effects of high pressure processing and temperature on selected beef quality attributes. *Meat Science*, 86, 629–634.

Mullen A.M. 2012. Developing novel convenient meat based products by application of high pressure processing (HP meat). Project number: 5580, Funding source: DAFF (R&D/TAFRC/521). Agriculture and Food Development Authority, Ireland.

Myers K., Montoya D., Cannon J., Dickson J., and Sebranek J. 2013. The effect of high hydrostatic pressure, sodium nitrite and salt concentration on the growth of *Listeria monocytogenes* on RTE ham and turkey. *Meat Science*, 93, 263–268.

Naugle A.L., Barlow K.E., Eblen D.R., Teter V., and Umholtz R. 2006. U.S. Food Safety and Inspection Service testing for Salmonella in selected raw meat and poultry products in the United States, 1998 through 2003: Analysis of set results. *Journal of Food Protection*, 69, 2607–2614.

O'Brien J.K. and Marshall R.T. 1996. Microbiological quality of raw ground chicken processed at high isostatic pressure. *Journal of Food Protection*, 59, 146–150.

Ohmori T., Shigehisa T., Taji S., and Hayashi R. 1991. Effect of high pressure on the protease activities in meat. *Agricultural and Biological Chemistry*, 55, 357–361.

Ohmori T., Shigehisa T., Taji S., and Hayashi R. 1992. Biochemical effects of high hydrostatic pressure on the lysosome and proteases involved in it. *Bioscience, Biotechnology, and Biochemistry*, 56, 1285–1288.

Patterson M.F. 2005. Microbiology of pressure treated foods. *Journal of Applied Microbiology*, 98, 1400–1409.

Pedras, M.M., Pinho, C.R.G., Tribst, A.A.L., Franchi, M.A., and Cristianini, M. 2012. The effect of high pressure homogenization on microorganisms in milk. *International Food Research Journal*, 19(1), 1–5.

Pietrzak D., Fonberg-Broczek M., Mucka A., and Windyga B. 2007. Effects of high pressure treatment on the quality of cooked pork ham prepared with different levels of curing ingredients. *High Pressure Research*, 27(1), 27–31.

Rubio R., Bover-Cid S., Martin B., Garriga M., and Aymerich T. 2013. Assessment of safe enterococci as bioprotective cultures in low acid fermented sausages combined with high hydrostatic pressure. *Food Microbiology* 33, 158–165.

Rubio B., Martinez B., Garcia-Gachan M.D., Rovira J., and Jaime I. 2007. Effect of high pressure preservation on the quality of dry cured beef. *Innovative Food Science & Emerging Technologies*, 8(1), 102–110.

Sentandreu M.A., Coulis G., and Ouali A. 2002. Role of muscle endopeptidases and their inhibitors in meat tenderness. *Trends Food Science Technology*, 13, 400–421.

Sikes A.L., Tobin A.B., and Tume R.K. 2009. Use of high pressure to reduce cook loss and improve texture of low-salt beef sausage batters. *Innovative Food Science & Emerging Technologies*, 10(4), 405–412.

Simonin H., Duranton F., and de Lamballerie M. 2012. New insights into the high-pressure processing of meat and meat products. *Comprehensive Reviews in Food Science and Food Safety*, 11, 285–306.

Sun X.D. and Holley R.A. 2010. High hydrostatic pressure effects on the texture of meat and meat products. *Journal of Food Science*, 75, 17–23.

Tewari G., Jayas D.S., and Holley R.A. 1999. High pressure processing of foods: An overview. *Science des Aliments*, 19, 619–661.

Tintchev F., Wackerbarth H., Kuhlmann U., Toepfl S., Knorr D., Hildebrandt P., and Heinz V. 2010. Molecular effects of high pressure processing on food studied by resonance Raman. *Annals of New York Academy of Science*, 1189, 34–42.

Tintchev F., Bindrich U., Toepfl S., Strijowski U., Heinz V., and Knorr D. 2013. High hydrostatic pressure/temperature modeling of frankfurter batters. *Meat Science*, 94, 376–387.

Toepfl S., Mathys A., Heinz V., and Knorr D. 2006. Review: Potential of high hydrostatic pressure and pulsed electric fields for energy efficient and environmentally friendly food processing. *Food Reviews International*, 22(4), 405.

Tokuşoğlu Ö. 2012. High pressure processing (HPP) strategies on polyphenolic bioactives and shelf life stability in foods and beverages. In *7th International Conference on High Pressure Bioscience and Biotechnology*. L1–14, Session 2, 1600–1645. Book of Abstracts. p. 40. October 29–November 2, 2012, Otsu, Kyoto, Japan. Keynote Lecture.

Torres J.A. and Velazquez G. 2005. Commercial opportunities and research challenges in the high pressure processing of foods. *Journal of Food Engineering*, 67, 95–112.

Troy D.J., Crehan C., Mullen A.M., and Desmond E. 2001. Development of value added beef products, *Teagasc Report*, The National Food Centre, Dunsinea, Castleknock, Dublin. ISBN 1841702617.

Villamonte G., Simonin H., Duranton F., Chéret R., and de Lamballerie M. 2013. Functionality of pork meat proteins: Impact of sodium chloride and phosphates under high-pressure processing. *Innovative Food Science and Emerging Technologies* 18, 15–23.

Vorst K.L., Todd E.C.D., and Ryser E.T. 2006. Transfer of *Listeria monocytogenes* during mechanical slicing of Turkey breast, Bologna, and Salami. *Journal of Food Protection*, 69(3), 619–626.

Vural H. and Tokuşoğlu Ö. 2012. High pressure processing on red, poultry meat and processed meat products. *Advanced Nonthermal Processing in Food Technology: Effects of Quality and Shelf Life of Food and Beverages*. 07–10 May, 2012, Kuşadasi-TURKEY.

Oral Presentation *In ANPFT2012 Proceeding Book*, page 109–113. ISBN: 978-975-8628-33-9, Celal Bayar University Publishing, Turkey.

Weiss J., Gibis M., Schuh V., and Salminen H. 2010. Advances in ingredient and processing systems for meat and meat products. *Meat Science*, 86, 196–213.

Xiong Y.L., Lou X., Wang C., Moody W.G., and Harmon R.J. 2000. Protein extraction from chicken myofibrils irrigated with various polyphosphate and NaCl solutions. *Journal of Food Science*, 65, 96.

Yoshioka K., Kage Y., and Omura H. 1992. Effect of high pressure on texture and ultrastructure of fish and chicken muscles and their gels. In: *High Pressure and Bioscience*, pp. 325–327. Balny, C., Hayashi, R., Heremans, K., and Masson, P., eds. John Libbey Eurotext; Montrouge, France.

Zhou G.H., Xu X.L., and Liu Y. 2010. Preservation technologies for fresh meat: A review. *Meat Science*, 86, 119–28.

6 Quality of High-Pressure Processed Pastes and Purees

Om Prakash Chauhan, L.E. Unni, and Stefan Toepfl

CONTENTS

6.1 INTRODUCTION

Quality and safety are the two major factors that drive the consumer acceptability of a processed food material. The conventional processing technologies utilizing heat suffer from the disadvantages of impaired organoleptic properties, nutritional damage, and other undesired effects. Increasing concerns over human health have led researchers and scientists to experiment with newer and better alternative technologies that can create foods with enhanced safety and better quality. Over the past few decades, a rapid progress in high-pressure (HP) research has enabled the development of a wide variety of value-added products by keeping quality better with fresh-like characteristics. In recent years, there is a great demand for minimally processed fruits and vegetable products with rich flavor and high quality. Fruits and vegetables are considered protective foods as they are rich in various bioactive compounds. The application of HP preserves the freshness and extends the shelf life of fruits and vegetables. Initially, pressure processing collapses the raw product (fruit/vegetable) and then it is suddenly "expanded." Therefore, a fruit or a vegetable in its natural form may change its structural properties, instead of remaining in its original shape after pressure processing. This is why fruits or vegetables are generally pressurized in juice, pulp nectar, paste, or puree form. Pastes and purees are a convenient way of preserving fruits and vegetables that are otherwise perishable and seasonal in nature. These fruit derivatives have an improved edibility and palatability. Fruit purees are used in a variety of products including jams, preserves, and smoothies, and contain many health-promoting antioxidants. Commercialization of pressure-treated pastes and purees enables the consumers to avail the benefits of fruits and vegetables in places distant from its production. Moreover, it provides shelf-life extension to the product, so that a greater number of consumers could have access to healthy and harmless food.

Quality is a measured degree of excellence. The overall quality may be broken down into component characteristics, including the physical attributes such as color, texture, flavor, nutrient content, freedom from microorganisms, and other undesirable substances. Small molecules that are responsible for flavor and nutrition are typically not changed by pressure. Studies also have proven that pressure-processed foods have better texture, nutrient retention, and color compared to heat-processed foods. The effect of high-pressure processing (HPP) on the quality characteristics of food has been mainly attributed to the stability of covalent bonds to HP.

6.2 HPP OF PASTES AND PUREES

HPP is a nonthermal technology for microbiologically safe and shelf-stable food products. Microorganisms and deteriorative enzymes can be inhibited or inactivated depending upon the amount of pressure and time applied to the product. In principle, any food material with sufficient moisture content can be subjected to HPP. In HPP, being a very predictable process, the pressure is uniformly distributed at all points within the pressure chamber. Thus, regardless of the product size, the pressure exposure period is identical. But, even though the pressure effects are uniform on the food, this technology is not devoid of the classical limitation of heat transfer particularly during pressure build up and decompression. An increase or decrease of pressure is associated with a proportional temperature change of the vessel contents, respectively, due to adiabatic heating or cooling temperature gradient. The effects of pressure and temperature on food components are governed by activation volume and activation energy. Differences in sensitivity of reactions toward pressure (activation volume) and temperature (activation energy) lead to the possibility of retaining or degrading some desired natural food quality attributes such as vitamins, pigments, and flavor or modifying the structure of food system and food functionality, while optimizing the microbial safety or minimizing the undesired food quality related enzymes. The pressure sensitivity of microorganisms and enzymes in these fruits and vegetable derivatives is dependent on both the type and amount of enzymes and microorganisms as well as food composition (Lakshmi et al. 2011b). In the case of puree products, the precooking process has to assure total inactivation of the endogenous polyphenol oxidase (PPO) enzyme, the enzyme mainly responsible for the development of undesirable oxidative browning (Weemaes et al., 1998). Water activity is a physical property that has a direct implication for microbiological safety of food. It influences the storage stability of foods as some deteriorative processes in foods are mediated by water. Lowering water activity by adding hydrocolloids further improves the efficacy of HP inactivation.

6.2.1 QUALITY OF HP-PROCESSED PASTES AND PUREES

Color, flavor, and texture are the important quality parameters of fruits and vegetables and also are the major factors affecting sensory perception and consumer acceptance of foods. The effectiveness of HP treatment on overall food quality and safety is influenced by the extrinsic (process) factors—such as treatment time, pressurization/decompression rate, pressure temperature levels, and the number of pulses—and also by the intrinsic factors of the product—such as food composition and physiological states of microorganisms (Knorr, 2001). Food is a complex system and the compounds responsible for sensory properties coexist with enzymes, metal ions, and so on. Cell wall/membrane disruption (Michel and Autio, 2001), enzyme catalyzed-conversion processes (Verlent et al., 2005), chemical reactions (Nguyen et al., 2003), modification of biopolymers including enzyme inactivation, protein denaturation, and gel formation (Kolakowski et al., 2001) can occur at the same time. Different pressure temperature combinations can be used to achieve the desired effects on color, texture, and flavor. The quality of fruits and vegetable products can, however, change during storage due to coexisting chemical reactions, such as oxidation and biochemical reactions when endogenous enzymes or microorganisms are partially inactivated. The incomplete inactivation of enzymes and microbes can be delayed using a combination of obstacles, such as refrigeration temperatures, low pH, low water activity, and antibrowning agents to enhance the shelf life of pastes and purees.

6.2.1.1 Color

The pigments that are responsible for color in fruits and vegetables (carotenoids, anthocyanins, lycopene, etc.) are quite stable on HP treatment (low and moderate temperatures). However, the color compounds may alter during storage owing to incomplete inactivation of enzymes and microorganisms, which can result in enzymatic and nonenzymatic reactions in the food matrix. For many fruits and vegetable products such as fruit jam, strawberries, tomato juice, guava puree, avocado puree, and banana puree, HP treatment was largely noted to preserve fresh color (Yen and Lin, 1996). The brightness (L—color value) and redness/greenness (a—color value) of pressure-treated products were found to be superior compared to thermally treated counterparts.

6.2.1.1.1 Carotenoids

Carotenoids impart an orange-yellow and red color to the fruits and vegetables. Studies show that they are quite stable toward pressure treatment, and HP treatment increases the extraction yield of carotenes from the plant matrix. The redness of tomato puree increased with pressure treatments as compared to unprocessed samples. Higher redness in pressure treatments (500 and 800 MPa/20°C/5 min) may be attributed to better extractability of carotenoids due to disintegration of chro-moplast. Pressure treatment induced rapid changes in the microstructure of tomato pulp. Any change in the microstructure of tomato pulp during processing involves changes in the exposition of hydrophilic structures or cell decompartmentalization influencing the disposition of internal membranes (Fernandez Garcia et al., 2001). Pressure-treated (345, 517, or 689 MPa for 10, 20, and 30 min at initial pH of 3.9, 4.1, or 4.3) avocado puree had a color equivalent to the freshly prepared avocado puree (Lopez Malo et al., 1998). de Ancos et al. (2000) reported that the carotenoid con-tent of the HP-treated tomato puree (600 MPa/25°C/10 min) was significantly higher than untreated purees. It has been suggested that food processing such as cooking or grinding might improve lycopene bioavailability, the major carotenoid present in tomatoes, by breaking down cell walls. Pressure-induced isomerization of all-*trans* lycopene in hexane was observed during HP treatment at 500 and 600 MPa at ambi-ent temperatures when processed for 12 min duration. But this phenomenon was not observed in tomato puree. Moreover, 500 MPa led to an increase of total lycopene content in tomato puree. It may be due to breaking down cell walls because of HPs weakening the binding force between lycopene and tissue matrix; thus, making lyco-pene more accessible. The presence of some macromolecules may also offer some protection for lycopene. HP may also lead to a release of other antioxidants found in tomatoes (carotenoids, ascorbic acid, and phenolic components). These active com-ponents could have a good retention and thus protect lycopene from degradation (Qiu et al., 2006). The color of tomato puree was found to be unchanged after HP treat-ment up to 700 MPa at 65°C even for 1 h (Rodrigo et al., 2007).

6.2.1.1.2 Anthocyanins

Anthocyanin pigments are responsible for the red, purple, and blue colors of many foods. They are water-soluble vacuolar flavonoids pigments and are not affected much by pressure treatments, but show deterioration in storage. However, anthocy-anin degradation in processed berry products has been reported to rise as a result of indirect oxidation by phenolic quinones generated by PPO and peroxidase (POD) (Kader et al., 1997). Mean anthocyanin levels for HP-treated samples at 400, 500, or 600 MPa were higher than for fresh samples, but the effect of HP was not significant. In general, other authors have also reported that anthocyanins are stable in HP treat-ment at moderate temperature. Garcia-Palazon et al. (2004) reported that pelargon-idin-3-glucoside and pelargonidin-3-rutinoside in red raspberry (*Rubus idaeus*) and strawberry (*Fragaria × ananassa*) were stable in HP treatment at 800 MPa when processed at 18–22°C for 15 min. Some reports also exist with regard to increased extractability of colored pigments in food components at extreme pressures (De Ancos et al., 2002; Patras et al., 2008). In general, anthocyanins were well retained at all pressure treatments and this was reflected in better retention of antioxidant activity of strawberry and blackcurrant purees.

6.2.1.2 Enzymatic Browning

In addition to color pigments, enzymatic browning reactions also greatly affect the color of fruit- and vegetable-based products such as pastes and purees. Immediately after the pressure treatment, no pronounced visual color differences (based on L^*, a^*

and $b*$ values) were observed. But in storage, discoloration due to enzymatic browning was visible. In guava and banana purees, the green color gradually decreased in storage because of browning as a result of residual PPO activity (Lopez Malo et al., 1998, Palou et al., 1999). Guerro-Beltran et al. (2005b) observed enzymatic browning in HP-treated (379–586 MPa/room temperature/0.033, 5, 10, 15, or 20 min) mango puree. The addition of ascorbic acid and cysteine inhibited the PPO activity, resulting in less browning. Further, the inhibition was enhanced by HP treatment.

The structure and pigmentation of food interact with each other to affect both color and translucency/opacity. Texture modification may result due to changes in the nature and extent of internally scattered light and the distribution of surface reflectance, which in turn may produce changes in color appearance that would be more expected rather than the changes in pigment concentration (Mac Dougall, 2002). Thus, color changes in fruits and vegetables may be related to changes in textural properties. This phenomenon has been observed in tomato-based products. HP treatment (400 °MPa/25°C/15 min) resulted in an increase in the $L*$ value of tomato puree, indicating lightening of the puree surface color. The CIE color parameters were significantly higher in the untreated and in the HP-treated tomato purees compared to the thermally treated ones (Sanchez Moreno et al., 2006). The reason could be the formation of a jelly-like translucent structure of tomato puree, as observed by Verlent et al. (2006) when pressure dropped below 400 MPa.

6.2.1.3 Flavor

Flavor is the sensory perception of a food that is determined mainly by chemical senses of taste and smell. For most fruit and vegetable products, the potential benefits of using HP mainly arise from the fact that fresh flavor can be maintained during pressure treatment as the structure of small molecular flavor compounds is not directly affected by HPs. HPP may indirectly alter the content of some flavor compounds and disturb the whole balance of flavor composition as it can enhance/retard enzymatic and chemical reactions. Many authors reported that trained sensory panels were unable to differentiate between fresh and pressurized fruit products made from the same raw material (Ogawa et al., 1990; Watanabe et al., 1991). On the contrary, studies of Lambert et al. (1999) on the effect of HP (200, 500, and 800 MPa, 20°C, 20 min) on the aromatic volatile profile [furaneol (2,5-dimethyl-4-hydroxy-furan-3-one) and nerolidol (3,7,11-trimethyl 1,6,10-dodecatrien-3-ol)] of strawberry puree indicated that pressure treatments of 200 and 500 MPa at 20°C for 20 min did not affect the aroma profile of the strawberry puree; whereas, a pressure of 800 MPa resulted in a slight decrease in hexanal content and induced the synthesis of new compounds such as 3,4-dimethoxy-2-methyl-furan and γ-lactone. For tomatoes and onions, however, some flavor defects caused by pressure treatment were observed; tomato had a rancid taste while onions smelled less intensely and more like fried onions (Butz et al., 1994; Poretta et al., 1995). In the former case, rancid flavor was attributed to a marked increase in n-hexanal, which is largely responsible for fresh tomato flavor in a concentration of 1–2 mg kg^{-1} and higher concentrations of the same impart rancid flavor. In the case of onions, pressure treatment was reported to diminish dipropylsulphide, a compound responsible for pungency and the characteristic odor of fresh onions and to increase transpropenyl disulfide and 3,4-dimethylthiophene concentrations leading to a flavor of bruised or fried onions.

6.2.1.4 Texture and Rheology

In general, pressures up to 350 MPa can be applied to plant systems without any major effect on overall texture and structure (Knorr, 1995). Several studies have revealed that pressure treatment of fruits and vegetables and their products can cause both firming and softening (Basak and Ramaswamy 1998), the effects being dependent on pressure level and pressurization time. In general, the softening curves revealed that texture changes caused by pressure occurred in two phases; a sudden loss as a result of the pulse action of pressure followed by further loss of gradual recovery during pressure holding phase. At low pressure (100 MPa), instantaneous pressure softening was caused by compression of cellular structures without disruption, while at higher pressure (>200 MPa), severe texture loss occurred owing to rupture of cellular membranes and consequent loss of turgor pressure. During pressure holding time, the instantaneous texture loss can be gradually recovered and some products become even firmer than their fresh counterparts. In many cases, pressure-treated vegetables do not soften during subsequent cooking, which is attributed to the action of PME that is only partially inactivated by pressure. Simultaneous disruption of cell structures allows interaction of the enzyme with the pectic substance. Therefore, the de-esterified cell wall pectin can crosslink with divalent ions leading to increased compactness of cellular structure. HP treatment can, however, affect the rheological properties of food products, such as crushed fruits and vegetables, purees, pulp, and juice. The observed effects are dependent on the conditions of the HP process and the type of fruit and vegetable (Oey et al., 2008).

The consistency of tomato puree expressed in viscosity and level of syneresis (serum separation) is an important quality parameter of tomato puree. HP treatments at ambient temperature resulted in a more jelly-like, homogenous structure of the tomato puree in comparison to other treatments. A shear thinning kind of behavior was observed for pressure processed tomato puree. As a result of heat/enzymatic degradation of pectin followed by HP pasteurization (300 MPa) at ambient temperature, the retention of viscosity of tomato puree was noted. With increasing pressure level, the viscosity increased linearly due to polygalacturanase inactivation, compacting effects or protein tissue coagulation (Krebbers et al. 2003). In guava puree, pressure processing showed no remarkable change in viscosity; whereas, heat treatment was found to increase the viscosity from 932 to 1090 cp. Viscosity reduction occurred apparent during storage after 10 days in untreated puree and 20 days in 400 MPa pressure-treated puree. This may be due to microbial growth and enzymatic degradation of cloud and pectin in the puree (Gow and Hsin, 1996). HP treatment increased the viscosity of acidified apple (cv. Granny Smith) puree. An increase in linearity of cell walls and volumes of particles due to the permeabilization of cell walls might have resulted in this increase (Oey et al., 2008). Keenan et al. (2011) studied the textural and rheological properties of heat and pressure-treated apple puree enriched with prebiotics and reported no significant difference in the texture (maximum force of back extrusion) of heat and pressure-treated apple purees. The samples exhibited shear thinning as shear rates increased, which is the characteristic of non-Newtonian fluids.

6.2.1.5 Vitamin Content

Bignon (1996) observed that Vitamins A, C, B_1, B_2, and E content in fruits and vegetable products was not significantly affected by pressure treatment in contrast to thermal treatments. In the case of strawberries and guava purees, the decrease in Vitamin C content during storage after pressure treatment (400–600 MPa/15–30 min) was found to be much lower compared to the fresh products (Sancho et al., 1999). de Ancos et al. (2000) showed that persimmon puree treated at 50, 300, and 400 MPa for 15 min at 25°C had higher levels of extractable carotenoids, which was related with the increase of Vitamin A. Furthermore, it has been reported that different HP combinations had different influence on the stability of Vitamin C in guava puree during storage (Yen and Lin, 1996). The ascorbic acid content of untreated and pressurized (400 MPa/room temperature/15 min) guava puree started to decline after 10 and 20 days, respectively. The latter could be caused by the inactivation of endogenous pro-oxidative enzyme during HP treatment at HP level.

6.3 MICROBIOLOGICAL QUALITY

There is growing (although still limited) evidence to suggest that HP treatments in combination with temperature can inactivate heat-resistant spore-forming microorganisms. HP treatments are effective in inactivating most vegetative pathogenic and spoilage microorganisms at pressures above 200 MPa at chilled or process temperatures less than 45°C, but the rate of inactivation is strongly influenced by the peak pressure. Commercially, HP is preferred as a means of accelerating the inactivation process, and the current practice is to operate at 600 MPa except for those products where protein denaturation needs to be avoided. The pressure resistance of vegetative microorganisms often reaches a maximum at ambient temperature so the initial temperature of the food prior to HPP can be reduced or elevated to improve inactivation at processing temperature.

In guava puree, HP treatment at 600 MPa and 25°C for 15 min showed a good inhibitory effect. The relatively low pH would have enabled yeasts and molds to better adapt to guava puree environment. This is clear from the fact that during initial ten days of storage, the total plate count and yeast and mold counts were numerically similar. Pressure treatment at 400 MPa possibly triggered spore germination, which is suggested by the increased microbial growth. HP pasteurization at 500 MPa and ambient temperature caused only minor inactivation of natural microflora in tomato puree; whereas, HP treatment at 700 MPa for 2 min reduced the natural microbiota (3.6 log cfu total aerobic plate count at 30°C/mL) to a level below the detection limit (<1.5 log cfu/mL) (Krebbers et al., 2003). HP-treated avocado puree prepared at reduced pH (<4.3) and 1.5% sodium chloride yielded a microbiologically stable (below level of detection) avocado purees during storage at 5, 15, and 25°C (Lopez Malo et al. 1998). The effect of HP on decimal reduction of microbes in various fruit and vegetable purees is given in Table 6.1.

TABLE 6.1

Effect of HP Treatment on Microbial Reduction in Some Fruit and Vegetable Pastes and Purees

Product	Microorganism	Treatment Conditions		Decimal Reductions	Reference
Banana puree	TPC	517 MPa 689 MPa	10 min	3.0	Palou et al. (1999)
	Yeast and mold	517 MPa 689 MPa	10 min	2.0	
Avocado puree	TPC	345 MPa	10 min 20 min 30 min	3.0	Lopez Malo et al. (1998)
	Yeast and mold	345 MPa	10 min	1.0	
Guacamole (avocado paste)	TPC	689 MPa	5 min	4.0	Palou et al. (1999)
	Yeast and mold	345 MPa	10 min	1.0	
Mango puree	TPC	345 MPa	2 s	1.62	Guerrero-Beltran et al. (2006)
	Yeast and mold	345 MPa	2 s	1.35	

6.4   BIOCHEMICAL QUALITY

When food products are microbiologically stable, the end of their shelf life is generally limited by changes in their sensory and nutritional properties, which are caused by deteriorative biochemical reactions (Hough et al., 2003). The enzymatic reactions are one of the key problem areas to address in HPP of fruit and vegetable products. HP treatment can fulfill the requirement of traditional hot water blanching diminishing mineral leaching and waste water.

6.4.1   ENZYME ACTIVITY

The effects of pressure on enzyme activity are expected to occur at the substrate–enzyme interaction and/or active-site conformational changes. Enzyme inactivation occurs at very HPs. Enzyme activation that could be present at relatively low pressures is attributed to reversible configuration and/or substrate molecules. Studies on catalase, phosphatase, lipase, pectin esterase, lipoxygenase, POD, PPO, and lactoperoxidase revealed that POD is the most baro-stable enzyme with 90% residual activity after 30 min treatment at 60°C and 600 MPa, suggesting that POD could be selected as an enzyme indicator for HP treatment (Terefe et al., 2010).

PPO is the enzyme responsible for many undesirable color changes during fruit product storage and it can be considered a HP target for fruit product processing. PPOs respond differently during and after HP treatments depending on the food source. Cano et al. (1997) studied the combination of HP and temperature on POD and PPO activities of fruit-derived products. Strawberry POD inactivation was obtained by combining 43°C and 230 MPa. A combination of HP and 35°C effectively reduced POD in orange juice. Fruit residual PPO activity after HP treatment suggests that undesirable enzymatic reactions such as browning require the combination of HP with one or more additional factors to inhibit activity. Low pH, blanching, and refrigeration temperatures can be used in combination with HP treatment for better inactivation. Lopez Malo et al. (1998) reported that avocado PPO activity can be reduced to less than 20% when treated at 689 MPa pressure and at a pH of 3.9, 4.1, or 4.3. This reduction in PPO activity is not enough to avoid browning during storage. Brown color development was delayed for more than 30 days in HP-treated avocado puree if stored at 5°C. Residual PPO activity in the range of 86.63% in guava puree treated at 400–600 MPa for 15 min and stored at 4°C caused a continuous decrease in lightness and greenness of guava puree, but maintained an acceptable quality for at least 20 days. When avocado pulp is removed from the intact fruit and processed into paste, the tissues undergo partial disruption of cellular organelles releasing PPO enzyme that reacts with its substrates (phenolic compounds) causing formation of o-quinones. These compounds undergo further polymerization producing brown pigments, undesirable flavors, and nutritional losses (Tomas Barberam and Espin, 2001; Weemaes et al., 1999). On the other hand, LOX catalyzes the degradation of lipids, generating off flavors and causing the coupled degradation of carotenoids and other nutrients during storage (Jacobo-Velázquez and Hernández-Brenes, 2010; Robinson et al., 1995).

Although enzymatic browning of HHP-treated avocado paste is an important parameter to evaluate the product quality, other sensory changes (not correlate with color) also take place during storage that affects the sensory shelf life. Jacobo-Velázquez and Hernández-Brenes (2010) reported on relevant physicochemical changes that take place in commercially HHP processed (600 MPa, 3 min, 23°C) avocado paste stored under refrigeration conditions (4°C) and concluded on their potential involvement on the deterioration mechanism of the product. Such biochemical changes included a decrease in pH values (from 4.9 to 5.42) and the reactivation of PPO and LOX enzymes during storage, which were only partially inactivated by the HP treatment. In the context of residual enzyme activity, reports suggest the activity of LOX and PPO as potential mechanisms of deterioration during the storage of pressurized avocado paste (Lopez Malo et al., 1998). LOX in the presence of oxygen transforms linoleic and linolenic acid into hydroperoxides. The subsequent action of enzymes such as hydroperoxide lyase and other cis–trans isomerases double bond reducing and alcohol dehydrogenases induces the conversion of these hydroperoxides into aldehydes and alcohols responsible for off-flavors such as rancid flavor (Hatanaka et al. 1987). The development of rancid flavor intensity during storage did not significantly affect acceptability of product by consumers. However, it confirms the previously reported LOX residual activity and reactivation during storage of HHP-treated avocado paste. Although stored in oxygen impermeable packs, LOX seemed to be capable of using the remaining oxygen of the food matrix. It is important to consider that the enzymatic activity of LOX and PPO has a detrimental effect on the nutraceutical properties of avocado. Both enzymes are capable of degrading some bioactive compounds that are present in avocado in significant amounts such as MUFA (linoleic and linolenic), carotenoids, Vitamin C, and phenolic compounds (Robinson et al., 1995).

Threshold pressures for inactivation at room temperature of PME from different sources have been reported to vary largely from about 150 to 1500 MPa depending on the origin and medium in which the inactivation is carried out (Van den Broeck et al. 2000a). Inactivation occurs faster in acid medium and is protected by an increased amount of soluble solids (Ogawa et al., 1990). Most studies report only partial inactivation of PME, which is ascribed to the presence of isozymes with difference pressure resistance. Application of low pressure increased the activity of PME, which became maximal at a pressure of 100–200 MPa in combination with a temperature of 60–65°C (Van den Broeck et al. 2000a,b). POD, generally considered to be the heat stable enzyme, is also pressure-resistant to some extent. In green beans, a pressure treatment of 900 MPa merely induced slight inactivation at room temperature, while in combination with elevated temperature enhanced the inactivation effect at 600 MPa (Quaglia et al., 1996). Contradictory results were found by Cano et al. (1997), who reported POD in strawberry puree to be increasingly inactivated at room temperature with pressures up to 300 and 400 MPa, respectively. At higher temperature (45°C), a decrease in activity was found for all different range of pressures (50–400 MPa).

6.4.2 Antioxidant Activity

In tomato puree, total scavenging activity (DPPH) in aqueous and organic fractions was not changed by a HP treatment at 400 MPa (25°C/15 min). The effect of additives (NaCl 0–0.8% w/w and citric acid 0–2% w/w) on total scavenging activity (DPPH) in aqueous and organic fractions of tomato puree has been studied at different pressure levels ranging from 50 to 400 MPa at 25°C for 15 min. Pressure increased the antioxidant activity of aqueous fraction of tomato puree in the absence of additives. However, combined treatments of pressure between 300–400 MPa and high citric acid concentration (1.2–2% w/w) decreased the antioxidant activity in tomato puree, while the opposite effect was observed at low pressures (50–150 MPa) and high citric acid concentrations (1.2–2% w/w). Sodium chloride (0–0.8% w/w) addition lowered the antioxidant activity. The latter effect was more pronounced at low pressures (50–150 MPa) than at HPs (300–400 MPa). In contrast, citric acid addition (1–2% w/w) increased the antioxidant capacity of tomato puree (organic portion). A slight increase in antioxidant activity occurred at a pressure of 200 MPa in the absence of additives, though the highest antioxidant capacity was found when HP treatment up to 400 MPa was combined with NaCl addition (Sanchez-Moreno et al. 2004). Figure 6.1 shows the effect of HPP on phytochemical composition and total antioxidant activity in ginger paste.

6.5 QUALITY OF PRESSURE VS. THERMAL TREATED PUREES

The retention of flavor compounds in strawberry puree was better in HP-treated (600 MPa/ambient temperature/5 min) samples rather than heat-treated (80°C, 5 min) ones (Patras et al. 2009). This has been supported by a study where an electronic nose detector was used to analyze the volatiles of the treated purees. HP-treated strawberry puree differed from heat-treated and unprocessed puree. Cross validation of the

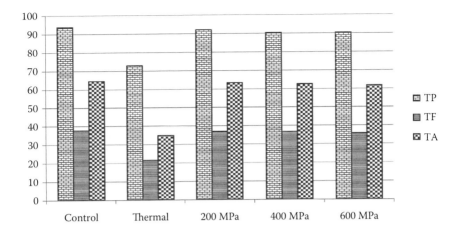

FIGURE 6.1 Effect of HPP on phytochemical composition and total antioxidant activity in ginger paste [TP: total phenolics (mg gallic acid equivalent/100 g), TF: total flavonoids (mg catechin equivalent/100 g), TA: total antioxidant activity (mM $FeSO_4$ equivalent/100 g)].

electronic nose data showed that heat treatment changed volatile compounds more than HPP. The corresponding results were reported for similarly processed raspberry and blackcurrant purees (Dalmadi et al. 2007). HP-processed purees had significantly higher antioxidant capacities when compared to thermally treated samples. This was reflected in better retention of carotenoids and ascorbic acid in HP-treated samples compared to thermally processed tomato and carrot puree samples. Unni et al. (2013) studied the effect of high pressure processing on garlic paste and observed retention of antioxidant activity in the order of 600 MPa (18.4%) > 400 MPa (18.2%) > 200 MPa (16.29 %) whereas a significant reduction ($p < 0.05$) of 49% in thermal treated garlic paste. The increase in antioxidant content or activity may be attributed to the increase in antioxidant components following HP treatment rather than an absolute increase. Unni et al. (2013) studied the effect of high pressure processing on garlic paste and observed retention of antioxidant activity in the order of 600 MPa (18.4%) > 400 MPa (18.2%) > 200 MPa (16.29 %) whereas a significant reduction ($p < 0.05$) of 49% in thermal treated garlic paste.

A better retention of phyto-chemicals such as phenolics, flavonoids, and antioxidant activities was observed in HP-treated ones as compared to thermally-treated samples. The tristimulus color (L^*, a^*, b^*) values and sensory acceptability, that is, sensory color, aroma, taste, consistency, and overall acceptability of ginger pastes were not much affected by HPs as compared to thermal processed ones (Chauhan et al. 2011). Figures 6.2–6.4 give a pictorial representation of tristimulus color (L^*, a^*, b^*) values of heat and pressure-treated ginger paste, tomato puree, and carrot puree.

6.6 SHELF LIFE AND STORAGE STABILITY

HP treatment is generally carried out at room/ambient temperature, which may not cause the complete inactivation of quality deteriorating enzymes and

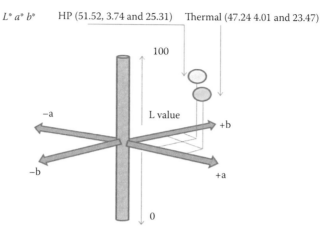

FIGURE 6.2 L^*, a^*, b^* values for thermal and pressure-treated ginger paste ((Adapted from Lakshmi, E. U. et al. 2011b. *International Journal of Food and Fermentation Technology* 1: 49–62.)

pressure-resistant bacterial spores that lead to undesirable changes in the HP-treated product during storage. Therefore, cold storage is needed to minimize detrimental quality modifications induced by residual enzyme activities and to ensure product safety. Navarro et al. (2002) observed that when HP-treated (400 MPa/ambient temperature/20 min) strawberry puree was stored for 30 days at 4°C, increase in contents of methyl butyrate, mesifurane, 2-methyl butyric acid, hexanoic acid, ethyl butyrate, ethyl hexanoate, 1-hexanol, and linalool were observed. During storage the content of 1-hexanol in the pressure-treated strawberry puree

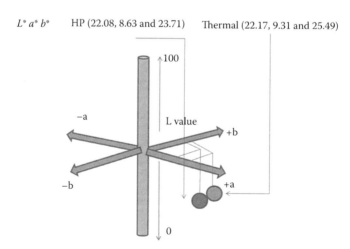

FIGURE 6.3 L^*, a^*, b^* values for thermal and pressure-treated tomato puree. (From Patras, A., Brunton, N. P., Da Pieve, S. and Butler, F. 2009. *Innovative Food Science and Emerging Technologies* 10(3): 308–313.)

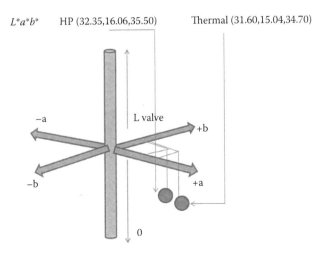

FIGURE 6.4 L^*, a^*, b^* values for thermal and pressure-treated carrot puree samples. (From Patras, A., Brunton, N. P., Da Pieve, S. and Butler, F. 2009. *Innovative Food Science and Emerging Technologies* 10(3): 308–313.)

increased probably due to the residual activity of lipoxygenase. In this case, POD was not considered responsible for flavor changes during storage, as the activity of POD was very low after HP treatment. Lopez Malo et al. (1998) pointed out that HP-processed (345–689 MPa) avocado puree (pH 3.9–4.3) lasted up to 100 days of storage at 5°C and 25°C with respect to microbial growth (<10 cfu/g). Avocado puree showed <45% PPO activity when stored at 5°C and had acceptable color up to 60 days of storage. Palou et al. (2000) treated guacamole at 689 MPa (5 min) for four times (cycles) and reported 50% residual PPO activity and a negligible microbial load of <10 cfu/g. However, browning during storage was observed and related with decrease in green color. In the case of HP-processed blanched (at selected times) banana puree (pH 3.4, a_w 0.97), synergistic effect was observed between longer blanching times and HPP and the samples were not spoiled (<10 cfu/g) after 15 days of storage at 25°C (Palou et al. 1999). The ginger, onion, and garlic paste samples added with ascorbic acid and citric acid and treated at 600 MPa pressure for 5 min, showed a shelf life of 6 months at low ($6 \pm 1°C$) temperature storage conditions on the basis of sensory microbiological attributes (Chauhan et al., 2011). Some studies undertaken on various HP-processed fruit and vegetable pastes and purees are given in Table 6.2.

6.7 CONCLUSION

HP, as a physical technique, has the potential for preservation of fruits and vegetable-based food products with improved quality. The main concerns about quality loss in food such as microorganisms, enzyme activity, chemical reactions, and physical changes may be effectively addressed using the novel HPP technology. A better understanding of the mechanism of inactivation/activation of enzymes, microbes, physicochemical changes in food matrix due to HP treatment would enhance the

TABLE 6.2

High-Pressure Processed Fruit and Vegetable Pastes and Purees

Paste/Puree	Process Conditions	Parameters Studied	Observation	Reference
Apple puree	400 and 600 MPa/5 min/20°C	Vitamin C, ascorbic acid, and total phenolic content	Total Vitamin C and ascorbic acid content were unaffected by 400 MPa. Phenolic content was not changed during 400 MPa treatment but was affected by 600 MPa	Landl et al. (2010)
Avocado puree	345,517 and 689 MPa for 10, 20 or 30 min at initial pH 3.9, 4.1 or 4.3	PPO color change and microbial inactivation	Standard plate as well as yeast and mold counts of HP-treated purees were <10 cfu/g during 100 days of storage at 5, 15, or 25°C. Significantly less ($p < 0.05$) PPO activity was observed with increasing pressure and decreasing initial pH	Lopez et al. (1998)
Avocado Guacamole	800 MPa, 25°C, pH 6–7	PPO	Inactivation of PPO	Weemaes et al. (1998)
	—	Enzyme activity, microbial, color	Standard plate and yeast and mold counts of high pressure-processed guacamole were <10 cfu/g. Sensory acceptability and color of high pressure guacamole were not significantly different ($p > 0.05$) from those of a guacamole control. Significantly less ($p < 0.05$) residual PPO and lipoxygenase (LOX) activities were obtained by increasing the process time and number of pressurization-decompression cycles	Palou et al. (2000)
Avocado paste	600 MPa/3 min	Enzyme activity	A decrease in PPO and LOX in avocado paste on storage for 45 days at 4°C. Reactivation of both enzymes was observed at 10–15 days of storage	Jacobo-Velázquez and Hernández-Brenes (2010)
Strawberry puree	200, 500, and 800 MPa, 20°C, 20 min	Aroma profile	Pressure treatments of 200 and 500 MPa at 20°C for 20 minutes did not affect the aroma profile of the strawberry puree; whereas, a pressure of 800 MPa significantly changed the aroma profile by inducing the synthesis of new compounds	Lambert et al. (1999)
Raspberry puree	200–800 MPa, 18–22°C, 15 min	Anthocyanin content and color stability	Highest stability of the anthocyanins was observed when processed at 200 and 800 MPa and stored at 4°C	Suthanthangjai et al. (2005)

Product	Treatment	Parameter measured	Results	Reference
Strawberry and blackberry purees	400–600 MPa/15 min/10–30°C	Total antioxidant activity and color	Ascorbic acid and anthocyanins contents or color in pressure treated purees were more or less similar to unprocessed samples	Patras et al. (2009)
Guava puree	600 MPa, 25°C, 15 min	Color, pectin, cloud and ascorbic acid content	No change in color, pectin, cloud and ascorbic acid content on storage. Inactivation of microorganisms to <10 cfu/mL	Gow and Hsin (1996)
Persimmon puree	50, 300, and 400 MPa, 15 min, 25°C	Carotenoids	Higher levels of extractable carotenoids, which was related to increase of vitamin A	de Ancos et al. (2000)
Tomato puree	400 MPa	Microbial and enzyme activity	4-log reduction of total microbial counts and an inactivation of polyphenoloxidase, peroxidase and pectinmethylesterase	Plaza et al. (2003)
Tomato puree	HP (400 MPa/25°C/15 min), low pasteurization (LPT) (70°C/30 s), High pasteurization (HPT) (90°C/1 min), freezing (F) (−38°C/15 min), and HPT plus freezing (HPT + F)	Color, Vitamin C, ascorbic acid content	CIE color parameters, total carotenoids, and provitamin A carotenoids were significantly higher in HP-processed tomato puree compared to other treatments; whereas, ascorbic acid and total Vitamin C content were lower in HP and thermal treatments compared to untreated and frozen tomato purees	Sanchez Moreno et al. (2006)
Mango puree	–	Enzymes and microbial activity	Pressure treatment at 207, 345, 483, and 552 MPa decreased the residual PPO activity. Total plate counts and yeasts were inactivated (<10 cfu/g) at pressure treatments of 483 or 552 MPa	Guerrero-Beltran et al. (2006)
Tomato puree	–	Rheological properties	Pressure treatment of less than 300 MPa caused significant losses in rheological properties; a combined HP/heat treatment (60°C, 500 MPa) improved the rheological properties of tomato puree but caused formation of a tomato gel structure; a pressure treatment of 500 MPa and temperatures higher than 60°C did not have any effect on the rheological properties and gel formation of the tomato puree	Verlent et al. (2006)
Tomato puree	100–600 MPa, 12 min, 20 ± 1°C	Lycopene content	The highest lycopene stability was reported in samples pressurized at 500 MPa and stored at 4 ± 1°C	Qiu et al. (2006)

continued

TABLE 6.2 (Continued)
High-Pressure Processed Fruit and Vegetable Pastes and Purees

Paste/Puree	Process Conditions	Parameters Studied	Observation	Reference
Tomato puree	300–700 MPa, 65°C, 60 min	Color	No color degradation was observed	Rodrigo et al. (2007)
Tomato puree	400–600 MPa/15 min/20°C	Color, phenolics antioxidant activity, and ascorbic acid	HP and thermal treatments did not affect the phenolic content in tomato puree but significantly affected color parameters	Patras et al. (2009)
Tomato puree	50–500 MPa/20–60°C/15 min	PPO, POD, and PME	A significant reduction of 32.5% of PME was observed when a combination of 150 MPa/30°C was employed, 25% reduction for 350 MPa/20°C treatment and 10% reduction in PPO for 200 MPa/20°C	Hernandez and Cano (1998)
Banana puree	517–689 MPa/10 min/21°C, pH 3.4, a_w 0.986–0.97	PPO activity and color	HHP treatments retained the initial color of the banana purees. Longer browning induction times and slower browning rates were observed when a longer blanching time was combined with a 689 MPa pressure treatment. A residual PPO activity <5% was observed in the puree when a 7 min blanch was followed by HHP treatment at 689 MPa for 10 min	Palou et al. (1999)
Carrot puree	400, 500, 600 MPa for 15 min/20°C and 70°C/2 min	Color, phenolics, antioxidant activity, ascorbic acid	Retention of antioxidant activity and ascorbic acid content was found to be better in pressure-treated products than untreated or thermally treated ones.	Patras et al. (2009)

Peach puree	103–517 MPa/5–25 min, with and without the addition of antibrowning agents (1000 ppm ascorbic acid or 300 ppm cysteine)	Enzyme activity	The higher the pressure treatment and the longer the treatment time for inactivating PPO, the less the discoloration of the peach puree with and without antibrowning agents during storage at low temperatures	Guerrero-Beltran et al. (2006)
Ginger, onion, and garlic paste	200, 400, and 600 MPa/5 min/30°C	Enzyme activity, microbial activity, color	Application of HP was found to decrease the enzymatic that is, PPO and POD activities with better retention of phytochemicals. The CIE color values and sensory attributes were not much affected by pressure treatment	Chauhan et al. (2011)
Strawberry puree	400 MPa/ambient temperature/20 min	Enzyme activity, microbial activity and color	HPP maintains the original aromatic distribution, but after 30 days of storage an increase in some volatile compounds was observed. After HP treatment and storage, a residual lipoxygenase activity was observed, but peroxidase was clearly inactivated	Navarro et al. (2002)
Apricot puree	300–900 MPa at 20°C for 1 min	Enzyme activity and microbial activity	Yeasts and molds inoculated into puree samples were completely inactivated by 400 MPa treatment at 20°C for 1 min. PPO was inactivated completely by preheating at 50°C before applying 700 MPa for 5 min	Rovere et al. (1994)

product quality and safety. Further, HP in combination with other processing techniques can be experimented to improve the stability. The challenge ahead is to find a feasible application so that HP products can be commercially made available on a larger scale.

REFERENCES

Basak, S. and Ramaswamy, H.S. 1998. Effect of high pressure processing on the texture of selected fruits and vegetables. *Journal of Texture Studies* 29: 587–601.

Bignon, J. 1996. Cold pasteurizers hyperbar for the stabilization of fresh fruit juices. *Fruit Processing* 2: 46–48.

Butz, P., Koller, W.D., Tauscher, B., and Wolf, S. 1994. Ultra-high pressure processing of onions: Chemical and sensory changes. *LWT-Food Science and Technology* 27: 463–467.

Cano, M.P., Hernandez, A. and De Ancos, B. 1997. High pressure and temperature effects on enzyme inactivation in strawberry and orange products. *Journal of Food Science* 62: 85–88.

Chauhan, O.P., Raju, P.S., Ravi, N. and Bawa, A.S. High pressure stabilization of onion ginger and garlic pastes. *iFood2011 – Innovation Food Conference*. Quakenbruck, Germany, October 11–14, 2011.

Dalmadi, I., Polyak Feher, K. and Farkas, J. 2007. Effects of pressure and thermal pasteurization on volatiles of some berry fruits. *High Pressure Research* 27: 169–172.

de Ancos, B., Gonzalez, E. and Cano, M.P. 2000. Effect of high-pressure treatment on the carotenoid composition and the radical scavenging activity of persimmon fruit purees. *Journal of Agricultural Food Chemistry* 48: 3542–3548.

de Ancos, B. Sgroppo, S., Plaza, L. and Cano, M.P. 2002. Possible nutritional and health related value promotion in orange juice preserved by high pressure treatment. *Journal of Science of Food and Agriculture* 82(8): 790–796.

Fernandez Garcia, A., Butz, P. and Taucher, B. 2001. Effect of high pressure processing on carotenoid extractability, antioxidant activity, glucose diffusion and water binding of tomato puree (*Lycopersicon esculentum* Mill.). *Journal of Food Science* 66(7): 1033–1038.

Garcia-Palazon, A., Suthanthangjai, W., Kajda, P. and Zabetakis, I. 2004. The effects of high hydrostatic pressure on β-glucosidase, peroxidase and polyphenoloxidase in red raspberry (*Rubus idaeus*) and strawberry (*Fragaria × ananassa*). *Food Chemistry* 88: 7–10.

Gow, C.Y. and Hsin, T.L. 1996. Comparison of high pressure treatment and thermal pasteurization effects on the quality and shelf-life of guava puree. *International Journal of Food Science and Technology* 31(2): 205–213.

Guerro-Beltran, J.A., Barbosa-Canovas, G.V., and Swanson, B.G. 2005. High hydrostatic pressure processing of fruit and vegetable products. *Food Reviews International* 21: 411–425.

Guerrero-Beltran, J.A., Barbosa-Canovas, G.V., Moraga-Ballesteros, G., Moraga-Ballesteros, M.J. and Swanson, B. G. 2006. Effect of pH and ascorbic acid on high hydrostatic pressure processed mango puree. *Journal of Food Processing and Preservation* 30(5): 582–596.

Hatanaka, A., Kajiawara, T., and Sekiya, J. 1987. Biosynthetic pathway for C6-aldehydes formation from linolenic acid in green leaves. *Chemistry and Physics of Lipids* 44: 341–361.

Hernandez, A. and Cano, M. 1998. High-pressure and temperature effects on enzyme inactivation in tomato puree. *Journal of Agricultural and Food Chemistry* 46: 266–270.

Hough, G., Langohr, K., Gomez, G., and Curia, A. 2003. Survival analysis applied to sensory shelf-life of foods. *Journal of Food Science* 68: 359–362.

Jacobo-Velázquez, D.A. and Hernández-Brenes, C. 2010. Biochemical changes during the storage of high hydrostatic pressure processed avocado paste. *Journal of Food Science* 75(6): S264–S270.

Kader, F., Rovel, B., Girardin, M., and Metche, M. 1997. Mechanism of browning in fresh highbushblueberry fruit (*Vaccinium corymbosum* L.). Role of blueberry polyphenol oxidase, chlorogenic acid and anthocyanins. *Journal of Science of Food and Agriculture* 74: 31–34.

Knorr, D. 1995. High pressure effects on plant derived foods. In: *High Pressure Processing of Foods*. Ledward, D.A., Johnston, D.E., Earnshaw, R.G., and Hasting, A.P.M. (eds.). Loughborough: Nottingham University Press, pp.123–35.

Knorr, D. 2001. High pressure processing for preservation, modification and transformation of foods, *Oral presentation in XXXIX European High Pressure Research Group Meeting*, Santander, Spain, 16-19 September, 2001.

Keenan, D.F., Brunton, N., Gormley, R., and Butler, F. 2011. Effects of thermal and high hydrostatic pressure processing and storage on the content of polyphenols and some quality attributes of fruit smoothies. *Journal of Agricultural and Food Chemistry* 59: 601–607.

Kolakowski, P., Dumay, E. and Cheftel, J.C. 2001. Effects of high pressure and low temperature on beta-lactoglobulin unfolding and aggregation. *Food Hydrocolloids* 15(3): 215–232.

Krebbers, B., Matser, A.M., Hoogerwerf, S.W., Moezelaar, R., Tomassen, M.M.M., and van den Berg, R.W. 2003. Combined high-pressure and thermal treatments for processing of tomato puree: evaluation of microbial inactivation and quality parameters. *Innovative Food Science and Emerging Technologies* 4: 377–385.

Lakshmi, E.U., Chauhan, O.P., Raju, P.S. and Bawa, A.S. 2011a. Comparative evaluation of thermal and high pressure processing methods for preservation of ginger paste. *International Symposium on Recent Trends in Processing and Safety of Specialty and Operational Foods at Defence Food Research Laboratory*, Mysore, November 22–24, 2011.

Lakshmi, E.U., Chauhan, O.P., Raju, P.S. and Bawa, A.S. 2011b. High pressure processing of foods: Present status and future strategies. *International Journal of Food and Fermentation Technology* 1: 49–62.

Lambert, Y., Demazeau, G., Largeteau, A., and Bovier, J. 1999. Changes in aromatic volatile composition of strawberry after high pressure treatment. *Food Chemistry* 67: 7–16.

Landl, A., Abadias, M., Sárraga, C., Viñas, I. and Picouet, P. A. 2010. Effect of high pressure on the quality of acidified Granny Smith apple puree product. *Innovative Food Science and Emerging Technologies* 11(4): 557–564.

Lopez, M.A., Palou, E., Barbosa-Canovas, G.V., Welti-Chanes, J. and Swanson, B.G. 1998. Polyphenoloxidase activity and color changes during storage of high hydrostatic pressure treated avocado puree. *Food Research International* 31(8): 549–556.

Mac Dougall, D.B. 2002. Color measurement of food: Principles and practice. In: *Color in Food Improving Quality*. Mac Dougall (ed.). Cambridge, UK: Woodhead Publishing Limited, pp. 33–63.

Michel, M. and Autio, K. 2001. Effect of high pressure on protein and polysaccharide based structures. In: *Ultra High Pressure Treatment of Foods*. M.E. Hendrickx and D. Knorr (eds.). New York: Kluwer Academic/Plenum Publishers, pp. 189–214.

Navarro, M., Verret, C., Pardon, P. and Moueffak, A. 2002. Changes in volatile aromatic compounds of strawberry puree treated by high pressure during storage. *High Pressure Research* 22: 693–696.

Nguyen, M.T., Indrawati, O. and Hendrickx, M.E. 2003. Model studies on the stability of folic acid and 5-methyltetrahydrofolic acids degradation during thermal treatment in combination with high hydrostatic pressure. *Journal of Agricultural Food Chemistry* 51(11): 3352–3357.

Oey, I., Lille, M., Van-Loey, A. and Hendrickx, M. 2008. Effect of high-pressure processing on color, texture and flavor of fruit- and vegetable-based food products: A review. *Trends in Food Science and Technology* 19(6): 320–328.

Ogawa, H., Fukuhisa, K., Kubo, Y. and Fukumoto, H. 1990. Pressure inactivation of yeasts molds and pectin esterase in Satsuma mandarin juice: Effects of juice concentration, pH and organic acids and comparison with heat sanitation. *Agricultural Biology and Chemistry* 54(5): 1219–1225.

Palou, E., Lopez-Malo, A., Barbosa-Canovas, G.V., Welti-Chanes, J. and Swanson, B.G. 1999. Polyphenoloxidase activity and colour of blanched and high hydrostatic pressure treated banana puree. *Journal of Food Science* 64(1): 42–45.

Palou, E., Hernández-Salgado, C., López-Malo, A., Barbosa-Cánovas, G.V., Swanson, B.G. and Welti-Chanes, J. 2000. High pressure-processed guacamole. *Innovative Food Science and Emerging Technologies* 1(1): 69–75.

Patras, A., Brunton, N., Da Pieve, S., Butler, F. and Downey, G. 2008. Effect of thermal and high pressure processing on antioxidant activity and instrumental color of tomato and carrot purées. *Innovative Food Science and Emerging Technologies* 10(1): 16–22.

Patras, A., Brunton, N.P., Da Pieve, S. and Butler, F. 2009. Impact of high pressure processing on total antioxidant activity, phenolic, ascorbic acid, anthocyanin content and color of strawberry and blackberry purees. *Innovative Food Science and Emerging Technologies* 10(3): 308–313.

Plaza, L., Munoz, M., de Ancos, B., and Cano, M.P. 2003. Effect of combined treatments of high-pressure, citric acid and sodium chloride on quality parameters of tomato puree. *European Food Research and Technology* 216: 514–519.

Porretta, S., Birzi, A., Ghizzoni, C., and Vicini, E. 1995. Effects of ultra-high hydrostatic pressure treatments on the quality of tomato juice. *Food Chemistry* 52: 35–41.

Quaglia, G.B., Gravina, R., Paperi, R., and Paoletti, F. 1996. Effect of high pressure treatments on peroxidase activity, ascorbic acid content and texture in green peas. *Lebensmittel-Wissenschaft und-Technology* 29: 552–555.

Qiu, W., Jiang, H., Wang, H. and Gao, Y. 2006. Effect of high hydrostatic pressure on lycopene stability. *Food Chemistry* 97: 516–523.

Robinson, C.R. and Sligar, S.G. 1995. Heterogeneity in molecular recognition by restriction endonucleases: Osmotic and hydrostatic pressure effects on *Bam*HI, *Pvu* II, and *Eco*RV specificity. *Proceedings of the National Academy of Science USA* 92: 3444–3448.

Rodrigo, D., Loey, A.V. and Hendrickx, M. 2007. Combined thermal and high pressure color degradation of tomato puree and strawberry juice. *Journal of Food Engineering* 79(2): 553–560.

Rovere, P., Carpi, G., Alessandra-Maggi-Gola, S. and Dall Aglio, G. 1994. Stabilisation of apricot puree by means of high pressure treatments. *Biotechnology Review* 32(4): 145–150.

Sancho, F., Lambert, Y., Demazeau, G., Largeteau, A., Bouvier, J., and Narbonne, J. 1999. Effect of ultra-high hydrostatic pressure on hydrosoluble vitamins. *Journal of Food Engineering* 39: 247–253.

Sanchez-Moreno, C., Plaza, L., de Ancos, B., and Cano, M.P. 2004. Effect of combined treatments of high-pressure and natural additives on carotenoid extractability and antioxidant activity of tomato puree (*Lycopersicum esculentum* Mill.). *European Food Research and Technology* 219: 151–160.

Sanchez Moreno, C., Plaza, L., De Ancos, B. and Cano, M.P. 2006. Impact of high-pressure and traditional thermal processing of tomato purée on carotenoids, vitamin C and antioxidant activity. *Journal of the Science of Food and Agriculture* 86: 171–179.

Suthanthangjai, W., Kajda, P. and Zabetakis, I. 2005. The effect of high hydrostatic pressure on the anthocyanins of raspberry (*Rubus idaeus*). *Food Chemistry* 90(1–2): 193–197.

Terefe, N.S., Yang, Y.H., Knoerzer, K., Buckow, R. and Versteeg, C. 2010. High pressure and thermal inactivation kinetics of polyphenoloxidase and peroxidase in strawberry puree. *Innovative Food Science and Emerging Technologies* 11: 52–60.

Tomas Barberam, F.A. and Espin, J.C. 2001. Phenolic compounds and related enzymes as determinants of quality in fruits and vegetables. *Journal of Science of Food and Agriculture* 81: 853–876.

Unni, L.E., Chauhan, O.P., and Raju, P.S. 2013. High pressure processing of garlic paste: Effect on the quality attributes. doi 10.1111/ijfs.12456.

Van den Broeck, I., Ludikhuyze, L.R., Van Loey, A.M., and Hendrickx, M.E. 2000a. Effect of temperature and/or pressure on tomato pectinesterase activity. *Journal of Agricultural and Food Chemistry* 48: 551–558.

Van den Broeck, I., Ludikhuyze, L.R., Van Loey, A.M., and Hendrickx, M.E. 2000b. Inactivation of orange pectinesterase by combined high pressure and temperature treatments: A kinetic study. *Journal of Agricultural and Food Chemistry* 48(5): 1960–1970.

Verlent, I., Smout, C., Duvetter, T., Hendrickx, M. E. and Van Loey, A. 2005. Effect of temperature and pressure on the activity of purified tomato polygalacturonase in the presence of pectins with different patterns of methyl esterification. *Innovative Food Science and Emerging Technologies* 6: 293–303.

Verlent, I., Hendrickx, M., Rovere, P., Moldenaers, P. and VanLoey, A. 2006. Rheological properties of tomato based products after thermal and high pressure treatment. *Journal of Food Science* 71(3): S243–S248.

Watanbe, M., Arai, E., Kumeno, K., and Honma, K. 1991. A new method for producing a non-heated jam sample: The use of freeze concentration and high-pressure sterilization. *Agricultural and Biological Chemistry* 55(8): 2175–2176.

Weemaes, C., Ludikhuyze, L., Broeck, I. Van den and Hendrickx, M. 1998. High-pressure inactivation of polyphenoloxidases. *Journal of Food Science* 63(5): 873–877.

Weemaes, C., Ludikhuyze, L.,Van den Broeck, I. and Hendrickx, M. 1999. Kinetic study of antibrowning agents and pressure inactivation of avocado polyphenoloxidase. *Journal of Food Science* 64(5): 823–827.

Yen, G.C. and Lin, H.T. 1999. Comparison of high pressure treatment and thermal pasteurisation on the quality and shelf life of guava puree. *International Journal of Food Science and Technology* 31: 205–213.

7 Fruit Juice Quality Enhancement by High-Pressure Technology

Özlem Tokuşoğlu, Barry G. Swanson, and Gustavo V. Barbosa-Cánovas

CONTENTS

7.1 INTRODUCTION

Fruit juice is a fruit liquid or liquid extract of the fruit that is commonly consumed as a beverage or used as an ingredient or flavoring in foods. Fruit juice is prepared by mechanically squeezing or macerating the fruit without the application of heat or solvents. Juice from fresh fruits may be prepared in the home using a variety of hand or electric juicers whereas many commercial juices are filtered to remove fiber or pulp. Fruit juices are globally accepted as healthy and nourishing foods owing to their richness in vitamins (especially ascorbic acid [vitamin C]), polyphenols, and flavonoids contributing to good antioxidant properties, their low sodium level, and minimal fat content (Anonymous, 2013; MEYED, 2013; Tokuşoğlu and Hall, 2011).

Fruit juice consumption overall in Europe, Australia, New Zealand, and the United States has increased in recent years, probably due to public perception of juices as a healthy natural source of nutrients and increased public interest in health issues. Especially, fruit phytochemicals are of significant interest for public health because of their protective and preventive effects in several chronic diseases and the pathogenesis of a definite class of cancers. Indeed, fruit juice intake has been consistently associated with the reduced risk of many cancer types, might be protective against stroke, and delays the onset of Alzheimer's disease (Anonymous, 2013; Kumar et al., 2009; Meskin et al., 2003; Patrignani et al., 2009; Tokuşoğlu and Hall, 2011).

7.2 SPOILAGE PROBLEM IN FRUIT JUICES

There is a great demand for freshly squeezed fruit juices due to their sensory and nutritional qualities, but fruit juices are susceptible to spoilage, thus having a limited shelf life. The spoilage of fruit juices is primarily owed to yeasts that are responsible for the fermented taste and carbon dioxide production, lactic acid bacteria which can produce a butter milk off-flavor, and molds that contribute to the spoilage by their surface growth. Studies show that unpasteurized juices can vehicle food-borne pathogens such as *Salmonella* spp and *Escherichia coli* O157:H7, in spite of the fact that fruit juices have always been considered safe because of their low pH properties (Kriskó and Roller, 2005; Lavinas et al., 2008; Tournas et al., 2006).

Unpasteurized fresh juices with low acidity (pH > 4.6) and high water activity (aw > 0.85) can support the growth of a variety of pathogenic microorganisms. Enteric food-borne pathogens including *E. coli* O157:H7 and *Salmonella enterica* serovar *Typhimurium*, and Gram-positive pathogens such as *Listeria monocytogenes* have all been reported to be capable of survival in raw fruit and vegetable juices (Pathanibul et al., 2009).

Traditionally, thermal processing is applied in fruit juice processing to improve safety and shelf life of commercial fruit juices. The heat treatment also inactivates the enzymes that may cause undesirable quality alterations (Jordan et al., 2001; MEYED, 2013). Heat-sensitive nutrients and volatile compounds may be negatively affected by the heat application and thermal processing causes some undesirable effects, such as nonenzymatic browning, off-flavor formation, and vitamin loss (Polydera et al., 2003; Plaza et al., 2006; Rattanathanalerk et al., 2005; Rodrigues and Narciso Fernandez, 2012).

There is an increasing demand for fresh juices obtained by alternative processing technologies to produce fruit juices with a minimum of nutritional, physicochemical, or organoleptic alterations and also without chemical preservatives.

7.3 *ALICYCLOBACILLUS* PROBLEM AND GUAIACOL FORMING IN FRUIT JUICE INDUSTRY

In the 1980s, an acidophilic *Bacillus* species was isolated from apple juice and was identified as a new type of spoilage bacterium. Originally named *Bacillus acidoterrestris,* this organism was later reclassified in a new genus, *Alicyclobacillus* as ω-alicyclic fatty acid was the major membrane fatty acid component of its cells (Deinhard et al., 1987; Sawaki, 2007; Wisotzkey et al., 1992).

Alicyclobacillus acidoterrestris (Figure 7.1) is a spore-forming, rod-shaped organism with a central, subterminal, or terminal oval spore and grows at pH values ranging from 2.5 to 6.0 at a temperature of 25–60°C (Murakami et al., 1998). *Alicyclobacillus* contains ω-alicyclic fatty acids as a major membrane fatty acid component. *Alicyclobacillus* spp have been isolated from soil, spoiled fruit juices, and several thermal environments (Murakami et al., 1998). All species metabolize sugars with acid production but no gas production (Sawaki, 2007). Water activity >0.9 is required for growth, and some species have been reported to grow in fruit juice with up to 18.2°Brix (Sprittstoesser et al., 1994).

 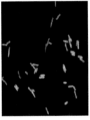 

Alicyclobacillus   *Alicyclobacillus 1*   *Alicyclobacillus 2*

FIGURE 7.1   *Acyclobacillus acidoterrestris.*

Alicyclobacillus is a major challenge for the fruit juice industry (Tribst et al., 2009). *A. acidiphilus, A. acidoterrestris, A. herbarius,* and *A. pomorum* have been reported to be associated with spoilage of fruit juices and beverages such as tea and herbal teas (Cerny et al., 1984; Goto et al., 2002, 2003; Matsubara et al., 2002). Some species of *Alicyclobacillus* spp, such as *A. acidoterrestris,* can cause commercially pasteurized apple juice to spoil (Lee et al., 2002; Matsubara et al., 2002).

A. acidoterrestris is an important spoilage organism of acidic foods because its spores are able to germinate and grow in highly acidic environments and produce guaiacol that causes "medicinal" or "antiseptic" off-flavors (Yamazaki et al., 1997). Guaiacol (2-methoxyphenol) (Figure 7.2) is a predominant metabolite associated with spoilage by *A. acidoterrestris* (Orr et al., 2000a). This off-flavor product has been widely reported (Corli Witthuhn et al., 2012; Savaş Bahçeci and Acar, 2007; Walls and Chuyate, 1998).

Figure 7.3 shows the guaiacol formation by *Alicyclobacillus* spp (Figure 7.3). In fruit juices, guaiacol (2-methoxyphenol) originates from phenolic acids (ferulic acid and vanillin) by *Alicyclobacillus* spp (Smit et al., 2011; Tokus̨og̃lu and Yamamoto, 2012).

The juice-associated outbreaks can be from *A. acidoterrestris* spoilage and guaiacol formation problem. According to the juice-hazards analysis and critical control point (HACCP) regulation-2001 by U.S. Food and Drug Administration (FDA), juice processors include in their HACCP plan measures to provide at least a 5-log reduction in the pertinent pathogens most likely to occur (FDA, 2001). The juice HACCP regulation only applies to pathogens, and there is no regulation for controlling juice spoilage. It is necessary to take measures to ensure the quality of their products for the juice and beverage industries. So far, traditional thermal pasteurization, using of chemical agents, and enzymatic methods are in progress as spoilage-control methods for *A. acidoterrestris* in the fruit juice sector. More than a 5-log reduction of spores

FIGURE 7.2   Guaiacol (2-methoxyphenol).

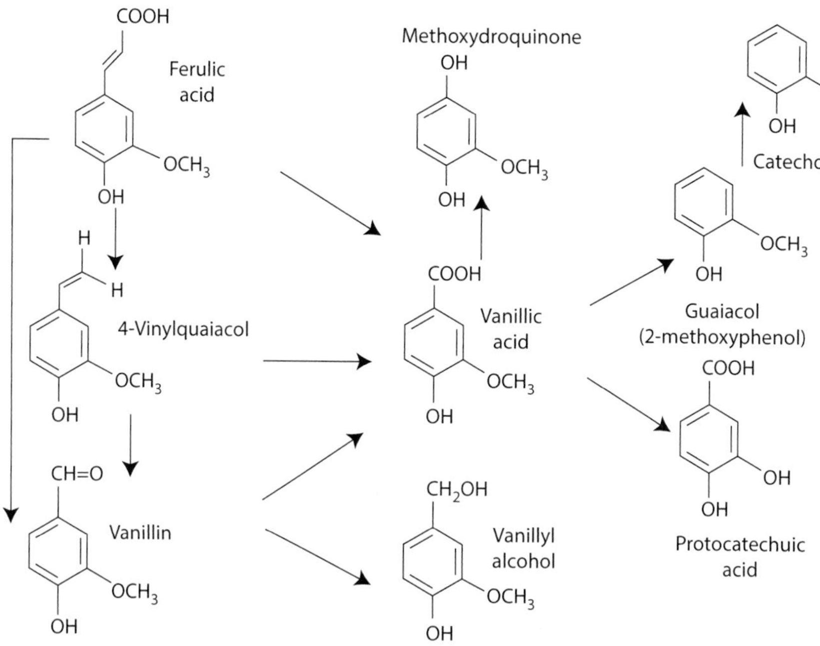

FIGURE 7.3 Guaiacol formation by *Alicyclobacillus* spp. (Adapted from Smit Y. et al. 2011. *Food Microbiology* 28, 331–349; Tokuşoğlu Ö., Yamamoto K. 2012. High pressure processing effects on guaiacol formation by *Alicyclobacillus* spp. from phenolic ferulic and vanillic acids in fruits juices, in fruity drinks. In *7th International Conference on High Pressure Bioscience and Biotechnology*. Poster Presentation. P1–19. *Book of Abstracts*. p. 63. October 29–November 2, 2012, Otsu, Kyoto, Japan.)

was observed after treament with 1000 ppm chlorine or 4% H_2O_2. The disinfectants were less effective in killing spores on the surface of apples. More than a 4-log reduction of spores was observed after treatment with 40 ppm chlorine dioxide for 5 min in aqueous suspension. For 4 min, 40 ppm free chlorine dioxide was applied to apple surfaces; the result was greater than a 4.8-log reduction of spores (Lee et al., 2004; Orr and Beuchat, 2000b; Tokuşoğlu and Yamamoto, 2012). Traditional thermal pasteurization is effective for inactivating vegetative cells of bacterial food-borne pathogens, but, as stated before, the current juice pasteurization treatment is not adequate to destroy spores of *A. acidoterrestris*. Heat treatment could have negative effects of the quality of juices, including loss of nutrients, and change in flavor, color, and texture (FDA, 2001; Tokuşoğlu and Yamamoto, 2012). There is a great demand for fresh juice free of chemical preservatives and obtaining fruit juices with improved quality and without nutritional, physicochemical, or organoleptic changes. Consumers expect premium juice not only to be safe, but also to have fresh, just-squeezed appearance, flavor, natural texture, and nutrition without additive compounds or preservatives. New emerging preservation technologies can be applied.

Novel nonthermal processes, such as high hydrostatic pressure (HHP), pulsed electric fields (PEFs), ionizing radiation, and ultrasonication require very high treatment intensities to achieve adequate microbial destruction in low-acid foods. Among these, HHP treatment is considered to be a promising alternative to thermal pasteurization for fruit and vegetable juices (Jordan et al., 2001; Ross et al., 2003). Currently, HHP-processed products are low-pH fruit juices (grapefruit juice, mandarin juice, apple, and orange juice), jams, jellies, and fruit dressing (Ohlsson, 2002).

7.4 HIGH-PRESSURE PROCESSING TECHNOLOGIES FOR FRUIT JUICES QUALITY

High-pressure processing (HPP) leaves juice tasting just like freshly squeezed. Extended refrigerated shelf life increases distribution opportunities, reduces returns and sensitivity to cool chain abuse, and allows more efficient production scheduling. HPP is a nonthermal, environmentally friendly process that allows the development of juices and beverages with fresh-like organoleptic quality and nutrition, while extending the shelf life of the product (Avure, 2013; Hiperbaric, 2013).

High pressure (HP) keeps the original fruit taste and color, allowing the creation of the highest-quality premium range of products. Nutritional and functional properties of the product remain intact, it can be obtained with natural, organic, preservative free, and functional products, and can be preserved with vitamins, antioxidants, and thermosensitive antimutagenic components, bringing a higher level of functionality to new products (Bull et al., 2004; Hiperbaric, 2013; Jordan et al., 2001; Sánchez-Moreno et al., 2009, 2005). It was reported that HP treatment caused the modification of several physical characteristics of the fruit juices, such as the juice viscosity (Campos and Cristianini, 2007; Donsì et al., 2009; Patrignani et al., 2010), and the mean particle size of the suspended solids (Betoret et al., 2009; Donsì et al., 2009; Stipp and Tsai, 1988).

7.5 RECENT STUDIES REGARDING HPP EFFECTS ON MICROBIAL SPOILAGE AND OVERALL JUICE QUALITY

The application of HPP in ensuring food safety and quality has been widely studied (Tokuşoğlu and Doona, 2011). HPP inactivates vegetative cells of microorganisms by breaking noncovalent bonds and causing damage to the cell membrane (Morris et al., 2007). The mechanism of inactivation of bacterial spores through HP was suggested to have two steps: HP will first induce spore germination and then inactivate the germinated spores (Gould and Sale, 1970). It is suggested that HPP can preserve certain foods better than heat by extending shelf life and inactivating microorganisms while retaining the inherent color, flavor, nutrients, and texture of the food. Pressure is transmitted instantaneously and is independent of mass; so, the treatment throughout the food is uniform (Zimmerman and Bergman, 1993).

Recent literatures show that fruit juices can be considered as an interesting and feasible field of application of HP treatments to achieve the microbial load reduction while preserving the quality of fruit juices (Donsì et al., 2009). It was reported that the potential of high-pressure homogenization (HPH) treatments for spoilage and pathogenic microorganisms inactivation is an alternative to heat treatment for

TABLE 7.1
Microbial Counts (log$_{10}$ CFU/g) in Mango Pulp after HHP Treatment and HTST

Pressure-Holding Time	Total Aerobic Bacteria					Yeasts and Molds				
	HHP				HTST	HHP				HTST
	300 MPa	400 MPa	500 MPa	600 MPa	110°C/8.6s	300 MPa	400 MPa	500 MPa	600 MPa	110°C/8.6 s
Control	4.64 ± 0.02[a]	4.88 ± 0.03[a]	4.39 ± 0.14[a]	4.54 ± 0.03[a]	4.75 ± 0.07[a]	3.87 ± 0.01[a]	3.52 ± 0.03[a]	3.74 ± 0.01[a]	3.89 ± 0.01[a]	3.65 ± 0.02[a]
1 min	4.14 ± 0.11[b]	3.24 ± 0.12[b]	1.11 ± 0.15[b]	ND[b]	ND[b]	1.97 ± 0.34[b]	1.56 ± 0.01[b]	ND[b]	ND[b]	ND[b]
2.5 min	3.40 ± 0.06[c]	ND[d]	ND[c]	ND[b]		1.85 ± 0.15[b]	1.22 ± 0.15[c]	ND[b]	ND[b]	
5 min	3.14 ± 0.05[d]	ND[d]	ND[c]	ND[b]		1.19 ± 0.24[c]	ND[d]	ND[b]	ND[b]	
10 min	1.80 ± 0.24[e]	ND[d]	ND[c]	ND[b]		ND[d]	ND[d]	ND[b]	ND[b]	
15 min	ND[f]	ND[d]	ND[c]	ND[b]		ND[d]	ND[d]	ND[b]	ND[b]	
20 min	ND[f]	ND[d]	ND[c]	ND[b]		ND[d]	ND[d]	ND[b]	ND[b]	

Source: Adapted from Liu F. et al. 2013. *Food Bioprocess Technology* (6), 2675–2684.

All data were the means ± standard deviation, *n*–3. Values with different letters within one column are significantly different ($p < 0.05$). HHP, high hydrostatic pressure; HTST, high-temperature short time; ND, not detectable.

improving fruit juice safety and shelf life (Bevilacqua et al., 2012; Briñez et al., 2006a,b, 2007; Patrignani et al., 2009, 2010, 2013).

In 2001, the European Commission authorized the sale of fruit juice pasteurized by HP (Anonymous, 2001). Fruit juices are normally treated at 400 MPa or greater for a few minutes at 20°C or below. This is adequate to significantly reduce the numbers of spoilage microorganisms, such as yeasts, molds, and lactic acid bacteria and extend refrigerated shelf life to about 30 days (Patterson, 2005).

According to Juice HACCP Hazards and Controls Guidance, low-acid juices, such as carrot juice, are not subjected to the low-acid canned foods regulation and they need to be distributed under refrigeration. Therefore, selection of pressure-resistant strains of food-borne pathogens, such as *E. coli* O157:H7, *Listeria* spp, *Salmonella* spp, or *Staphylococcus* spp, will also apply for low-acid juices. Carrot juice is one of the most popular vegetable juices (pH > 6.0). As carrot juice is a low-acid food, it has a higher risk of bacterial contamination than the other acidic foods (Zhou et al., 2009).

It was stated that the effect of HHP on microorganisms in cantaloupe juice was studied and demonstrated that they achieved 5-log cycles reductions of *E. coli* within 8 min at 500 MPa at room temperature in melon juice. Sensory evaluation indicated that there was no significant aromatic difference between HHP-treated and HHP-untreated cantaloupe juice samples (Ma et al., 2010).

Liu et al. (2013) stated the effects of HHP treatments at pressures of 300–600 MPa/1–20 min and of high-temperature, short-time (HTST) treatment on the inactivation of natural microorganisms in blanched mango pulp (BMP) and unblanched mango pulp (UBMP) (Table 7.1).

It was found that no yeasts, molds, or aerobic bacteria were detected in BMP or UBMP after HHP treatments at 300 MPa/15 min, 400 MPa/5 min, 500 MPa/2.5 min, and 600 MPa/1 min and HTST treatment at 110°C/8.6 s (Liu et al., 2013). These conditions were selected by Li et al. (2013) to detect the effects of HHP and HTST treatments on pectin methylesterase (PME) activity, water-soluble pectin (WSP) levels, and the rheological characteristics of UBMP and BMP (Li et al., 2013).

It was detected that HHP treatment at 600 MPa/1 min significantly reduced PME activity in UBMP and significantly activated PME in BMP, whereas pressures of 300–500 MPa activated PME regardless of blanching (Figure 7.4) (Li et al., 2013). It was found that PME activity was reduced by 97% in UBMP while it was entirely inactivated in BMP by HTST treatment (Li et al., 2013).

Varela-Santos et al. (2012) reported the effect of HHP processing (350–550 MPa for 30/90/150 s) on microbial quality, physicochemical, and bioactive compounds of pomegranate juices during 35 days of storage at 4°C. It was found that the microbiological results that showed HHP at or over 350 MPa/150 s resulted in a microbial load reduction around 4.0 log cycles, and were adequate to keep microbial populations below the detection limit during the pomegranate juice storage period (Table 7.2) (Varela-Santos et al., 2012). The above-mentioned treatments were able to extend the microbiological shelf life of pomegranate juice stored at 4°C for more than 35 days (Varela-Santos et al., 2012).

It was determined that the total aerobic mesophilic bacteria (TAMB) count on untreated samples increased up to the 13th day, reaching counts >6.0 $\log_{10}$ CFU mL^{-1}

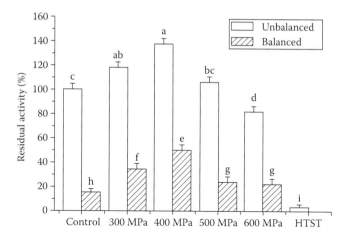

FIGURE 7.4 Effects of HHP and HTSTs on PME in UBMP and BMP. (Adapted from Liu F. et al. 2013. *Food Bioprocess Technology* (6), 2675–2684.)

TABLE 7.2

Effect of HHP Treatment on Microbial Populations Studied (log$_{10}$ CFU mL) in Pomegranate Juice

Treatments (MPa/s)	Microbial Populations	
	Aerobic Mesophilic Bacteria	**Molds and Yeasts**
Control	2.98 (0.18)[a]	3.79 (0.32)[a]
350/30	1.18 (0.25)[b]	1.67 (0.30)[b]
350/90	1.00 (0.00)[b]	1.33 (0.27)[b]
350/150	N.D[c]	N.D[c]
450/30	N.D[c]	N.D[c]
450/90	N.D[c]	N.D[c]
450/150	N.D[c]	N.D[c]
550/30	N.D[c]	N.D[c]
550/90	N.D[c]	N.D[c]
550/150	N.D[c]	N.D[c]

Source: Adapted from Liu F. et al. 2013. *Food Bioprocess Technology* (6), 2675–2684.
Note: N.D = not determined (levels of microbial populations studied were below the limit of detection at the corresponding treatment). Mean values with different letters (a through c) in the same column are significantly different ($p < 0.05$; LSD). Values in parentheses represent standard deviations.

(Figure 7.5a) that is considered as the upper acceptable limit by Patrignani et al. (2009). It was found that for HHP-treated samples at 350 MPa/30 and 90 s, TAMB value was not exceeded, and was found to be <3.0 log$_{10}$ CFU mL^{-1} at the end of the storage period, whereas for pressurized samples at or above 350 MPa/150 s, aerobic mesophilic bacteria count remained below the detection limit (<1.0 log CFU mL^{-1}) during 35 days of storage (Varela-Santos et al., 2012). It was stated that the yeast and mold counts (Figure 7.5b) in untreated samples exceeded the upper acceptable limit (6.0 log$_{10}$ CFU mL^{-1}) around the 8th day, and were found to be >10 log$_{10}$ CFU mL^{-1} at the end of the storage period (Figure 7.5b). Varela-Santos et al. (2012) stated that for treated samples at 350 MPa (30/90 s), the upper acceptable limit was not reached

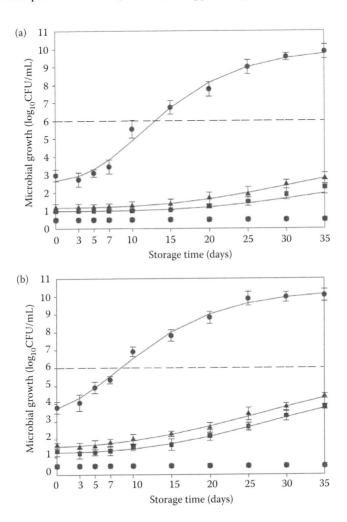

FIGURE 7.5 Growth curve of (a) aerobic mesophilic and (b) molds and yeasts in untreated and HHP-treated pomegranate juice during storage at 4°C/35 days. Symbols and lines represent observed and modeled (reparameterized version of the Gompertz equation) values, respectively. Symbols are means of three measurements ± D.S. control (●), 350 MPa/30 s (▲), 350 MPa/90 s (■), 350 MPa/150 s (▼), 450 MPa/30 s (△), 450 MPa/90 s (□), 450 MPa/150 s (▽), 550 MPa/30 s (◆), 550 MPa/90 s (×), and 550 MPa/150 s (★). The dotted line shows the upper acceptable limit. (Adapted from Varela-Santos E. et al. 2012. *Innovative Food Science and Emerging Technologies* 13, 13–22.)

to <4.5 $\log_{10}$ CFU mL^{-1} at the end of the storage period, whereas molds and yeasts were not detected immediately after pressure treatment at or above 350 MPa/150 s, and survivors were kept below the detection limit throughout the cold-storage period (Figure 7.5b).

It was shown that HHP treatment gave similar results regarding the microbiological shelf-life extention of fruit products, including tomato juice (Hsu et al., 2008), tomato purée (Krebbers et al., 2003), peach purée (Guerrero-Beltrán and Barbosa-Cánovas, 2004), mango purée (Guerrero-Beltrán et al., 2006), and cashew apple juice (Lavinas et al., 2008).

In the study reported by Varela-Santos et al. (2012), the ΔE values, which are an indicator of total color difference, showed that there were significant differences ($p < 0.05$) in color between untreated and treated samples and a significant decrease ($p < 0.05$) in ΔE values during storage time. The highest color difference was obtained at day 35 for 550 MPa/90 s and it was concluded that the color stability of pomegranate juice depended on the processing conditions (Varela-Santos et al., 2012).

HHP (400 MPa/5 min) and high temperature short time (HTST) (110°C/8.6 s) treatment were comparatively qualified by determining their influences on pH, total soluble solids (TSS), titratable acidity (TA) (Figure 7.6), on microorganisms, color, total phenols, anthocyanins, antioxidant capacity and shelf-life characteristics of 90 days at 4°C. It was shown that the inactivation effect of microorganisms by HHP fitted theWeibull model well and HHP at 400 MPa/5 min inactivated microorganisms effectively by Chen et al. (2013). Figure 7.7 shows the Weibull model of microbial inactivation of cloudy pomegranate juice treated by HHP (Varela-Santos et al., 2012).

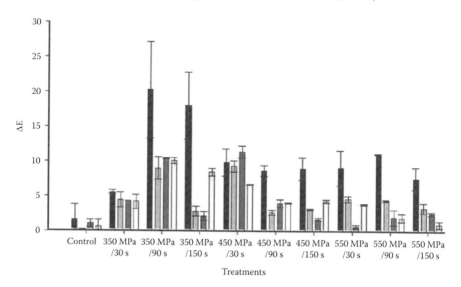

FIGURE 7.6 The effect of HHP processing on pH (a), total titratable acidity (TA) (b), and total soluble solids (TSS) (c) for processed pomegranate juice during storage after 35 days at 4°C (■ Day 0 ▨ Day 15 ■ Day 25 ▨ Day 35 days: 3, 5, 7, 10, 20, and 30, data not shown). (Adapted from Chen D. et al. 2013. *Innovative Food Science and Emerging Technologies* 19, 85–94.)

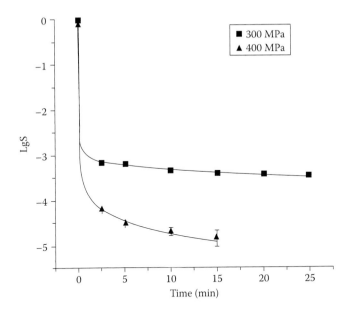

FIGURE 7.7 The Weibull model of microbial inactivation of cloudy pomegranate juice treated by HHP. (Adapted from Varela-Santos E. et al. 2012. *Innovative Food Science and Emerging Technologies* 13, 13–22.)

McKay et al. (2011) stated the microbiota of raw apple juice treated by HHP and HPH during storage at ideal (4°C) and abusive (12°C) temperatures. It was found that after the HPH, only low numbers of microorganisms were detected after treatment at 300 MPa (typically 2–3 log.mL^{-1}) (Figure 7.8) and these were identified as *Streptomyces* spp; the numbers had not increased during storage of the juice for 35 days, irrespective of storage temperature (McKay et al., 2011). In the study reported by McKay et al. (2011), after 500–600 MPa/1 min HHP treatment, the total aerobic counts were also reduced significantly ($p < 0.05$) (Figure 7.9); the counts were not increased significantly during storage at 4°C whereas they increased during storage at 12°C and by day 14, counts at 500 MPa were not significantly different from the control juice. It was concluded that good temperature control was important for the full benefits of HHP treatment and it was found that *Frateuria aurantia* dominated the microbiota of the HHP-treated apple juice stored at 12°C along with low levels of *Bacillus* and *Streptomyces* spp. It was stated that both the HPH and HHP juices turned brown during storage, indicating that neither treatment was sufficient to inactivate polyphenol oxidase (PPO) and PPO enzyme was pressure resistant and this discoloration was controlled by a heat treatment (70°C for 1 min) used in commercial practice and given prior to HP treatment (McKay et al., 2011).

It is stated that *F. aurantia* resembles the acetic acid group of bacteria in its ability to oxidize ethanol to acetic acid and the fact that it can grow at pH 3.6. Despite these similarities with the acetic acid bacteria, the genus *Frateuria* is classified within the γ-proteobacteria. Characteristics of *F. aurantia* that support this classification

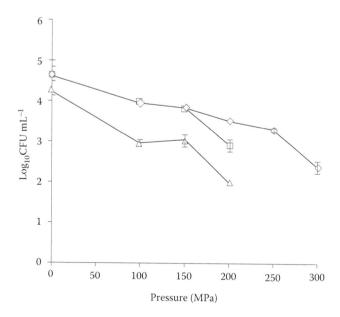

FIGURE 7.8   The effect of high pressure homogenisation (HPH) on the microbial counts in apple juice. -◊- total aerobic count (PCA), □-yeasts and moulds count (OGYE),-△- lactic acid bacteria count (MRSSA), Error bars represent ± standard error. The lactic acid bacteria count and the yeasts & moulds count were both below the limit of detection after HPH at 250 MPa and 300 MPa. (Adapted from McKay A.M. et al. 2011. *Food Microbiology* 28, 1426–1431.)

include similarity values for the 16S rRNA gene sequence. *F. aurantia* has been isolated from the fruit of raspberries (*Rubus parvifolius*) (Yamada et al., 1976).

McKay et al. (2011) reported that the predominance of *F. aurantia* in the microbiota of apple juice treated with HHP at 500 and 600 MPa may result from its initial presence on the apple fruit and in processed juice, followed by survival of HHP treatment and growth during storage of the apple juice. Table 7.3 shows the HHP resistance of *F. aurantia* in apple juice (McKay et al., 2011). It was stated that the resistance of *F. aurantia* to HHP was not exceptionally high (Table 7.3); the study on the presence or absence of growth after HHP revealed that the microorganism can survive at low numbers in apple juice (McKay et al., 2011). On the basis of the study described by McKay et al. (2011), this microorganism dominated the microbiota under conditions of growth at low temperature in an acidic environment following HHP treatment (McKay et al., 2011). Suarez-Jacobo et al. (2010) stated the ability of unidentified aerobic mesophilic bacteria to survive HPH at 300 MPa, probably as spores. It was reported that the aerobic mesophilic count did not increase during refrigerated storage of the HPH-treated apple juice in the study described by Suarez-Jacobo et al. (2010). The bacteria that behaved in a similar manner in the study reported by McKay et al. (2011), have been identified as *Streptomyces* spp (McKay et al., 2011).

McKay et al. (2011) concluded that HPH, a technique that is not currently exploited commercially can control microbial spoilage of apple juice as effectively

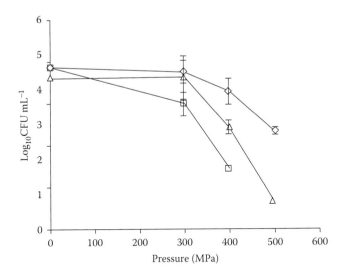

FIGURE 7.9 The effect of high hydrostatic pressure (HHP) (1 min hold time at 20°C) on the microbial counts in apple juice. -◇- total aerobic count (PCA), -□- yeasts and moulds count (OGYE), -△- lactic acid bacteria count (MRSSA). error bars represent ± standard error. The yeasts & moulds count was below the limit of detection after HHP treatment at 500 Mpa and 600 MPa. The lactic acid bacteria and total aerobic count were below the limit of detection after HHP treatment at 600 Mpa. (Adapted from McKay A.M. et al. 2011. *Food Microbiology* 28, 1426–1431.)

as a commercial HHP process. It was also stated that completion of the comparison between HHP and HPH requires further research of the organoleptic and nutritional properties of apple juice treated by HPH at different pressures to correspond to results on HHP (McKay et al., 2011).

Lee et al. (2002) reported that the spores of *A. acidoterrestris* were successfully destroyed by a treatment combining the HHP technique with mild-to-high temperatures. The microorganism had been suspended in commercial apple juice. It was stated that 3.5-log microbial reductions were achieved by applying 207 MPa pressure at 45°C for 10 min, 4-log reductions with the parameter of 621 MPa applied pressure at 71°C for 1 min, and over 5.5-log reductions were achieved by applying 621 MPa pressure at 71°C for 10 min (Lee et al., 2002).

The effect of HHP on inactivation of *A. acidoterrestris* vegetative cells in BAM broth model system and in orange, apple, and tomato juices was studied. After treating with 350 MPa at 50°C for 20 min, a 4.7-log reduction of cells was achieved in BAM broth while in all juices, over a 4-log reduction was achieved (Alpas et al., 2003). It was reported that the effect of HPH on the rheological properties of tomato juice HPH processing decreased the viscosity of the juice serum (Augusto et al., 2012) and increased the consistency, thixotropy, viscous, and elastic behavior of the tomato juice (Augusto et al., 2013). It was concluded that according to rheological analysis, it indicated that this technology could be used to increase the consistency

TABLE 7.3

HHP Resistance of *F. aurantia* in Apple Juice

HHP (1 min at 18°C)	Log$_{10}$ CFU mL^{-1}	Presence/Absence in 5 mL	
		Immediately after HHP	48 h after HHP (Stored at 4°C)
Untreated	8.36	Not done	Not done
300 MPa	6.74	Not done	Not done
400 MPa	2.22	Present 4/4 samples	Present 4/4 samples
500 MPa	< 1.30	Present 2/4 samples	Present 2/4 samples
600 MPa	< 1.30	Not done	Not done

Source: Adapted from McKay A.M. et al. 2011. *Food Microbiology* 28, 1426–1431.
Note: Results are the mean of four replicates (LSD 1/4 2.025).

of tomato juice, improving its sensory acceptance, reducing the need for adding hydrocolloids, and reducing particle sedimentation and serum separation (Augusto et al., 2013).

Apple juices (17.5°, 35°, and 70°Brix) inoculated with *A. acidoterrestris* spores were subjected to three pressure treatments (207, 414, and 621 MPa) at 22°C, 45°C, 71°C, and 90°C. Results showed that the effectiveness of treatment increased as pressure and temperature increased. It was found that the effectiveness of HHP was affected by soluble solids content, with reduction in inhibition observed when the concentration of juice was increased (Chen, 2011). Over 5- and 4-log reductions were found in juice of 17.5° and 35°Brix, respectively, at 90°C; however, there was no significant reduction of spores at the highest concentration (70°Brix) (Lee et al., 2006). The effectiveness of treatment increased as pressure and temperature increased. At 25°C of temperature, there was no significant reduction of spores in all juice samples for all pressures. Spores suspended in apple juice were successfully destroyed by combining HP with a mild or high temperature (45°C, 71°C, or 90°C) (Lee et al., 2006).

Chen (2011) stated that HPP and dimethyl dicarbonate (DMDC) treatments significantly ($p < 0.05$) inhibited growth of vegetative cells. HPP caused a 1-2 log reduction in vegetative cell populations at 300 MPa for four strains (N-1100, N-1108, N-1096, and OS-CAJ), whereas HPP caused only 0.5 log reduction for the most resistant strain of *A.acidoterrestris* (SAC strain).

Silva Filipa et al. (2012) stated that the use of HHP (200 –600 MPa) in combination with mild temperature (45 –65°C) for 1–15 min, inactivated the spores in orange juices (Filipa et al., 2012). It was stated that treatment with 207 MPa at 45°C for 10 min or at 71°C for 1 min resulted in more than 3.5-log reductions in the number of viable spores in apple juices (Tokuşoğlu and Yamamoto, 2012).

Gobbi et al. (2010) reported on the early detection of *A. acidoterrestris* in different fruit juices, and Concina et al. (2010) tested the electronic nose as a diagnostic tool toward *Alicyclobacillus* spp in three commercial soft drinks.

Hartyáni et al. (2013) stated the survival rate of *A. acidoterrestris* DSMZ 2498 in apple and orange juices that were treated by combined heat (20°C, 50°C, 60°C) and

HHP (200, 400, and 600 MPa for 10 min) and stored for 0–4 weeks (Hartyáni et al., 2013). It was stated that the different levels of HP treatment at each temperature levels were applied and, in each case, the treatment time was set to 10 min. It was found that in the case of *A. acidoterrestris*-inoculated apple juice, the application of 50°C treatment temperature and 200 MPa pressure level resulted in 2-log reduction in the number of spores. It was stated that by applying 400 and 600 MPa pressure level, 3-log reduction was achieved. Concerning the vegetative cell number, the survival rate decreased to a detection limit with the application of increasing pressure levels (Hartyáni et al., 2013). Hartyáni et al. (2013) stated that in the case of inoculated orange juice, the treatment at 60°C using 200, 400, and 600 MPa/10 min caused 1.1-log, 2.5-log, and 3-log reduction, respectively, in the spore number. It was also stated that at 60°C the survival rate of vegetative cells decreased to a detection limit in the case of all applied pressure levels (Hartyáni et al., 2013). It was concluded that the combined heat and HHP treatment followed by storage of the samples at 4°C proved to be a successful strategy. Hartyáni et al. (2013) reported that the electronic nose is a useful tool for tracing sensor changes after preservation treatments and storage in connection with the aroma alterations as an effect of inactivation of the problematic microorganism in fruit juices (Hartyáni et al., 2013).

Sokołowskaa et al. (2013) stated factors influencing the inactivation of *A. acidoterrestris* spores exposed to HHP in apple juice. It was concluded that spores were examined in concentrated apple juice (71.1Bx) by 200 MPa/50 C/45 min pressurization, and no significant alterations were observed. In the apple juices with a soluble solids content of 35.7°, 23.6°, and 11.2Bx, the reduction in spores was 1.3–2.4 log, 2.6–3.3 log, and 2.8–4.0 log, respectively. No clear effect of the spore's age on the survival under HP conditions was found. Spores surviving pressurization and subjected to subsequent HHP treatment showed increased resistance to pressure, by even as much as 2.0 log (Sokołowskaa et al., 2013).

Mert et al. (2013) stated that red and white grape juices were treated with HHP at three different pressures, temperature, and time values to determine the HHP effects on natural microflora and some quality properties of the grape juices (Mert et al., 2013).

It was reported that increased pressure, temperature, and time showed significant effect on the microbial reduction and no microbial growth was observed in HHP-treated grape juices up to 90 days. It was found that the ΔL^* values obtained for red grape juice were negative under all pressure, temperature, and time applications, which indicates a darker color tendency, whereas positive values were obtained for white grape juice, indicating a lighter color tendency after HHP processing. The color change values (ΔE) were between 1 and 7 units and were between 0.2 and 0.6 for HHP-treated red and white grape juices, respectively (Figure 7.10) (Mert et al., 2013). HHP had little or no effect on pH and color of the juices. 5-hydroxymethyl-furfural (HMF) formation was observed in heat pasteurized samples while no HMF was detected in HHP-treated grape juices (Mert et al., 2013).

Laura Espina et al. (2013) reported that the inactivation was achieved with *E. coli* O157:H7 and *L. monocytogenes* EGD-e by combined processes of HHP and essential oils (EOs) or their chemical constituents (CCs) (Laura Espina et al., 2013). In the study described by Laura Espina et al. (2013), HHP treatments

(175–400 MPa/20 min) were combined with 200 µL/L of each EO (*Citrus sinensis* L., *Citrus lemon* L., *Citrus reticulata* L., *Thymus algeriensis* L., *Eucalyptus globulus* L., *Rosmarinus officinalis* L., *Mentha pulegium* L., *Juniperus phoenicea* L., and *Cyperus longus* L.) or each CC ((+)-limonene, α-pinene, β-pinene, p-cymene, thymol, carvacrol, borneol, linalool, terpinen-4-ol, 1,8-cineole, α-terpinyl acetate, camphor, and (+)-pulegone) in buffer of pH 4.0 or 7.0 (Laura Espina et al., 2013). It was stated that the tested combinations achieved different degrees of inactivation, the most effective being (+)-limonene, carvacrol, *C. reticulata* L. EO, *T. algeriensis* L. EO, and *C. sinensis* L. EO that were capable of inactivating about 4–5 $\log_{10}$ cycles of the initial cell populations in combination with HHP and, therefore, showed outstanding synergistic effects. (+)-Limonene was also capable of inactivating 5 $\log_{10}$ cycles of the initial *E. coli* O157:H7 population in combination with HHP (300 MPa for 20 min) in orange and apple juices, and a direct relationship was established between the inactivation degree caused by the combined process with (+)-limonene and the occurrence of sublethal injury after the HHP treatment (Laura Espina et al., 2013). It was concluded that the potential of EOs and CCs in the inactivation of food-borne pathogens in combined treatments with HP, proposes their possible use in liquid food such as fruit juices (Laura Espina et al., 2013).

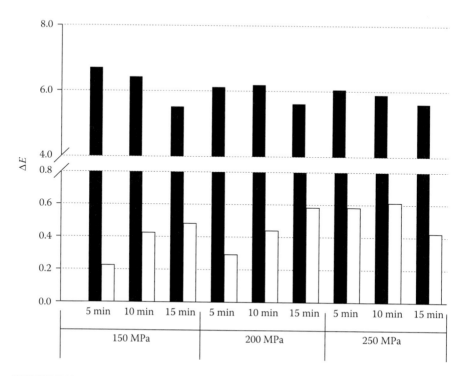

FIGURE 7.10   The ΔE values of red (black color) and white (white color) grape juices at different pressure and time values at 40°C. (Adapted from Mert M., Buzrul S., Alpas H. 2013. *High Pressure Research* 33(1), 55–63.)

As a conclusion, HHP processing improved the microbial stability, and eliminated the pathogen microorganisms whereas an effective tool for nonpathogen *A. acidoterrestris* in fruit juices could be used in the fruit juice industry.

REFERENCES

Alpas H., Alma L., Bozoğlu F. 2003. Inactivation of *Alicyclobacillus acidoterrestris* vegetative cells in model system, apple, orange and tomato juices by high hydrostatic pressure. *World Journal of Microbiology and Biotechnology* 19(6), 619–623.

Augusto P.E.D., Ibarz A., Cristianini M. 2012. Effect of high pressure homogenization (HPH) on the rheological properties of a fruit juice serum model. *Journal of Food Engineering* 111(2), 474–477.

Augusto P.E.D., Ibarz A., Cristianini M. 2013. Effect of high pressure homogenization (HPH) on the rheological properties of tomato juice: Viscoelastic properties and the Cox–Merz rule. *Journal of Food Engineering* 114, 57–63.

Anonymous. 2013. Research Report.West Europe Fruit Juice Market Research, Trends, AnalysisTOC.

Anonymous. 2001. Commission decision. 2001/424/EC, authorising the placing on the market of pasteurised fruit-based preparations produced using high pressure pasteurisation under Regulation (EC) no 258/97 of the European Parliament and of the Council.

Avure. 2013. High pressure applications. *Avure Technologies Service and Sales—Americas*, Ohio, OH, USA.

Betoret E., Betoret N., Carbonell J.V., Fito P. 2009. Effects of pressure homogenization on particle size and the functional properties of citrus juices. *Journal of Food Engineering* 92(1), 18–23.

Bevilacqua A., Corbo M.R., Sinigaglia M. 2012. Use of natural antimicrobials and high pressure homogenization to control the growth of *Saccharomyces bayanus* in apple juice. *Food Control* 24, 109–115.

Briñez W.J., Roig-Sagues A.X., Herrero M.M.H., Lopez B.G. 2006a. Inactivation by ultra-high-pressure homogenization of *Escherichia coli* strains inoculated into orange juice. *Journal of Food Protection* 69(5), 984–989.

Briñez W.J., Roig-Sagues A.X., Herrero M.M.H., Lopez B.G. 2006b. Inactivation of *Listeria innocua* in milk and orange juice by ultrahigh-pressure homogenization. *Journal of Food Protection* 69(1), 86–92.

Briñez W.J., Roig-Sagues A.X., Herrero M.M.H., Lopez B.G. 2007. Inactivation of *Staphylococcus* spp. strains in wholemilk and orange juice using ultra high pressure homogenisation at inlet temperatures of 6 and 20°C. *Food Control* 18(10), 1282–1288.

Bull M.K., Zerdin K., Howe E., Goicoechea D., Paramanandhan P., Stockman R., Sellahewa J., Szabo E.A., Johnson R.L., Stewart C.M. 2004. The effect of high pressure processing on the microbial, physical and chemical properties of Valencia and navel orange juice. *Innovative Food Science and Emerging Technologies* 5, (2004), 135–149.

Campos F.P., Cristianini M. 2007. Inactivation of *Saccharomyces cerevisiae* and *Lactobacillus plantarum* in orange juice using ultra high-pressure homogenisation. *Innovative Food Science and Emerging Technologies* 8(2), 226–229.

Cerny G., Hennlich W., Poralla K. 1984. Fruchtsaftverderb Durch Bacillen: Isolierung Und Charakterisierung Des Verderbserregers (Spoilage of fruit juice by bacilli. Isolation and characterization of spoiling microorganism). *Z. Lebens Unters Forsch* 179, 224–227 (in German).

Chen W. 2011. Inactivation of *Alicyclobacillus Acidoterrestris* using high pressure homogenization and dimethyl dicarbonate. Master Thesis. University of Tennessee, Knoxville, USA. http://trace.tennessee.edu/utk_ gradthes/863

Chen D., Xi H., Guo X., Qin Z., Pang X., Hu X., Liao X., Wu J. 2013. Comparative study of quality of cloudy pomegranate juice treated by high hydrostatic pressure and high temperature short time. *Innovative Food Science and Emerging Technologies* 19, 85–94.

Concina I., Bornsek M., Baccelliere S., Falasconi M., Gobbi E., Sbarveglieri G. 2010. *Alicyclobacillus* spp.: Detection in soft drinks by electronic nose. *Food Research International* 43, 2108–2114.

Corli Witthuhn R., van der Merwe E., Venter P., Cameron M. 2012. Short communication. Guaiacol production from ferulic acid, vanillin and vanillic acid by *Alicyclobacillus acidoterrestris*. *International Journal of Food Microbiology* 157, 113–117.

Deinhard G., Blanz P., Poralla K., Altan E. 1987. *Bacillus acidoterrestris* sp. nov., a new thermotolerant acidophile isolated from different soils. *Systematic Applied Microbiology* 10, 47–53.

Donsì F., Esposito L., Lenza E., Senatore B., Ferrari G. 2009. Production of shelfstable Annurca apple juice with pulp by high pressure homogenization. *International Journal of Food Engineering* 5(4), 1556–3758.

FDA. 2001. Hazard analysis and critical control point (HACCP); procedures for the safe and sanitary processing and importing of juice: Final rule (21 CFR Part 120). *Federal Register* 66, 6137–6202.

Gobbi E., Falasconi M., Concina I., Mantero G., Bianchi F., Mattarozzi M., Musci M., Sbarveglieri G. 2010. Electronic nose and *Alicyclobacillus* spp. spoilage of fruit juices: An emerging diagnostic tool. *Food Control* 21, 1374–1382.

Goto K., Matsubara H., Mochida K., Matsumura T., Hara Y., Niwa M., Yamasato K. 2002. *Alicyclobacillus herbarius* sp. nov., a novel bacterium containing cycloheptane fatty acids, isolated from herbal tea. *International Journal of Systematic and Evolutionary Microbiology* 52, 109–113.

Goto K., Mochida K., Asahara M., Suzuki M., Kasai H., Yokota K. 2003. *Alicyclobacillus pomorum* sp. nov., a novel thermoacidophilic, endospore-forming bacterium that does not possess alicyclic fatty acids, and emended description of the genus *Alicyclobacillus*. *International Journal of Systematic and Evolutionary Microbiology* 53, 1537–1544.

Gould G.W., Sale J.H. 1970. Initiation of germination of bacterial spores by hydrostatic pressure. *Journal of General Microbiology* 60, 335–346.

Guerrero-Beltrán J.A., Barbosa-Cánovas G.V. 2004. High hydrostatic pressure processing of peach puree with and without antibrowning agents. *Journal of Food Processing and Preservation* 28, 69–85.

Guerrero-Beltrán J.A., Barbosa-Cánovas G.V., Moraga-Ballesteros G., Moraga-Ballesteros M.J., Swanson B.G. 2006. Effect of pH and ascorbic acid on high hydrostatic pressure processed mango puree. *Journal of Food Processing and Preservation* 30, 582–596.

Hartyáni P., Dalmadi I., Knorr D. 2013. Electronic nose investigation of *Alicyclobacillus acidoterrestris* inoculated apple and orange juice treated by high hydrostatic pressure. *Food Control* 32 (2013), 262–269.

Hiperbaric. 2013. Juices applications. *High Pressure Processing Technology*. Hiperbaric, Miami, FL, USA; Hiperbaric, Burgos, Spain.

Hsu K., Tan F., Chi H. 2008. Evaluation of microbial inactivation and physicochemical properties of pressurized tomato juice during refrigerated storage. *LWT—Food Science and Technology* 41, 367–375.

Jordan S.L., Pascual C., Bracey E., Mackey B.M. 2001. Inactivation and injury of pressure resistant strains of *Escherichia coli* O157 and *Listeria monocytogenes* in fruit juices. *Journal of Applied Microbiology* 91(3), 463–469.

Kriskó G., Roller S. 2005. Carvacrol and p-cymene inactivate *Escherichia coli* O157:H7 in apple juice. *BMC Microbiology* 5, 36.

Kumar S., Thippareddi H., Subbiah J., Zivanovic S., Davidson P.M., Harte F. 2009. Inactivation of *Escherichia coli* K-12 in apple juice using combination of high pressure homogenization and chitosan. *Journal of Food Science* 74(1), 8–14.

Krebbers B., Matser A.M., Hoogerwerf S.W., Moezelaar R., Tomassen M.M.M., Van den Berg R.W. 2003. Combined high pressure and thermal treatments for processing of tomato puree: Evaluation of microbial inactivation and quality parameters. *Innovative Food Science and Emerging Technology* 4, 377–385.

Laura Espina L., García-Gonzalo D., Laglaoui A., Mackey B.M., Pagán R. 2013. Synergistic combinations of high hydrostatic pressure and essential oils or their constituents and their use in preservation of fruit juices. *International Journal of Food Microbiology* 161, 23–30.

Lavinas F.C.L., Miguel M.A., Lopes M.L.M., Valentemesquita V.L. 2008. Effect of high hydrostatic pressure on cashew apple (*Anacardium occidentale* L.) juice preservation. *Journal of Food Science* 73(6), 273–277.

Lee S.Y., Dougherty R.H., Kang D.H. 2002. Inhibitory effects of high pressure and heat on *Alicyclobacillus acidoterrestris* spores in apple juice. *Applied and Environmental Microbiology* 68, 4158–4161.

Lee S.Y., Gray P.M., Dougherty R.H., Kang D.H. 2004. The use of chlorine dioxide to control *Alicyclobacillus acidoterrestris* pores in aqueous suspension and on apples. *International Journal of Food Microbiology* 92, 121–127.

Liu F., Wang Y., Bi X., Guo X., Fu S., Liao X. 2013. Comparison of microbial inactivation and rheological characteristics of mango pulp after high hydrostatic pressure treatment and high temperature short time treatment. *Food Bioprocess Technology* 2013 (6), 2675–2684.

Ma Y., Hu X., Chen J., Zhao G., Liao X., Chen F., Wu J., Wang Z. 2010. Effect of UHP on enzyme, microorganism and flavor in cantaloupe (*Cucumis melo* L.) juice. *Journal of Food Process Engineering* 33, 540–553.

Matsubara H., Goto K., Matsumura T., Mochida K., Iwaki M., Niwa M., Yamasato K. 2002. *Alicyclobacillus acidiphilus* sp. nov., a novel thermo-acidophilic, *N*-alicyclic fatty acid containing bacterium isolated from acidic beverages. *International Journal of Systematic and Evolutionary Microbiology* 52, 1681–1685.

McKay A.M., Linton M., Stirling J., Mackle A., Patterson M.F. 2011. A comparative study of changes in the microbiota of apple juice treated by high hydrostatic pressure (HHP) or high pressure homogenisation (HPH). *Food Microbiology* 28, 1426–1431.

Mert M., Buzrul S., Alpas H. 2013. Effects of high hydrostatic pressure on microflora and some quality attributes of grape juice. *High Pressure Research* 33(1), 55–63.

Meskin M.S., Bidlack W.R., Davies A.J., Lewis D.S., Randolph R.K. 2003. Phytochemicals: Mechanisms of Action. BOOK. ISBN-10: 0849316723; ISBN-13: 978-0849316722. 224 page. CRC Press, Taylor & Francis Group, Boca Raton, Florida, USA.

MEYED. 2013. Fruit Juice Industry Association, Turkey.

Morris C., Brody A.L., Wicker L. 2007. Non-thermal food processing/preservation technologies: A review with packaging implications. *Packaging Technology and Science* 20, 275–286.

Murakami M., Tedzuka H., Yamazaki K. 1998. Thermal resistance of *Alicyclobacillus acidoterrestis* spores in different buffers and pH. *Food Microbiology* 15, 577–582.

Ohlsson T. 2002. Minimal processing of foods with non-thermal methods. In: Ohlsson, T. and Bengtsson, N. (eds.), *Minimal Processing Technologies in the Food Industry*, 1st ed., CRC Press, Boca Raton, FL, pp. 34–47.

Orr R.V., Shewfelt R.L., Huang C.J., Tefera S., Beuchat L.R. 2000a. Detection of guaiacol produced by *Alicyclobacillus acidoterrestris* in apple juice by sensory and chromatographic analysis, and comparison with spore and vegetative cell population. *Journal of Food Protection* 63, 1517–1522.

Orr R.V., Beuchat L.R. 2000b. Efficacy of disinfectants in killing spores of *Alicyclobacillus acidoterrestris* and performance of media for supporting colony development by survivors. *Journal of Food Protection* 63, 1117–1122.

Pathanibul P., Taylor M.T., Davidson P.M., Harte F. 2009. Inactivation of *Escherichia coli* and *Listeria innocua* in apple and carrot juices using high pressure homogenization and nisin. *International Journal of Food Microbiology* 129, 316–320.

Patrignani F., Tabanelli G., Siroli L., Gardini F., Lanciotti R. 2013. Combined effects of high pressure homogenization treatment and citral on microbiological quality of apricot juice. *International Journal of Food Microbiology* 160, 273–281.

Patrignani F., Vannini L., Kamdem S.L.S., Lanciotti R., Guerzoni M.E. 2010. Potentialities of high-pressure homogenization to inactivate *Zygosaccharomyces bailii* in fruit juices. *Journal of Food Science* 75(2), 116–120.

Patrignani F., Vannini L., Sado-Kamdem S.L., Lanciotti R., Guerzoni M.E. 2009. Effect of high pressure homogenization on *Saccharomyces cerevisiae* inactivation and physicochemical features in apricot and carrot juices. *International Journal of Food Microbiology* 136, 26–31.

Patterson M.F. 2005. Microbiology of pressure-treated foods. *Journal of Applied Microbiology* 98, 1400–1409.

Plaza L., Sánchez-Moreno C., Elez-Martínez P., De Ancos B., Martín-Belloso O., Cano M.P. 2006. Effect of refrigerated storage on vitamin C and antioxidant activity of orange juice processed by high-pressure or pulsed electric fields with regard to low pasteurization. *European Food Research and Technology* 223, 487–493.

Polydera A.C., Stoforos N.G., Taoukis P.S. 2003. Comparative shelf life study and vitamin C loss kinetics in pasteurised and high pressure processed reconstituted orange juice. *Journal of Food Engineering* 60(1), 21–29.

Rattanathanalerk M., Chiewchan N., Srichumpoung W. 2005. Effect of the thermal processing on the quality loss of pineapple juice. *Journal of Food Engineering* 66(2), 259–265.

Rodrigues S., Narciso Fernandez F.A. 2012. *Advances in Fruit Processing Technologies*. ISBN 978-1-4398-5152-4, Contemporary Food Engineering Series. 454 p., CRC Press, Taylor & Francis, Boca Raton, FL, USA.

Ross A.I.V., Griffiths M.W., Mittal G.S., Deeth H.C. 2003. Combining nonthermal technologies to control foodborne microorganisms. *International Journal of Food Microbiology*. 89(2–3), 125–138.

Sánchez-Moreno C., Plaza L., Elez-Martínez P., de Ancos B., Martín-Belloso O., Cano P. 2005. Impact of high pressure and pulsed electric fields on bioactive compounds and antioxidant capacity of orange juice in comparison with traditional thermal processing. *Journal of Agricultural and Food Chemistry* 53(11), 4403–4409.

Sánchez-Moreno C., De Ancos B., Plaza L., Elez-Martínez P., Cano M.P. 2009. Nutritional approaches and health-related properties of plant foods processed by high pressure and pulsed electric fields. *Critical Reviews in Food Science and Nutrition* 49, 552–576.

Savas Bahceci K., Acar J. 2007. Determination of guaiacol produced by *Alicyclobacillus acidoterrestris* in apple juice by using HPLC and spectrophotometric methods, and mathematical modeling of guaiacol production. *European Food Research Technology* 225, 873–878.

Sawaki T. 2007. Introduction to *Alicyclobacillus*. In: *Alicyclobacillus Thermophilic Acidophilic Bacilli*. Akira Y., Tateo F., Keiichi G., (eds.), Springer-Verlag GmbH, Berlin, Heidelberg. ISBN: 978-4-431-69849-4 (Print) 978-4-431-69850-0.

Silva Filipa V.M., Tan E.K., Farid M. 2012. Bacterial spore inactivation at 45–65°C using high pressure processing: Study of *Alicyclobacillus acidoterrestris* in orange juice. *Food Microbiology* 32(1), 206–211.

Sokołowskaa B., Skąpskaa S., Fonberg-Broczekb M., Niezgodaa J., Chotkiewicza M., Dekowskaa A., Rzoskab S.J. 2013. Factors influencing the inactivation of *Alicyclobacillus acidoterrestris* spores exposed to high hydrostatic pressure in apple juice. *High Pressure Research* 33(1), 73–82.

Smit Y., Cameron M., Venter P., Witthuhn R.C. 2011. *Alicyclobacillus* spoilage and isolation. A review. *Food Microbiology*.28, 331–349.

Sprittstoesser D.F., Churey J.J., Lee C.Y. 1994. Growth characteristics of aciduric sporeforming *Bacilli* isolated from fruit juices. *Journal of Food Protection* 57, 1080–1083.

Stipp G.K., Tsai C.H. 1988. Low viscosity orange juice concentrates useful for high Brix products having lower pseudoplasticity and greater dispersibility. Vol. U.S. Patent No. 4,946,702, The Procter & Gamble Company.

Suarez-Jacobo A., Gervilla R., Guamis B., Roig-Sagues A.X., Saldo J. 2010. Effect of UHPH on indigenous microbiota of apple juice. A preliminary study of microbial shelf-life. *International Journal of Food Microbiology* 136, 261–267.

Tokuşoğlu Ö., Hall C. 2011. *Fruit and Cereal Bioactives: Sources, Chemistry & Applications.* ISBN: 9781439806654; ISBN-10: 1439806659. 459 page. CRC Press, Taylor & Francis Group, Boca Raton, Florida, USA.

Tokuşoğlu Ö., Yamamoto K. 2012. High pressure processing effects on guaiacol formation by *Alicyclobacillus* spp. from phenolic ferulic and vanillic acids in fruits juices, in fruity drinks. In *7th International Conference on High Pressure Bioscience and Biotechnology.* Poster Presentation. P1–19. *Book of Abstracts.* p. 63. October 29–November 2, 2012, Otsu, Kyoto, Japan.

Tribst A.A., Sant'Ana Ade S., de Massaguer P.R. 2009. Review: Microbiological quality and safety of fruit juices—Past, present and future perspectives. *Critical Reviews in Microbiology* 35, 310–339. ISSN 1549-7828; 1040-841X. doi: 10.3109/10408410903241428.

Tokuşoğlu Ö., Doona C. 2011. High pressure processing technology on Bioactives in Fruits & cereals (Chapter 21—Part IV. Functionality, Processing, Characterization and Applications of Fruit & Cereal Bioactives). In: Özlem Tokuşoğlu and Clifford Hall (eds.), *Fruit and Cereal Bioactives: Sources, Chemistry and Applications.* ISBN: 9781439806654; ISBN-10: 1439806659. 459 page. CRC Press, Taylor & Francis Group, Boca Raton, Florida, USA.

Tournas V.H., Heeres J., Burgess L. 2006. Moulds and yeasts in fruit salads and fruit juices. *Food Microbiology* 23(7), 684–688.

Varela-Santos E., Ochoa-Martinez A., Tabilo-Munizaga G., Reyes J.E., Pérez-Won M., Briones-Labarca V., Morales-Castro J. 2012. Effect of high hydrostatic pressure (HHP) processing on physicochemical properties, bioactive compounds and shelf-life of pomegranate juice. *Innovative Food Science and Emerging Technologies* 13 (2012), 13–22.

Yamada Y., Okada Y., Kondo K. 1976. Isolation and characterisation of polarly flagellated intermediate strains in acetic acid bacteria. *Journal of Genetic Applied Microbiology* 22, 237–245.

Yamazaki K., Okubo T., Inoue N., Shinano H. 1997. Randomly amplified polymorphic DNA (RAPD) for rapid identification of the spoilage bacterium *Alicyclobacillus acidoterrestris.* *Bioscience and Biotechnology Biochemistry* 61, 1016–1018.

Walls I., Chuyate R. 1998. *Alicyclobacillus*—Historical perspective and preliminary characterization study. *Dairy Food and Environment Sanitation* 18, 499–503.

Wisotzkey J.D., Jurtshuk J.R.P., Fox G.E., Deinhard G., Poralla K. 1992. Comparative sequence analyses on the 16S rRNA (rDNA) of *Bacillus acidocaldarius*, *Bacillus acidoterrestris*, and *Bacillus cycloheptanicus* and proposal for creation of a new genus, *Alicyclobacillus* gen. nov. *International Journal of Systematic Bacteriology* 42, 263–269.

Zhou L., Wang Y., Hu X., Wu J., Liao X. 2009. Effect of high pressure carbon dioxide on the quality of carrot juice. *Innovative Food Science and Emerging Technologies* 10, 321–327.

Zimmerman F., Bergman C. 1993. Isostatic high-pressure equipment for food preservation. *Food Technology* 47:162–163.

8 Mild High-Pressure Treatments as an Alternative to Conventional Thermal Blanching
A Case Study on Pepper Fruits

Sónia Marília Castro, Jorge Alexandre Saraiva, and Ivonne Delgadillo

CONTENTS

8.1 GENERAL INTRODUCTION

There is a food market tendency for more convenient, fresher, less heavily processed (e.g., processed with less heat), more natural, and healthier products. This can be achieved with minimal processing methods that preserve food but also retain to a greater extent their nutritional quality and sensory characteristics by reducing the reliance on heat as the main preservation action. Traditionally, fermented foods have many of these characteristics; irradiation has been adopted in some countries as a minimal method of food preservation, and chilling together with controlled or modified atmosphere are now the more widely adopted methods to suppress microbial growth. But there has also been an increasing interest in developing other combinations of existing and novel methods to achieve mild preservation. And currently, nonthermal food-processing techniques are regarded with special interest by the food industry. Among nonthermal techniques, high-pressure processing, also described as high hydrostatic pressure, or ultra-high-pressure processing, has been recently introduced in some food sectors.

Even though referred as a new technology, the use of high pressure in food processing is just an extension of a technology that is commonly employed in other industrial processes. Its commercial interest in the food industry has occurred since the early 1990s, albeit the effect of high pressure in inactivating microorganisms has been known for more than a century. Hite (1899) demonstrated the potential of high pressure as a means of preserving milk, fruit juices, and meat. During the next 15 years, Hite and his co-workers examined a variety of foods and showed that high pressure could be used to decrease the microbial load of foods and, thus, preserves several fruits and vegetables (Hite et al., 1914). Initial emphasis on high-pressure processing was directed toward food preservation with the goal of extending product shelf life with minimum impact on product quality. Subsequently, the great potential of food and food constituents for physical modification of structure and function as well as the possibility for new processes developments (i.e., pressure-assisted freezing or thawing) has been recognized (Urrutia-Benet et al., 2004).

Portuguese cuisine is rich in using fresh vegetables, such as sweet bell peppers (*Capsicum annuum* L.), on soups, salads, sauces, packet food, and many convenience foods. However, their processed form is scarce in the market. During processing of peppers, quality attributes and nutritional values are inevitably affected. Among others, texture is one of the key sensorial parameters associated with *Capsicum* fruits.

The relation between texture and pectic substances is well known, since textural changes on fruit and vegetable tissues are accompanied by significant changes in the characteristics of these compounds. Taking into account this consideration, to improve the texture or reduce softening, the respective pectic degrading enzymes should be considered as the target for suppression or control.

The aim of this work was to evaluate the possible use of high-pressure treatments to substitute the conventional thermal blanching applied to bell pepper fruits prior to freezing, to obtain a final product with better quality. In order to compare both blanching and pressure processes, the soluble protein content, the activity of the enzymes PME, POD, and PPO, the ascorbic acid content, and the firmness were quantified. These are the main parameters quantified at the industrial level, to evaluate the adequacy of thermal blanching of peppers fruits. The effect of freezing on firmness, as carried out in industrial conditions to produce commercial frozen peppers, was also evaluated for both thermally blanched and high-pressure-treated peppers. The physicochemical characterization of the peppers used was also carried out.

8.2 THE FRUIT *C. ANNUUM*

The genus *Capsicum* is a member of the Solanaceae family, which also includes tomato, potato, and eggplant. *Capsicum* is a perennial small shrub. Although botanically they are berries, *Capsicums* are considered vegetables and usually classified within fruit characteristics, that is, pungency, color, shape, flavor, size, and their use (Bosland, 1992; Smith et al., 1987). The botanical classification of *Capsicum* members showed at least two major species, that is, *C. annuum* and *Capsicum frutescens* (Bailey and Bailey, 1976). The varieties that belong to the first group are classified as mild-flavored sweet-peppers, large and less hot, or even lacking this characteristic property, while the other species include smaller and hotter fruits. More recently, the genus *Capsicum* is considered to have approximately 22 wild species and five domesticated ones, that is, *C. annuum*, *C. baccatum*, *C. chinense*, *C. frutescens*, and *C. pubescens* (Bosland, 1994). Despite their vast peculiarity differences, most of the cultivars commercially cultivated in the world belong to the species *C. annuum* (Salunkhe and Desay, 1984).

8.2.1 QUALITY AND SAFETY PARAMETERS OF *CAPSICUM* FRUIT

The most appreciated characteristics in fruits and vegetables, and *C. annuum* without exception, are color, flavor, texture, shape, and absence of external defects, which are influenced by chemical and biochemical changes that occur during pre- and post-harvesting conditions. Good quality sweet bell peppers should be of uniform shape, size, and color typical of the variety. The flesh should be firm, relatively thick with a bright skin color and sweet flavor, and free from defects such as cracks, decay, and sunburn.

Color changes occur during ripening, storage, transport, and sale of fresh *Capsicum* fruits as well as in several pepper-based minimally processed products. Green *Capsicum* fruits are often harvested before they ripen and are usually less expensive because they can withstand transport and tend to last long. But when allowed to ripen, *Capsicum* fruits can turn from green to yellow, orange, red, purple, or even black (Simmone et al., 1997). The intense red color of *C. annuum* fruits is due to carotenoid pigments, aliphatic or alicyclic structures composed of isoprene units, which are normally fat-soluble (Bunnell and Bauernfeind, 1962), and synthesized massively during fruit ripening.

Another important attribute in *Capsicum* fruits and most appreciated in many foods is flavor. Green maturation stages were characterized by attributes like grassy, cucumber, and green bell pepper aroma whereas the ripe stages had a distinct red bell pepper aroma (Luning et al., 1994a). During the ripening of bell peppers, the volatiles associated with the "green" aroma notes seemed to disappear (Chitwood et al., 1983; Luning et al., 1994a,b,c). The other attributes, besides aroma characteristics, appear to be important for the perceived flavor. It seems that the perceived sharpness in different varieties of green *Capsicum* fruits is associated with a certain type of alkaloid compounds (Luning et al., 1994a), the capsaicinoids, which are only found in this genus (Govindarajan et al., 1987; Hoffman et al., 1983). The presence of these compounds is directly related with another quality characteristic in *Capsicum* fruits, pungency.

The texture of *Capsicum* fruits, and in particular their crispiness, is an important quality attribute to consumers. In *C. annuum* fruits, an excessive softening is a major post-harvest problem, and ripe fruits are likely to become flaccid more quickly than green fruits. Fruit softening is primarily due to changes in cell-wall carbohydrate metabolism that result in a net decrease of certain structural components of cell wall in most fruits (Bartley and Knee, 1982), mainly pectins. Pectins can be degraded by either an enzymatic and nonenzymatic way. The nonenzymatic way, β-elimination, is a chemical reaction that takes place at elevated temperatures and alkaline conditions (Keijbets and Pilnik, 1974), while the enzymatic reactions are catalyzed by a combined action of pectic enzymes, mainly PME and polygalacturonase (PG), which can lead to drastic textural changes of fruits and vegetables (i.e., Whitaker, 1984). The final degradation products are short and de-esterified pectin chains that cause an increase of pectin solubility, loosening of cell walls, and softening of tissues (Van Buren, 1979) and, therefore, care should be taken after post-harvesting, during storage and processing.

Besides the attractive characteristics mentioned above regarding the different varieties, fresh *Capsicum* fruits are also an excellent source of vitamins (A, C, and E) as well as neutral and acidic phenolic compounds, known as important antioxidants in a variety of plant defense responses, and xanthophylls. However, and as far as vitamin C is concerned, there have been some contradictory results regarding its content during ripening. Luning (1995) reported that the content in vitamin C increases during the first developmental stages of *Capsicum* fruit and slowly decreases in the red maturation stage, while later Yahia and his coworkers (2001) observed a dramatic loss of vitamin C much earlier than the full color intensity stage, which seemed to be related with the increase in the activity of a related enzyme, the ascorbic acid oxidase. It seems that the levels of these compounds can vary by genotype and maturity and are influenced by growing conditions and losses after processing and analytical methods (i.e., Daood et al., 1996; Howard et al., 1994, 2000; Lee et al., 1995; Mejia et al., 1988; Osuna-Garcia et al., 1998; Simmone et al., 1997).

8.2.2 Enzymes in *Capsicum* Fruit

8.2.2.1 Pectin Methylesterase and Polygalacturonase

The changes in cell-wall carbohydrate composition result from the action of hydrolytic enzymes produced in the plant tissue, mainly PME and PG (Kays, 1997), both considered to be implicated due to an increase in their activities as ripening continues and the relation with changes in the cell-wall pectin content (Fischer and Bennett, 1991). The enzymatic degradation by PME and PG due to successive demethoxylation and depolymerization reduces intercellular adhesiveness and tissue rigidity (Alonso et al., 1997), and therefore plays an important role in process-induced textural changes.

Although peppers, like tomatoes, belong to the Solanaceae family, there is little information in the literature on the pectolytic enzymes and their relationship with biochemical cell-wall changes of bell peppers and its texture (Gross et al., 1986; Jen and Robinson, 1984; Sethu et al., 1996). The lack of information is probably due to the anticipated low-enzyme activities in peppers, as the texture degradation in peppers is a slow process (Jen and Robinson, 1984). And still, contradictory information has been published regarding these two pectic enzymes with respect to their activity during the *Capsicum* fruit development. During the ripening, it seemed to be an overall decrease in PME activity (Jen and Robinson, 1984; Prabha et al., 1998; Sethu et al., 1996). In the early stages of ripening, there was an initial increase to a maximal PME activity, followed by a decrease (Jen and Robinson, 1984; Sethu et al., 1996). In accordance with previous studies, such as those of Castro et al. (2008, 2011), the PME activity was only detected in green peppers, even though increased amounts of enzymatic extract were used and longer reaction times were studied. Castro et al. (2008) suggested that the absence of PME activity in red peppers could be due to absence of the enzyme, or due to the occurrence of a well-known PME inhibitor, a glycoprotein that usually appears or increases its amount with ripening (An et al., 2008; Giovane et al., 2004). In contrast, the PG activity increased to a maximal level at the turning stage, with an overall increase of activity (Sethu et al., 1996). According to Sethu et al. (1996), the texture of the fruit declined concomitantly with the increase in PG activity. Softening in peppers did not correlate with PME activity since this enzyme decreased as the softening increased. Castro et al. (2003) also reported that the PG activity was not detected in green bell peppers. It seems that firm-fruited domesticated pepper varieties lack expression of endo-PG (Rao and Paran, 2003).

8.2.2.2 Peroxidase and Polyphenol Oxidase

Both peroxidase (POD) and PPO are oxidative enzymes that catalyze the oxidation of phenolic substrates (Gomes and Ledward, 1996; Jiang et al., 2004). PODs are implicated in a number of higher plant processes and their catalytic reaction is involved in the oxidation of a large number of aromatic structures at the expense of H_2O_2. In vegetables, POD induces negative flavor changes during storage. Recently, PODs have also been considered to play an important role in the metabolism of alkaloids. Bernal et al. (1993a,b) suggested that POD is involved in the degradation of capsaicinoids, the oxidation of capsaicin, and dihydrocapsaicin. It seemed that oxidation of dihydrocapsaicin by *Capsicum* POD was strictly dependent on the presence of H_2O_2 (Bernal et al. 1993b). In addition, *Capsicum* POD, like capsaicin, is mainly located in the placenta and the outermost epidermal cell layers (Bernal et al. 1993c, 1994). Later, Contreras-Padilla and Yahia (1998) demonstrated that there was an inverse relationship between the evolution of capsaicinoids and POD activity during development stages, maturation, and senescence of two hot varieties of *Capsicum* fruits. A basic POD of high p*I* appeared to be directly related to capsaicinoid metabolism since capsaicin, dihydrocapsaicin, and their phenolic precursors were easily oxidized by this enzyme (Bernal et al., 1993a,b; 1995). More recently, Pomar and co-workers (1997) demonstrated that capsaicin oxidation was also possible by an acidic POD found in *Capsicum* fruits. Even though there is an increase of these isoenzymes during the development and maturation of the fruit, it is only in the last stage of maturation, when loss of compartmentalization occurs, that these acidic PODs intervene in the oxidation of capsaicin because, until then, they are both found in different sub-cellular compartments (Estrada et al. 2000).

PPO is predominantly located in cell organelles such as chloroplasts thylakoid, mitochondria, and peroxisomes where it is firmly bound to the membrane and may also be found in the soluble fraction of the cell (Barbagallo et al., 2009; Scuderi et al., 2011). In young and unripe fruits, it is mostly present under conjugated forms, while in ripe ones it is present in the soluble fraction (Conforti et al., 2007).

8.2.2.3   Other Enzymes

In addition to PME and PG, a variety of glycanases and glycosidases have been assigned roles in fruit cell-wall metabolism (Fischer and Bennett, 1991). The behavior of glycanases – namely cellulase, xylanase, mannanase, glucanase, and galactanase – and glycosidases, like α-D-mannosidase and β-D-galactosidase, were monitored during the ripening of *Capsicum* fruits (Sethu et al., 1996). The activities of cellulase and xylanase increased up to a maximum in the early stages of ripening, followed by a decrease as ripening proceeded; while mannanase, glucanase, and galactanase were present constantly in a significant amount in all stages. As glycosidases are concerned, α-D-mannosidase increased during the ripening process and β-D-galactosidase in the early stages, and thereafter remained constant. According to Ogasawara et al. (2007), β-galactosidase plays a significant role in bell pepper fruit ripening. β-hexoaminidase, another glycosidase which may have an important function in fruit ripening by way of deglycosylation and generating free *N*-glycans (Priem et al., 1993), was found in *Capsicum* fruits (Jagadeesh and Prabha, 2002). This enzyme activity seemed to increase during development followed by a further increase during ripening.

Lipoxygenases constitute a group of enzymes that catalyzes lipid oxidation and pigment bleaching found ubiquitously in plants. Pinski et al. (1971) determined the LOX activity in several fruit and vegetable tissues and found that in red bell peppers it was very low as compared to other vegetables, such as eggplant, which belong to the same Solanaceae family. Later, some evidences for the presence of LOX in red pepper seeds was given by Daood and Biacs (1986). Mínguez-Mosquera et al. (1993) reported that LOX activity decreased during ripening of some *C. annuum* varieties, but suggested that seasonal influences leveled down this phenomenon. The relation between some lipid-derived volatile compounds in bell peppers, formed upon tissue disruption, and LOX activity was suggested by Wu et al. (1986). Besides differences in flavor that might result from the action of LOX, this enzyme seemed to be involved in color changes during ripening. *In vitro* studies revealed that the products formed during the enzymatic reaction catalyzed by pepper LOX have a strong destructive action on the carotenoids pigments of the fruits (Jarén-Galán and Mínguez-Mosquera, 1999).

8.3 CONVENTIONAL TREATMENTS APPLIED TO FRESH *CAPSICUM* FRUITS

As previously mentioned, the different varieties of *Capsicum* fruit have a wide range of food applications (Govindarajan, 1985; Govindarajan et al., 1987), due to their distinct colors, pungency, and flavor. Bell peppers are used to produce dehydrated products (such as paprika), pickled peppers, and sliced or diced frozen peppers to be used in pizzas or to be eaten raw in salads. However, *Capsicum* fruits are considered to be a perishable product, like any other fruit and vegetable and, therefore, not suitable for long-term storage (Senesi et al., 2000). Since the presence of *Alternaria* black rot, especially on the stem end, is a symptom of chilling injury, the best way to control it is to store the *Capsicum* fruits at 7–10°C (Barth et al., 2002; Paull, 1990). At these temperatures, chilling injury and ripening are minimal (Barth et al., 2002), and soft rot decay is delayed but not prevented (Sherman et al., 1982). Hot water dips at 53–55°C (4 min) can effectively control *Botrytis* rot without causing fruit injury. The operations involved in minimal processing (i.e., washing, coring, slicing, or cutting) of horticultural products disrupt the plant tissues, and the products become more perishable than the intact ones (Watada et al., 1996). Consequently, fresh-cut products (i.e., salad or salsa) require very special attention because of the magnitude of enzymatic and respiratory factors as well as microbiological concerns that will have an impact on safety. Fresh-cut bell peppers should be stored at lower temperatures than uncut fruits, between 0 and 5°C, in order to maintain either visual quality as compositional characteristics (Barth et al., 2002). As a consequence of processing, changes in several quality parameters can occur. The changes regarding pungency (Govindarajan, 1985), color pigments (Ihl et al., 1998; Mínguez-Mosquera et al., 1994, 2000; Ramakrishnan and Francis, 1973; Sgroppo et al., 2001), flavor (Luning et al., 1995; van Ruth and Roozen, 1994), and texture (Chang et al., 1995; Domínguez et al., 2001; Gu et al., 1999; Senesi et al., 2000; Villarreal-Alba et al., 2004) have already been reported.

The demand for sliced and diced frozen raw peppers has been increasing considerably in the last few years, due to consumers' willingness to eat raw, minimally processed vegetable products, as part of healthier food habits. Freezing operations can also result in severe changes in the food product, such as loss of turgor, weakening of the cell wall, and some degree of cell separation due to destruction of the cytoplasmatic structure. As a consequence, it will produce a severe effect on the texture of frozen products (Préstamo et al., 1998). Prior to freezing, foods of vegetable origin are submitted to thermal blanching to reduce the microbial load and inactivate deleterious enzymes (Cano, 1996), since many of the quality changes that occur during distribution and storage of these foods are due to detrimental reactions catalyzed by enzymes, such as POD, PPO, and PME (Bahçeci et al., 2004; Cano, 1996). However, heating also causes losses of sensorial (texture, taste, flavor, and color) and nutritional quality attributes, such as reduction of ascorbic acid content (Howard et al., 1994; Rao et al., 1981). Matthews and Hall (1978) concluded that unblanched peppers were superior in flavor, texture, and appearance to the blanched (100°C, 2.5 min) frozen peppers. However, studies related with mild thermal pre-treatments, between 50 and 60°C, reduced canning-induced softening in several plant tissues (Domínguez et al., 2001; Lee and Howard, 1999; Ng et al., 1998; Villarreal-Alba et al., 2004; Wu and Chang, 1990). It has been suggested that the firming effect obtained from mild heat treatments alone or even combined with calcium chloride treatments (Hoogzand and Doesburg, 1961; Hsu et al., 1964; Stolle-Smits et al., 1998; Vu et al., 2004) may be attributed to PME and/or to increased Ca^{2+} diffusion into the tissue. Blanching treatment conditions should be kept at levels strictly sufficient to cause inactivation of the deleterious enzymes, to minimize quality losses. This is particularly relevant for frozen sweet bell peppers intended to be stored frozen and eaten raw after thawing. This limitation has been a driving force for food processors to seek for other processing technologies that can substitute the conventional thermal blanching and, at the same time, cause less damaging effects on bell pepper texture. Treatments that increase the PME activity would in particular have an effect on improving the texture of fruits and vegetables (Knorr, 1993; Stute et al., 1996).

8.4   MATERIALS AND METHODS

8.4.1   MATERIALS

Sweet green, yellow, and red bell peppers (*C. annuum* L.) were supplied by a local company (Friopesca Refrigeração Aveiro, Lda., Aveiro, Portugal) and were brought to the laboratory immediately after harvesting, where they were stored at $4 \pm 1°C$, until further use. All chemicals used in this study were of analytical grade.

8.4.2   ASCORBIC ACID CONTENT

Five grams of pepper fruit were homogenized with 50 mL of 4% (w/v) solution of metaphosphoric acid for 15 min; the mixture was filtered and diluted to 100 mL and divided into several aliquots, that were then frozen in liquid nitrogen and stored at −20°C until quantification of ascorbic acid (AsA) by HPLC, based on the method described by Daood and co-workers (1994). All the operations were carried out with protection from light, using aluminum foil, to avoid oxidation of AA. An aliquot of AA extract was thawed and filtered through a 0.45 μm millipore filter prior to injection onto the chromatographic column. The HPLC apparatus that is used consisted of an L-6200A pump, with a 20 μL injection loop and an L-4250 UV–vis detector, with a D-2500 Chromato-integrator. The column was a LiChrosorb 100 RP-18 column (250 × 4.6 mm), with particle size of 5 μm (Merck). The mobile phase was constituted by 0.1 M phosphate and methanol (97:3), containing 0.75 mM ammonium tetrabutylhydroxide at pH 2.75, and a flow rate of 1.0 mL min⁻¹ was used. The detection was performed at 254 nm for AA, which was first identified and further quantified by comparing retention time, absorption spectra, and peak areas with those of AsA standard.

8.4.3 ENZYMATIC ACTIVITY

8.4.3.1 Crude Extract Preparation

Activity of soluble and ionically bound forms of PPO, PME, and POD were quantified separately, since these two forms can show different behavior during maturation, as found for POD (Silva et al., 1990), and different properties like stability to temperature and pressure. Fifty grams of thermal and pressure treated and unprocessed pepper samples were thawed at 4°C, homogenized with 75 mL of 0.2 M sodium phosphate buffer (pH 6.5), a low ionic strength buffer, and 4% (w/w) polyvinylpyrrolidone (PVP) in a Warring blender, followed by agitation during 1 h at 4°C. The homogenate was filtered through several layers of cheesecloth and then centrifuged (10000g, 20 min) at 4°C. The supernatant, further designated enzymes soluble fraction (SF) extract, was collected and the precipitate was re-suspended in 30 mL of 0.2 M sodium phosphate buffer (pH 6.5), followed by agitation for 30 min, and the extraction is performed again alike, to ensure that all enzymes from SF were extracted. The resulting pellet was mixed with 50 mL of 0.2 M sodium phosphate buffer (pH 6.5) with 1 M NaCl for 2 h, a high ionic strength buffer, in order to obtain the enzymes ionically bound fraction (IF) extract. The extraction was carried out twice as described for the SF. All the extracts were divided into aliquots, frozen in liquid nitrogen, and stored at −20°C until quantification of the enzymatic activities. All procedures to obtain the enzymatic extracts were carried out at 4°C and the enzymatic activities were determined in triplicate. No measurable activity was found for the enzymes studied in the second extraction of SF and IF, revealing that first extraction was complete in both cases. Protein content was quantified using the same extracts used for quantification of enzymatic activities and no protein was found for the second extraction of SF and IF, revealing that the first extraction was also complete. The total enzymatic activity for each enzyme and total protein content was calculated as the sum of the values obtained for SF and IF fractions. All enzymatic determinations were performed in triplicate.

8.4.3.2 Polyphenol Oxidase

A spectrophotometric assay using cathecol as substrate at pH 6.5, was used to quantify PPO activity, as described by Ramírez et al. (2002), using a 6405 UV/Vis JenWay spectrophotometer. The reaction mixture consisted of 1000 µL of substrate at 25°C and 40 µL of enzyme extract. The PPO activity was determined from the slope of the linear portion of the curve, relating absorbance at 411 nm with time, and was expressed as ΔAbs_{411}/min/100 g fresh weight.

8.4.3.3 Peroxidase

Activity was determined based on the method described by Worthington (1978). The substrate solution was composed by hydrogen peroxide (0.975 mM), phenol (83.1 mM), and 4-aminoantipyrine (1.19 mM) in 0.1 M sodium phosphate buffer (pH

7.0). To 1.450 mL of daily prepared substrate solution, incubated at 25°C, 50 μL of enzymatic extract were added and the increase in absorbency was recorded at 510 nm. The slope of the linear portion of the curve relating absorbance at 510 nm to time was used to calculate the enzyme activity, expressed in ΔAbs_{510}/min/100 g fresh weight.

8.4.3.4 Pectin Methylesterase

Activity determination was carried out according to the method described by Hagerman and Austin (1986). In order to achieve a constant starting pH for the reaction, all the solutions (pectin, indicator dye, and water) were adjusted to pH 7.5 with 2 M NaOH. The enzymatic extracts were also adjusted to pH 7.5 with 0.5 M NaOH. Four milliliters of citrus pectin solution (0.5%, w/v) were mixed with 300 μL of bromothymol blue (0.01%, w/v), and distilled water and enzyme extract, up to 6 mL. The PME activity was expressed as ΔAbs_{620}/min/100 g fresh weight.

8.4.3.5 Polygalacturonase

Polygalacturonase activity determination was carried out at 276 nm after incubating an appropriated amount of enzyme extract with 200 μL of polygalacturonic acid 0.4% (w/v) (Sigma, P-3889), 75 μL of 200 mM of acetate buffer (pH 4.4) and deionized water, in a final volume of 400 μL, at 30°C, for a period of time up to 48 h. The quantification of reducing sugars was done using cyanoacetomide. In the conditions used in this work, no PG activity was detected for unprocessed peppers and the applied temperature/pressure conditions (Gross, 1982). Additionally, no PG activity was detected in hot pepper, at several maturation stages, ranging from immature green to 100% red surface color (Gross et al., 1986).

8.4.4 TEXTURE MEASUREMENTS

The texture of the peppers was measured using a texture analyzer (TA-HDplus, Stable Micro System) with a 7 mm diameter hole, and the following parameters: 5 kg force load cell, 2 mm diameter aluminum cylinder probe, and 2.0 mm/s test speed. The property "firmness" (hardness), the maximum force applied to puncture the pepper tissue, was measured as an indicator of texture. The measurements were done on both sides of the pepper tissue, that is, from the skin and the flesh sides. Rupture of the skin from the flesh side required a lower force when compared with the same action from the skin side. An average value of firmness from 5 puncture measurements was calculated for each experimental condition. Texture analysis of pre-treated pepper samples was carried within one hour after the thermal and pressure treatments have been applied, and the samples were kept at 4°C during this period. The results are presented as relative firmness (%), calculated from the ratio between the firmness of the control (untreated sample) and the thermally blanched or pressure-treated samples.

8.4.5 THERMAL BLANCHING AND PRESSURE TREATMENTS

Temperature (70, 80, and 98°C) and time duration (60 or 150 s) of the thermal blanching treatments were the same as those typically used for the industrial production of frozen peppers, depending on the ripening stage of the peppers used. The pepper samples were cut in slices of 15 × 75 mm and placed in a plastic bag that was vacuum-heat sealed. For the blanching treatments, the packaged samples were immersed in a thermostatic water bath (Grant, Y28), pre-set at the adequate temperature, for 60 or 150 s, immediately cooled in a water bath at 4°C for 5 min. For the pressure treatments, the packaged samples were then pressurized using an Autoclave Engineers (Erie, PA, USA) isostatic press (Model IP3-23-30), with a cylindrical pressure chamber (i.d. 76 mm, height 610 mm), containing a pressure medium consisting of water with 2% hydraulic fluid (Hydrolubric 142; E. F. Houghton and Co., Valley Forge, PA, USA). The pressure was built up at room temperature (18–20°C), up to 100 or 200 MPa, which took ca. 45 and 60 s, respectively, maintained during 10 or 20 min, followed by decompression (ca. 45 s). Temperature increment during pressurization was very small, due to the low range of pressures and the slow pressurization rate used. The maximum temperature reached during pressurization was 22–24°C for 100 MPa and 24 – 26°C for 200 MPa and decreased during the pressurization holding time to 18–20°C. After each thermal/pressure treatment, samples were frozen in liquid nitrogen and stored frozen until further use. The designation for the different pressure and temperature conditions were: nonprocessed (Control, C), 70°C/60 s (BI.1), 70°C/150 s (BI.2), 80°C/60 s (BII.1), 80°C/150 s (BII.2), 98°C/60 s (BIII.1), 98°C/150 s (BIII.2), 100 MPa/10 min (PI.1), 100 MPa/20 min (PI.2), 200 MPa/10 min (PII.1), and 200 MPa/20 min (PII.2), according to Table 8.1. For the determination of enzymatic activities and determination of ascorbic acid content, thermally blanched and pressure-treated peppers were frozen in liquid nitrogen and stored frozen at −20°C until used.

8.4.6   THE EFFECT OF FREEZING ON FIRMNESS

Thermally blanched and pressure-treated peppers were kept at 4°C, transported to the company that supplied the peppers and frozen in a tunnel freezer at −30°C, with

TABLE 8.1

Temperature, Pressure, and Time Conditions used for Thermal Blanching and Pressure Treatments of Green and Red Peppers

Sample Code	Operating Conditions
C	Control, unprocessed sample
BI.1	Blanching at 70°C for 60 s
BI.2	Blanching at 70°C for 150 s
BII.1	Blanching at 80°C for 60 s
BII.2	Blanching at 80°C for 150 s
BIII.1	Blanching at 98°C for 60 s
BIII.2	Blanching at 98°C for 150 s
PI.1	Pressurizing at 100 MPa for 10 min
PI.2	Pressurizing at 100 MPa for 20 min
PII.1	Pressurizing at 200 MPa for 10 min
PII.2	Pressurizing at 200 MPa for 20 min

an air-blast freezing system, following the same procedure used to freeze commercial peppers. The frozen samples were stored at −20°C and prior to the texture measurements the samples were thawed and equilibrated at room temperature.

8.4.7 Data Analysis

ANOVA and bilateral Tukey's tests were carried out to determine significant differences ($P < 0.05$).

8.5 RESULTS

8.5.1 Ascorbic Acid Content

The obtained results for AsA content confirm that peppers are in fact an excellent source of AsA (Howard et al., 1994) (Figure 8.1). Also, the AsA content is within the ranges found in other studies (Guil-Guerrero et al., 2006; Howard et al., 1994; Rhaman et al., 2007; Yahia et al., 2001). For unprocessed peppers, AsA content was significantly higher ($P < 0.05$) for yellow peppers, a result that is in accordance with other studies (Guil-Guerrero et al., 2006). Nevertheless, AsA is very susceptible to chemical and enzymatic oxidation during processing and storage of produce. On one hand, blanching and pasteurization prevent the action of AsA oxidase and other plant enzymes, indirectly responsible for AsA loss (Lee and Kader, 2000; Yahia et al., 2001), but on the other hand, AsA is a heat-sensitive bioactive compound in the presence of oxygen and pH > 7. Thus, high temperatures during processing can greatly affect the rates of its degradation through an aerobic pathway (Odriozola-Serrano et al., 2008).

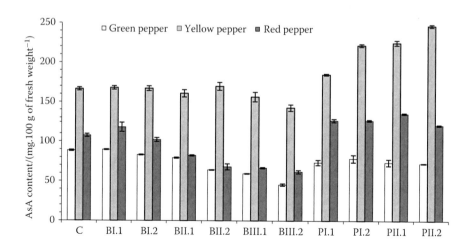

FIGURE 8.1 Effect of thermal blanching and pressure treatments, on ascorbic acid (AsA) content (mg of AsA/100 g fresh weight) of peppers. The bars represent the standard deviation ($n = 3$). The meaning of the abbreviations used in the abscissa axis is indicated in Table 8.1.

AsA content decreased progressively as blanching conditions were more severe, to about 45% and 30% of the initial value, for green and red peppers, respectively, whereas for treated yellow peppers it remained constant ($P > 0.05$), with the exception of the more severe treatment (BIII.2) (Figure 8.1). According to several authors, thermal sensitivity of AsA can depend on many factors, such as the individual crop (Lee et al., 1976; Roig et al., 1995), the maturity stage, the blanching method, and the time–temperature processing conditions (Howard et al., 1994; Lee and Howard, 1999; Matthews and Hall, 1978; Oey et al., 2006; Quaglia et al., 1996) that can cause different degrees of inactivation of AsA oxidase and removal of residual oxygen from vegetable tissue (Selman, 1994). Pressure-treated yellow and red peppers showed an increase of about 11–48% and 10–20% (Castro et al., 2008, 2011), respectively, while green bell peppers showed a decrease of about 15–20% of AsA content (Castro et al., 2011), when compared to unprocessed samples. The reason for the augmentation of AsA for pressurized yellow and red peppers is unknown. It has been reported that pressure can induce an increase of extractability (Sancho et al., 1999), possibly by decompartmentalization (Ludikhuyze et al., 2003; Rastogi et al., 2007). Lee et al. (1995) also reported higher stability toward pressure of AsA and/or other components, for example, oxidation inhibitors, which can inhibit the oxidation of AsA.

Globally, pressurized green bell peppers showed a similarity to higher retention of AsA when compared to the blanching treatments of 80 and 98°C, respectively, while pressurized red peppers showed a higher content of around 50–100%, when compared to the same blanching treatments. AsA was clearly better retained in pressurized red peppers than in green peppers, indicating an effect of the color maturation stage, as already observed for the soluble protein content.

8.5.2 Enzymatic Activity

8.5.2.1 Polyphenol Oxidase

While PPO is considered a moderately heat-stable enzyme, its baroresistance varies greatly depending on the enzyme source. For example, mushroom, potato, and avocado PPO are very pressure stable, since treatments at 800–900 MPa are needed to reduce enzyme activity at room temperature (Gomes and Ledward, 1996), while apricot and strawberry PPO were reported to be inactivated by pressures exceeding 100 and 400 MPa, respectively (Amati et al., 1996; Jolibert et al., 1994). The PPO activity (Figure 8.2) of unprocessed green peppers was found to be about 26 and 50% higher than that of unprocessed yellow and red peppers, respectively, due to a lower activity of SF extract of yellow and red peppers (activity of the SF extract represented 85, 80 and 70% of total activity for green, yellow, and red peppers, respectively). The PPO activity in eggplant (*Solanum melongena* L.) was also found to be mostly in the SF (Concellón et al., 2004).

As it can be seen in Figure 8.2 for green peppers, all treatments decreased the PPO activity, with the thermal blanching treatments causing, generally, a 25–75% decrease on activity, with increasing severity of the treatments, while all pressure treatments caused a decrease of activity of about 50%. The PPO activity in yellow peppers decreased progressively ($P < 0.05$) as the temperature and time of blanching

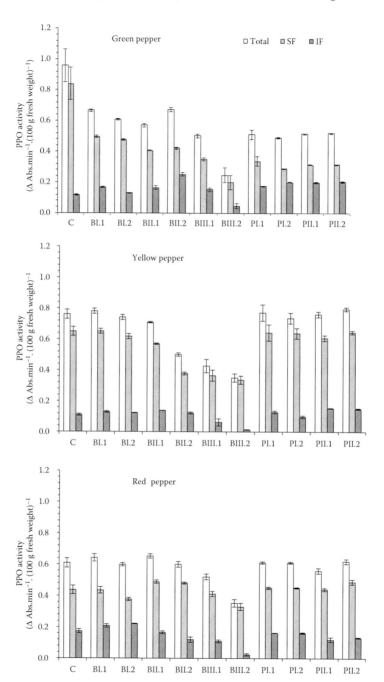

FIGURE 8.2 Effect of thermal blanching and pressure treatments on polyphenol oxidase activity of peppers (SF—activity of the soluble fraction; IF—activity of the ionically bound fraction; Total—activity of SF + IF). The bars represent the standard deviation ($n = 3$). The meaning of the abbreviations used in the abscissa axis is indicated in Table 8.1.

increased, till reaching about 50% the value of fresh peppers. The applied pressure treatments showed no effect ($P > 0.05$) on the PPO activity regarding yellow peppers. PPO was found to be more stable to both thermal blanching and pressure treatments in red peppers with only the blanching treatments at 98°C causing significant ($P < 0.05$) reduction of activity (Figure 8.2). Globally, pressure-treated peppers showed a PPO activity level similar to thermally blanched peppers, except for the three most severe blanching treatments.

8.5.2.2 Peroxidase

For all unprocessed samples, the POD activity was found to be mainly present in SF and no significant ($P > 0.05$) variation was found between them (data not shown). In tomato (Thomas et al., 1981) and strawberry (Civello et al., 1995), the POD activity decreased with ripening. In other fruits, the POD activity has been found to increase with ripening (Silva et al., 1990; Thomas et al., 1981).

In general, POD is well known as a heat-stable enzyme, and for this reason is used to evaluate the adequacy of fruits and vegetables thermal blanching (Barrett and Theerakulkait, 1995). For instance, Bahçeci et al. (2004) found that a blanching treatment at 90°C for 3 min was necessary to inactivate 90% of the activity of green bean POD. However, as it can be seen in Figure 8.3, all the applied blanching treatments reduced significantly ($P < 0.05$) the POD activity, with the exception of treatment BI.1 for red peppers (where a reduction of 70% was verified). For yellow peppers, the most severe treatment (BIII.2) led to a complete POD inactivation. These results indicate that pepper POD has a low stability to temperature. Therefore, using absence of POD activity as an indicator of adequate thermal blanching is inadequate for peppers, since POD shows lower heat stability than PPO and PME, as it can be seen in Figures 8.2 and 8.4.

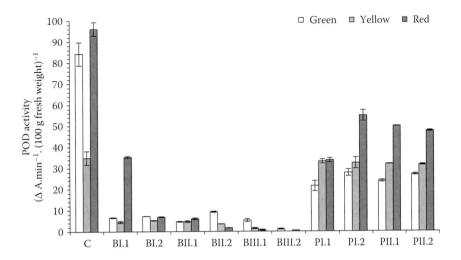

FIGURE 8.3 Effect of thermal blanching and pressure treatments on total peroxidase activity of peppers. The bars represent the standard deviation ($n = 3$). The meaning of the abbreviations used in the abscissa axis is indicated in Table 8.1.

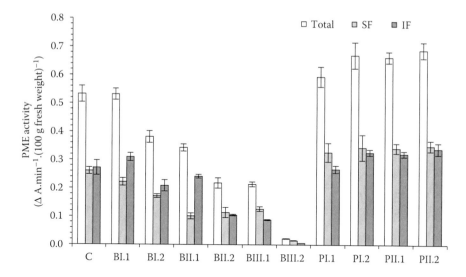

FIGURE 8.4   Effect of thermal blanching and pressure treatments on pectin methylesterase activity of green peppers (SF—activity of the soluble fraction; IF—activity of the ionically bound fraction; Total—activity of SF + IF). The bars represent the standard deviation ($n = 3$). The meaning of the abbreviations used in the abscissa axis is indicated in Table 8.1.

Pressure-treated peppers showed a reduction on POD activity of about 10, 40, and 70%, for yellow, red, and green peppers, respectively. The results found in the literature indicate that POD can show a wide spectrum in what concerns pressure stability. Pressure-treated green beans at 900 MPa for 10 min at room temperature, showed an 88% inactivation of POD and combination with temperature treatments enhanced the inactivating effect at 600 MPa (Quaglia et al., 1996). Green bean POD showed 75% residual activity after being pressure treated at 500 MPa for 60 s at room temperature (Krebbers et al., 2002). Lower pressure conditions (300 and 400 MPa) could inactivate POD from strawberry puree and orange juice at room temperature (Cano et al., 1997). In tomato puree, an increase in POD activity was reported for pressure treatments below 350 MPa at room temperature, while a significant inactivation was obtained above 350 MPa (Hernández and Cano, 1998). The obtained results pointed out that pepper POD activity has a low stability to pressure, with POD from red peppers presenting a higher stability to pressure, followed by yellow and green peppers. Overall, pressure treatments caused much lower inactivation of POD activity than the thermal blanching treatments.

8.5.2.3   Pectin Methylesterase

The PME activity was only detected in green peppers, even when increased amounts of enzymatic extract were used and longer reaction times were studied (absence of PME activity in yellow and red peppers indicates again an effect of the color maturation stage). No detected PME activity can be due to absence of the enzyme, or due to the occurrence of a well-known PME inhibitor, a glycoprotein that usually appears or increases its amount with ripening (Giovane et al., 2004). A gene in *C. annuum*

L. has been isolated and functionally characterized, which encodes a PME inhibitor protein in pepper leaves when submitted to bacterial stress (An et al., 2008). Both Jen and Robinson (1984), and Sethu et al. (1996) reported a decrease in PME activity in *C. annuum* L. fruits during ripening. In green peppers the PME activity was, generally, equally distributed between SF and IF extracts (Figure 8.4). Although PME is mainly located in the cell walls of higher plants and ionically bound, it has been suggested that soluble-protein extraction procedures can increase their recovery dramatically. These observations suggest that some of the failure to detect acidic isoforms, weakly adsorbed onto the cell wall components in higher plants, might be related to the experimental conditions used for protein extraction. Although Castro et al. (2004) have used a high ionic strength buffer to extract PME from the cell wall of green *C. annuum* L. fruits, they obtained three alkaline isoelectric points (p*I*s), and two acidic ones, indicating the possibility that acidic PME isoforms might be present. The different PME isoforms also seemed to be related with the different behavior toward temperature and pressure conditions as well as varied model systems (Castro et al., 2004, 2005).

The PME activity of green peppers declined progressively, as blanching temperature and time increased, until reaching almost absence of activity for treatment BIII.2 (Figure 8.4). Castro et al. (2005) concluded that green pepper PME was completely inactivated after heating at 80°C for 5 min, both in crude extract and in Tris buffer (pH 5.6), results that are in agreement with those obtained in this work. Pressure-treated green peppers showed a slight increase in activity when compared to the control samples. Shook et al. (2001) observed a significant increment of PME activity, when diced tomatoes were pressurized at 400 MPa and 45°C compared to nonpressurized samples. Again, Castro et al. (2005) found that after pressurization green pepper pieces up to 500 MPa and at room temperature, the PME activity increased. Augmentation of PME activity of tomato cell cultures, caused by pressure treatments up to 150 MPa, was also ascribed by Dörnenburg and Knorr (1998) to a more effective extraction of the enzyme, due to damage of plant cell wall/membrane and changes in cell wall association state of the enzyme. The same phenomena might be responsible for the increment of PME activity found in this work for pressure-treated green peppers.

8.5.3 Texture

Firmness (kg force) measured from the skin side (green, 1.12 ± 0.13; yellow: 0.86 ± 0.05; red, 1.03 ± 0.06) was about threefold higher ($P < 0.05$) than firmness measured from the flesh side (green, 0.446 ± 0.043; yellow: 0.339 ± 0.034; red, 0.414 ± 0.074) (Castro et al., 2008, 2011). A more notorious and significant ($P < 0.05$) effect between the different treatments was observed for firmness from the skin side of peppers (Castro et al., 2008), with the exception of yellow peppers, since puncture tests from the flesh side were more difficult to perform, due to the presence of uneven tissue, and illustrated in the higher standard deviation bars (Castro et al., 2008). After thermal blanching and the pressure treatments, firmness of yellow peppers was better retained when measured from the flesh side than from the skin side (Castro et al., 2011).

Generally, the results showed a trend for firmness of red peppers to be more sensitive to both the thermal blanching and the pressure treatments, than the firmness of green peppers. These results might be related to the absence of PME activity in yellow and red peppers, previously mentioned. For green peppers, PME activity can cause de-methylation of pectin molecules in the middle lamella (Alvarez et al., 2001). The de-esterified pectins are consequently less susceptible to β-eliminative degradation and, therefore, more heat resistant and less soluble, which is generally thought to increase the cell–cell adhesion (Ng and Waldron, 1997), and can crosslink with calcium ions, forming calcium pectates that contribute to increase firmness (Lee and Howard, 1999). This hypothesis is supported by observation of the results shown in Figure 8.5 for green peppers: decrement of firmness caused by the blanching treatments follows a pattern similar to the decrease of PME activity caused by the same treatments, while for pressure treatments that caused no reduction of PME activity, no decrease of firmness was observed. According to Castro et al. (2005), pepper PME seemed to be rather thermal and pressure stable under the studied conditions for the different models systems (purified, crude extract, and pepper pieces). Besides, in products where PG activity is negligible, it has been reported that thermal stimulation of cell-wall-bound PME within the optimal range of PME catalytic activity (generally between 50 and 80°C) reduces the vulnerability of the products to thermal softening (Canet et al., 2005; Fuchigami et al., 1995; Ni et al., 2005; Stanley et al., 1995; Zhang and Chen, 2006). Also, Sila et al. (2004) accounted that PME activity due to pressure activation (treatments higher than 100 MPa) caused beneficial effects on carrot firmness. The maximum activity for purified PME has been reported at 200 MPa and 55°C by Castro and coworkers (2006). Also, Basak and Ramaswamy (1998) stated that there is a dual effect on the texture of fruits and vegetables, as characterized by an initial loss in texture, due to the instantaneous pulse action of pressure, followed by a more gradual change as a result of pressure-hold. The extent of the initial loss of texture was less prominent at lower pressures and partial recovery of texture was more prominent.

For yellow and red peppers, it is clear that enzymes other than PME and PG are also involved in fruit softening during ripening. As previously mentioned, β-galactosidase and β-hexoaminidase seemed to play a significant role in bell pepper fruit ripening (Jagadeesh and Prabha, 2002; Ogasawara et al., 2007; Priem et al., 1993). Therefore, it might be worthwhile to screen for the presence of these enzymes in further studies concerning process activity and stability in *C. annuum* fruits or other fruit and vegetables in order to clear up their role in processing-related softening.

Freezing in itself caused a considerable decrease in firmness (Castro et al., 2008; 2011). However, an inversion of what was previously mentioned was also observed: a more notorious and significant ($P < 0.05$) effect between the different treatments was observed for firmness from the flesh side of peppers (Castro et al., 2008), with the exception of yellow peppers. A decrease of about 27% and 33% on firmness (Figure 8.5), measured from the flesh side, was observed for frozen, unprocessed green and red peppers, respectively, after thawing, while firmness quantified from the skin side was not affected (Castro et al., 2008). Moreover, freezing also caused a higher decrease on firmness, for the thermally blanched green and red peppers, when it was measured from the flesh side, indicating, possibly, the occurrence of

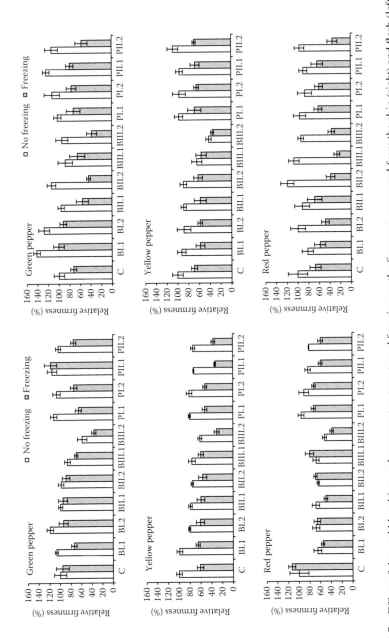

FIGURE 8.5 Effect of thermal blanching and pressure treatments and freezing on the firmness, measured from the skin (right) and flesh (left) side of peppers. The bars represent the standard deviation ($n = 5$). The meaning of the abbreviations used in the abscissa axis is indicated in Table 8.1.

more detrimental effects on pepper flesh cells, caused by the treatments. Globally, thawed green and red peppers that were pressure treated showed a similar to better texture, compared to the thermally blanched peppers. For yellow peppers, processed samples showed similar relative decreases in firmness, with the exception of peppers processed at 200 MPa and when texture was measured from the skin side, for which the decrease in firmness was higher and comparable to that found for peppers blanched at 98°C for 150 s.

8.6 FINAL CONCLUSIONS

Pressure treatments applied to peppers (100 and 200 MPa, 10 and 20 min) caused a lower reduction on ascorbic acid contents than thermal blanching, particularly for yellow and red peppers, which showed an increase in the amount of ascorbic acid content when compared to the unprocessed peppers.

In general, pressure-treated peppers showed a level of residual PPO activity similar to that of thermally blanched peppers, except for the most severe blanching treatments. PPO of green peppers showed lower pressure and temperature stability than PPO of yellow and red peppers. Peroxidase was more stable to pressure than to temperature, showing a lower stability towards blanching treatments than PPO and PME. PME activity was only detected in green peppers and its activity declined progressively, as blanching temperature and time increased, while pressure treatments caused a slight increase in its activity. No PG was detected in fresh and processed samples. Firmness was equal or better retained in pressure-treated peppers than in thermally blanched peppers, before and after freezing, with red peppers showing higher sensitivity to lose firmness.

Globally, the quality parameters of peppers showed a different behavior to thermal blanching and pressure treatments; and pressure treatments of 100 and 200 MPa constitute potential alternatives to the thermal blanching of bell peppers, yielding peppers with higher ascorbic acid content and better firmness.

ACKNOWLEDGEMENTS

The authors acknowledge the Glass and Ceramics Engineering Department (Aveiro University, Portugal) for the use of the high-pressure equipment; António Oliveira and Daniela Marques from Friopesca Refrigeração Aveiro, Lda. (Aveiro, Portugal), and Dr. José A. L. da Silva for the use of the texture analyzer. Thanks are also due to Fundação para a Ciência e a Tecnologia (FCT, Portugal), European Union, QREN, FEDER, and COMPETE for funding the QOPNA research unit (project PEst-C/QUI/UI0062/2011).

REFERENCES

Alonso, J., Canet, W., Rodriguez, T. 1997. Thermal and calcium pretreatment affects texture, pectinesterase and pectic substances of frozen cherries. *Journal of Food Science* 62: 511–515.

Alvarez, M. D., Canet, W., Tortosa, M. E. 2001. Kinetics of thermal softening of potato tissue (cv. Monalisa) by water heating. *European Food Research and Technology* 212: 588–596.

Amati, A., Castellari, M., Matricardi, L. et al. 1996. Modificazione indotte in mosti d'uva da trattamenti con alte pressione idrostatische (Effects of high pressure on grape musts composition). *Industrie delle Bevande* 25: 324–328.

An, S. H., Sohn, K. H., Choi, H. W. et al. 2008. Pepper pectin methylesterase inhibitor protein CaPMEI1 is required for antifungal activity, basal disease resistance and abiotic stress tolerance. *Planta* 228: 61–78.

Bahçeci, S. K., Serpen, A., Gökmen, V. et al. 2004. Study of lipoxygenase and peroxidase as indicator enzymes in green beans: Change of enzyme activity, ascorbic acid and chlorophylls during frozen storage. *Journal of Food Engineering* 66: 187–192.

Bailey, L. A. and Bailey, E. Z. 1976. *Hortus Third: A Concise Dictionary of Plants Cultivated in the United States and Canada, 322, 673.* New York: MacMillan Publishing Co.

Barbagallo R. N., Chisari M., Spagna, G. 2009. Enzymatic browning and softening in vegetable crops: Studies and experiences. *Italian Journal of Food Science* 21: 3–16.

Barrett, D. M. and Theerakulkait, C. 1995. Quality indicators in blanched, frozen, stored vegetables. *Food Technology* 49: 62, 64–65.

Barth, M. M., Zhuang, H., Salveit, M. E. 2002. Fresh-cut vegetables. In *The Commercial Storage of Fruits, Vegetables and Florist and Nursery Stocks*, K. C. Gross, C. Y. Wang and M. Salveit (eds), 50–72. Washington, DC: US Department of Agriculture.

Bartley, I. M. and Knee, M. 1982. The chemistry of textural changes in fruit during storage. *Food Chemistry* 9: 47–58.

Basak, S. and Ramaswamy, H. S. 1998. Effect of high pressure processing on the texture of selected fruits and vegetables. *Journal of Texture Studies* 29: 587–601.

Bernal M. A., Calderón A. A., Ferrer M. A. et al. 1995. Oxidation of capsaicin and capsaicin phenolic precursors by the basic peroxidase isoenzyme B6 from hot pepper. *Journal of Agriculture and Food Chemistry* 43: 352–355.

Bernal, M. A., Calderón, A. A., Pedreño, M. A. et al. 1994. Purification of a basic peroxidase isoenzyme from *Capsicum* fruits and the immunoinhibition of its capsaicin oxidation capacity by antibodies raised against horseradish peroxidase. *Zeitschrift für Lebensmittel-Untersuchung und –Forschung* 199: 240–242.

Bernal, A. M., Calderon, A. A., Pedreño, M. A. et al. 1993c. Capsaicin oxidation by peroxidase from *Capsicum annuum* (var. *annuum*) fruits. *Journal of Agriculture and Food Chemistry* 41: 1041–1044.

Bernal, A. M., Calderon, A. A., Pedreño, M. A. et al. 1993b. Dihydrocapsaicin oxidation by *Capsicum annuum* (var. annuum) peroxidase. *Journal of Agriculture and Food Chemistry* 58: 611–613, 679.

Bernal, M. A., Pedreño, M. A., Calderon, A. A. et al. 1993a. The subcellular localization of isoperoxides in *Capsicum annuum* leaves and their different expression in vegetative and flowered plants. *Annals of Botany* 72: 415–421.

Bosland, P. W. 1992. Chiles: A diverse crop. *HortTechnology* 2: 6–10.

Bosland, P. W. 1994. Chiles: History, cultivation, and uses. In: *Spices, Herbs and Edible Fungi (Herbs)*, G. Charalambous (ed.), 347–366. Amsterdam: Elsevier Science Publishers.

Bunnell, R. H. and Bauernfeind, J. C. 1962. Chemistry, uses, and properties of carotenoids in foods. *Food Technology* 16: 36–43.

Canet, W., Alvarez M. D., Luna, P. et al. 2005. Blanching effects on chemistry, quality and structure of green beans (cv. Moncayo). *European Food Research and Technology* 220: 421–30.

Cano, M. P. 1996. Vegetables. In: *Freezing Effects on Food Quality*, L. E. Jeremiah (ed.), 247–298. New York: Marcel Dekker.

Cano, M. P., Hernandez, A., De Ancos, B. 1997. High pressure and temperature effects on enzyme inactivation in strawberry and orange products. *Journal of Food Science* 62: 85–88.

Castro, S. M., Saraiva J., Delgadillo I. et al. 2003. Effect of pressure and blanching treatments on endogenous enzymatic activity and vitamin C content on bell pepper fruit. *High-Pressure Research* 23: 97–100.

Castro, S. M., Saraiva, J. A., Domingues, F. M. J. et al. 2011. Effect of mild pressure treatments and thermal blanching on yellow bell peppers (*Capsicum annuum* L.). *Lebensmittel-Wissenschaft und-Technologie* 44: 363–369.

Castro, S. M., Saraiva, J. A., Lopes-da-Silva, J. A. et al. 2008. Effect of thermal blanching and of high pressure treatments on sweet green and red bell pepper fruits (*Capsicum annuum* L.). *Food Chemistry* 107: 1436–1449.

Castro, S. M., Van Loey, A., Saraiva J. et al. 2004. Activity and process stability of purified green pepper (*Capsicum annuum*) pectin methylesterase. *Journal of Agriculture and Food Chemistry* 52: 5724–5729.

Castro, S. M., Van Loey, A., Saraiva J. et al. 2005. Process stability of *Capsicum annuum* pectin methylesterase in model systems, pepper puree and intact pepper tissue. *European Food Research and Technology* 221: 452–458.

Castro, S. M., Van Loey, A., Saraiva J. et al. 2006. Identification of pressure/temperature combinations for optimal pepper (*Capsicum annuum*) pectin methylesterase activity. *Enzyme and Microbial Technology* 38: 831–838.

Chang, C. V., Lai, L. R., Chang, W. H. 1995. Relationships between textural changes and the changes in the linkages of pectic substances of sweet pepper during cooking processes, and the applicability of the models of interactions between pectin molecules. *Food Chemistry* 53: 409–416.

Chitwood, R. L., Pangborn, R. M., Jenning, W. 1983. GC/MS and sensory analysis of volatiles from three cultivars of capsicum. *Food Chemistry* 11: 201–216.

Civello, P. M., Martínez, G. A., Chaves, A. R. et al. 1995. Peroxidase from strawberry fruit (*Fragaria ananassa* Dutch): Partial purification and determination of some properties. *Journal of Agriculture and Food Chemistry* 43: 2596–2601.

Concellón, A., Añon, M. C., Chaves, A. R. 2004. Characterization and changes in polyphenol oxidase from eggplant fruit (*Solanum melongena* L.) during storage at low temperature. *Food Chemistry* 88: 17–24.

Conforti, F., Statti, G. A., Manichini, F. 2007. Chemical and biological variability of hot pepper fruits (*Capsicum annuum* var. *acuminatum* L.) in relation to maturity stage. *Food Chemistry* 102: 1096–1104.

Contreras-Padilla, M. and Yahia, E. M. 1998. Changes in Capsaicinoids during development, maturation, and senescence of Chile peppers and relation with peroxidase activity. *Journal of Agriculture and Food Chemistry* 46: 2075–2079.

Daood, H. G., Biacs, P. A., Dakar, M. A. et al. 1994. Ion pair chromatography and photodiode-array detection of vitamin C and organic acids. *Journal Chromatography Science* 12: 481–487.

Daood, H. G. and Biacs, P. A. 1986. Evidence for the presence of lipoxygenase and hydroperoxide-decomposing enzyme in red pepper seeds. *Acta Alimentaria* 15: 307–318.

Daood, H. G., Vinkler, M., Markus, F. et al. 1996. Antioxidant vitamin content of spice red pepper (paprika) as affected by technological and varietal factors. *Food Chemistry* 55: 365–372.

Domínguez, R., Quintero-Ramos, A., Bourne, M. et al. 2001. Texture of rehydrated dried bell peppers modified by low-temperature blanching and calcium addition. *International Journal of Food Science and Technology* 36: 523–527.

Dörnenburg, H. and Knorr, D. 1998. Monitoring the impact of high-pressure processing on the biosynthesis of plant metabolites using plant cell cultures. *Trends in Food Science and Technology* 9: 355–361.

Estrada, B., Bernal, M. A., Díaz, J. et al. 2000. Fruit development in *Capsicum annuum*: Changes in capsaicin, lignin, free phenolics, and peroxidase patterns. *Journal of Agriculture and Food Chemistry* 48: 6234–6239.

Fischer, R. L. and Bennett, A. B. 1991. Role of cell wall hydrolases in fruit ripening. *Annual Review of Plant Physiology and Plant Molecular Biology* 42: 675–703.

Fuchigami, M., Miyazaki, K. Hyakumoto, N. 1995. Frozen carrots texture and pectic components as affected by low-temperature-blanching and quick freezing. *Journal of Food Science* 60: 132–136.

Giovane, A., Servillo, L., Balestrieri, C. et al. 2004. Pectin methylesterase inhibitor. *Biochimica and Biophysica Acta* 1696: 245–252.

Gomes, M. R. and Ledward, D. A. 1996. Effect of high-pressure treatment on the activity of some polyphenol oxidases. *Food Chemistry* 56: 1–5.

Govindarajan, V. S. 1985. *Capsicum* production, technology, chemistry, and quality. Part I. History, botany, cultivation and primary processing. *Critical Reviews in Food Science and Nutrition* 22: 109–175.

Govindarajan, V. S., Rajalakshmi, D., Chand, N. 1987. Capsicum production, technology, chemistry, and quality. Part IV. Evaluation of quality. *Critical Reviews in Food Science and Nutrition* 25: 185–283.

Gross, K. C. 1982. A rapid and sensitive spectrophotometric method for assaying polygalacturonase using 2-cyanoacetamide. *HortScience* 17: 933–934.

Gross, K. C., Watada, A. E., Kang, M. S. et al. 1986. Biochemical changes associated with the ripening of hot peppers fruit. *Physiologia Plantarum* 66: 31–36.

Gu, Y. S., Howard, L. R., Wagner, A. B. 1999. Firmness and cell wall characteristics of pasteurised jalapeno pepper rings as affected by calcium chloride and rotary processing. *Journal of Food Science* 64: 494–497.

Guil-Guerrero, J. L., Martínez-Guirado, C., Rebolloso-Fuentes, M. et al. 2006. Nutrient composition and antioxidant activity of 10 pepper (*Capsicum annuum*) varieties. *European Food Research and Technology* 224: 1–9.

Hagerman, A. E. and Austin, P. J. 1986. Continuous spectrophotometric assay for plant pectin methyl esterase. *Journal of Agriculture and Food Chemistry* 34: 440–444.

Hernández, A. and Cano, M. P. 1998. High pressure and temperature effects on enzyme inactivation in tomato puree. *Journal of Agriculture and Food Chemistry* 46: 266–270.

Hite, B. H. 1899. The effect of pressure in the preservation of milk. *West Virginia University Agricultural Experiment Station Bulletin* 58: 15–35.

Hite, B. H., Giddings, N. J., Weakley Jr., C. E. 1914. The effect of pressure on certain microorganisms encountered in the preservation of fruits and vegetables. *West Virginia University Agricultural Experiment Station Bulletin* 146: 3–67.

Hoffman, P. G., Lego, M. C., Galetto, W. G. 1983. Separation and quantitation of red pepper major heat principles by reverse-phase high pressure liquid chromatography. *Journal of Agriculture and Food Chemistry* 31: 1326–1330.

Hoogzand, C. and Doesburg, J. J. 1961. Effect of blanching on texture and pectin canned cauliflower. *Journal of Food Technology* 15: 160–163.

Howard, L. R., Smith, R. T., Wagner, A. B. et al. 1994. Provitamin A and ascorbic acid content of fresh pepper cultivars (*Capsicum annuum*) and processed Jalapeños. *Journal of Food Science* 59: 362–365.

Howard, L. R., Talcott, S. T., Brenes, C. H. et al. 2000. Changes in phytochemical and antioxidant activity of selected pepper cultivars (*Capsicum* species) as influenced by maturity. *Journal of Agriculture and Food Chemistry* 48: 1713–1720.

Hsu, C. P., Deshpande, S. N., Desrosier, N. W. 1964. Role of pectin methylesterase in firmness of canned tomatoes. *Journal of Food Science* 30: 583–588.

Ihl, M., Monslaves, M., Bifani, V. 1998. Chlorophyllase inactivation as a measure of blanching efficacy and colour retention of Artichokes (*Cynara scolymus* L.). *Lebensmittel-Wissenschaft und-Technologie* 31: 50–56.

Jagadeesh, B. H. and Prabha, T. N. 2002. β-hexosaminidase, an enzyme from ripening bell pepper capsicum (*Capsicum annuum* var. *variata*). *Phytochemistry* 61: 295–300.

Jarén-Galán, M. and Mínguez-Mosquera, M. I. 1999. Quantitative and qualitative changes associated with heat treatments in carotenoids content of paprika oleoresins. *Journal of Agriculture and Food Chemistry* 47: 4379–4383.

Jen, J. J. and Robinson, M. L. 1984. Pectolytic enzymes in sweet bell peppers (*Capsicum annuum* L.). *Journal of Food Science* 49: 1085–1087.

Jiang, Y., Duan, X., Joyce, D. et al. 2004. Advance in understanding of enzymatic browning in harvested litchi fruit. *Journal of Food Chemistry* 88: 443–446.

Jolibert, F., Tonello, C., Sagegh, P., et al. 1994. Les effets des hautes pressions sur la polyphénol oxydase des fruits. *Bioscience et Boissons* 251: 27–35.

Kays, S. J. 1997. *Postharvest Physiology of Perishable Plant Product*. Georgia: Exon Press.

Keijbets, M. J. H., Pilnik, W. 1974. β-Elimination of pectin in the presence of anions and cations. *Carbohydrates Research* 33: 359–362.

Knorr, D. 1993. Effects of high-hydrostatic pressure processes on food safety and quality. *Food Technology*, 47: 156–161.

Krebbers, B., Matser, A. M., Koets, M. et al. 2002. Quality and storage-stability of high-pressure preserved green beans. *Journal of Food Engineering* 54: 27–33.

Lee, S. K. and Kader, A. A. 2000. Preharvest and postharvet factors influencing vitamin C content of horticultural crops. *Postharvest Biology and Technology* 20: 207–220.

Lee, Y., Howard, L., R., Villalon, B. 1995. Flavonoids and antioxidant activity of fresh pepper (*Capsicum annuum*) cultivars. *Journal of Food Science* 60: 473–476.

Lee, Y. and Howard, L. 1999. Firmness and phytochemical losses in pasteurized yellow banana peppers (*Capsicum annuum*) as affected by calcium chloride and storage. *Journal of Agriculture and Food Chemistry* 47: 700–703.

Lee, C. Y., Downing, D. L., Iredale, H. D. et al. 1976. The variations of ascorbic acid content in vegetable processing. *Food Chemistry* 1: 15–22.

Ludikhuyze, L., Van Loey, A., Indrawati, et al. 2003. Effects of combined pressure and temperature on enzymes related to quality of fruits and vegetables: From kinetic information to process engineering aspects. *Critical Reviews in Food Science and Nutrition* 43: 527–586.

Luning, P., Carey, A. T., Roozen, J. P. et al. 1995. Characterization and occurrence of lipoxygenase in bell peppers at different ripening stages in relation to the formation of volatile flavor compounds. *Journal of Agriculture and Food Chemistry* 43: 1493–1500.

Luning, P., de Rijk, T., Wichers, H. J. et al. 1994c. Gas chromatography, mass spectrometry and sniffing port analysis of volatile compounds of fresh bell peppers (*Capsicum annuum*) at different ripening stages. *Journal of Agriculture and Food Chemistry* 42: 977–983.

Luning, P., Vries, R. V., Yuksel, D. et al. 1994b. Combined instrumental and sensory evaluation of flavour of fresh bell peppers (*Capsicum annuum*) harvested at three different maturation stages. *Journal of Agriculture and Food Chemistry* 42: 2855–2861.

Luning, P., Yuksel, D., Roozen, J. P. 1994a. Sensory attributes of bell peppers (*Capsicum annuum*) correlated with composition of volatile compounds. In: *Trends in Flavour Research*, H. Maarse, and D. G. van der Heij (ed.), 241–248. Amsterdam: Elsevier Science Publishers.

Luning, P. A. 1995. Characterisation of the flavour of fresh bell peppers and its changes after hot-air drying, and instrumental and sensory evaluation. PhD Thesis, Agrotechnological Research Institute ATO-DLO, Wageningen, The Netherlands.

Matthews, R. F. and Hall, J. W. 1978. Ascorbic acid, dehydroascorbic acid and diketogulonic acid in frozen green peppers. *Journal of Food Science* 43: 32–534.

Mejia, L. A., Hudson, E., Gonzalez de Mejia, E. et al. 1988. Carotenoid content and vitamin A activity of some common cultivars of Mexican peppers (*Capsicum annuum*) as determined by HPLC. *Journal of Food Science* 53: 1448–1451.

Mínguez-Mosquera, M. I. and Hornero-Mendez, D. 1994. Formation and transformation of pigments during the fruit ripening of *Capsicum annuum* cv Bola and Agridulce. *Journal of Agriculture and Food Chemistry* 42: 38–44.

Mínguez-Mosquera, M. I., Jarén-Galán, M., Garrido-Fernandéz, J. 1993. Lipoxygenase activity during pepper ripening and processing of paprika. *Phytochemistry* 32: 1103–1108.

Mínguez-Mosquera, M. I., Pérez-Gálvez, A., Garrido-Fernández, J. 2000. Carotenoid content of the varieties *Jaranda* and *Jariza* (*Capsicum annuum* L.) and response during the industrial slow drying and grinding steps in paprika processing. *Journal of Agriculture and Food Chemistry* 48: 2972–2976.

Ng, A., Harvey, A. J., Parker, M. L. et al. 1998. Effect of oxidative coupling on the thermal stability of texture and cell wall chemistry of beet root (*Beta vulgaris*). *Journal of Agriculture and Food Chemistry* 46: 3365–3370.

Ng, A. and Waldron, K. W. 1997. Effect of cooking and pre-cooking on cell-wall chemistry in relation to firmness of carrot tissues. *Journal of Science and Food Agriculture* 73: 503–512.

Ni, L., Lin, D., Barrett, D. M. 2005. Pectin methylesterase catalyzed firming effects on low-temperature blanched vegetables. *Journal of Food Engineering* 70: 546–56.

Odriozola-Serrano, I., Soliva-Fortuny, R., Martín-Belloso, O. 2008. Phenolic acids, flavonoids, vitamin C and antioxidant capacity of strawberry juices processed by high-intensity pulsed electric fields or heat treatments. *European Food Research Technology* 228: 239–248.

Oey, I., Verlinde, P., Hendrickx, M. et al. 2006. Temperature and pressure stability of L-ascorbic acid and/or [6s] 5-methyltetrahydrofolic acid: A kinetic study. *European Journal of Food Technology* 223: 71–77.

Ogasawara, S., Abe, K., Nakajima, T. 2007. Pepper β-galactosidase 1 (PBG1) plays a significant role in fruit ripening in bell pepper (*Capsicum annuum*). *Bioscience and Biotechnology Biochemistry* 71: 309–322.

Osuna-Garcia, J. A., Wall, M. M., Waddell, C. A. 1998. Endogenous levels of tocopherols and ascorbic acid during fruit ripening of New Mexican-Type Chile. *Journal of Agriculture and Food Chemistry* 46: 5093–5096.

Paull, R. E. 1990. Chilling injury of crops of tropical and subtropical origin. In *Chilling injury of horticultural crops*, C. Y. Wang (ed.), 17–36. Boca Raton: CRC Press.

Pinski, A., Grossman, S., Trop, M. 1971. Lipoxygenase content and antioxidant activity in some fruits and vegetables. *Journal of Food Science* 36: 571–572.

Pomar, F., Bernal, M. A., Díaz, J. et al. 1997. Purification, characterization and kinetic properties of pepper fruit acidic peroxidase. *Phytochemistry* 46: 1313–1317.

Prabha, T., Neelwarne, B., Tharanathan, R. 1998. Carbohydrate changes in ripening *Capsicum annuum* in relation to textural degradation. *European Food Research and Technology* 206: 121–125.

Préstamo, G., Fuster, C., Risueno, M. C. 1998. Effect of blanching and freezing on the structure of carrot cells and their implications for food processing. *Journal of Science Food and Agriculture* 77: 223–229.

Priem, B., Gitti, R., Bush, C. A. et al. 1993. Structure of ten free *N*-glycans in ripening tomato fruit. *Plant Physiology* 102: 445–458.

Quaglia, G. B., Gravina, R., Paperi, R. et al. 1996. Effect of high pressure treatments on peroxidase activity, ascorbic acid content and texture in green peas. *Lebensmittel-Wissenschaft und-Technologie* 29: 552–555.

Ramakrishnan, T. V. and Francis, F. J. 1973. Color and carotenoid changes in heated paprika. *Journal of Food Science* 38: 25–28.

Ramírez, E. C., Whitaker, J. R., Virador, V. M. 2002. Polyphenol oxidase. In *Handbook of Food Enzymology*, J. R. Whitaker, A. G. J. Voragen, D. W. S. Wong (eds), 50–523. New York: Marcel Dekker, Inc.

Rao, G. U. and Paran, I. 2003. Polygalacturonase: A candidate gene for the soft flesh and deciduous fruit mutation in Capsicum. *Plant Molecular Biology* 51: 135–141.

Rao, M. A., Lee, C. Y., Katz, J. et al. 1981. Kinetic study of the loss of vitamin C, color, and firmness during thermal processing of canned peas. *Journal of Food Science* 46: 636–637.

Rastogi, N. K., Raghavarao, K. S. M. S., Balasubramaniam, V. M. et al. 2007. Opportunities and challenges in high pressure processing of foods. *Critical Reviews in Food Science and Nutrition* 47: 69–112.

Rhaman F. M. M., Buckle K. A., Edwards, R. A. 2007. Changes in total solids, ascorbic acid and total pigment content of capsicum cultivars during maturation and ripening. *International Journal of Food Science and Technology* 13: 445–450.

Roig, M. G., Rivera, Z. S., Kennedy, J. F. 1995. A model study on rate of degradation of L-ascorbic acid during processing using home-produced juice concentrates. *International Journal of Food Sciences and Nutrition* 46: 107–115.

Salunkhe, B. B. and Desay, D. K. 1984. Pepper (Capsicum sp.). In *Post-harvest Biotechnology of Vegetables*, B. B. Salunkhe and D. K. Desay (eds), 49–58. Boca Raton: CRC Press, Inc.

Sancho, F., Lambert, Y., Demazeau, G. et al. 1999. Effect of ultra-high hydrostatic pressure on hydrosoluble vitamins. *Journal of Food Engineering* 39: 247–253.

Selman, J. D. 1994. Vitamin retention during blanching of vegetables. *Food Chemistry* 49: 137–147.

Scuderi, D., Restuccia, C., Chisari, M. et al. 2011. Salinity of nutrient solution influences the shelf-life of fresh-cut lettuce grown in floating system. *Postharvest Biology Technology* 59: 132–137.

Senesi, E., Prinzivalli, C., Sala, M. et al. 2000. Physicochemical and microbiological changes in fresh-cut green bell peppers as affected by packing and storage. *Italian Journal of Food Science* 12: 55–64.

Sethu, K. M. P., Prabha, T. N., Tharanathan, R. N. 1996. Post-harvest biochemical changes associated with the softening phenomenon in *Capsicum annuum* fruits. *Phytochemistry* 42: 961–966.

Sgroppo, S. C., Fusco, A. J. V., Avanza, J. R. 2001. Influencia de la temperatura de secado de pimientos en trozos. *Informacion Tecnologica* 12: 53–58.

Sherman, M., Kasmire, R. F., Shuler, K. D. et al. 1982. Effect of precooling methods and peduncle lengths on soft rot decay of bell peppers. *Horticultural Science* 17: 251–252.

Shook, C. M., Shelhammer, T. H., Schwartz, S. J. 2001. Polygalacturonase, pectinemethylesterase, and lipoxygenase activities in high-pressure processed diced tomatoes. *Journal of Agriculture and Food Chemistry* 49: 664–668.

Sila, D. N., Smout, C., Vu T. S. et al. 2004. Effects of high-pressure pretreatment and calcium soaking on the texture degradation kinetics of carrots during thermal processing. *Journal of Food Science* 69: 205–11.

Silva, E., Lourenço, E. J., Neves, V. A. 1990. Soluble and bound peroxidase from papaya fruit. *Phytochemistry* 29: 1051–1056.

Simmone, A. H., Simmone, E. H., Eitenmiller, R. R. et al. 1997. Ascorbic acid and provitamin A contents in some unusually colored bell peppers. *Journal of Food Composition and Analysis* 10: 299–311.

Smith P. G., Villalon, B., Villa, P. L. 1987. Horticultural classification of peppers grown in the United States. *HortScience* 22: 11–13.

Stanley, D. W., Bourne, M. C., Stone, A. P. et al. 1995. Low-temperature blanching effects on chemistry, firmness and structure of canned green beans and carrots. *Journal of Food Science* 60: 327–333.

Stolle-Smits, T., Donkers, J., van Dijk, C. et al. 1998. An electron microscopy study on the texture of fresh, Blanched and sterilized green bean pods (*Phaseolus vulgaris* L.). *Lebensmittel-Wissenschaft und-Technologie* 31: 237–244.

Stute, R., Eshtiaghi, M. N., Boguslawaski, S. et al. 1996. High pressure treatments of vegetables. In: *High Pressure Chemical Engineering, Process Technology Proceedings*, P. R. Rohrvon and C. Trepp (eds), 271–276. Amsterdam: Elsevier Science.

Thomas, R. L., Jen, J. J., Morr, C. V. 1981. Changes in soluble and bound peroxidase-IAA oxidase during tomato fruit development. *Journal of Food Science* 47: 158–61.

Urrutia-Benet, G., Schlüter, O., Knorr, D. 2004. High pressure–low temperature processing. Suggested definitions and terminology. *Innovative Food Science and Emerging Technologies* 5: 413–427.

van Buren, J. P. 1979. The chemistry of texture in fruits and vegetables. *Journal of Texture Studies* 10: 1–23.

van Ruth, S. M. and Roozen, J. P. 1994. Gas chromatography/sniffing port analysis and sensory evaluation of commercially dried bell peppers (*Capsicum annuum*) after rehydration. *Food Chemistry* 51: 165–170.

Villarreal-Alba, E. G., Contreras-Esquivel, J. C., Aguilar-Gonzalez, C. N. et al. 2004. Pectinesterase activity and texture of jalapeno peppers. *European Food Research and Technology* 218: 164–166.

Vu, T. S., Smout, C., Sila, D. N. et al. 2004. Effect of preheating on thermal degradation kinetics of carrot texture. *Innovative Food Science and Emerging Technologies* 5: 37–44.

Watada, A. E., Ko, N. P., Minott, D. A. 1996. Factors affecting quality of fresh-cut horticultural products. *Postharvest Biology and Technology* 9: 115–125.

Whitaker, J. R. 1984. Pectic substances, pectic enzymes and haze formation in fruit juices. *Enzyme Microbial Technology* 6: 341–9.

Worthington Biochemical Corporation. 1978. *Enzymes and Related Biochemicals*, Millipore Corporation, Bedford, MA, USA.

Wu, A. and Chang, H. 1990. Influence of precooking on the firmness and pectic substances of three stem vegetables. *International Journal of Food Science and Technology* 25: 558–565.

Wu, C.-M., Liou, S.-E., Wang, M.-C. 1986. Changes in volatile constituents of bell peppers immediately and 30 minutes after stir frying. *Journal of the American Oil Chemists' Society* 63: 1172–1175.

Yahia, E. M., Contreras-Padilla, M., Gonzalez-Aguilar, G. 2001. Ascorbic acid content in relation to ascorbic acid oxidase activity and polyamine content in tomato and bell pepper fruits during development, maturation and senescence. *Lebensmittel-Wissenschaft und-Technologie* 34: 452–457.

Zhang, D. and Chen, M. 2006. Effects of low temperature soaking on color and texture of green eggplants. *Journal of Food Engineering* 74: 54–59.

9 High-Pressure Processing for Improving Digestibility of Cooked Sorghum Protein

*Jorge Alexandre Saraiva, Ivonne Delgadillo,
Alexandre Nunes, and Ana Isabel Loureiro Correia*

CONTENTS

9.1 INTRODUCTION

High-pressure treatments applied in sorghum porridges avoid/revert the decrease of sorghum protein digestibility. In fact, cooking deleterious effects on sorghum protein digestibility are avoided or reverted when pressure is applied either before or after cooking suspensions of sorghum flour in water, although best results are obtained when pressure is applied before the cooking process. Digestibility of cooked sorghum proteins increases from 16.1% to 35.3% or 25.4% when pressure at the level of 300 MPa was applied 15 min before/after cooking, respectively. When 300 MPa were applied for 5 min before cooking, similar results were obtained for digestibility (36.0%).

Analysis of sodium dodecyl sulfate polyacrylamide gel electrophoresis (SDS-PAGE) of sorghum prolamins, revealed that high-molecular-weight aggregates and a 45 kDa dimer, which usually increase with cooking and are related to protein digestibility decrease, did not significantly change when high pressure was applied.

A relationship between infra-red (IR) spectra and protein digestibility by means of a partial least-square (PLS1) regression was assessed, showing changes in proteins and also on lipids and starch.

Pressurization of sorghum flour, before or after cooking, particularly the former, is a suitable process to greatly improve cooked sorghum protein digestibility.

9.2 PROBLEM OF REDUCED DIGESTIBILITY OF SORGHUM PROTEIN

Sorghum (*Sorghum bicolor* [L.] Moench) is one of the most important crops in Africa, Asia, and Latin America (Dicko et al., 2006). Sorghum, together with millet, is a major staple food for around 60 million people concentrated in the inland areas of tropical Africa, for who, it is an important source of proteins (FAO, 2004). A study carried out with urban residents of Polokwane (South Africa), revealed that while sorghum is easily purchasable and widely consumed, increasing consumer demand for sorghum-based products (namely for breakfast meals and for inclusion in infants' cereals used as weaning foods) is occurring (Bichard et al., 2005).

Among sorghum proteins, kafirins (sorghum prolamins) are the most abundant, making up 70–80% of the total endosperm protein. α-Kafirin, which comprises 80% of the total kafirins, is located in light-staining areas, while β- and γ-kafirins are found in dark-staining areas; the former is found in the interior and the latter is found inside and at the edge of the protein bodies (Shull et al., 1992).

Generally, sorghum cultivars present low protein digestibility when compared to other cereals (Hamaker et al., 1986; MacLean et al., 1981). Sorghum is unique among the plant food proteins in that it becomes obviously less digestible after cooking (Axtell et al., 1981; Eggum et al., 1983; Mitaru et al., 1985; Oria et al., 1995). In fact, porridges protein digestibility is lower than raw sorghum grain protein digestibility (Weaver et al., 1998).

Processing methods applied to prepare a wide array of sorghum-based foods usually involve cooking (Murty and Kumar, 1995). These processing methods promote modifications on the protein fraction, considerably reducing its digestibility and the nutritional value of the foods so prepared (Hamaker et al., 1986; MacLean et al., 1981).

During decades, *in vivo* experiments were developed that demonstrated the reduction of the sorghum protein digestibility when cooked. Kurien et al. (1960) found that rice proteins fed to seven 11–12-years-old boys were 75% digestible; if rice was substituted by sorghum, the digestibility would lower to 55%. Similar results were found by Daniel et al. (1966) in young girls. MacLean et al. (1981) fed two normal and two high-lysine sorghum to children who were 6–24-months old, and the apparent digestibility value of cooked sorghum gruels was 46%, compared with 81% for wheat, 73% for maize, and 66% for rice. Posterior studies with malnourished Peruvian children fed with cooked cereal porridges showed values of apparent protein digestibility that confirm the results presented above (MacLean et al., 1983). Compared to the casein diet control, sorghum protein was only 57% digestible (Weaver et al., 1998). To confirm these findings, other animals were also used to develop other *in vivo* studies. In nonruminant farm animals, protein digestibility of sorghum is slightly less when compared with other cereals. Rats fed with low-tannin sorghum that was cooked to gruel digested 7% less protein than those fed with the uncooked flour (Eggum et al., 1983). Mitaru and Blair (1984) also developed a study with rats; they reported decreases in protein digestibility of cooked sorghum compared with uncooked sorghum. Mitaru et al. (1985) found that chickens digested 31.5% less protein from cooked, whole-grain, low-tannin sorghum than from the uncooked grain. In swine, protein digestibility is very similar to the protein digestibility of one variety of maize (Rooney, 1990).

Fortunately, *in vitro* digestibility studies were developed and routinely implemented and *in vivo* experiments stopped.

Chibber et al. (1980) tested an *in vitro* pepsin method to determine protein digestibility values. This *in vitro* protein digestibility method followed some essential steps that other researchers applied with some minor modifications in time of digestibility, extraction time, and type of enzyme used (Aboubacar et al., 2001; Axtell et al., 1981; Chibber et al., 1980; Dahlin and Lorenz 1993; Duodu et al., 2002a, b; Elkin et al., 1996; Hamaker et al., 1986, 1987; Mertz et al., 1984; Oria et al., 1995; Rom et al., 1992; Weaver et al., 1998). This type of protein digestion procedure has been largely applied and the results obtained with this method were quite similar to the results obtained in *in vivo* digestibility studies.

Hamaker et al. (1987) studies reported that protein digestibility values decreased by 22.5% with cooking. By comparison, cooked maize flour was either digestible or slightly more digestible than the uncooked flour. Rom et al. (1992) had determined digestibility values in both cooked and uncooked sorghum flour. According to this study, after 5 min of *in vitro* pepsin digestibility of uncooked sorghum, it was 37% and 79% after 120 min. Conversely, cooked sorghum was only 18% digested after 5 min and 58% digested after 120 min. Oria et al. (1995) showed that *in vitro* protein digestibility decreases from 69.2% to 43.6% with cooking.

Elkin et al. (1996) introduced a major improvement into the digestibility method. After the *in vitro* digestion process they studied, by SDS-PAGE, the nondigested proteins of the resulting residue allowing to identify the proteins that resist enzyme digestion.

As sorghum is the main staple food for the world's poorest people and is consumed in different traditional forms in various geographic areas (Eggum et al., 1982), avoiding or reducing these deleterious cooking effects would be of great importance to produce cooked sorghum-based foods, with improved protein nutritional value.

9.3   EFFECTS INDUCED BY DIFFERENT PROCESSING METHODS ON SORGHUM PROTEINS

Sorghum is processed into a wide array of foods. Processing methods applied to the preparation of sorghum products usually involve wet or dry heat (Murty and Kumar, 1995), germination, and fermentation (Gadaga et al., 1999). These processing methods promote modifications in sorghum protein structure, affecting its nutritional quality.

It has been shown that, when sorghum is submitted to cooking some changes occur in the protein fraction, reducing its digestibility (Nunes et al., 2004). Some authors suggest that cooking promotes the formation of disulfide-bonded protein polymers, leading to a change in protein structure (Hamaker et al., 1986). This is believed to induce the reduction of protein digestibility in cooked sorghum products.

As opposed to wet cooking, processing methods such as germination and fermentation alter protein structures to make it more digestible and bioavailable.

Germination causes activation of intrinsic amylases, proteases, phytases, and fiber-degrading enzymes, thereby increasing nutrient digestibility (Taylor et al., 1985). According to Elkhalil et al. (2001), the activity of intrinsic proteases in germinated seeds leads to an increase in *in vitro* protein digestibility. Several indigenous fermented foods prepared from sorghum are common in India and many parts of Africa. Taylor and Taylor (2002) found that sorghum fermentation increases insoluble protein digestibility, suggesting that fermentation causes structural changes in sorghum storage proteins (prolamins and glutelins), making them more accessible to pepsin attack. According to Yousif and El Tinay (2001), this increase in protein digestibility can be attributed to the partial degradation of complex storage proteins into more simple and soluble products.

Popped sorghum, consumed as a snack, is traditionally prepared by dry heating grain in a hot pan or bowl over a steady fire (Murty and Kumar, 1995). Parker et al. (1999) reported that dry thermal treatment in the form of popping does not decrease sorghum protein digestibility, unlike cooked sorghum. These authors suggested that popping is an explosive process that leads to a fragmentation of the cell walls of the vitreous endosperm. This effect appeared to improve the accessibility of the starch and protein components of the endosperm foam to enzymes in the digestive tract that is beneficial for protein digestibility. Duodu et al. (2001) found that the extent of secondary structural changes promoted by popping is lower than that promoted by wet cooking.

Correia et al. (2010) showed that popping, an explosive process that drastically changes starch structure, has no consequences in protein digestibility. Changes in protein structure occur slightly more in popped samples than in cooked samples. In popped samples, proteins may suffer denaturation and, consequently, became less extractable. However, digestive enzymes accessibility is not altered. The same study demonstrated that dry-heated samples practically present the same digestibility values as unprocessed samples and protein extractability was not compromised when this treatment was applied, leading the authors to conclude that when water is used in heating process, with concomitant starch gelatinization, protein digestibility value is compromised.

9.4   EFFECT OF HIGH-PRESSURE TREATMENTS ON COOKED SORGHUM PROTEIN DIGESTIBILITY

High hydrostatic pressure (HP) is an emerging technology that is being increasingly applied to processed foods, being very effective for preservation of foods by pasteurization, as it has little influence on food-quality parameters (Castro et al., 2008). HP is also increasingly studied for the modification of functional properties of foods.

Studies on the effects of HP on seed storage proteins were done only on soybean (Alvarez et al., 2008; Puppoa et al., 2008; Wang et al., 2008), lupine (Chapleau and De Lamballerie-Anton, 2003; De Lamballerie-Anton et al., 2002), and red-kidney-bean protein isolates (Yin et al., 2008) and only a few works were done on the effects of HP on protein digestibility (Elgasin and Kennich, 1980; Klepacka et al., 1996; Yin et al., 2008).

9.4.1 HIGH-PRESSURE PARAMETERS OPTIMIZATION

Whole-sorghum grains ground to pass through a 0.4-mm sieve were mixed with water in a proportion of one part ground flour with 10 parts water. For cooked samples preparation, the mixture was boiled in a water bath for 20 min. Cooked and uncooked samples were subjected to HP treatments, freeze dried, and ground to obtain cooked/HP and uncooked/HP samples, respectively. To study the effect of HP prior to cooking, HP/cooked, a sorghum-uncooked sample was submitted to HP, and afterwards cooked, freeze dried, and ground.

All sorghum samples were vacuum conditioned in heat-sealed polyethylene bags, using a laboratory-scale vacuum- packaging machine. High-pressure treatments were carried out in a HP (Unipress Equipment, Model U33, Poland), with a pressure vessel of 100 mL (35 mm diameter and 100 mm height), surrounded by an external jacket, connected to a thermostatic bath to control the temperature. Prior to the experiments, the temperature of the pressure vessel and the samples were equilibrated to 20.0°C.

High-pressure treatments were conducted first at 300 MPa during 15 min. To observe the effect of different pressure levels and time under high pressure at 300 MPa, another set of experiments was conducted at 100 and 450 MPa for 15 min and at 300 MPa during 5 and 30 min. Compression was done at a rate of about 200 MPa/min and the temperature inside the vessel increased to a maximum of 23°C, due to adiabatic heating when compression ended, but was then kept at 20°C by the external jacket.

9.4.2 EFFECT OF HIGH-PRESSURE TECHNOLOGY ON SORGHUM PROTEIN STRUCTURE

When sorghum flour was cooked, its protein digestibility was reduced from 42.1% to 16.1% (Table 9.1). The decrease in sorghum proteins digestibility promoted by cooking, has been attributed to the formation of disulfide cross-linkages between β-kafirins and γ-kafirins, located on the surface of sorghum protein bodies (Duodu et al., 2002a,b; Emmanbux and Taylor, 2009; Wong et al., 2009).

Uncooked sorghum flour treated at 300 MPa of pressure for 15 min showed a protein digestibility value (43.5%, Table 9.1) similar to that of the uncooked sorghum not submitted to HP. However, the protein digestibility of cooked sorghum, subjected to the same HP treatment before or after cooking, was much higher (Table 9.1), respectively, 35.3% (1.6-fold) and 25.4% (2.2-fold), compared to the cooked sorghum not HP treated (16.1%). These results indicate that the 300 MPa HP treatment applied can largely avoid (applied before cooking) or even revert (applied after cooking), the deleterious effect of cooking on sorghum protein digestibility. A slight increase (around more than 1.1-fold) of digestibility of extracted kafirins was observed when sorghum was pressure cooked (Emmanbux and Taylor, 2009). The higher effect observed when HP was applied prior to cooking can be related to the higher water mobility in the raw sample during the high-pressure treatment. It is known that HP can lead to protein

TABLE 9.1

In Vitro **Protein Digestibility Results of Sorghum Samples Not Processed by Pressure (Uncooked and Cooked) and Processed by Pressure at 300 MPa during 15 min—Uncooked (Uncooked/HP), First Cooked and Then Processed (Cooked/HP), and First Processed and Then Cooked (HP/Cooked)**

Processed Samples	Digestibility Values (%)[a]
Uncooked	$42.1 \pm 0.6a$
Cooked	$16.1 \pm 0.7b$
Uncooked/HP	$43.5 \pm 1.0a$
HP/cooked	$35.3 \pm 1.4c$
Cooked/HP	$25.4 \pm 1.4d$

[a] Mean of three replicates ± standard deviation. Same letters show nonsignificant differences ($p > 0.05$).

unfolding, due to changes on noncovalent bonds and disruption of SS bonds and to the increase of free SH groups (Messens et al., 1997; Wang et al., 2008).

The enhancement on sorghum flour protein digestibility caused by HP can be a result of changes on noncovalent bonds within protein molecules and disruption of SS bonds and increase of free SH groups, leading to better pepsin accessibility. Another work done with lupine also revealed an increase in protein digestibility of lupine flour samples submitted to HP processing (De Lamballerie-Anton et al., 2002).

To try to understand the effect of HP on sorghum proteins that leads to increased protein digestibility, proteins of these samples were extracted and analyzed by SDS-PAGE. The electrophoregram of prolamins from sorghum uncooked flour presented bands that correspond to high molecular weight (HMW) aggregates, a 66 kDa trimer, a 45 kDa dimer, and γ-, α-, β-monomers, with the respective areas presented in Table 9.2 (similar results were obtained by Nunes et al., 2004).

Cooked sorghum flour showed an increase in HMW aggregates and 45 kDa dimer areas, a typical behavior also reported previously by Nunes et al. (2004), which has

TABLE 9.2

Quantification of Each Area of the Electrophoretic Profiles of Sorghum Prolamins

	Uncooked (%)	Cooked (%)	Uncooked/ HP (%)	HP/Cooked (%)	Cooked/ HP (%)
HMW (%)	3.6	6.9	3.4	3.5	3.7
66 (%)	3.7	4.7	2.1	3.0	3.7
45 (%)	34.9	41.9	37.3	36.9	34.3
γ + α (%)	47.0	37.7	45.6	45.0	48.3
β1 + β2 (%)	10.8	9.4	11.6	11.6	10.0

been associated with the decrease in protein digestibility value caused by cooking. Simultaneously, the disappearance of the β2 monomer, already observed before by Nunes (2004), was detected together with a decrease in bands that correspond to the γ + α-monomers (Table 9.2). These results pointed toward the involvement of the β2 monomer and not the β1 monomer in the formation of HMW aggregates and 45 kDa dimers that increased in prolamin fraction of sorghum after cooking.

No significant differences were observed in the electrophoretic profile areas of uncooked sorghum prolamins, compared to pressure-treated uncooked sorghum (Table 9.2), a result that is in accordance with the results of protein digestibility. Also, cooked sorghum samples, submitted to high pressure either before or after cooking, showed no remarkable changes in the electrophoretic prolamins profile, compared to uncooked non-pressure-treated sorghum flour (a result that is also in agreement with the results of protein digestibility). It is of relevance to highlight the great similarity of the electrophoretic area profiles of these three samples. In the presence of high pressure, the HMW aggregates and the 45 kDa dimer, which usually increase with cooking, and the γ + α-monomers, which decrease with cooking, did not significantly change when high pressure was additionally applied to cooked sorghum samples (Table 9.2).

Using the 1800–800 cm^{-1} spectral region preprocessed with standard normal deviate (SNV), three latent variables (LVs) were found to give a calibration model with predictive power (Helland 2001). The cross-validated coefficient of determination (Q^2) was found to be 0.98, the root mean square error of cross-validation (RMSECV) was 1.2%, and the root mean square error of prediction (RMSEP) was 9.9%. The relationships between actual and predicted digestibility values, as well as the corresponding b vector profile, are plotted in Figure 9.1. The b vector profile (Figure 9.1b) showed that the increase in digestibility values of samples was positively related to proteins characteristic bands (1644 and 1536 cm^{-1}). This supports the findings observed by SDS-PAGE that high pressure promotes changes on proteins, leading to an increase in protein digestibility. A shift was also noticed from a triglyceride characteristic peak, at 1743 cm^{-1}, to a free fatty acids peak at 1710 cm^{-1}. Another shift was also obscured, starch's characteristic peak moved from 1032 to 987 cm^{-1}. Several authors reported that high pressure can promote gelatinization of starch granules at room temperature (Blaszczak et al., 2007; Buckow et al., 2007).

Since the most beneficial effect of the 300 MPa pressure treatment on sorghum protein digestibility was observed when HP was applied before cooking, the influence of different pressure levels (100 and 450 MPa during 15 min) and HP treatment times (300 MPa during 5 and 30 min), was studied by application to samples before cooking, on a second set of experiments.

In all cases, a significant increase of protein digestibility was achieved when compared to the cooked sample without HP treatment (16.1%), with values ranging from 27.6% (450 MPa, 15 min) to 36.0% (300 MPa, 5 min). When samples were submitted to a lower pressure (100 MPa), the high-pressure treatment was less efficient on avoiding the decrease of protein digestibility caused by cooking, resulting in a lower digestibility value (27.6%). A similar result was observed when a higher pressure (450 MPa) was applied, resulting in a digestibility value of 24.3% (Table 9.3). A possible explanation for these results, supported by conclusions of other authors (Wang

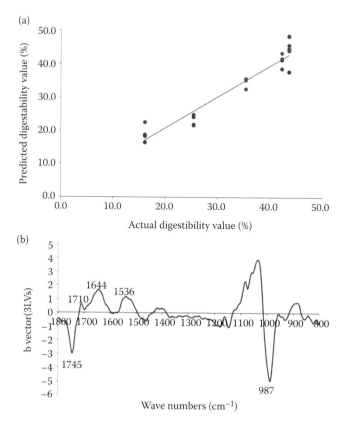

FIGURE 9.1 (a) Relationship between actual and predicted digestibility values and (b) PLS1 regression b vector profile for digestibility prediction.

TABLE 9.3

In Vitro Protein Digestibility Results of Cooked Sorghum Samples Submitted to HP Treatment, at Different Pressure Levels and Times, before Cooking

Processed Samples	Digestibility Values (%)[a]
100 MPa, 15 min	27.6 ± 1.3b
450 MPa, 15 min	24.3 ± 0.8d
300 MPa, 5 min	36.0 ± 0.5c
300 MPa, 30 min	28.7 ± 1.1b

[a] Mean of three replicates ± standard deviation. Same letters show nonsignificant differences ($p > 0.05$).

et al., 2008; Yin et al., 2008), is that up to certain levels of pressure, protein unfolding and disruption of SS bonds occur (which may be responsible to a higher accessibility of pepsin to prolamins). However, at higher-pressure levels, a subsequent aggregation/reassociation of unfolded proteins takes place due to hydrophobic interactions, thus decreasing protein digestibility. The same negative effect seems to occur when duration time of HP treatment is increased to 30 min.

9.5 POSSIBLE BENEFICIAL EFFECTS ON SORGHUM PROTEINS DIGESTIBILITY

High-pressure treatments, applied before/after cooking of sorghum flour, largely avoid/revert deleterious effects promoted by cooking on sorghum proteins digestibility. It was found that the pressurization of sorghum flour at 300 MPa during 5 and 15 min before cooking gives the best results, yielding cooked sorghum samples with protein digestibility values of 36.0% and 35.3%, respectively, that compare with a value of 16.1% for cooked sorghum not submitted to pressure and 42.1% for uncooked sorghum. In accordance, electrophoretic sorghum protein profiles of the pressure-treated samples were found to be similar to that of uncooked sorghum. By PLS1 regression, it was possible to relate IR signals with protein digestibility values and to associate proteins digestibility improvement with changes in proteins, while changes in lipids and starch caused by the pressure treatments were also detected. The results of this work indicate pressure treatments of sorghum flour as a suitable way to greatly improve cooked sorghum protein digestibility.

REFERENCES

Aboubacar, A., Axtell, J.D., Huang, C.-P. and Hamaker, B.R. 2001. A rapid protein digestibility assay for identifying highly digestible sorghum lines. *Cereal Chemistry* 78(2), 160–165.

Alvarez, P.A., Ramaswamy, H.S. and Ismail, A.A. 2008. High pressure gelation of soy proteins: Effect of concentration, pH and additives. *Journal of Food Engineering* 88, 331–340.

Axtell, J.D., Kirleis, A.W., Hassen, M.M., Mason, N.d.C., Mertz, E.T. and Munck, L. 1981. Digestibility of sorghum proteins. *Proceedings of the National Academy Science* 78(3), 1333–1335.

Bichard, A., Dury, S., Schonfeldt, H.C., Moroka, T., Motau, F. and Bricas, N. 2005. Access to urban markets for small-scale producers of indigenous cereals: A qualitative study of consumption practices and potential demand among urban consumers in Polokwane. *Development Southern Africa* 22(1), 125–141.

Blaszczak, W., Fornal, J., Kiseleva, V.I., Yuryev, V.P., Sergeev, A.I. and Sadowska, J. 2007. Effect of high pressure on thermal, structural and osmotic properties of waxy maize and Hylon VII starch blends. *Carbohydrate Polymers* 68, 387–396.

Buckow, R., Heinz, V. and Knorr, D. 2007. High pressure phase transition kinetics of maize starch. *Journal of Food Engineering* 81, 469–475.

Castro, S.M., Saraiva, J.A., Lopes-da-Silva, J.A., Delgadillo, I., Van Loey, A., Smout, C. and Hendrickx, M. 2008. Effect of thermal blanching and of high pressure treatments on sweet green and red bell pepper fruits (*Capsicum annuum* L.). *Food Chemistry* 107, 1436–1449.

Chapleau, N. and De Lamballerie-Anton, M. 2003. Improvement of emulsifying properties of lupin proteins by high pressure induced aggregation. *Food Hydrocolloids* 17, 273–280.

Chibber, B.A.K., Mertz, E.T. and Axtell, J.D. 1980. *In vitro* digestibility of high tannin sorghum at different stages of dehulling. *Journal of Agriculture Food Chemistry* 28, 160–161.

Correia, I., Nunes, A., Barros, A. and Delgadillo, I. 2010. Comparison of the effects induced by different processing methods on sorghum proteins. *Journal of Cereal Science* 51, 146–151.

De Lamballerie-Anton, M., Delépine, S. and Chapleau, N. 2002. High pressure effect on meat and lupin protein digestibility. *High Pressure Research* 22, 649–652.

Dahlin, K. and Lorenz, K. 1993. Protein digestibility of extruded cereal grains. *Food Chemistry* 48, 13–18.

Daniel, V.A., Leela, R., Doraiswamy, T.R., Rajalakshmi, D., Rao, S.V., Swaminathan, M. and Parpia, H.A.B. 1966. The effort of supplementing a poor Kaffir corn (Sorghum vulgare) diet with L-lysine and D-L-theonine of the digestibility coefficient, biological value and net utilization of protein retention of nitrogen in children. *Journal of Nutrition and Dietetics* 3, 10–14.

Dicko, M.H., Gruppen, H., Traoré, A.S., Voragen, A.G.J. and van Berkel, W.J.H. 2006. Sorghum grain as human food in Africa: Relevance of content of starch and amylase activities. *African Journal of Biotechnology* 5(5), 384–395.

Duodu, K.G., Nunes, A., Delgadillo, I. and Belton, P.S. 2002a. Low protein digestibility of cooked sorghum—Causes and needs for further research. *Proceedings of AFRIPRO, Workshop on the Proteins of Sorghum and Millets: Enhancing Nutritional and Functional Properties for Africa*, Pretoria.

Duodu, K.G., Nunes, A., Delgadillo, I., Parker, M.L., Mills, E.N.C., Belton, P.S. and Taylor, J.R.N. 2002b. Effect of grain structure and cooking on sorghum and maize *in vitro* protein digestibility. *Journal of Cereal Science* 35, 161–174.

Duodu, K.G., Tang, H., Grant, A., Wellner, N., Belton, P.S., and Taylor, J.R.N., 2001. FTIR and solid state 13C NMR spectroscopy of proteins of wet cooked and popped sorghum and maize. *Journal of Cereal Science* 33, 261–269.

Eggum, B.O., Bach Knudsen, K.E., Munck, L., Axtell, J.D. and Mukuru, S.Z. 1982. Milling and nutritional value of sorghum in Tanzânia. *Journal of Cereal Science* 1, 127–137.

Eggum, B.O., Bach Knudsen, K.E., Munck, L., Axtell, J.D. and Mukuru, S.Z. 1982. Milling and nutritional value of sorghum in Tanzânia. In *Proceedings of the International Symposium on Sorghum Grain Quality* (L.W. Rooney and D.S. Murty, eds.) ICRISAT, Patancheru, A P, Índia, pp. 211–225.

Elgasin, E.A. and Kennich, W.H. 1980. Effect of pressurization of pre-rigor beef muscles on protein quality. *Journal of Food Science* 45, 1122–1124.

Elkin, R.G., Freed, M.B., Hamaker, B.R., Zhang, Y. and Parsons, C.M. 1996. Condensed tannins are only partially responsible for variations in nutrient digestibilities of sorghum grain cultivars. *Journal of Agriculture Food Chemistry* 44, 848–853.

Elkhalil, E.A.I., El Tinay, A.H., Mohamed, B.E., and Elsheikh, E.A.E. 2001. Effect of malt pretreatment on phytic acid and in vitro protein digestibility of sorghum flour. *Food Chemistry* 72, 29–32.

Emmanbux, M.N. and Taylor, R.N. 2009. Properties of heat-treated sorghum and maize meal and their prolamin proteins. *Journal of Agricultural and Food Chemistry* 57, 1045–1050.

FAO, 2004. Food outlook. Available at ftp://ftp.fao.org/docrep/fao/006/j2084e/j2084e00.pdf, consulted in November 25 of 2008

Gadaga, T.H., Lehohla, M., and Ntuli, V. 1999. Traditional fermented foods of *Lesotho*. *Journal of Microbiology, Biotechnology and Food Sciences* 2(6), 2387–2391.

Hamaker, B.R., Kirleis, A.W., Mertz, E.T. and Axtell, J.D. 1986. Effect of cooking on protein profiles and *in vitro* digestibility of sorghum and maize. *Journal of Agricultural and Food Chemistry* 34, 647–649.

Hamaker, B.R., Kirleis, A.W., Butler, L.G., Axtell, J.D. and Mertz, E.T. 1987. Improving the *in vitro* protein digestibility of sorghum with reducing agents. *Proceedings of the National Academy Science* 84, 626–628.

Helland, I.S. 2001. Some theoretical aspects of partial least squares regression. *Chemometrical and Intelligent Laboratory Systems* 58, 97–107.

Klepacka, M., Porzucek, H., Piecyk, M. and Salanski, P. 1996. Effect of high pressure on solubility and digestibility of legume proteins. *Polish Journal of Food Nutrition* 47(2), 41–49.

Kurien, P.P., Narayanarao, M., Swaminathan, M. and Subrahmanyan, M. 1960. The metabolism of nitrogen, calcium and phosphorus in undernourished children. *British Journal of Nutrition* 13, 213–238.

MacLean, W.C., Lopez de Romana, G., Placko, R.P. and Graham, G. 1981. Protein quality and digestibility of sorghum in preschool children: Balance studies and plasma free amino acids. *Journal of Nutrition* 111, 1928–1936.

MacLean, W.C., Lopez de Romana, G., Gastanaduy, A. and Graham, G.G. 1983. The effect of decortication and extrusion on the digestibility of sorghum by preschool children. *Journal of Nutrition* 113, 2171–2177.

Mertz, E.T., Hassen, M.M., Cairns-Whittern, C., Kirleis, A.W., Tu, L. and Axtell, J.D. 1984. Pepsin digestibility of proteins in sorghum and other major cereals. *Proceedings of the National. Academy of Science* 81, 1–2.

Messens, W., van Camp, J. and Huyghebaert, A. 1997. The use of high-pressure to modify the functionality of food proteins. *Food Hydrocolloids* 15, 263–269.

Mitarau, B.N., Blair, R., Reichert, R.D., and Roe, W.E.A. 1984. Dark and yellow rapeseed hulls, soybean hulls and a purified fiber source; their effects on dry matter, energy, protein and amino acid digestibilities in cannulated pigs. *Journal of Animal Science* 59, 1510–1518.

Mitaru, B.N., Reichert, R.D. and Blair, R. 1985. Protein and amino acid digestibilities for chickens of reconstituted and boiled sorghum grain varying in tannin contents. *Poultry Science* 64, 101–106.

Murty, D.S. and Kumar, K.A. 1995. Traditional uses of sorghum and millets. In *Sorghum and Millets: Chemistry and Technology*, Dendy, D.A.V., ed. American Association of Cereal Chemists Inc., St. Paul, Minnesota, pp. 185–221.

Nunes, A. 2004. Study of the interaction between components of flour of Sorghum bicolor (L.) Moench (Estudo de interacção entre componentes de farinha de *Sorghum bicolor* (L.) Moench). PhD thesis. (In Portuguese) Department of Chemistry, Universidade de Aveiro Portugal.

Nunes, A., Correia, I., Barros, A. and Delgadillo, I. 2004. Sequential *in vitro* pepsin digestion of uncooked and cooked sorghum and maize samples. *Journal of Agricultural and Food Chemistry* 52(7), 2052–2058.

Oria, M.P., Hamaker, B.R. and Shull, J.M. 1995. Resistance of sorghum alfa-, beta- and gamma-kafirins to pepsin digestion. *Journal of Agriculture Food Chemistry* 43, 2148–2153.

Parker, M.L., Grant, A., Rigby, N.M., Belton, P.S. and Taylor, J.R.N. 1999. Effects of popping on the endosperm cell walls of sorghum and maize. *Journal of Cereal Science* 30, 209–216.

Puppoa, M.C., Beaumalb, V., Chapleauc, N., Speronia, F., de Lamballeriec, M., Anón, M.C. and Anton, M. 2008. Physicochemical and rheological properties of soybean protein emulsions processed with a combined temperature/high-pressure treatment. *Food Hydrocolloids* 22, 1079–1089.

Rom, D.L., Shull, J.M., Chandrashekar, A. and Kirleis, A.W. 1992. Effects of cooking and treatment with sodium bisulfite on *in vitro* protein digestibility and microstructure of sorghum flour. *Cereal Chemistry* 69(2), 178–181.

Rooney, L.W. 1990. Processing methods to improve nutritional value of sorghum livestock. In *Proceedings of the International Conference on Sorghum Nutritional Quality*, G. Ejeta, ed. Purdue University, West Lafayette, IN.

Shull, J.M., Watterson, J.J. and Kirleis, A.W. 1992. Purification and immunocytochemical localization of kafirins in *Sorghum bicolor* (L.) Moench endosperm. *Protoplasma* 171, 64–74.

Taylor, J. and Taylor, J.R.N. 2002. Alleviation of the adverse effect of cooking on sorghum protein digestibility through fermentation in traditional African porridges. *International Journal of Food Science Technology* 37(2), 129–137.

Taylor, J.R.N., Noveille, L. and Liebenberg, N.W. 1985. Protein body degradation in the starchy endosperm of germinating sorghum. *Journal of Experimental Biology* 36, 1287–1295.

Wang, X.-S., Tang, C.-H., Li, B.-S., Yang, X.-Q., Li, L. and Ma, C.-Y. 2008. Effects of high pressure treatment on some physicochemical and functional properties of soy protein isolates. *Food Hydrocolloids* 22, 560–567.

Weaver, C.A., Hamaker, B.R. and Axtell, J.D. 1998. Discovery of grain sorghum germ plasma with high uncooked and cooked *in vitro* protein digestibilities. *Cereal Chemistry* 75(15), 665–670.

Wong, J.H., Lau, T., Cai, N., Singh, J., Pedersen, J.F., Vensel, W.H., Hurkman, W.J., Wilson, J.D., Lemaux, P.G. and Buchanan, B.B. 2009. Digestibility of protein and starch from sorghum (*Sorghum bicolor*) is linked to biochemical and structural features of grain endosperm. *Journal of Cereal Science* 49, 73–82.

Yin, S.-W., Tang, C.-H., Wen, Q.-B., Yang, X.-Q. and Li, L. 2008. Functional properties and *in vitro* trypsin digestibility of red kidney bean (*Phaseolus vulgaris* L.) protein isolate: Effect of high-pressure treatment. *Food Chemistry* 110, 938–945.

Yousif, N.E. and El Tinay, A.H., 2001. Effect of fermentation on sorghum protein fractions and in vitro protein digestibility. *Plant Foods for Human Nutrition* 56, 175–182.

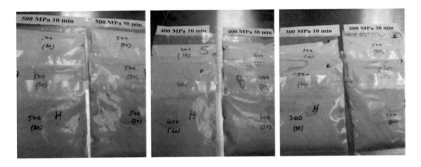

FIGURE 3.11 HHP-processed ice-cream mixes at 300/400/500 MPa, 10/30 min. (Photo: Tokuşoğlu, 2010, WSU.)

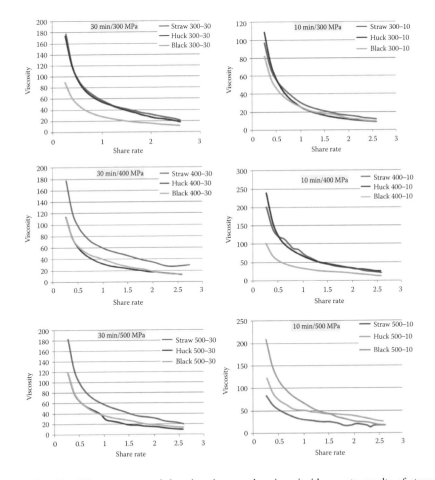

FIGURE 3.12 HP treatment and duration time on the viscosity/shear rate results of strawberry, huckleberry, and BIMs. (Adapted from Tokuşoğlu Ö. et al. 2011. Rheological, microbial quality characteristics and antioxidant activity of high hydrostatic pressure (HHP) processed berry ice-cream mixes: Weibull modeling and response surface methodology approach for treatment pressure and holding time effects. *Nonthermal Processing Division Workshop 2011—Innovation Food Conference.* October 12–14, Osnabrück, Germany. Oral Presentation.)

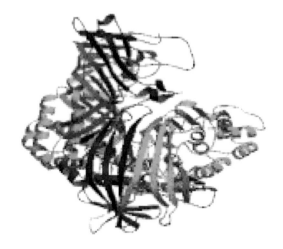

FIGURE 12.5   Phosvitin structure (Adapted from Anonymous, 2013a).

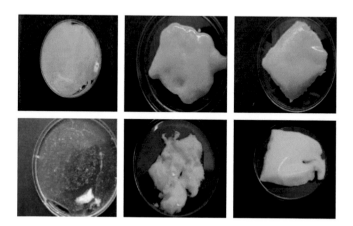

FIGURE 12.13   Texture profiles of egg yolk and egg white. (Control, 400, 600 MPa of HHP treatment). (Adapted from Singh A., Ramaswamy H.S. 2010. Effect of HPP on physicochemical properties of egg components. In: Annual Meeting of Institute of Food Technologists (IFT), Chicago, USA, July 2010.)

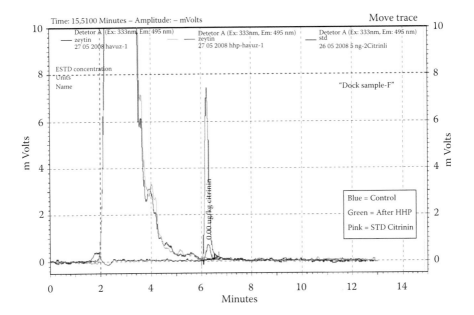

FIGURE 12.19 HPLC chromatogram of citrinin (CIT) occurrence in control "dock-F" and HHP-treated "dock-F" olives and standard citrinin separation. (From Tokuşoğlu Ö., Alpas H., and Bozoğlu F.T. 2010. *Innovative Food Science and Emerging Technologies*, 11(2), 250–258.)

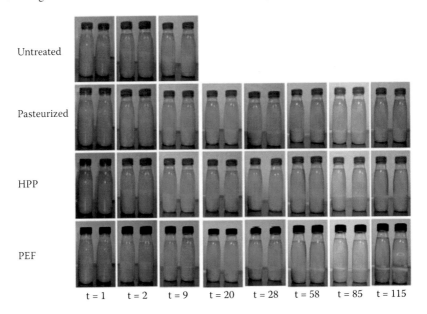

FIGURE 17.8 Observed sedimentation and cloud loss of untreated, mild heat-pasteurized, HPP-, and PEF-processed orange juice bottles during the first 115 days of storage at 4°C. (Adapted from Timmermans R.A.H. et al. 2011. *Innovative Food Science and Emerging Technologies*, 12, 235–243.)

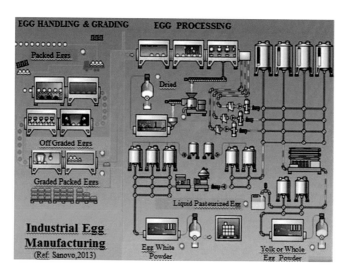

FIGURE 18.2 Schematized industrial egg-manufacturing process. (Adapted from Sanovo 2013. *Egg Production. Sanovo Egg Group.* SANOVO International A/S, Denmark.)

10 Modeling and Simulating of the High Hydrostatic Pressure Inactivation of Microorganisms in Foods

Sencer Buzrul

CONTENTS

10.1 INTRODUCTION

There is a growing concern among consumers that thermally preserved foods may be overprocessed and, hence, their nutritive value and overall quality may decrease to some extent. The demand from the consumers for safe and "fresh" food products has increased the interest to nonthermal food preservation methods, notably high hydrostatic pressure (HHP) (Buzrul et al., 2005a). HHP treatment may allow the production of high-quality foods that are minimally processed, additive free, and microbiologically safe (Alpas and Bozoglu, 2000). Over the last 20 years, research about HHP has been explored (Avsaroglu et al., 2006; Buzrul and Alpas, 2004; Buzrul et al., 2005b,c; Erkan et al., 2010, 2011; García-Graells et al., 1999; Gervilla et al., 1997; Hayakawa et al., 1994; Palou et al., 1997; Pilavtepe-Çelik et al., 2009, 2011; Tay et al., 2003) and there is a large body of literature on the subject. Moreover, several commercial products (such as fruit juices, seafoods, meat, and vegetable products) treated by HHP are now available in the market in different countries (Buzrul et al., 2008a).

Since the first-order kinetics is widely accepted, it is also extended to nonthermal preservation methods such as HHP and pulsed electric fields (FDA/USDA, 2000); however, to determine the safety of nonthermal methods and compare them to that of thermal methods, one would need a concept that should account for the frequently observed deviations from the standard first-order kinetics model. Furthermore, in light of the growing importance of microbial safety issues and the availability of mathematical tools to deal with complex phenomena, it seems appropriate to re-examine the currently held concepts of microbial inactivation kinetics and develop new approaches to its description, interpretation, and eventually utilization (Peleg, 2003a).

A number of nonlinear models for describing the HHP inactivation of microbial cells such as the (modified) Gompertz equation (Chen and Hoover, 2003; Patterson and Kilpatrick, 1998; Yamamoto et al., 2005), the Fermi equation (Donsì et al., 2003), log–logistic equation (Buzrul and Alpas, 2004; Chen and Hoover, 2003; Yamamoto et al., 2005), the modified Baranyi model (Koseki and Yamamoto, 2007a), and the Weibull model (Avsaroglu et al., 2006; Buzrul and Alpas, 2004; Chen and Hoover, 2003; Yamamoto et al., 2005) have been proposed in literature (Figure 10.1). While all these models—in their original and modified forms—could be used to fit survival curves of microbial cells inactivated by HHP, none of them can be considered as unique. Nevertheless, among them the Weibull model is, per-haps, the most simple and flexible one as it is characterized by only two parameters and it can accurately fit linear, concave upward, and concave downward survival curves (Figure 10.1).

This chapter presents some issues in microbial modeling during HHP treatment and also demonstrates the application of the Weibull model to the published survival data of *Salmonella typhimurium* KUEN 1357 in tryptone soy broth and *Escherichia coli* ATCC 11229 in aqueous whey protein.

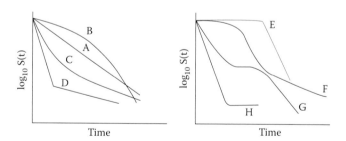

FIGURE 10.1   Commonly observed types of survival curves. Left plot: Linear (A), down-ward concave (B), upward concave (C), and biphasic (D). Right plot: Linear with a preced-ing shoulder (E), sigmoidal I—starting with a downward concavity and ending up with an upward concavity (F), sigmoidal II—starting with an upward concavity and ending up with a downward concavity (G), and linear with tailing (H). (Adapted from Xiong, R et al. 1999. *International Journal of Food Microbiology* 46: 45–55; Peleg, M. 2003b. *Critical Reviews in Food Science and Nutrition* 43: 645–658; Geeraerd, A.H., Valdramidis, V.P., Van Impe, J.F. 2005. *International Journal of Food Microbiology* 102: 95–105.)

10.2   SOME ISSUES IN MICROBIAL MODELING DURING HHP TREATMENT

10.2.1   ATTAINMENT OF ISOBARIC–ISOTHERMAL CONDITIONS

The treatment of foods or food ingredients by using HHP is accompanied with an increase of the temperature in the treated volume and subsequent heat transfer among pressure medium, pressure chamber, and the samples. In the past, HHP systems were seldom equipped with sensors to measure temperature within the pressure vessel; hence, the effect of temperature is not considered due to the inability to monitor and control the temperature of the product (Guan et al., 2006). Since the models (such as first-order kinetics and the Weibull models) used for describing the inactivation of microorganisms are applicable at "only" iso-conditions; attainment of isobaric–isothermal conditions during HHP is an important issue.

Figure 10.2 demonstrates the temperature increases of water and ethylene glycol (commonly used as pressure-transmitting fluids) during HHP treatment. Both were initially at 20°C; however, upon compression to 200 MPa in 1 min, about 5°C and 8.3°C of temperature increases were observed for water and ethylene glycol, respectively. Therefore, the target temperature of water rises to 25°C during HHP treatment (or during holding time period) while it reaches up to 28.3°C for ethylene glycol (if

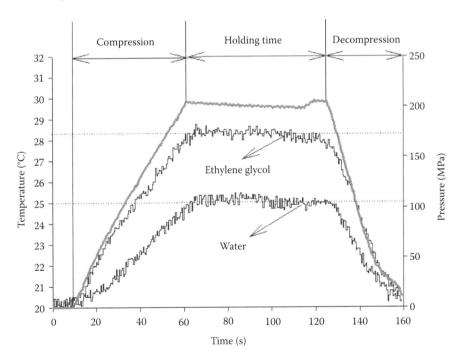

FIGURE 10.2   Pressure (gray solid line) and temperature (black solid lines) profiles of water and ethylene glycol. Dashed lines indicate the targeted process temperature values for water (25°C) and ethylene glycol (28.3°C) that were initially at 20°C. (Adapted from Buzrul, S. et al. 2008b. *Journal of Food Engineering* 85: 466–472.)

the heat loss through the pressure vessel is minimized and uniform temperature profile throughout the vessel is attained) (Figure 10.2).

Attainment of an isobaric–isothermal HHP treatment is mainly dependent on the initial temperature of the sample at the beginning of pressurization, type of pressure-transmitting fluid, pressure vessel design, pressurization type, and come-up rate. The targeted process temperature is equal to the starting temperature plus temperature increase due to compression heating ($T_{process} = T_{initial} + T_{heat}$). However, it should be noted that food products (i.e., samples) and pressure-transmitting fluids could have different physical properties and therefore different compression-heating values (Patazca et al., 2007) and this difference should be taken into account during HHP treatments. Furthermore, it should be noted that inactivation of microorganisms at nonisobaric–nonisothermal conditions (during compression and decompression periods) should be quantified. Figure 10.3 demonstrates these phenomena; the difference ($\log_{10}N_{diff}$) between the initial number of microorganisms ($\log_{10}N_{init}$) and after the stabilization of treatment conditions ($\log_{10}N_0$) can be mathematically described as (Figure 10.3)

$$\log_{10} N_{diff} = \log_{10} N_{init} - \log_{10} N_0 \tag{10.1a}$$

Depending on the target pressure and temperature, $\log_{10}N_{diff}$ could reach a value as high as 4.60 $\log_{10}$ for *E. coli* at 350 MPa, 40°C indicating the importance of quantifying the nonisobaric–nonisothermal effect on microbial inactivation (Valdramidis et al., 2007; Van Opstal et al., 2005). In fact, most of the studies indicated that $\log_{10}N_{init}$ is due to only a compression (come-up) period (see, e.g., Palou et al., 1997; Valdramidis et al., 2007); however, it is due to compression followed by decompression (HHP treatment with no holding time) since it is not possible to take samples from the HHP equipment before the pressure is reduced to 1 atm (0.1 MPa). It is known that the decompression rate (especially rapid decompression) also has an impact on

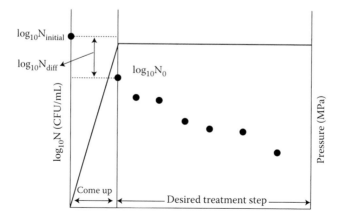

FIGURE 10.3 A demonstration of the impact of pressure come-up period on microbial inactivation during HHP. (Adapted from Valdramidis, V.P. et al. 2007. *Journal of Food Engineering* 78: 1010–1021.)

microbial inactivation (Hayakawa et al., 1998; Noma et al., 2002); hence, it should not be neglected. The assumption made by using Equation 10.1a has been that inactivation due to compression followed by decompression due to holding time is additive.

10.2.2 INJURY

Bacterial cells exposed to different lethal treatments suffer injury that could be reversible in food materials during storage (Bozoglu et al., 2004). Injury has also been observed for many bacterial cells exposed to HHP (Alpas et al., 2000; Erkmen and Dogan, 2004; García-Graells et al., 2000). Since the published predictive models of HHP-induced microbial inactivation are based on the notion that a cell exposed to high pressure can only be in one of the two stages; alive and countable or dead and irrecoverable [it should be noted that the recovery takes place usually in a medium and under conditions that favor growth and incubations for up to several days to allow injured cells to form visible colonies; however, the result would most probably not be the same for the bacteria in a food matrix]. These models did not take into account the recovery of injured bacteria induced by HHP (Corradini and Peleg, 2007; Koseki and Yamamoto, 2007b). As a result, the effect of microbial inactivation induced by HHP could be overestimated, leading to an increased risk of food poisoning or spoilage and could be a critical issue in terms of the safety of HHP-treated foods (Koseki and Yamamoto, 2007b).

New survival models should be used that specifically account for the fact that at least some of the survivors might be injured and perhaps die at a later time. These models should also take into account the possibility that some of the cells considered dead, are actually only injured and might become viable at a later time after the damage repair (Corradini and Peleg, 2007). Two important recent modeling studies (Corradini and Peleg, 2007; Koseki and Yamamoto, 2007b) that consider injury can be good examples for these new models.

10.2.3 SECONDARY MODELS: PRESSURE AND TEMPERATURE DEPENDENCY OF PRIMARY MODEL PARAMETERS

Primary models describe the change in bacterial counts with time under particular environmental conditions. Secondary models describe one or more parameters of a primary model under different environmental conditions such as pH, water activity, temperature, and pressure. Since both pressure and temperature are responsible for inactivation of microorganisms in HHP treatments, describing the primary model parameters in terms of pressure and temperature is another important issue.

In literature, it is possible to find empirical secondary models that describe the parameter(s) of primary models in terms of pressure or temperature. For example, Peleg (2002) used two different empirical models to describe the parameters of the Weibull model (Equation 10.2) in terms of pressure at a constant temperature. Similarly, Buzrul et al. (2005b) used polynomial (linear) function to describe the rate parameter (b) of the Weibull model in terms of temperature at constant pressure values. Valdramidis et al. (2007) used a polynomial function to describe the rate constant of the first-order model in terms of both pressure and temperature.

The polynomial function (response surface-type model) is not the only option to describe the model parameters in terms of pressure and temperature and their interactions. The square root model (Ratkowsky et al., 1983; Skinner et al., 1994), the Bigelow-type model (Bigelow, 1921; Leguerinel et al., 2000; Mafart, 2000), and the linear Arrhenius–Davey model (Davey et al., 1995) could be the other alternatives. It is also possible to derive new empirical functions that can be used as secondary models. As no model is able to accurately define microbial responses under all circumstances, it is best to compare several models (van Gerwen and Zwietering, 1998); therefore, both primary and secondary model selection—considering the isothermal–isobaric experimental setup and injury issues—is a critical step for safety design and optimization of HHP treatments in food processing.

10.3 APPLICATION OF THE WEIBULL MODEL

Different forms of the Weibull model were presented in literature (Mafart, 2002; Peleg, 1999; van Boekel, 2002), but they all have the same good fit (Buzrul, 2007) and it was expressed by Peleg and Cole (1998) as

$$\log_{10} S(t) = -b(P,T) \cdot t^{n(P,T)} \tag{10.1b}$$

where b is the pressure- and temperature-dependent rate parameter and n is the pressure- and temperature-dependent shape parameter. Such a model presents the main advantage of being very simple and sufficiently robust to describe both monotonic downward concave (shoulder) survival curves ($n > 1$) and monotonic upward concave (tailing) survival curves ($n < 1$). The traditional first-order model is then a special case ($n = 1$) of the Weibull model.

Moreover, if the distribution of the lethal events (heat, pressure, disinfectant, etc.) is indeed Weibullian, then $n(P, T)$ is expected to be either P and T independent or only a weak function of P and T (Peleg et al., 2005). When the power $n(P, T)$ is fixed, that is, $n(P, T) = n$, then Equation 10.1b becomes

$$\log_{10} S(t) = -b(P,T) \cdot t^{n} \tag{10.2}$$

A second-order polynomial function (Equation 10.3) was proposed to describe P and T dependency of the rate parameter (b):

$$\ln\left[b(P,\ T)\right] = a_0 + a_1 P + a_2 P^2 + a_3 T + a_4 T^2 + a_5 PT \tag{10.3}$$

where $a_0, a_1, a_2, a_3, a_4,$ and a_5 are the coefficients of Equation 10.3.

If the shape parameter (n) is also dependent on pressure and temperature, the same secondary model could be used for $n(P, T)$:

$$\ln\left[n(P,\ T)\right] = a_0 + a_1 P + a_2 P^2 + a_3 T + a_4 T^2 + a_5 PT \tag{10.4}$$

A backward regression procedure may be applied to remove the coefficients that are not significant $(P > 0.05)$.

Let us assume that under constant pressure and temperature conditions, survival curves of *S. typhimurium* and *E. coli* obey the Weibull model (Equation 10.1) and let us also assume that recovery from injury is not a factor and that the process is short enough so that these microorganisms have no time to adapt physiologically (Peleg, 2006).

10.4   MODELING HHP INACTIVATION OF *S. TYPHIMURIUM*

In the first attempt, the Weibull model (Equation 10.1b) was used to describe the high-pressure inactivation data of *S. typhimurium* (at 200, 250, 300, and 350 MPa; 15°C, 25°C, 35°C, and 45°C) published by Erkmen (2009). Figure 10.4 shows the fit of Equation 10.1 for survival data of *S. typhimurium* at 300 MPa, 15°C, 25°C, and 45°C. Visual inspection of Figure 10.4 indicated that Equation 10.1b produced reasonable fits for the HHP inactivation of *S. typhimurium* (Figure 10.4).

The shape parameters obtained from Equation 10.1b for the 16 survival curves (4 pressure values × 4 temperature values) of *S. typhimurium* were all less than one (ranged from 0.43 to 0.85) indicating that sensitive members of the population are destroyed at a relatively fast rate, leaving behind the survivors of higher and higher resistance (14). The fixed shape parameter value was calculated by averaging 16 shape parameter values obtained from Equation 10.1b that was 0.59. The second attempt was done with Equation 10.2 where $n = 0.59$.

Figure 10.5 shows the fit of the Weibull (Equation 10.1b) and the reduced Weibull (Equation 10.2) models for survival data of *S. typhimurium* at 350 MPa, 15°C, 25°C,

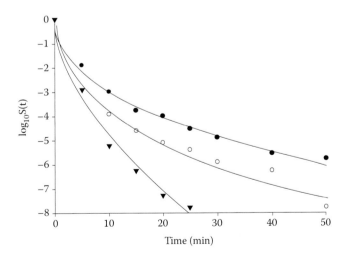

FIGURE 10.4   Survival curves of *S. typhimurium* in tryptone soy broth at 300 MPa; closed circles: 15°C, open circles: 25°C, and closed reversed triangle: 45°C. Solid lines indicate that data were fitted with the Weibull model (Equation 10.1b). (The original data are from Erkmen, O. 2009. *Food and Bioproducts Processing* 87: 68–73.)

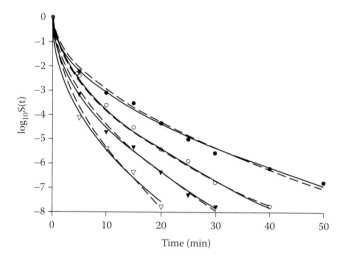

FIGURE 10.5 Survival curves of *S. typhimurium* in tryptone soy broth at 350 MPa; closed circles: 15°C, open circles: 25°C, closed reversed triangle: 35°C, and open reversed triangle: 45°C. Solid lines indicate that the data were fitted with the Weibull model (Equation 10.1b) with variable *n* and dashed lines indicate that the data were fitted with the reduced Weibull model (Equation 10.2) with a fixed *n* value (*n* = 0.59). (The original data are from Erkmen, O. 2009. *Food and Bioproducts Processing* 87: 68–73.)

35°C, and 45°C. Table 10.1 shows the obtained parameters and mean square error (MSE) values of both models. Although, there was slight loss of good fits by fixing the shape parameters of the survival curves of *S. typhimurium,* fixing of *n* causes a better stability of *b* estimates and leads to an improvement of the robustness of the model (Couvert et al., 2005). Normally, choosing the average value to evaluate a single *n* value may not be suitable since the number of data in each kinetic is not equal, that is, each kinetic does not have the same weight on *n* value evaluation. However, in this case, average value and the value considering the number of data of each kinetic gave the same result; therefore, the average value was used for the fixed *n* value. Fixing *n* would be useful when warranted by the results (Figure 10.5 and Table 10.1).

Erkmen (2009) has defined the same survival curves with a modified Gompertz equation that has three parameters excluding $\log_{10}N_0$. However, this study has shown that it could be possible to define these survival curves (with tailing) with only one parameter (*b*) since the shape parameter (*n*) was fixed. One of the criteria of selecting a model is simplicity, that is, the simplest model is the best one to describe the data adequately, this is known as the rule of parsimony.

For the secondary model, Equation 10.5 was obtained:

$$\ln\left[b(P,T)\right] = (-10.4 \pm 0.8^a) + (0.055 \pm 0.006^a)$$
$$\cdot P - (7.8 \times 10^{-5} \pm 1 \times 10^{-5a}) \cdot P^2 + (0.22 \pm 0.02^a) \cdot T \tag{10.5}$$

TABLE 10.1

Parameters of the Weibull Model (Equation 10.1b), the Reduced Weibull Model (Equation 10.2), and MSE Values for HHP Inactivation of *S. typhimurium* in Tryptone Soy Broth

		Weibull Model (Equation 10.1)			Reduced Weibull Model (Equation 10.2)		
Pressure (MPa)	Temperature (°C)	$b(P,T)$	$n(P,T)$	MSE	$b(P,T)$	n	MSE
200	15	0.06	0.79	0.008	0.11	0.59	0.01
250		0.11	0.85	0.07	0.27		0.11
300		1.07	0.44	0.03	0.65		0.15
350		0.97	0.50	0.02	0.72		0.07
200	25	0.15	0.60	0.004	0.16		0.004
250		0.63	0.49	0.12	0.45		0.13
300		1.39	0.43	0.07	0.80		0.28
350		1.01	0.56	0.009	0.90		0.02
200	35	0.21	0.54	0.005	0.17		0.005
250		0.82	0.45	0.12	0.51		0.18
300		1.40	0.44	0.24	0.83		0.43
350		1.43	0.50	0.03	1.09		0.08
200	45	0.31	0.53	0.01	0.25		0.02
250		0.68	0.51	0.01	0.52		0.003
300		1.38	0.55	0.11	1.23		0.09
350		1.89	0.46	0.05	1.35		0.14

Source: Original data are from Erkmen, O. 2009. *Food and Bioproducts Processing* 87: 68–73.

where superscript a is the standard error and MSE of the fit was 0.011. Note that only the significant coefficients were retained after the backward regression procedure. Figure 10.6 shows the three-dimensional surface of Equation 10.5.

When the secondary model (Equation 10.5) is integrated into the primary model (Equation 10.2), Equation 10.6 is obtained:

$$\log_{10} S(t) = -\exp\left[-10.4 + 0.055 \cdot P - 7.8 \times 10^{-5} \cdot P^2 + 0.22 \cdot T\right] \cdot t^{0.59} \qquad (10.6)$$

This equation was used to simulate the survival curves at other pressure and temperature values (throughout the interpolation region). As indicated by Figure 10.7, generated survival curves of *S. typhimurium* at 275 MPa at 30°C and 40°C and 325 MPa at 30°C and 40°C are all upward concave since the shape parameter was fixed to 0.59.

10.5   MODELING HHP INACTIVATION OF *E. COLI*

The high-pressure inactivation data of *E. coli* (between 207 and 438.9 MPa; 30°C, 40°C, and 50°C) were published by Doona et al. (2005), Feeherry et al. (2005), and

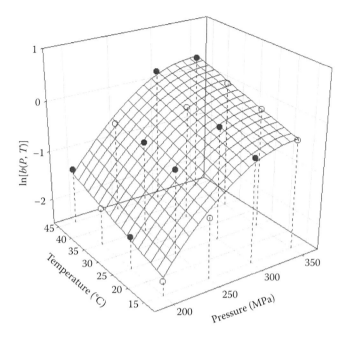

FIGURE 10.6   The three-dimensional surfaces of the quadratic polynomial model (Equation 10.5) with significant coefficients for rate parameter (b) of *S. typhimurium* that obtained from the fit of the reduced Weibull model (Equation 10.2). Data points: under the surface (open circles) and above the surface (closed circles). MSE of the fit was 0.011.

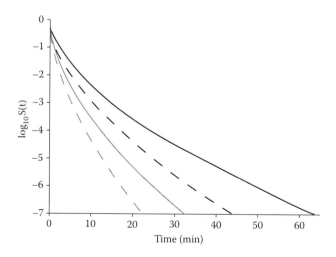

FIGURE 10.7   Generated survival curves of *S. typhimurium* using Equation 10.6; black solid line: 275 MPa, 30°C, black dashed line: 275 MPa, 40°C, gray solid line: 325 MPa, 30°C, and gray dashed line: 325 MPa, 40°C.

Ross et al. (2005) and the same mathematical model, that is, the Weibull model (Equation 10.1) was also used by Doona et al. (2007) for the same published data; however, they proposed different secondary models—see below.

Figure 10.8 shows the fit of the Weibull model (Equation 10.1) for survival data of *E. coli* at 335.5, 370, and 438.9 MPa at 30°C. Table 10.2 shows the obtained parameters and MSE values of the model. It should be noted that upward and downward concavity as well as (almost) linear survival curves were observed for *E. coli* (see Figure 10.8 and Table 10.2). Downward concavity ($n > 1$) indicates that the remaining members become increasingly damaged and if the survival curve is linear, then microorganisms have time-independent and same (equal) probability of mortality (Avsaroglu et al., 2006) (Figure 10.8).

It is not possible to fix the shape parameter as was done for *S. typhimurium*. Therefore, Equation 10.4 was used to describe the $n(P,T)$ in terms of pressure and temperature:

$$\ln[n(P,T)] = (1.88 \pm 0.26^{a}) - (9.7 \times 10^{-6} \pm 1 \times 10^{-6a}) \\ \cdot P^{2} - (0.019 \pm 0.006^{a}) \cdot T \tag{10.7}$$

where superscript a is the standard error and MSE of the fit was 0.04. Figure 10.9 shows the three-dimensional surface of Equation 10.7 (Figure 10.9).

For the rate parameter, the following equation was obtained:

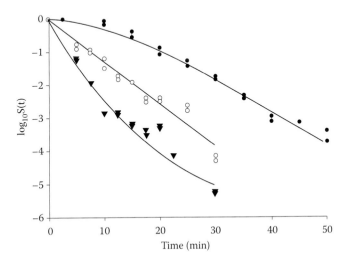

FIGURE 10.8 Survival curves of *E. coli* aqueous whey protein at 30°C; closed circles: 335.5 MPa, open circles: 370 MPa, and closed reversed triangle: 438.9 MPa. Solid lines indicate that the data were fitted with the Weibull model (Equation 10.1b). (The original data are from Doona, C.J. et al. 2007. *High Pressure Processing of Foods*. Blackwell, NY. pp. 115–144.)

TABLE 10.2

Parameters of the Weibull Model (Equation 10.1b) and MSE Values for HHP Inactivation of *E. coli* Aqueous Whey Protein

Pressure (MPa)	Temperature (°C)	$b(P,T)$	$n(P,T)$	MSE
213.6	30	2.7×10^{-8}	3.37	0.003
241.2		5.0×10^{-4}	1.63	0.14
301.1		0.006	1.37	0.10
335.5		0.012	1.47	0.04
370		0.13	0.99	0.07
438.9		0.48	0.70	0.07
207.4	40	4.0×10^{-4}	1.58	0.04
266.6		3.0×10^{-4}	2.24	0.02
301.1		0.047	1.19	0.13
335.5		0.18	0.89	0.38
370		0.44	0.75	0.05
385.8		0.62	0.69	0.09
438.9		2.16	0.36	0.68
206.7	50	0.003	1.81	0.005
266.6		0.08	1.21	0.09
301.1		0.14	1.16	0.10
335.5		0.64	0.85	0.24
370		0.75	0.72	0.14
438.9		2.65	0.48	0.93

Source: Original data are from Doona et al. (2007). *High Pressure Processing of Foods.* Blackwell, NY. pp. 115–144.

$$\ln\left[b(P,T)\right] = (-64.9 \pm 11.3^a) + (0.22 \pm 0.05^a) \cdot P - (1.6 \times 10^{-4} \pm 7.3$$
$$\times 10^{-5a}) \cdot P^2 + (0.77 \pm 0.22^a) \cdot T - (0.0017 \pm 0.0007^a) \cdot P \cdot T \quad (10.8)$$

where superscript a is the standard error and MSE of the fit was 2.97. Figure 10.10 shows the three-dimensional surface of Equation 10.8 (Figure 10.10).

When the secondary models (Equations 10.7 and 10.8) are integrated into the primary model (Equation 10.1b), Equation 10.9 is obtained:

$$\log_{10} S(t) = -\exp\left[-64.9 + 0.22 \cdot P - 1.6 \times 10^{-4} \cdot P^2 + 0.77 \cdot T - 0.0017 \cdot P \cdot T\right]$$
$$\cdot t^{\exp\left[1.88 - 9.7 \times 10^{-6} \cdot P^2 - 0.019 \cdot T\right]} \quad (10.9)$$

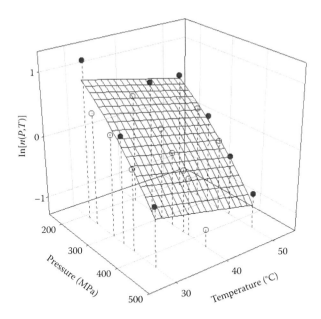

FIGURE 10.9   The three-dimensional surfaces of the quadratic polynomial model (Equation 10.7) with significant coefficients for shape parameter (n) of *E. coli* that obtained from the fit of the Weibull model (Equation 10.1b). Data points: under the surface (open circles) and above the surface (closed circles). MSE of the fit was 0.04.

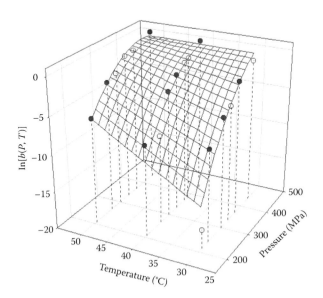

FIGURE 10.10   The three-dimensional surfaces of the quadratic polynomial model (Equation 10.7) with significant coefficients for rate parameter (b) of *E. coli* that obtained from the fit of the Weibull model (Equation 10.1b). Data points: under the surface (open circles) and above the surface (closed circles). MSE of the fit was 2.97.

This equation could be used to simulate the survival curves at other pressure and temperature values (throughout the interpolation region)—see Figure 10.11. The survival curve generated by Equation 10.9 at 350 MPa, 30°C is almost linear and the survival curve generated at 350 MPa, 40°C is downward concave; however, at 450 MPa at both 30°C and 40°C, upward survival curves are generated. The concavity direction can be reversed when the treatment's intensity is increased or decreased (Peleg, 2003b) (Figure 10.11).

Doona et al. (2007) used the following equations [which were initially proposed by Peleg (2006)] for $n(P,T)$ and $b(P,T)$ and obtained reasonable fits as judged by the statistical criteria:

$$n(P)\big|_{T=const} = C_{0P}(T) \cdot \exp\left[-C_{1P}(T) \cdot P\right] \tag{10.10}$$

in which $C_{0P}(T)$ and $C_{1P}(T)$ are temperature-dependent coefficients, and

$$n(T)\big|_{P=const} = C_{0T}(P) \cdot \exp\left[-C_{1T}(P) \cdot T\right] \tag{10.11}$$

in which $C_{0T}(P)$ and $C_{1T}(P)$ are pressure-dependent coefficients.

$$b(P)\big|_{T=const} = \ln\left[1 + \exp\left\{-K_P(T) \cdot \left[P - P_c(T)\right]\right\}\right] \tag{10.12}$$

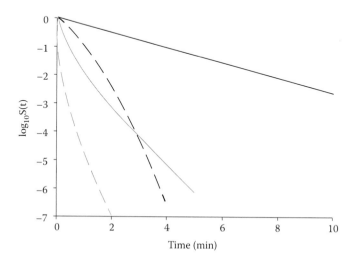

FIGURE 10.11 Generated survival curves of *E. coli* using Equation 10.9; black solid line: 350 MPa, 30°C, black dashed line: 350 MPa, 40°C, gray solid line: 450 MPa, 30°C, and gray dashed line: 450 MPa, 40°C.

in which $K_P(T)$ and $P_c(T)$ are temperature-dependent coefficients, and

$$b(T)\big|_{P=const} = \ln\left[1 + \exp\left\{-K_T(P)\cdot\left[T - T_c(P)\right]\right\}\right] \tag{10.13}$$

in which $K_T(P)$ and $T_c(P)$ are pressure-dependent coefficients.

10.6   CONCLUDING REMARKS

There is now enough evidence that thermal inactivation of microorganisms does not follow first-order kinetics (van Boekel, 2002) and the same can be said about nonthermal inactivation methods such as HHP. This chapter demonstrates that the Weibull model could be successfully used to describe the survival curves of *S. typhimurium* and *E. coli* treated with HHP.

A second-order polynomial function was proposed as the secondary model. The proposed secondary model, although the coefficients do not have a physical meaning, can also be used as an alternative to the models proposed by Doona et al. (2007) and Peleg (2006). The coefficients can be reduced by retaining only the significant ones in the model. It should be noted that pressure and temperature values in the secondary model are constant; therefore, attainment of the isobaric–isothermal conditions is important during HHP treatment. Furthermore, inactivation of microorganisms at nonisobaric–nonisothermal conditions (during compression and decompression periods) should be quantified.

Combined pressure and temperature, also known as pressure-assisted thermal sterilization (PATS), was recently approved by the FDA (Food and Drug Administration) in 2009 for commercial sterilization (Bermúdez-Aguirre and Barbosa-Cánovas, 2011). In this case, most probably, a high number of microorganisms would be inactivated during the compression period (see, e.g., Ahn and Balasubramaniam, 2007) in which both pressure and temperature vary concomitantly. In such a situation, the proposed secondary model is inadequate. Nevertheless, at moderate pressures (200–500 MPa) and temperatures (20–50°C), this model would be useful. The prediction capability survival curves generated by the use of the integrated model (secondary model integrated into the primary one) can be easily validated experimentally.

More studies should be carried out with different spoilage and pathogenic microorganisms in different foods at various pressure and temperature values that would supply the data needed for the modeling issues. These data accumulation would also help modelers to construct new secondary models, that is, to determine the pressure and temperature dependency of the primary model's parameters.

REFERENCES

Ahn, J., Balasubramaniam, V.M. 2007. Effects of inoculum level and pressure pulse on the inactivation of *Clostridium sporogenes* spores by pressure-assisted thermal processing. *Journal of Microbiology and Biotechnology* 17: 616–623.

Alpas, H., Bozoglu, F. 2000. The combined effect of high hydrostatic pressure, heat and bacteriocins on inactivation of foodborne pathogens in milk and orange juice. *World Journal
of Microbiology and Biotechnology* 16: 387–392.

Alpas, H., Kalchayanand, N., Bozoglu, F., Ray, B. 2000. Interactions of high hydrostatic
pressure, pressurization temperature and pH on death and injury of pressure-resistant
and pressure-sensitive strains of foodborne pathogens. *International Journal of Food
Microbiology* 60: 33–42.

Avsaroglu, M.D., Buzrul, S., Alpas, H., Akcelik, M., Bozoglu, F. 2006. Use of the Weibull
model for lactococcal bacteriophage inactivation by high hydrostatic pressure.
International Journal of Food Microbiology 108: 78–83.

Bermúdez-Aguirre, D., Barbosa-Cánovas, G.V. 2011. An update on high hydrostatic pressure,
from the laboratory to industrial applications. *Food Engineering Reviews* 3: 44–61.

Bigelow, W. 1921. The logarithmic nature of thermal death time curves. *Journal of Infectious
Diseases* 29: 528–536.

Bozoglu, F., Alpas, H., Kaletunç, G. 2004. Injury recovery of foodborne pathogens in high
hydrostatic pressure treated milk during storage. *FEMS Immunology and Medical
Microbiology* 40: 243–247.

Buzrul, S., Alpas, H. 2004. Modeling the synergistic effect of high pressure and heat on the
inactivation kinetics of *Listeria innocua*: A preliminary study. *FEMS Microbiology
Letters* 238: 29–36.

Buzrul, S., Alpas, H., Bozoglu, F. 2005a. Effect of high hydrostatic pressure on quality parameters of lager beer. *Journal of the Science of Food and Agriculture* 85:
1672–1676.

Buzrul, S., Alpas, H., Bozoglu, F. 2005b. Use of Weibull frequency distribution model to
describe the inactivation of *Alicyclobacillus acidoterrestris* by high pressure at different
temperatures. *Food Research International* 38: 151–157.

Buzrul, S., Alpas, H., Bozoglu, F. 2005c. Effects of high hydrostatic pressure on shelf life of
lager beer. *European Food Research and Technology* 220: 615–618.

Buzrul, S. 2007. On the use of Weibull model for isothermal and nonisothermal heat treatments. *Molecular Nutrition and Food Research* 51: 374–375.

Buzrul, S., Alpas, H., Largeteau, A., Demazeau, G. 2008a. Inactivation of *Escherichia coli*
and *Listeria innocua* in kiwifruit and pineapple juices by high hydrostatic pressure.
International Journal of Food Microbiology 124: 275–278.

Buzrul, S., Alpas, H., Largeteau, A., Bozoglu, F., Demazeau, G. 2008b. Compression heating
of selected pressure transmitting fluids and liquid foods during high hydrostatic pressure
treatment. *Journal of Food Engineering* 85: 466–472.

Chen, H., Hoover, D.G. 2003. Pressure inactivation kinetics of *Yersinia enterocolitica* ATCC
35669. *International Journal of Food Microbiology* 87: 161–171.

Corradini, M.G., Peleg, M. 2007. A Weibullian model for microbial injury and mortality.
International Journal of Food Microbiology 119: 319–328.

Couvert, O., Gaillard, S., Savy, N., Mafart, P., Leguérinel, I. 2005. Survival curves of heated
bacterial spores: Effect of environmental factors on Weibull parameters. *International
Journal of Food Microbiology* 101: 73–81.

Davey, K.R., Hall, R.F., Thomas, C.J. 1995. Experimental and model studies of the combined
effect of temperature and pH on the thermal sterilization of vegetative bacteria in liquid.
Transactions of Chemistry E 73: 127–132.

Donsì, G., Ferrari, G., Maresca, P. 2003. On the modeling of the inactivation kinetics of
Saccharomyces cerevisiae by means of combined temperature and high pressure treatments. *Innovative Food Science and Emerging Technologies* 4: 35–44.

Doona, C.J., Feeherry, F.E., Ross, E.W. 2005. A quasi-chemical model for growth and death
of microorganisms in foods by non-thermal and high pressure processing. *International
Journal of Food Microbiology* 100: 21–32.

Doona, C.J., Feeherry, F.E., Ross, E.W., Corradini, M.G., Peleg, M. 2007. The quasi-chemical and Weibull distribution models of non linear inactivation kinetics of Escherichia coli ATCC 11229 by high pressure processing. In: Doona, C.J. and Feeherry, F.E. (Eds.) *High Pressure Processing of Foods*. Blackwell, NY. pp. 115–144.

Erkan, N., Alpas, H., Üretener, G., Selçuk, A., Buzrul, S. 2010. Changes in the physicochemical properties of high pressure treated rainbow trout. *Archiv für Lebensmittelhygiene* 61: 183–188.

Erkan, N., Üretener, G., Alpas, H., Selçuk, A., Özden, Ö., Buzrul, S. 2011. The effect of different high pressure conditions on the quality and shelf life of cold smoked fish. *Innovative Food Science and Emerging Technologies* 12: 104–110.

Erkmen, O., Dogan, C. 2004. Effects of ultra high hydrostatic pressure on *Listeria monocytogenes* and natural flora in broth, milk and fruit juices. *International Journal of Food Science and Technology* 39: 91–97.

Erkmen, O. 2009. Mathematical modeling of *Salmonella typhimurium* inactivation under high hydrostatic pressure at different temperatures. *Food and Bioproducts Processing* 87: 68–73.

Feeherry, F.E., Doona, C.J., Ross, E.W. 2005. The quasi-chemical kinetics model for the inactivation of microbial pathogens using high pressure processing. *Acta Horticulturae* 674: 245–251.

García-Graells, C., Masschalck, B., Michiels, C.W. 1999. Inactivation of *Escherichia coli* in milk by high-hydrostatic-pressure treatment in combination with antimicrobial peptides. *Journal of Food Protection* 62: 1248–1254.

García-Graells, C., Valckx, C., Michiels, C.W. 2000. Inactivation of *Escherichia coli* and *Listeria innocua* in milk by combined treatment with high hydrostatic pressure and the lactoperoxidase system. *Applied and Environmental Microbiology* 66: 4173–4179.

Geeraerd, A.H., Valdramidis, V.P., Van Impe, J.F. 2005. GInaFiT, a freeware tool to assess non-log-linear microbial survivor curves. *International Journal of Food Microbiology* 102: 95–105.

Gervilla, R., Capellas, M., Ferragut, V., Guamis, B. 1997. Effect of high hydrostatic pressure on *Listeria innocua* 910 CECT inoculated into ewe's milk. *Journal of Food Protection* 60: 33–37.

Guan, D., Chen, H., Ting, E.Y., Hoover, D.G. 2006. Inactivation of *Staphlyococcus aureus* and *Escherichia coli* O157:H7 under isothermal-endpoint pressure conditions. *Journal of Food Engineering* 77: 620–627.

Hayakawa, I., Kanno, T., Yoshiyama, K., Fujio, Y. 1994. Oscillatory compared with continuous high pressure sterilization on *Bacillus stearothermophilus* spores. *Journal of Food Science* 59: 164–167.

Hayakawa, I., Furukawa, S., Midzunaga, A., Horiuchi, H., Nakashima, T., Fujio, Y., Yano, Y., Ishikura, T., Sasaki, K. 1998. Mechanism of inactivation of heat-tolerant spores of *Bacillus stearothermophilus* IFO 12550 by rapid decompression. *Journal of Food Science* 63: 371–374.

Koseki, S., Yamamoto, K. 2007a. A novel approach to predicting microbial inactivation kinetics during high pressure processing. *International Journal of Food Microbiology* 116: 275–281.

Koseki, S., Yamamoto, K. 2007b. Modelling the bacterial survival/death interface induced by high pressure processing. *International Journal of Food Microbiology* 116: 136–143.

Leguerinel, I., Couvert, O., Mafart, P. 2000. Relationship between the apparent heat resistance of *Bacillus cereus* spores and the pH and NaCl concentration of the recovery medium. *International Journal of Food Microbiology* 55: 223–227.

Mafart, P. 2000. Taking injuries of surviving bacteria into account for optimising heat treatments. *International Journal of Food Microbiology* 55: 175–179.

Mafart, P., Couvert, O., Gaillard, S., Leguerinel, I. 2002. On calculating sterility in thermal preservation methods: Application of the Weibull frequency distribution model. *International Journal of Food Microbiology* 72: 107–113.

Noma, S., Shimoda, M., Hayakawa, I. 2002. Inactivation of vegetative bacteria by rapid decompression treatment. *Journal of Food Science* 67: 3408–3411.

Palou, E., López-Malo, A., Barbosa-Cánovas, G.V., Welti-Chanes, J., Swanson, B.G. 1997. Kinetic analysis of *Zygosachharomyces bailii* inactivation by high hydrostatic pressure. *Lebensmittel-Wissenschaft und-Technologie* 30: 703–708.

Patazca, A., Koutchma, T., Balasubramaniam, V.M. 2007. Quasi-adiabatic temperature increase during high pressure processing of selected foods. *Journal of Food Engineering* 80: 199–205.

Patterson, M.F., and Kilpatrick, D.J. 1998. The combined effect of high hydrostatic pressure and mild heat on inactivation of pathogens in milk and poultry. *Journal of Food Protection* 61: 432–436.

Peleg, M., Cole, M.B. 1998. Reinterpretation of microbial survival curves. *Critical Reviews in Food Science* 38: 353–380.

Peleg, M. 1999. On calculating sterility in thermal and non-thermal preservation methods. *Food Research International* 32: 271–278.

Peleg, M. 2002. Simulation of E. coli inactivation by carbon dioxide under pressure. *Journal of Food Science* 67: 896–901.

Peleg, M. 2003a. Microbial survival curves: Interpretation, mathematical modeling, and utilization. *Comments on Theoretical Biology* 8: 357–387.

Peleg, M. 2003b. Calculation of the non-isothermal inactivation patterns of microbes having sigmoidal isothermal semi-logarithmic survival curves. *Critical Reviews in Food Science and Nutrition* 43: 645–658.

Peleg, M., Normand, M.D., Corradini, M.G. 2005. Generating microbial survival curves during thermal processing in real time. *Journal of Applied Microbiology* 98: 406–417.

Peleg, M. 2006. *Advanced Quantitative Microbiology for Foods and Biosystems: Models for Predicting the Growth and Inactivation.* Boca Raton, FL: CRC Press.

Pilavtepe-Çelik, M., Buzrul, S., Alpas, H., Bozoğlu, F. 2009. Development of a new mathematical model for inactivation of *Escherichia coli* O157:H7 and *Staphylococcus aureus* by high hydrostatic pressure in carrot juice and peptone water. *Journal of Food Engineering* 90: 388–394.

Pilavtepe-Çelik, M., Buzrul, S., Alpas, H., Largeteau, A., Demazeau, G. 2011. Multi-pulsed high hydrostatic pressure treatment for inactivation and injury of *Escherichia coli*. *Journal für Verbraucherschutz und Lebensmittelsicherheit (Journal of Consumer Protection and Food Safety)* 6: 343–348.

Ratkowsky, D.A., Lowry, R.K., McMeekin, T.A., Stokes, A.N., Chandler, R.E. 1983. Model for the growth rate throughout the entire biokinetic temperature range. *Journal of Bacteriology* 154: 1222–1226.

Ross, E.W., Taub, I.A., Doona, C.J., Feeherry, F.E., Kustin, K. 2005. The mathematical properties of the quasi-chemical model for microorganism growth/death kinetics in foods. *International Journal of Food Microbiology* 99: 157–171.

Skinner, G.E., Larkin, J.W., Rhodehamel, E.J. 1994. Mathematical modelling of microbial growth—A review. *Journal of Food Safety* 14: 175–217.

Tay, A., Shellhammer, T.H., Yousef, A.E., Chism, G.W. 2003. Pressure death and tailing behavior of *Listeria monocytogenes* strains having different barotolerances. *Journal of Food Protection* 66: 2057–2061.

U.S. FDA. Kinetics of microbial inactivation for alternative food processing technologies. 2000. U.S. Food and Drug Administration Center for Food Safety and Applied Nutrition Web site, http://vm.cfsan.fda.gov/ ~ comm/ift-toc.html.

Valdramidis, V.P., Geeraerd, A.H., Poschet, F., Ly-Nguyen, B., Van Opstal, I., Van Loey, A.M., Michiels, C.W., Hendrickx, M.E., Van Impe, J.F. 2007. Model based process design of the combined high pressure and mild heat treatment ensuring the safety and quality of a carrot simulant system. *Journal of Food Engineering* 78: 1010–1021.

van Boekel, M.A.J.S. 2002. On the use of the Weibull model to describe thermal inactivation of microbial vegetative cells. *International Journal of Food Microbiology* 74: 139–159.

van Gerwen, S.J.C., Zwietering, M.H. 1998. Growth and inactivation models to be used in quantitative risk assessments. *Journal of Food Protection* 61: 1541–1549.

Van Opstal, I., Vanmuysen, S.C.M., Wuytack, E.Y., Masschalck, B., Michiels, C.W. 2005. Inactivation of *Escherichia coli* by high hydrostatic pressure at different temperatures in buffer and carrot juice. *International Journal of Food Microbiology* 98: 179–191.

Xiong, R., Xie, G., Edmondson, A.E., Sheard, M.A. 1999. A mathematical model for bacterial inactivation. *International Journal of Food Microbiology* 46: 45–55.

Yamamoto, K., Matsubara, M., Kawasaki, S., Bari, M.L., Kawamoto, S. 2005. Modeling the pressure inactivation dynamics of *Escherichia coli*. *Brazilian Journal of Medical and Biological Research* 38: 1253–1257.

11 Phytochemical Quality, Microbial Stability, and Bioactive Profiles of Berry-Type Fruits, Grape, and Grape By-Products with High-Pressure Processing

Özlem Tokuşoğlu, Barry G. Swanson, and Gustavo V. Barbosa Cánovas

CONTENTS

11.1 INTRODUCTION TO BERRY FRUITS

Berry fruits, commonly called aggregate fruits, have clusters of one-seeded drupelets, each cluster of drupelets developing from a single flower. The drupelets are typically eaten as a cluster and not individually. It is stated that the origin of berries is very complicated and there are numerous cultivated varieties that have been developed through the centuries (Tokuşoğlu and Stoner, 2011).

Berry is a term used for a fruit having succulent pericarp. Berry fruits are a simple type of fruit that are fleshy or succulent at maturity. Cranberry, blueberry, red, black and white currant, and grape are examples of berry fruits (Simpson, 2010; Tokuşoğlu and Stoner, 2011). Although strawberry is an achenecetum type of aggregate fruit, and blackberry and raspberry are a drupacetum type of aggregate fruit; these fruits are accepted as berry fruits by most of the people (Simpson, 2010).

A multitude of phenolic compounds in berries and berry-type fruits has been detected (Häkkinen et al., 1999; Kähkäonen et al, 2003; Moyer et al., 2002; Stöhr and Hermann, 1975; Tokuşoğlu and Stoner, 2011) their content being highly variable in different berries.

Figure 11.1 shows major bioactives in berry fruits, and it is stated that specific bioactives can be found in specific berries (Tokuşoğlu and Stoner, 2011). It is reported that genotype variety is the major factor in determining a fruit's nutritional quality, but it is also affected by crop conditions, including environmental and cultivation methodology, ripening season, preharvest and postharvest conditions, shelf-life, and processing (Connor et al., 2002; Prior et al., 1998; Tokuşoğlu and Stoner, 2011; Wang and Lin, 2000; Wang, 2007). Table 11.1 shows some phenolic compounds found in different fruits.

Phenolic compounds are one of the plant's secondary metabolites, which are generally synthesized through the shikimic acid pathway. Phenolics have one or more hydroxyl groups attached directly to a benzene ring or to another complex aromatic ring structure (Crozier et al., 2006; Dixon and Paiva, 1995; Tokuşoğlu, 2001; Tokuşoğlu and Hall, 2011; Vasco, 2009). Polyphenols are divided into several subgroups, including flavonoids, hydroxybenzoic and hydroxycinnamic acids, lignans, stilbens, tannins, and coumarins that have specific physiological and biological effects (Andersen and Markham 2006; Meskin et al., 2003; Tokuşoğlu, 2001; Tokuşoğlu and Hall, 2011) (Figure 11.2).

Phenolic compounds play some important roles in fruits, such as visual appearance, taste, and aroma. For example, these compounds are responsible for most of

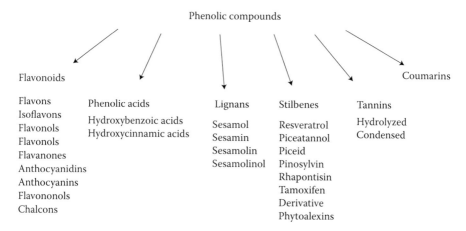

FIGURE 11.1 Family of phenolics. (Compiled by Tokuşoğlu Ö. 2001. The Determination of the Major Phenolic Compounds (Flavanols, Flavonols, Tannins) and Aroma Properties of Black Teas. PhD Thesis. Department of Food Engineering, Bornova, Izmir, Turkey: Ege University.)

TABLE 11.1
Some of the Phenolic Compounds Found in Various Fruits

Phenolic Compound	Source	Fruit Type
Hydroxybenzoic acids	Blackberry	Aggregate fruits–Drupecetum
Protocatechuic acid	Raspberry	Aggregate fruits–Drupecetum
Gallic acid	Black currant	Simple fruits–Berry
p-Hydroxybenzoic acid	Strawberry	Aggregate fruits–Achenecetum
Hydroxycinnamic acids	Blueberry	Simple fruits–Berry
Caffeic acid	Kiwi fruit	Simple fruits–Berry
	Apricot	Simple fruits–Drupe
	Plum	Simple fruits–Drupe
Chlorogenic acid	Cherry	Simple fruits–Drupe
	Cranberry	Simple fruits–Berry
	Apricot	Simple fruits–Drupe
Neochlorogenic acid	Plum	Simple fruits–Drupe
p-Coumaric acid	Plum	Simple fruits–Drupe
	Apricot	Simple fruits–Drupe
	Cranberry	Simple fruits–Berry
	Strawberry	Aggregate fruits–Achenecetum
Ferulic acid	Apricot	Simple fruits–Drupe
Sinapic acid	Apple	Simple fruits–Pome
	Pear	Simple fruits–Pome
Anthocyanins		
Cyanidin	Blackberry	Aggregate fruits–Drupecetum
	Apple	Simple fruits–Pome
	Plum	Simple fruits–Drupe
	Blueberry	Simple fruits–Berry
	Cranberry	Simple fruits–Berry
	Black currant	Simple fruits–Berry
	Grape	Simple fruits–Berry
	Orange	Simple fruits–Hesperidium
	Pear	Simple fruits–Pome
	Pomegranate	Simple fruits–Berry
Pelergonidin	Black currant	Simple fruits–Berry
Peonidin	Blueberry	Simple fruits–Berry
Delphinidin	Black grape	Simple fruits–Berry
	Blueberry	Simple fruits–Berry
	Pomegranate	Simple fruits–Berry
Malvidin	Cherry	Simple fruits–Drupe
	Strawberry	Aggregate fruits–Achenecetum
	Plum	Simple fruits–Drupe
	Blueberry	Simple fruits–Berry
Flavonols		
Quercetin	Cranberry	Simple fruits–Berry
	Raspberry	Aggregate fruits–Drupecetum
	Blueberry	Simple fruits–Berry
	Black currant	Simple fruits–Berry

continued

TABLE 11.1 (continued)
Some of the Phenolic Compounds Found in Various Fruits

Phenolic Compound	Source	Fruit Type
	Orange juice	Simple fruits–Hesperidium
Quercetin 3-rutinoside	Apricot	Simple fruits–Drupe
Kaempferol	Strawberry	Aggregate fruits–Achenecetum
Kaempferol 3-rutinoside	Apricot	Simple fruits–Drupe
Myricetin	Cherry tomato	Simple fruits–Berry
	Blueberry	Simple fruits–Berry
	Black currant	Simple fruits–Berry
	Apricot	Simple fruits–Drupe
	Apple	Simple fruits–Pome
	Black grape	Simple fruits–Berry
	Cranberry	Simple fruits–Berry
Flavanones	Orange juice	Simple fruits–Hesperidium
Hesperidin	Grapefruit juice	Simple fruits–Hesperidium
	Lemon juice	Simple fruits–Hesperidium
	Orange juice	Simple fruits–Hesperidium
Naringenin	Lemon juice	Simple fruits–Hesperidium
	Grapefruit	Simple fruits–Hesperidium
Monomeric Flavanols		
Catechin	Cranberry	Simple fruits–Berry
	Apricot	Simple fruits–Drupe
	Grape	Simple fruits–Berry
	Strawberry	Aggregate fruits–Achenecetum
	Plum	Simple fruits–Drupe
Epicatechin	Apricot	Simple fruits–Drupe
	Cherry	Simple fruits–Drupe
	Grape	Simple fruits–Berry
	Peach	Simple fruits–Drupe
	Blackberry	Aggregate fruits–Drupecetum
	Apple	Simple fruits–Pome
	Apricot	Simple fruits–Drupe

Source: Altuner, E.M. and Tokuşoğlu, Ö. 2013. *International Journal of Food Science and Technology* (IJFST). In Press; Tokuşoğlu, Ö. and Hall, C. 2011. *Fruit and Cereal Bioactives: Sources, Chemistry & Applications.* ISBN: 9781439806654; ISBN-10: 1439806659. 459 page. CRC Press, Taylor & Francis Group, Boca Raton, Florida, USA; Vasco, C. 2009. Phenolic Compounds in Ecuadorian Fruits. PhD. Thesis, Swedish University of Agricultural Sciences, Uppsala; Tomas-Barberan, F.A. and Espin, J.C. 2001. *Journal of the Science of Food and Agriculture,* 81, 853–876; Chen, H., Zuo, Y. and Deng, Y. 2001. *Journal of Chromatography A,* 913, 387–395; Häkkinen, S. 2000. *Flavonols and Phenolic Acids in Berries and Berry Products.* PhD Doctoral dissertation. University of Kuopio, Kuopio.

the blue, purple, red, and intermediate colors of fruits. An increase in the amount of these compounds will cause more colorful fruits. On the other hand, the bitter, sweet, pungent, or astringent tastes of fruits are also caused by some phenolic compounds. Phenolic compounds also have some health-promoting benefits. In addition, previous research has shown that phenolic compounds are relevant to the quality of plant-derived foods and beverages (Beecher, 1999; Crozier et al., 2006; Tokuşoğlu, 2001; Tomas-Barberan and Espin, 2001).

Flavonoids are a chemically defined family of polyphenols that includes several thousand compounds. The main subclasses of flavonoids include flavan-3-ols (catechins), flavonols, flavones, flavanones, isoflavones, anthocyanidins, anthocyanins, flavononols, and chalcons (Figure 11.2) that are distributed in plants and foods of plant origin (Crozier et al., 2006; Tokuşoğlu, 2001).

Anthocyanins are types of phenolic compounds classified under the flavonoids group of phenolic compounds, which are water-soluble glycosides of anthocyanidins. Anthocyanins commonly present in berry juices are based on cyanidin, delphinidin, malvidin, petunidin, and peonidin (Andersen and Markham, 2006; Kähkäonen et al., 2003; Kong et al., 2003; Tokuşoğlu and Stoner, 2011; Tokuşoğlu and Yıldırım, 2012).

Flavonols are also phenolic compounds classified under the flavonoids group. Flavonols are poorly soluble substances and they are present in berries as kaempferol, quercetin, and myricetin (Andersen and Markham, 2006; Häkkinen, 2000; Kühnau, 1976; Rajnarayana et al., 2001; Tokuşoğlu, 2001; Tokuşoğlu and Hall, 2011; Tokuşoğlu et al., 2003).

Phenolic contents of the fruits obviously vary from fruit to fruit. This difference may depend on both the methods used for the extraction of the phenolic compounds

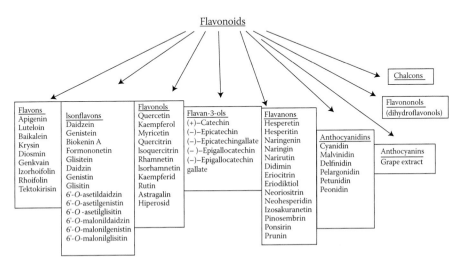

FIGURE 11.2 Flavonoid family in food plants. (Compiled by Tokuşoğlu Ö. 2001. The Determination of the Major Phenolic Compounds (Flavanols, Flavonols, Tannins) and Aroma Properties of Black Teas. PhD Thesis. Department of Food Engineering, Bornova, Izmir, Turkey: Ege University.)

and the methods used for analysis. Besides, the phenolic compound profile in fruits is affected by some intrinsic factors, such as using different genus, species, or cultivars, and extrinsic factors, such as the time when the collecting the fruits, location, environmental factors, and storage. In addition to these intrinsic and extrinsic factors, some food-processing technologies can also affect the plant phenolics' composition (Crozier et al., 2006; Tokuşoğlu, 2001; Tokuşoğlu and Hall, 2011; Tomas-Barberan and Espin, 2001).

There has been great interest shown in the research of flavonoids from dietary sources, owing to growing evidence of the versatile health benefits of flavonoids, including anti-inflammatory, antioxidant, antiproliferative and anticancer activity, freeradical scavenging capacity, antihypertensive effects, coronary heart disease prevention, and anti-human immunodeficiency virus functions (Xiao et al., 2011).

The major phenolics and phytochemicals in berry fruits were reported by Tokuşoğlu and Stoner (2011) (Figure 11.3). Tokuşoğlu and Stoner (2011) stated, in detail, the major phytochemical bioactives in blackberry (*Rubus fruticosus* sp.), blueberry, "highbush" blueberry (*V. corymborsum and V. ashei*), "lowbush" blueberry (*V. augustifolium*), red and black raspberry (*Rubus idaeus* L.), strawberry (*Fragaria X ananassa* Duch), bayberry (*Myrica rubra Sieb.* et Zucc.), chokeberry (*Aronia melanocarpa*), black currant (*Ribes rubrum, R. Petraeum* and *R. sativum*), cranberry (*Vaccinium macrocarpon*), elderberry (*Sambucus nigra* or *Sambucus simpsonii*), and in gooseberry (*Ribes uva-crispa* or *Ribes grossularia*).

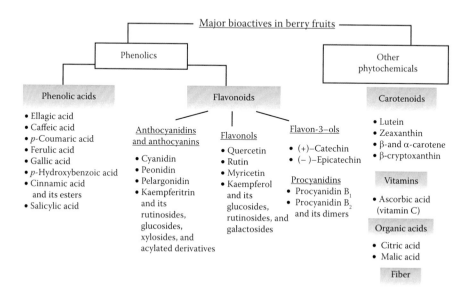

FIGURE 11.3 Major bioactives in berry fruits. (Adapted from Tokuşoğlu, Ö. and Stoner, G. 2011. *Fruit and Cereal Bioactives: Sources, Chemistry & Applications.* Boca Raton, Florida, USA: CRC Press, Taylor & Francis Group, p. 459 ISBN: 9781439806654; ISBN-10: 1439806659.)

Mechanistic studies have shown that the berries positively modulate the expression levels of genes associated with proliferation, apoptosis, inflammation, angiogenesis, cell cycling and adhesion, differentiation, and multiple metabolic processes (Stoner et al., 2008; Wang et al., 2011).

There are some strategies for preserving or increasing phenolic-related quality of foods. Genetic modification, agronomic treatments, and postharvest treatments are some of these strategies. There are also some new processing technologies that produce milder treatments without affecting the content of phenolics, used to obtain food products with a better quality. Edible coatings, thermal treatments, high-pressure treatments, high field electric pulses, microwaves, and ultrasound can be given as examples of these new processing technologies (Altuner and Tokuşoğlu, 2013; Tokuşoğlu and Hall, 2011; Tomas-Barberan and Espin, 2001).

It was put forward that high hydrostatic pressure (HHP) processing affected the extraction, retention, and the stability of anthocyanins and flavonols of berry fruits and juices (Tokuşoğlu and Doona, 2011).

11.2 HHP PROCESSING

HHP processing is a novel nonthermal food preservation process having reduced the effects on the nutritional properties and quality of processed foods when compared to thermal food treatment processes (Rastogi et al., 2007; Tiwari et al., 2009; Tokuşoğlu and Doona, 2011). High pressure is a cold, isostatic, super-high hydraulic pressure that ranges from 100 to 800 MPa or even up to 1000 MPa. HHP has some advantages such as applied pressure that is transmitted uniformly and in all directions independent of the geometry of the food. In addition, pressure can be immediately transferred through a pressure transmitting medium (Oey et al., 2008; Rawson et al., 2011).

It is also known that pressurized cells show increased permeability owing to the mass transfer assessment and it is stated that HHP increases the rate of dissolution (Tokuşoğlu et al., 2010a,b; Zhang et al., 2005). A rapid permeation is observed under high pressure due to the large differential pressure between the cell interior and the exterior of cell membranes. The increasing penetration through the broken membranes into cells increases the mass transfer rate owing to increased permeability. This means the higher the hydrostatic pressure is, the more solvent can enter into the cell (Tokuşoğlu and Doona, 2011; Tokuşoğlu et al., 2011a,b; Zhang et al., 2005). As more solvent enters, more compounds can permeate the cell membrane, which could cause a higher yield of extraction (Ahmed and Ramaswamy, 2006; Dornenburg and Knorr, 1993; Zhang et al., 2005).

HHP can effectively be used to extract active ingredients from plant materials (Altuner et al., 2012a,b; Altuner and Tokușoğlu, 2013; Tokușoğlu et al. 2010; Zhang et al., 2005). High-pressure extraction (HPE) is considered an alternative method for the extraction of active ingredients from plant products, which is proven to be fast and more effective than most of the other extraction methods (Altuner et al., 2012a; US Food and Drug Administration Center [USFDAC], 2000; Zhang et al., 2004). Recently, some authors (Altuner et al., 2012b; Corrales et al., 2008a,b,c; Tokușoğlu et al., 2012, 2011a,b; Zhang et al., 2004, 2005) have reported that HPE techniques can reduce the processing time and obtain a higher extraction yield, and HPE has some advantages for the extraction of natural products (Tokuşoğlu, 2012; Tokuşoğlu and Doona, 2011). Moreover, this technology has been used successfully for the extraction of flavonoids from propolis (Zhang et al., 2005), anthocyanins from grape skin (Corrales et al., 2008a), ginsenosides from the roots of *Panax ginseng* (Zhang et al., 2007), flavonoids from grape pomaces (GPs) (Tokuşoğlu et al., 2012, 2011a), from berry-based foods (Tokuşoğlu et al., 2011b), and so on. High pressure can cause some structural alterations, including cellular deformation, cellular membrane damage and protein denaturation, owing to the pressure applied (Richard, 1992; Zhang et al., 2005).

11.3 EFFECT OF HHP ON THE BERRY FRUIT PHYTOCHEMICALS

Several researchers have conducted examinations about the effect of high hydrostatic pressure processing (HHPP) on the retention of total phenolic content, total monomeric anthocyanins (TMAs), kaempferol, myricetin, and quercetin content of berry fruits and juices.

Corrales et al. (2008c) reported an insignificant reduction of cyanidin-3-glucoside at HHPP by applying 600 MPa at 20°C for 30 min. They also determined that increasing pressure increases anthocyanin reduction at higher temperatures.

Giacarini (2008) tested the effect of HHP on cranberry juices. According to the results, TMA content of cranberry juice increased immediately after HHPP; but TMA content was negatively affected by storage conditions such as time and temperature. Anthocyanin pigment values were observed to decrease after a month of storage, and the samples stored at 37°C had lower TMA values than the samples stored at 22°C. HHPP at 278 MPa for 15 min is proposed to be an alternative processing method to the pasteurization of cranberry juices. Oey et al. (2008) have reported that the anthocyanin content of various fruits and vegetables is minimally affected by HHP treatment at room temperature.

Patras et al. (2009) observed the impact of HHPP on total phenols of strawberry and blackberry purées. They compared the amount of total phenols in HHPP at room temperature with the amounts identified in both unprocessed and thermally processed samples. As a result of the study, they found that the amount of total phenols increases as the pressure increases in HHPP, whereas the amount of total phenols is lower in thermally processed samples than the amount in unprocessed samples. The results of this study are given in Table 11.2.

In another research, Cao et al. (2011) tested the change in kaempferol, quercetin, and myricetin in strawberry pulps with HHPP applications at 400, 500, and 600 MPa for different processing times. According to the results, the change in the kaempferol amount is not very significant when compared to the control sample. The myricetin amount falls below the control level for all pressures and times, but the quercetin amount is higher than the control level at 500 and 600 MPa (Table 11.3).

Barba et al. (2013) reported the physicochemical and nutritional characteristics of blueberry juice after high-pressure processing. The total phenolic content in the blueberry juice was observed to increase, mainly after HHP at 200 MPa for all treatment times such as 5, 9, and 15 min. The total and monomeric anthocyanin was

TABLE 11.2

Comparison of the Amount of Total Phenols in Unprocessed, HHPP, and Thermally Processed Strawberry and Blackberry Samples

Process	Strawberry (mg GAE/100 g Dry Weight)	Blackberry (mg GAE/100 g Dry Weight)
Unprocessed	855.02 ± 6.52	1694.19 ± 3.0
Thermal	817.01 ± 5.26	1633.62 ± 8.4
HHPP, 400 MPa	859.03 ± 6.56	1546.26 ± 8.0
HHPP, 500 MPa	926.00 ± 5.93	1724.65 ± 0.7
HHPP, 600 MPa	939.01 ± 0.99	1778.44 ± 6.0

Source: Adapted from Patras, A. et al. 2009. *Innovative Food Science and Emerging Technologies*, 10, 308–313.

TABLE 11.3

Phenolic Componds (mg kg⁻¹) in Strawberry Pulps Treated with High Hydrostatic Pressure

Conditions	Kaempferol	Quercetin	Myricetin
Control	3.6 ± 0.2	9.5 ± 0.5	6.4 ± 0.2
400 MPa, 5 min	3.3 ± 0.1	8.7 ± 0.2	6.0 ± 0.3
400 MPa, 10 min	3.4 ± 0.1	8.4 ± 0.3	5.9 ± 0.2
400 MPa, 15 min	3.5 ± 0.1	9.3 ± 0.3	6.1 ± 0.3
400 MPa, 20 min	3.6 ± 0.1	9.5 ± 0.4	6.2 ± 0.3
400 MPa, 25 min	3.6 ± 0.1	9.4 ± 0.5	6.2 ± 0.3
500 MPa, 5 min	3.5 ± 0.1	9.6 ± 0.4	5.8 ± 0.2
500 MPa, 10 min	3.6 ± 0.1	9.5 ± 0.3	6.0 ± 0.3
500 MPa, 15 min	3.4 ± 0.1	9.9 ± 0.1	5.9 ± 0.2
500 MPa, 20 min	3.5 ± 0.1	10.1 ± 0.5	5.8 ± 0.3
500 MPa, 25 min	3.4 ± 0.1	9.8 ± 0.3	6.1 ± 0.3
600 MPa, 5 min	3.3 ± 0.1	10.0 ± 0.3	6.1 ± 0.2
600 MPa, 10 min	3.3 ± 0.1	9.3 ± 0.2	6.0 ± 0.5
600 MPa, 15 min	3.4 ± 0.1	9.5 ± 0.4	6.2 ± 0.4
600 MPa, 20 min	3.5 ± 0.2	10.0 ± 0.3	6.2 ± 0.5
600 MPa, 25 min	3.4 ± 0.1	9.8 ± 0.3	6.3 ± 0.3

Source: Extracted from Cao, X. et al. 2011. *Journal of the Science of Food and Agriculture*, 91, 877–885.

similar or higher than the value estimated for the fresh juice, being maximum at 400 MPa for 15 min (Barba et al., 2013).

The antioxidant capacity values were not statistically different for treatments at 200 MPa for 5–15 min in comparison with fresh juice, however, for 400 MPa/15 min and 600 MPa for all times (8–16% reduction) (Barba et al., 2013).

Suthanthangjai et al. (2005) reported the impact of high pressure on the color molecules in raspberries. Fruit samples were pressured under 200, 400, 600, and 800 MPa for 15 min at a temperature controlled between 18°C and 22°C. After the application of pressure, the high-pressure-treated samples were kept at refrigerator temperature (4°C), room temperature (20°C), and at 30°C. Two pigments were identified and quantified: cyanidin-3-glucoside and cyanidin-3-sophoroside, and the highest stability of the anthocyanins was found when raspberries were pressured under 200 and 800 MPa and stored at 4°C.

Zabetakis et al. (2000) stated the effect of HHP on the strawberry anthocyanins. In the study described by Zabetakis et al. (2000), the color stability of fruit juice made from strawberries (*Fragaria × ananassa*, cv. Elsanta) that were subjected to HHP was studied by measuring the anthocyanin content. The samples were pressurized under 200, 400, 600, and 800 MPa for 15 min at a temperature controlled between 18 and 22°C and the high-pressure-treated samples were kept at refrigerator temperature (4°C), room temperature (20°C), and 30°C. Two pigments were identified and quantified as pelargonidin 3-glucoside and pelargonidin 3-rutinoside, and it was found that the highest stability of the anthocyanins is achieved when strawberries were stored at a temperature of 4°C (Zabetakis et al., 2000). It was stated that HP treatment at 800 MPa led to the lowest losses at 4°C (Zabetakis et al., 2000).

Suthanthangjai et al. (2005) analyzed the HHP effects on the anthocyanins of raspberry (*Rubus idaeus*). Raspberries were pressured under 200, 400, 600, and 800 MPa for 15 min at a temperature controlled between 18°C and 22°C. After HHP, the high-pressure-treated samples were kept at refrigerator temperature (4°C), room temperature (20°C), 30°C and then analyzed after 1, 2, 4, 7, and 9 days of storage by high-performance liquid chromatography (HPLC)-UV using an isocratic elution system. Two pigments were identified and quantified as cyanidin-3-glucoside (C3G) and cyanidin-3-sophoroside (C3S). It was found that greater stability was achieved at 800 MPa for C3G and C3S at storage temperature of 4°C.

Kouniaki et al. (2004) reported the effect of HHP on anthocyanins and ascorbic acid in black currants (*Ribes nigrum*). The stability of two anthocyanins compound and ascorbic acid found in fruit juices made from black currants subjected to high-pressure processing (HPP) was presented. In the study performed by Kouniaki et al. (2004), the HHP impacts on the nutritionally important anthocyanins, cyanidin-3-rutinoside, delphinidin-3-rutinoside, and also ascorbic acid (Vitamin C) in black currants was assessed. Fruit samples were treated at 200, 400, 600, and 800 MPa for 15 min under controlled temperature (20–22.5°C) and after high-pressure treatment the samples were stored at refrigerator temperature (5°C), room temperature (20°C) and at 30°C. The best stability for cyanidin-3-rutinoside pigment was found when black currants were treated at a HP of 600 MPa and were stored at 5°C, whereas delphinidin-3-rutinoside had the lowest losses when treated at 800 MPa and stored at

5°C. Kouniaki et al. (2004) found that high pressure at 600 MPa gave the best preservation for all the samples stored at 5°C and 20°C, and the rates of losses observed for anthocyanins were linked to those observed for the ascorbic acid (Kouniaki et al., 2004). HP processing up to 600 MPa and ambient temperature did not significantly affect the Vitamin C level (Sancho et al., 1999).

It was reported that HP processing at pressures of up to 500 MPa and ambient temperature did not affect the aroma profile of strawberries (Lambert et al., 1999).

Zabetakis et al. (2000) stated the effect of HP processing at 200–800 MPa/15 min at 18–22°C on strawberry anthocyanins and no alteration was found in the concentration of pelargonidin-3-glucoside and pelargonidin-3-rutinoside following HP treatment. Zabetakis et al. (2000) and Kouniaki et al. (2004) identified that the highest storage stability of anthocyanins content in strawberry and black currant was established after applying a pressure of 800 MPa for 15 min. The reason for the stability of pelargonidin-3-glucoside and pelargonidin-3-rutinoside types of anthocyanins found in strawberry and red raspberry at 800 MPa for 15 min at moderate temperature (18°C–22°C) was explained by Garcia-Palazon et al. (2004). This stability was maintained due to complete inactivation of polyphenoloxidase under this condition. Cano et al. (1997) reported that enzymes such as polyphenoloxidase, peroxidase, and β-glucosidase act important roles in the degradation of anthocyanins.

Hilz et al. (2006) stated the effect of HPP on cell wall polysaccharides in berries. It was found that the degree of methyl esterification (DM) decreased by high-pressure processing due to the possible activation of pectin methyl esterase improving the extractability of pectins. According to the data obtained by Hilz et al. (2006), the activity of rhamnose-releasing enzymes in minor quantities might be enhanced after HPP, resulting in a decrease of rhamnose in the polymeric cell wall material. Based on the data by Hilz et al. (2006), high pressure can be the possible processing technique for polysaccharide modification.

Engmann et al. (2013) stated the high-pressure processing effect on anthocyanins composition of mulberry (Morus moraceae) (*Morus alba* L.) juice (Engmann et al., 2013). In the study performed by Engmann et al. (2013), the anthocyanin constituents (Figure 11.4) were determined after treatments of 200, 400, and 600 MPa for 20 min. Anthocyanins were identified and characterized using HPLC and electrospray ionization–mass spectrometry (ESI/MS). Engmann et al. (2013) found the cyanidin-3-*O*-glucopyranoside (55.56%) and cyanidin-3-*O*-coumaroylglucoside (44.44%) in the untreated sample, while two new anthocyanins [pelargonidin-3-*O*-coumaroylglucoside (0.46%) and delphinidin-3-*O*-coumaroylglucoside (5.8%)] were identified in the sample treated at 200 MPa for 20 min. One new anthocyanin, delphinidin-3-*O*-coumaroylglucoside (5.38%), was found in the juice treated at 400 MPa for 20 min (Figure 11.5), whereas no new anthocyanins were detected at 600 MPa for 20 min (Engmann et al., 2013).

Tokuşoğlu et al. (2011b) stated that instrumental viscosity for HHP-treated berry ice creams was significantly greater than that of the untreated berry ice creams ($p < 0.05$). It was found that the viscosity of HP-treated ice cream mix increased with increasing pressure and treatment time and with increasing phenolic content ($p < 0.05$) (Tokuşoğlu et al. 2011b).

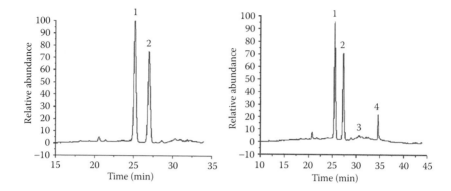

FIGURE 11.4 HPLC-DAD recorded at 520 nm for untreated mulberry juice (left figure) and for mulberry juice treated at 200 MPa for 20 min (right figure). (1,2,3,4 numbered peaks correspond to cyanidin-3-*O*-glucoside, cyanidin-3-*O*-rutinoside, pelargonidin-3-*O*-coumaroyl glucoside, delphinidin-3-*O*-coumaroylglucoside, respectively.) (Adapted from Engmann, N.F. et al. 2013. *Czech Journal of Food Science*, 31(1), 72–80.)

In the study described by Tokuşoğlu et al. (2011b), prior to HP processing, each ice cream mix and samples were weighed, gently transferred to 7.6 × 15.2 cm plastic pouches (Whirl-Pak bags, ~110 mL; Cole-Parmer Instrument Co., Vernon Hills, IL, USA) and vacuum-sealed by hand-pressure plastic bag sealing machine (Sealer Sales, Inc., CA, USA) (Figure 11.6).

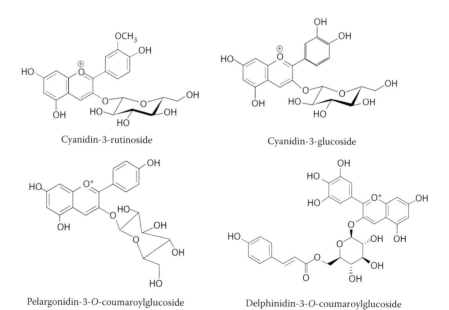

FIGURE 11.5 Mulberry (*Morus moraceae*) juice anthocyanins. (Compiled by Tokuşoğlu, Ö. 2014. Functional and MultiPurpose Foods. *Lecture Notes*. Celal Bayar University, Manisa. Unpublished.)

FIGURE 11.6   Prior to HPP, vacuum sealed mix samples through hand-pressure plastic bag sealing machine. (Adapted from Tokuşoğlu Ö. et al. 2011a. High pressure processed black grape pomaces: Multivariate analysis of major flavan-3-ol phenolics, antioxidant activity and microbiological quality profiles. Nonthermal Processing Section Technical Research Paper. 2011-TRP-2908-IFT 2011 IFT Annual Meeting+Food Expo, New Orleans, Louisina, USA.)

The strawberry, huckleberry, and blackberry ice cream mixes were pressurized with HHP at 300, 400, 500 MPa and held during 10/30 min at $25 \pm 1°C$ in sealed packages and equipped with a cylindrical pressure chamber (of 0.1 m diameter, 0.25 m height) at WSU FSHN Pilot Plant (Tokuşoğlu et al. 2011b). The rheology, overall quality, and antioxidant activity tests were performed (see Chapter 3). After HHP processing of mixes, the manufactured ice ceam samples were analyzed by HPLC (Biological Systems Engineering, Pullman, WSU) with gradient elution using three mobile phase conditions as shown below (Figure 11.7) (Tokuşoğlu et al. 2011b). Figure 11.8 shows the simultaneous HPLC chromatogram of the major phenolics (procyanidin B_1, catechin, quercetin) of manufactured huckleberry ice cream (Figure 11.8).

A 2.9-fold increase in quercetin levels for huckleberry ice creams was detected (Tokuşoğlu et al., 2011b). It is stated that cell permeability and diffusion of secondary metabolites increased by using HPP. Therefore, the extraction capacity of phenolic constituents of berries in mixes increased after HHP and higher levels of bioactive compounds were preserved in HPP-treated ice creams. In the study described by Tokuşoğlu et al. (2011b), quercetin levels highly increased while other detected phenolics were considerably preserved in HHP-treated samples. HHP is a cold-processing technique for quality in berry ice cream mixtures, and it was determined that higher levels of bioactives were retained in HPP-treated samples (Tokuşoğlu et al., 2011b).

11.4   EFFECT OF HHP ON THE MICROBIAL STABILITY OF BERRY FRUITS AND DRINKS

Berry fruits like strawberries are some of the most delicate fruits with an extremely short postharvest life. Recently, the microbial safety of strawberries have been raised owing to the implication of strawberries and some other berry fruits in several large foodborne outbreaks. It was stated that strawberries and

Time (min)	Flow rate (ml/min)	A%	B%
Initial	0.80	90	10
5.01	0.80	90	10
40.00	0.80	70	30
55.00	1.0	0	100
60.00	1.0	0	100
60.01	1.0	90	10
65.00	1.0	90	10

A:	3% acetic acid
B:	80% ACN, 19.6% H_2O, 0.4% acetic acid

FIGURE 11.7 The high performance liquid chromatographic analyses conditions of HHP processed samples (Varian, Photodiode Array Detector (PDA) model 330-HPLC; Washington State University, Biological Systems Eng., Analytical Chemistry Service Center, Pullman, WA) by Tokuşoğlu, WSU-2010.

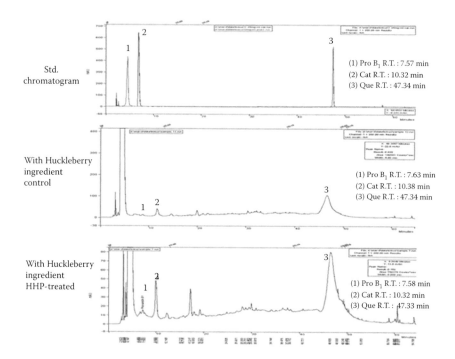

FIGURE 11.8 HHP effects on major polyphenols (procyanidin B_1, catechin, quercetin) in Huckleberry ice cream. ((Adapted from Tokuşoğlu Ö. et al. 2011b. Rheological, Microbial Quality Characteristics and Antioxidant Activity of High Hydrostatic Pressure (HHP) Processed Berry Ice-Cream Mixes: Weibull Modeling and Response Surface Methodology Approach for Treatment Pressure and Holding Time Effects. *Nonthermal Processing Division Workshop 2011-Innovation Food Conference. October 12-14, Osnabrück*, Germany. Oral Presentation.)).

raspberries have been associated with outbreaks of hepatitis A virus and human norovirus (Niu et al., 1992; Ransay and Upton, 1989; Reid and Robinson, 1987 Sarvikivi et al., 2012).

It was reported that fresh strawberries from a farm in Oregon were linked with an *Escherichia coli* O157:H7 outbreak, that caused at least 15 people sick including one death (ODA, 2011). U.S. Food and Drug Administration (FDA) conducted a survey in 1999 and it was found that 1 out of 143 imported strawberry samples was tested positive for *Salmonella* (FDA, 2001).

It is known that strawberries are usually sold as fresh or processed into frozen berries or puree. Strawberry puree is an intermediate product and is an important industrial ingredient that is widely used in a variety of foods, including yogurt, jams or jellies, popsicle, smoothies, snacks, and different drinks. For food safety on strawberry and strawberry products, the extending shelf life, and inactivating pathogens, a thermal treatment (88°C for 1.5–2 min) is usually required. It is reported that thermal pasteurization processes are sometimes not adequate to pathogen elimination (Baert et al., 2008).

Huang et al. (2013) stated the inactivation of *Escherichia coli* O157:H7 and *Salmonella* spp. in strawberry puree (SP) by HHP with/without subsequent frozen storage. In the study described by Huang et al. (2013), fresh SP was inoculated with high (~6 log CFU/g) and low (~3 log CFU/g) levels of *E.coli* O157:H7 or *Salmonella* spp. and stored at −18°C for 12 weeks. It was found that pressure treatment of 450 MPa/2 min at 21°C was able to eliminate both pathogens in SP (Figures 11.9 and 11.10) (Huang et al., 2013). It was determined that 4–8 days of frozen storage was able to reduce the pressure level needed for the elimination of both pathogens to 250–300 MPa, whereas natural yeasts and molds in SP were effectively reduced by pressure of 300 MPa/2 min at 21°C. No adverse influences on physical characteristics was reported, including color, soluble solids content, pH, and viscosity of HPP-treated SPs (Huang et al., 2013). Table 11.4 shows the physical properties and yeasts and molds counts of the control and HHP processed SP during 8 days frozen storage (Huang et al., 2013). It was put forward that 300 MPa/2 min treatment at 21°C followed by 4 days frozen storage at −18°C was recommended for the minimal processing of SP with a great retention for fresh-like sensory properties. It was concluded that HPP could be a promising alternative to traditional thermal processing for berry purees (Huang et al., 2013).

11.5 EFFECTS OF HHP ON THE PHYTOCHEMICALS IN GRAPES, GRAPE-BASED DRINKS, GRAPE BY-PRODUCTS AND GRAPE POMACES

The grape is one of the most grown fruit in the world, and consequently, it is used as raw material for various food products. Grape fruits have a large amount of polyphenols that have antioxidative properties and also include organic acids and sugar. These substances are very sensitive in conventional food processes and can be degraded easily. Recently, the utilization of novel food-processing techniques, including high-pressure processing (HHP) on grape and grape-derived food products, has been

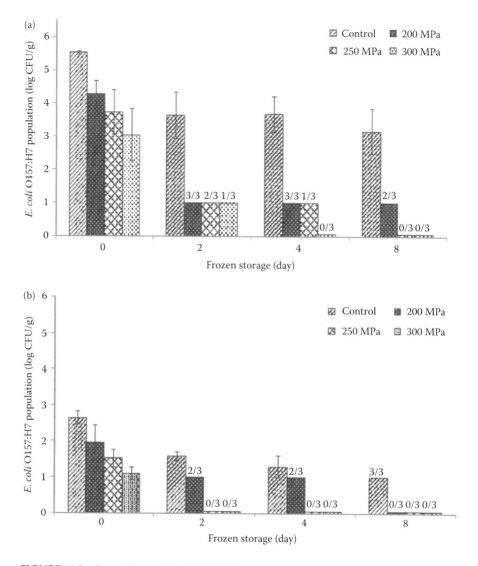

FIGURE 11.9 Populations of *E. coli* O157:H7 in the control and pressure-treated strawberry puree with high (a) and low (b) inoculation levels during 8 days frozen storage. Inoculated samples were treated at 200–300 MPa/2 min at 21°C and then stored at −18°C. Error bars shown in the figures represent one standard deviation. Enrichment was conducted when the bacterial counts were below the detection limit by the plating method (1 log CFU/g). Numbers in fraction represent the number of samples testing positive after enrichment out of a total of 3 trials. (Adapted from Huang, Y., Ye, M. and Chen, H. 2013. *International Journal of Food Microbiology*, 160, 337–343.)

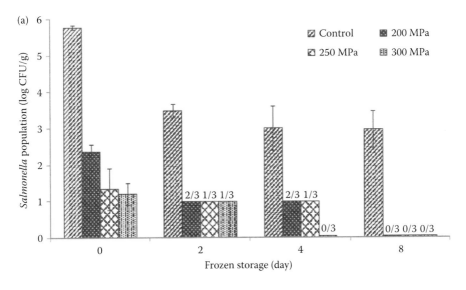

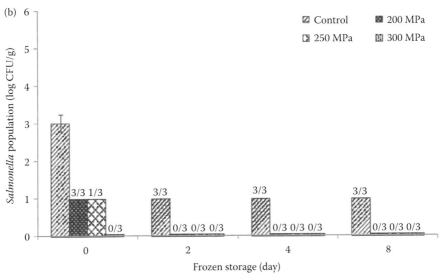

FIGURE 11.10 Populations of *Salmonella* spp. in the control and pressure treated strawberry puree with high (a) and low (b) inoculation levels during 8 days of frozen storage. Inoculated samples were treated at 200–300 MPa/2 min at 21°C and then stored at −18°C. Error bars shown in figures represent one standard deviation. Enrichment was conducted when the bacterial counts were below the detection limit by the plating method (1 log CFU/g). Numbers in fraction represent the number of samples testing positive after enrichment out of a total of 3 trials. (Adapted from Huang, Y., Ye, M. and Chen, H. 2013. *International Journal of Food Microbiology,* 160, 337–343.)

TABLE 11.4

Physical Properties and Yeast and Mold Counts of the Control and Pressure-Treated Strawberry Puree (SP) During 8 Days Frozen Storage. Uninoculated SP was Treated at 250–300 MPa/2 min at 21°C and Then Stored at −18°C

Treatment	Day	Yeast and Mold (log CFU/g)	Viscosity (Pa s)	SSC (°Brix)	pH	Color L*	Color a*
Control	0	4.3 ± 0.0[aA]	5.1 ± 0.0[cA]	8.8 ± 0.3[aA]	3.57 ± 0.03[aA]	30.7 ± 0.2[aA]	17.3 ± 0.5[aA]
	4	4.1 ± 0.1[aAB]	4.7 ± 0.0[cB]	8.6 ± 0.3[aA]	3.59 ± 0.02[aA]	30.4 ± 1.6[aA]	16.4 ± 2.9[aA]
	8	3.8 ± 0.2[aB]	4.7 ± 0.0[cB]	8.9 ± 0.1[aA]	3.60 ± 0.01[aA]	31.1 ± 0.3[aA]	17.5 ± 0.8[aA]
250 MPa	0	1.9 ± 0.3[bA]	5.8 ± 0.0[bA]	9.0 ± 0.1[aA]	3.60 ± 0.01[aA]	31.8 ± 0.6[aA]	19.6 ± 1.1[aA]
	4	2.1 ± 0.1[bA]	4.9 ± 0.0[bB]	8.4 ± 0.3[aB]	3.61 ± 0.01[aA]	31.0 ± 0.3[aA]	17.3 ± 0.5[aB]
	8	1.9 ± 0.3[bA]	4.8 ± 0.0[bB]	8.7 ± 0.2[aAB]	3.62 ± 0.02[aA]	31.1 ± 0.1[aA]	17.4 ± 0.2[aB]
300 MPa	0	< 1[cA]	6.1 ± 0.0[aA]	8.6 ± 0.1[aA]	3.59 ± 0.01[aA]	31.7 ± 0.6[aA]	19.5 ± 0.9[bA]
	4	< 1[cA]	5.3 ± 0.0[aB]	8.5 ± 0.1[aA]	3.59 ± 0.01[aA]	31.4 ± 0.8[aA]	18.4 ± 1.3[aA]
	8	< 1[cA]	5.3 ± 0.0[aB]	8.6 ± 0.2[aA]	3.62 ± 0.01[aB]	31.8 ± 1.0[aA]	18.8 ± 1.8[aA]

Source: Adapted from Huang, Y., Ye, M., Chen, H. 2013. *International Journal of Food Microbiology* 160, 337–343.

Note: The limits of detection were 1 log CFU/g for the plating method. Data represent mean value of three replicates ± 1 standard deviations. Data in the same column and storage day followed by the same lowercase letter are not significantly different ($p > 0.05$). Data in the same column and treatment category followed by the same uppercase letter are not significantly different ($p > 0.05$).

extensively studied (Güler and Tokuşoğlu, 2012). Several researchers conduct studies about the effect of HHP on the grape phenolics, and polyphenols of grape-based foods. There are limited studies concerning high pressure processed grape products.

Corrales et al. (2008a) reported the extraction of anthocyanins from grape skins assisted by HHP. It was found that the antioxidant capacity of the samples extracted at 200, 400, and 600 MPa (expressed as TROLOX equivalents (TE) g^{-1} DM) was up to three-fold higher than control samples (Corrales et al., 2008a). Figure 11.11 (left) shows the effect of HHP (600 MPa) at different temperatures on the antioxidant capacity of Dornfelder *Vitis vinifera* L. grape skin extracts. It was determined that the antioxidant capacity of the extracts was higher in heat-pressurized samples than in the control samples (Figure 11.11, left) (Corrales et al., 2008a).

It was known that temperature played an important role in the extraction process because high extraction temperatures increased the solubility of phenolics and the diffusion coefficient, which thereby reduced the extraction times (Cacace and Mazza, 2003; Lou et al., 1997). Corrales et al. (2008a) detected that the extraction yields at 600 MPa/70°C were two-fold higher than the controls, but the antioxidant capacity obtained at 90°C was significantly lower than at 70°C (Figure 11.11, left) (Corrales et al., 2008a). This situation may be due to the thermal degradation and oxidation of extract constituents that are sensitive when applied at high temperatures (Kirca and Cemeroglu, 2003).

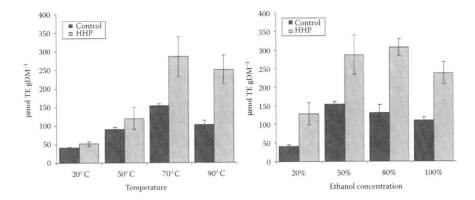

FIGURE 11.11 The effect of HHP (600 MPa) at different temperatures on the antioxidant capacity of Dornfelder *Vitis vinifera* L. grape skin extracts expressed as 1 molTEg⁻¹DM (left figure). The effect of HHP and different ethanol concentrations on the antioxidant capacity of these grape skin extracts (right figure). Extractions were carried out for 30 min using an ethanol concentration of 50%. Bars represent mean ± standard deviation, $n = 3$. (Adapted from Corrales, M. et al. 2008a. *Journal of Food Engineering*, 90, 415–421.)

Figure 11.11 (right) shows the effect of HHP and different ethanol concentrations on the antioxidant capacity of studied grape skin extracts. It was detected that HHP increased the antioxidant capacity by more than double when compared to control extracts and the highest recovery achieved at ethanol concentrations of 50% and 80% were 286.08 and 306.61 µmolTE g⁻¹DM, respectively (Figure 11.11, right) (Corrales et al., 2008a).

Corrales et al. (2008b) showed that the extraction of anthocyanins from grape by-products increased with the application of 600 MPa pressure, when compared to the conventional extraction methods as shown in Table 11.5.

Figure 11.12 shows the total phenolic content (µmol GAE g⁻¹ DM) and the antioxidant capacity (µmol TE g⁻¹ DM) level for grape by-products extracted by ultrasonics, HHP, and pulsed electrical field (PEF) (Corrales et al., 2008b).

In the study reported by Corrales et al. (2008b), the total phenolic content of grape by-products extracted with different methods ultrasonics, HHP, and PEF were not significantly different ($p < 0.05$) in comparison to the control samples (Figure 11.12, left). It was found that the antioxidant activity was significantly different among the various treatments utilized ($p < 0.05$), and the highest antioxidant content (784.34 ± 150.41 µmol TE g⁻¹ DM) was obtained in the (PEF) treated samples followed by HHP (548.49 ± 47.97 µmol TE g⁻¹ DM) (Figure 11.12, right), Corrales et al. (2008b)). Barbosa-Canovas et al. (1998) stated that HHP increased the extraction yields owing to its aptitude to deprotonate charged groups and disrupt salt bridges and hydrophobic bonds in cell membranes that may lead to a higher permeability.

Corrales et al. (2008c) reported the cyanidin-3-glucoside in a model solution at processing conditions of 600 MPa, 20°C, and 30 min. A 25% loss was reported at 600 MPa,

TABLE 11.5
Anthocyanin Content (mg Cyanidin-3-Glucoside eq.g⁻¹ DM) From Grape By-Products

Compound	Control	HHP
Anthocyanins Monoglucosides		
Delphinidin-3-glucoside	0.40 ± 0.016	0.47 ± 0.229
Cyanidin-3-glucoside	0.30 ± 0.002	0.33 ± 0.073
Petunidin-3-glucoside	0.48 ± 0.002	0.66 ± 0.034
Peonidin-3-glucoside	4.22 ± 0.017	2.90 ± 0.333
Malvidin-3-glucoside	2.06 ± 0.061	1.68 ± 0.260
Subtotal	7.46 ± 0.098	6.05 ± 0.929
Acylated Anthocyanins Monoglucosides		
Delphinidin-3-acetylglucoside	0.28 ± 0.001	0.74 ± 0.597
Petunidin-3-acetylglucoside	0.30 ± 0.008	0.32 ± 0.032
Peonidin-3-acetylglucoside	0.35 ± 0.002	0.39 ± 0.097
Malvidin-3-acetylglucoside	0.50 ± 0.005	0.62 ± 0.326
Cyanidin-3-*p*-coumaroylglucoside	0.30 ± 0.006	0.32 ± 0.02
Petunidin-3-*p*-coumaroylglucoside	0.38 ± 0.004	0.43 ± 0.053
Peonidin-3-*p*-coumaroylglucoside	0.53 ± 0.012	0.59 ± 0.168
Malvidin-3-*p*-coumaroylglucoside	1.17 ± 0.053	1.70 ± 0.222
Subtotal	0.48 ± 0.091	5.15 ± 1.518
Total content	7.93 ± 0.189	11.21 ± 2.447

Source: Extracted from Corrales, M. 2008b. *Innovative Food Science and Emerging Technologies*, 9, 85–91.

70°C/30 min compared to a 5% loss at 70°C for 30 min; this indicated the HHP accelerated anthocyanin reduction at elevated temperatures (Corrales et al. 2008c). Barbagallo et al. (2007) reported that enzymatic degradation of anthocyanins by β-glucosidase is mainly owed to the loss of glycosidic moiety, leading to the formation of aglycon (anthocyanidin) consequently affecting juice color (Barbagallo et al., 2007).

Del Pozo-Insfran et al. (2007) stated the HHP effects on anthocyanins and ascorbic acid levels of muscadine grape juices. In the study reported by Del Pozo-Insfran et al. (2007), the effect of polyphenol oxidase (PPO) activity on the phytochemical stability of an ascorbic acid-fortified muscadine grape juice following HHP treatment (400 and 550 MPa for 15 min) and after 21 days of storage at 25°C was analyzed. After 400 and 550 MPa HHP processing for 15 min, delphinidin 3,5-diglycoside, petunidin 3,5-diglycoside, peonidin 3,5-diglycoside, and malvinidin 3,5-diglycoside profiles were studied. Total anthocyanin loss of 70% and 46% after 400 and 550 MPa treatment was found, respectively (Del Pozo-Insfran et al., 2007).

Del Pozo-Insfran et al. (2007) stated that the addition of rosemary and thyme polyphenolic extracts (copigmentation) were evaluated as a means to stabilize anthocyanins

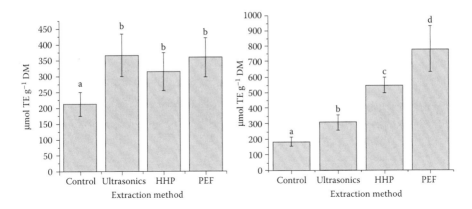

FIGURE 11.12 The total phenolic content (μmol GAE g^{-1} DM) (left figure) and antioxidant capacity (μmol TE g^{-1} DM) from grape by-products extracted by ultrasonics, HHP and PEF. Different letters above bars indicate significant differences between mean values ($p < 0.05$). (Adapted from Corrales, M. et al. 2008b. *Innovative Food Science and Emerging Technologies*, 9, 85–91.)

and ascorbic acid during pressurization and subsequent storage. Figure 11.13 shows total anthocyanin content of muscadine grape juice as affected by HHP processing and copigmentation with rosemary or thyme copigments in the absence (A) or presence (B) of ascorbic acid (450 mg/L) (Figure 11.13). It was found that copigmentation increased anthocyanin retention in reference to pressurized controls and 3 and 3.2-fold for rosemary and thyme treatments were determined, respectively. In stored juices, a higher anthocyanin content (>2 fold) and antioxidant capacity (>1.5 fold)

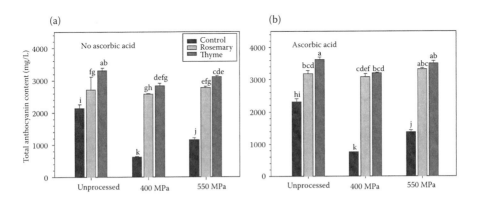

FIGURE 11.13 Total anthocyanin content of muscadine grape juice as affected by HHP processing and copigmentation with rosemary or thyme copigments in the absence (a) or presence (b) of ascorbic acid (450 mg/L). Bars with different letters for each processing treatment are significantly different (LSD test, $p < 0.05$). (Adapted from Del Pozo-Insfran, D., et al. 2007. *Journal of Food Science*, 72(4), 247–253.)

were detected for copigmented treatments when compared to control grape juices (Del Pozo-Insfran et al., 2007).

It is known that black grape juice is considered to have health-promoting properties due to the presence of rich amounts of antioxidants, phenolics, and flavonoids in black grapes (Alcolea et al., 2002; Larrauri et al., 1997).

Chauhan et al. (2011) stated the effect of high-pressure processing on total antioxidant activity, phenolic, and flavonoid content of black grapes juice. It was found that the optimum levels were 550 MPa, 44°C, and 2 min for pressure, temperature, and processing time, respectively, for getting the maximum retention of total antioxidant activity, phenolics, and flavonoids in the juice (Chauhan et al., 2011).

Ju and Howard (2003) reported the pressurized liquid extraction for extracting anthocyanins from the freeze-dried skin of a highly pigmented red wine grape with six solvents at 50°C, 10.1 MPa, and 3×5 min extraction cycles. The highest levels of total monoglucosides and total anthocyanins by the extraction with acidified methanol were found. In the study stated by Ju and Howard (2003), the highest levels of total phenolics and total acylated anthocyanins by methanol/acetone/water HCl (40:40:20:0.1) were also found. It was also identified that pressure-aided extraction of anthocyanins was effective by using acidified water at 80–100°C (Ju and Howard, 2003).

Zhao et al. (2009) studied the extraction of polyphenolis from grape skins under ultra-high pressure. It was found that the optimum extraction parameters for polyphenols from grape skins were 70% of alcohol, extraction pressure of 550 MPa and time 10 min, by adding 2000:1 (v/v) HCl. It was revealed that the extracting effectiveness of polyphenols by alcohol solution from the grape skins could be improved by HHP processing (Zhao et al., 2009).

Lafka et al., (2007) have shown the high efficiency of natural phenolic extracts from GP as potent antioxidants. Recent studies have shown that HHP preserves the phytochemicals and minor food bioactives without the quality, quantity, and damage caused by heat treatments. It has shown that HPE can shorten the processing times, and can provide higher extraction yields while having less negative effects on the structure and antioxidant activity of bioactive constituents (Dornenburg and Knorr, 1993; Tokuşoğlu et al. 2010). HHP enables food processing at ambient temperatures; even at lower temperatures, it causes microbial death while virtually eliminating heat damage and the use of chemical preservatives/additives, thereby leading to improvements in the overall quality of foods. It can be utilized to create ingredients with novel functional properties. Microbial growth and reproduction are delayed at moderate pressures (up to 200–300 MPa) (San Martin et al. 2002).

Flavan-3-ols and procyanidins have received much attention because of their antioxidant capacity, free-radical-scavenging ability, and cardio-protective effects. The most common procyanidins are mixtures of oligomers and polymers consisting of (+)-catechin and/or (−)-epicatechin units linked mainly through C4 f B8 bonds (B type) (Gu et al., 2002). Bourzeix et al. (1986) analyzed the (+)-catechin (Cat), epicatechin (Epicat), procyanidin dimmers (B_1–B_4), and trimers in grape skin and seed and reported that Cat and Epicat were about the same level in grape seed while B_2 dimer is dominant (38%) and following that B_1 (29%) and B_4 (24%) in grape seed. It has been determined that B_1 dimer is dominant (64%) in grape skin. Besides, it

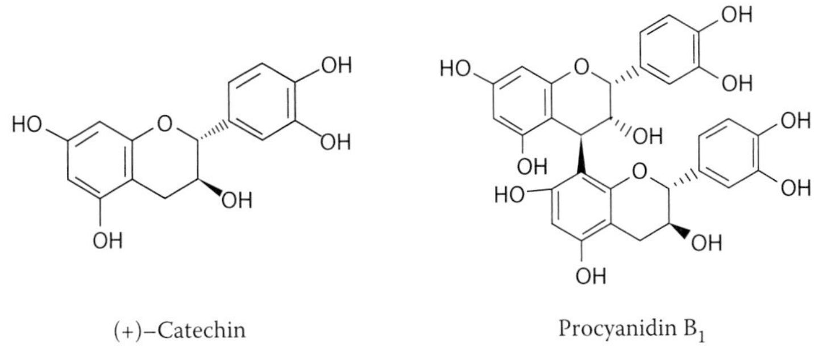

(+)–Catechin

Procyanidin B$_1$

FIGURE 11.14 The chemical structure of (+)-catechin and procyanidin B$_1$ in GPs.

was detected that the (+)-catechin (Cat) level was 4-fold more than the epicatechin (Epicat) amount in grape skin (Bourzeix et al., 1986). Hence, catechin and procyanidin B$_1$ levels are the most important for pomace phenolic quality and antioxidant activity.

Tokuşoğlu et al. (2012) reported the high-pressure processing (HHP) effects on antioxidant activity, total phenolics, and the major flavan-3-ols (catechin and procyanidin B$_1$) (Figure 11.14) of 10 black and white GPs (Figure 11.15) (Tokuşoğlu et al., 2012). Figure 11.15 shows the manufacturing flow for one of the GPs (black/red grape-var. *Alicante Bouschet*).

In the study described by Tokuşoğlu et al. (2012), prior to HP processing, each GP from 10 *Vitis vinifera* L. types (var. *Merlot,* var. *Cabernet-sauvignon,* var. *Bogazkere,* var. *Alicanthe-bouche,* var. *Öküzgözü,* var. *Narince,* var. *Ugni-bls,* var. *Emir,* var. *Syrah,* var. *Kalecik-karası*) were weighed, gently transferred to 7.6 × 15.2 cm plastic pouches (Whirl-Pak bags, ~110 mL; Cole-Parmer Instrument Co.,Vernon Hills, IL, USA), and vacuum-sealed by hand-pressure plastic bag sealing machine (Sealer

FIGURE 11.15 The sample manufacturing from grape to pomace. (a) The harvesting of black/red grape var. *Alicante Buschet,* (b) Maceration form of grapes (Including Skin and Seeds), (c) Pomace form of grapes (GP) (Moisture level: <4%). (Adapted from Tokuşoğlu, Ö. et al. 2012. High pressure processing on black and white grape pomaces. *Advanced Nonthermal Processing in Food Technology: Effects of Quality and Shelf Life of Food and Beverages.* 07–10 May, 2012, Kuşadasi-Turkey. Oral Presentation. *In ANPFT2012 Proceeding Book*, pp. 228–234. ISBN: 978–975-8628-33-9, Celal Bayar University Publishing, Turkey.)

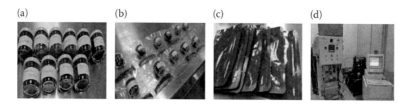

FIGURE 11.16 Preparing of grape pomace (GPs) samples to HHP processing. (a) Grape pomaces prior to preparing. (b) Vacuum sealed bags of grape pomaces. (c) Laminated vacuum pouches of sealed grape pomaces. (d) Processing by HHP apparatus.

Sales, Inc., CA, USA). Then laminated vacuum pouches were performed via a vacuum chamber packaging machine (UltraVac™ 500, Ultravac Solutions LLC, Kansas City, MO, USA). The pomace samples in laminated pouches were held in cold room (1–2°C) until HHP processing or analyzing. Pomaces were pressurized with HHP at 300/400/500 MPa and held during 30 min in sealed packages equipped with a cylindrical pressure chamber (of 0.1 m diameter, 0.25 m height) (FSHN/Pilot Plant,WSU) (Figure 11.16) (Tokuşoğlu et al., 2012).

After HHP processing, the pomaces were analyzed by HPLC (Biological Systems Engineering, Pullman, WSU) with gradient elution. Procyanidin B_1 and catechin were analyzed by simultaneous detection using two mobile phase mixtures (A: 3% acetic acid; B: 80% ACN + 19.6% H_2O + 0.4% acetic acid). Figure 11.17 shows standard and sample HPLC chromatograms of catechin and procyanidin B_1 for control and HHP-treated GP *Alicante Buschet* at different pressures (Figure 11.17) (Tokuşoğlu et al., 2012).

In the study described by Tokuşoğlu et al. (2012), antioxidant activity (AA) increased 1.22–1.98-fold after HHP processing, AA changed between 281.14 and 637.72 IC_{50} µg/g in 10 GPs, while 375.3–1165.87 IC_{50} µg/g in HHP-processed GPs (Figure 11.18, upper figure) (Tokuşoğlu et al., 2012). It was found that the total phenolics (TPs) increased 1.35–2.16-fold after HHP processing. TPs changed between 82.45 and 568.89 mg GAE/g DW in 10 GPs, while 133.65–1092.26 mg GAE/gDW in HHP-processed GPs (Figure 11.18, bottom figure). The correlation between the TP control and TP-HHP processed was found very high for all GP samples ($R^2 = 0.9635$) ($y = 2.1386x - 78.103$) (Tokuşoğlu et al., 2012).

Tokuşoğlu et al. (2012) found that (+)-catechin (CAT) phenolic increased 1.11–2.42-fold after HHP processing ($y = 0.4523x - 0.0461$, $R^2 = 0.9978$). CAT changed between 1267.87 and 8856.32 µg/g in 10 GPs, while 2087.16–17592.23 µg/g in HHP-processed GPs. It was also found that Procyanidin B_1 (Pro B_1) phenolic decreased 1.27–2.34-fold after HHP processing. Pro B_1 changed between 1584.83 and 12575.97 µg/g in 10 GPs, while 808.57–6383.22 µg/g in HHP-processed GPs (Figure 11.19). Procyanidin B_1 levels decreased depending upon increased pressure, whereas catechin levels increased depending on increased pressure (Tokuşoğlu et al., 2012). It was detected that catechin amounts of GPs at 500 MPa were much higher than that of at 300 MPa. It was concluded that the convertion of Procyanidin B_1 compounds to catechin units, with catechin concentration increased by HHP

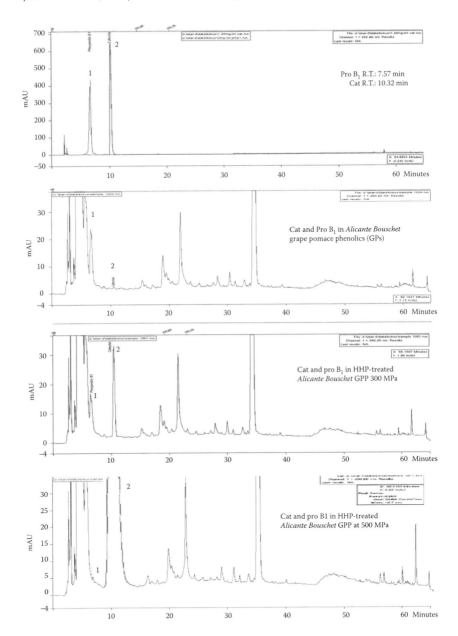

FIGURE 11.17 Standard and sample HPLC chromatograms of simultaneous catechin (Peak No 1) and procyanidin B1 (Peak No 2) for control and HHP-treated grape pomaces (GPs) at different HHP treatments (300/400/500 MPa). (Adapted from Tokuşoğlu, Ö. et al. 2012. High pressure processing on black and white grape pomaces. *Advanced Nonthermal Processing in Food Technology: Effects of Quality and Shelf Life of Food and Beverages.* 07–10 May, 2012, Kuşadasi-Turkey. Oral Presentation. *In ANPFT2012 Proceeding Book*, pp. 228–234. ISBN: 978–975-8628-33-9, Celal Bayar University Publishing, Turkey.)

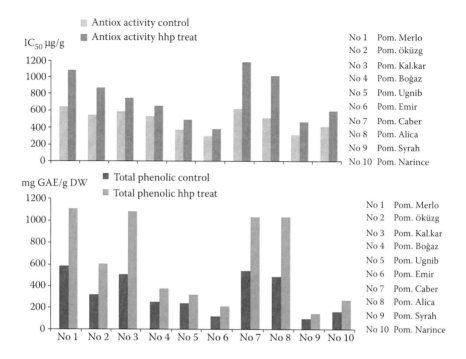

FIGURE 11.18 Antioxidant activity (AA) (upper figure) and total phenolic levels (bottom figure) of control and HHP-treated grape pomaces. (Adapted from Tokuşoğlu, Ö. et al. 2012. High pressure processing on black and white grape pomaces. *Advanced Nonthermal Processing in Food Technology: Effects of Quality and Shelf Life of Food and Beverages.* 07–10 May, 2012, Kuşadasi-Turkey. Oral Presentation. *In ANPFT2012 Proceeding Book*, pp. 228–234. ISBN: 978–975-8628-33-9, Celal Bayar University Publishing, Turkey.)

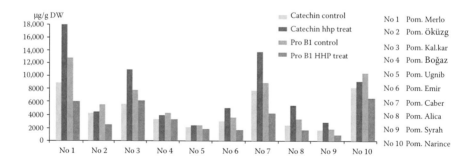

FIGURE 11.19 Catechin/ProB1 levels of control and HHP-treated grape pomaces. (Adapted from Tokuşoğlu, Ö. et al. 2012. High pressure processing on black and white grape pomaces. *Advanced Nonthermal Processing in Food Technology: Effects of Quality and Shelf Life of Food and Beverages.* 07–10 May, 2012, Kuşadasi-Turkey. Oral Presentation. *In ANPFT2012 Proceeding Book*, pp. 228–234. ISBN: 978–975-8628-33-9, Celal Bayar University Publishing, Turkey.)

FIGURE 11.20.   The Microbial Quality Detection of HHP-Treated Grape Pomace Samples (via total mold-yeast and total plate count (TPC) based on USA-FDA Microbiological Methods Manual and Expressed as cfu/g). (Adapted from Tokuşoğlu, Ö. et al. 2012. High pressure processing on black and white grape pomaces. *Advanced Nonthermal Processing in Food Technology: Effects of Quality and Shelf Life of Food and Beverages.* 07–10 May, 2012, Kuşadasi-Turkey. Oral Presentation. *In ANPFT2012 Proceeding Book*, pp. 228–234. ISBN: 978-975-8628-33-9, Celal Bayar University Publishing, Turkey; Tokuşoğlu, Ö. et al. 2011a. High pressure processed black grape pomaces: Multivariate analysis of major flavan-3-ol phenolics, antioxidant activity and microbiological quality profiles. Nonthermal Processing Section Technical Research Paper. 2011-TRP-2908-IFT2011IFT Annual Meeting+Food Expo, New Orleans, Louisina, USA.)

effects (Figure 11.19) (Tokuşoğlu et al., 2012). HHP is a nonthermal technology for black/red and white GPs, and it can be used for microbial quality and extending the shelf life of GPs (Figure 11.20). The antioxidative phenolic flavan-3-ol profiles improved by HHP while higher levels of bioactives retained in HHP-treated samples (Tokuşoğlu et al., 2012).

11.6   EFFECT OF HHP ON THE MICROBIAL STABILITY OF GRAPE POMACES, GRAPE-BASED DRINKS, AND GRAPE MUST

Tokuşoğlu et al. (2012, 2011a) reported the high-pressure processing (HHP) effects on the microbial quality of 10 black and white GPs after HHP.

In the study reported by Tokuşoğlu et al. (2012, 2011a), with HHP application of pomaces, total mold and yeast load was reduced more than 95% at 25°C, whereas total plate count was reduced more than 95% and high-level microbial stability was obtained (Figures 11.21–11.23) (Tokuşoğlu et al., 2012, 2011a).

Wine is one of the most popular fermented beverages and contains 9–13% alcohol. If not properly pasteurized, low alcohol in wine may cause quality changes, spoilage, sourness, and off-flavor during storage due to the growth of undesirable microorganisms such as *Acetobacter* and lactic acid bacteria. The traditional thermal treatment may be undesirable because wine's taste, flavor, and color are very sensitive to heat (Mermelstein, 1998). Since wine is very sensitive to heat treatment and is usually packaged in bulk for aging, the HHP treatment as nonthermal preservation technology is very suitable for wine preservation. Delfini et al. (1994) stated the antiseptic effect of HHP on wine quality. Puig et al. (2003) reported that 500 MPa/5 min HHP

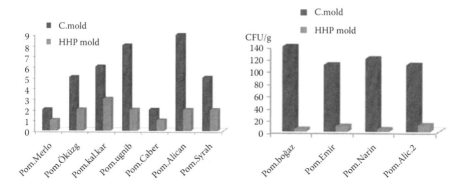

FIGURE 11.21 Mold levels of control and HHP-treated pomaces. (Adapted from Tokuşoğlu, Ö. et al. 2012. High pressure processing on black and white grape pomaces. *Advanced Nonthermal Processing in Food Technology: Effects of Quality and Shelf Life of Food and Beverages.* 07–10 May, 2012, Kuşadasi-Turkey. Oral Presentation. *In ANPFT2012 Proceeding Book*, pp. 228–234. ISBN: 978–975-8628-33-9, Celal Bayar University Publishing, Turkey; Tokuşoğlu, Ö. et al. 2011a. High pressure processed black grape pomaces: Multivariate analysis of major flavan-3-ol phenolics, antioxidant activity and microbiological quality profiles. Nonthermal Processing Section Technical Research Paper. 2011-TRP-2908-IFT2011IFT Annual Meeting + Food Expo, New Orleans, Louisina, USA.)

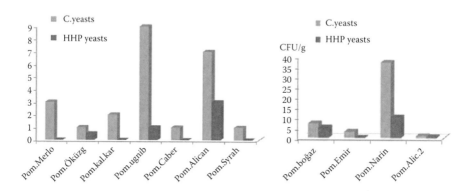

FIGURE 11.22 Yeast levels of control and HHP-treated pomaces. (Adapted from Tokuşoğlu, Ö. et al. 2012. High pressure processing on black and white grape pomaces. *Advanced Nonthermal Processing in Food Technology: Effects of Quality and Shelf Life of Food and Beverages*, 07–10 May, 2012, Kuşadasi-Turkey. Oral Presentation. *In ANPFT2012 Proceeding Book*, pp. 228–234. ISBN: 978–975-8628-33-9, Celal Bayar University Publishing, Turkey; Tokuşoğlu, Ö. et al. 2011a. High pressure processed black grape pomaces: Multivariate analysis of major flavan-3-ol phenolics, antioxidant activity and microbiological quality profiles. Nonthermal Processing Section Technical Research Paper. 2011-TRP-2908-IFT2011IFT Annual Meeting + Food Expo, New Orleans, Louisina, USA.)

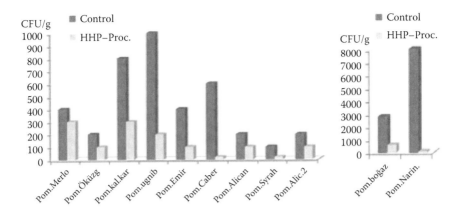

FIGURE 11.23   Total plate count of control and HHP-treated grape pomaces. (Adapted from Tokuşoğlu, Ö. et al. 2012. High pressure processing on black and white grape pomaces. *Advanced Nonthermal Processing in Food Technology: Effects of Quality and Shelf Life of Food and Beverages.* 07–10 May, 2012, Kuşadasi-Turkey. Oral Presentation. *In ANPFT2012 Proceeding Book*, pp. 228–234. ISBN: 978–975-8628-33-9, Celal Bayar University Publishing, Turkey; Tokuşoğlu, Ö. et al. 2011a. High pressure processed black grape pomaces: Multivariate analysis of major flavan-3-ol phenolics, antioxidant activity and microbiological quality profiles. Nonthermal Processing Section Technical Research Paper. 2011-TRP-2908-IFT2011IFT Annual Meeting + Food Expo, New Orleans, Louisina, USA.)

treatment resulted in a 99.99% reduction in the initial microbial population without any alterations in the chemical or organoleptic properties of the wine (Puig et al., 2003).

Mok et al. (2006) presented the effects of HHP treatments on the physiochemical properties (alcohol, pH, acidity, total sugar) and microbiological quality (aerobic bacteria, yeast, and lactic acid bacteria). It was found that the pasteurization effect of the HHP treatments increased with treatment pressure and time (Mok et al., 2006). The relationship between the treatment pressure and the inactivation rate constants k_1 and k_2 are shown in Figure 11.24. As shown in semilogarithmic plotting, the linear relationships were observed, indicating the constants k_1 and k_2 increased exponentially with increasing pressure (Figure 11.24) (Mok et al., 2006).

In the study reported by Takush and Osborne (2011), a Pinot noir grape must was inoculated with *Saccharomyces cerevisiae, Brettanomyces bruxellensis, Kluveromyces thermotolerans, Lactobacillus hilgardii, Oenococcus oeni,* and *Acetobacter aceti* at similar levels ($1 \times 10(5)$ cfu/mL) and subjected to HHP treatment (551 MPa/10 min) for investigating HHP processing as a tool for studying yeast during red winemaking. No viable cells were detected in the grape must after HHP treatment. It was determined that no significant differences in color or hue were obtained, but wine produced from HHP-treated grapes contained higher total phenolics. The produced wines was evaluated by a trained sensory panel and a slight increase in overall fruit aroma was found while there were no significant differences between any aroma and flavor descriptors (Takush and Osborne, 2011).

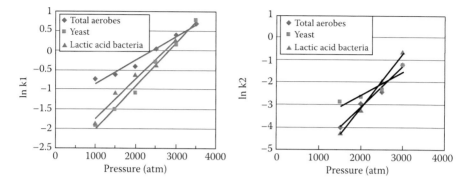

FIGURE 11.24 Semilogarithmic plots of reduction rate constants k_1 (left figure) and k_2 (right figure) of microorganisms in wine compared with hydrostatic pressure. (Adapted from Mok, C. et al. 2006. *Journal of Food Science*, 71(8), 265–269.)

Morata et al. (2012) stated the effect of HHP on wines contaminated with Dekkera/Brettanomyces populations of 10^4 and 10^6 cfu mL^{-1} growing at either pH 3.2 or 3.6 (both normal pHs for red wine) and at room temperature (25°C). It was determined that HHP treatment at 100 MPa/24 h was highly effective in controlling the growth of all combinations of starting yeast populations and pH, and scant modification of thermosensitive wine molecules such as pigments and volatile compounds that greatly influence wine quality was detected. The potential use of HHP was put forward as a means of cold pasteurized wines to control Dekkera/Brettanomyces (Morata et al., 2012).

It is stated that HPP is the ideal process for preserving cold pressed juices owing to HHP doesn't heat the juice beyond room temperature. HPP extends the shelf life of the juice from days to weeks by preserving the juice using high pressure (Tokuşoğlu, 2013).

As a conclusion, HHP processing improved the phytochemical properties and microbial quality of berry-type fruits, grape and grape by-products. HHP provided the microbial stability and the enhanced bioactive profiles for berry and grape fruits, and these by-products. As nonthermal processing, HHP could be utilized in the industry of berry-type fruits, grape fruit, berry and grape-based drinks.

REFERENCES

Ahmed, J. and Ramaswamy, H.S. 2006. High pressure processing of fruits and vegetables. *Stewart Postharvest Review*, 1, 1–10.

Alcolea, J.F., Cano, A., Acosta, M. and Arnao, M.B. 2002. Hydrophilic and lipophilic antioxidant activities of grapes. *Food/Nahrung*, 46, 353–356.

Altuner, E.M., İşlek, C., Çeter, T. and Alpas, H. 2012a. High hydrostatic pressure extraction of phenolic compounds from *Maclura pomifera* fruits. *African Journal of Biotechnology*, 11, 930–937.

Altuner, E.M., İşlek, C., Çeter, T. and Alpas, H. 2012b. High hydrostatic pressure: A method which increases efficiency of phenolic compounds extraction from fruits having adhesive nature. *Proceedings Book of ANPFT 2012—Advanced Nonthermal Processing in Food Technology: Effects on Quality and Shelf Life of Food & Beverages, 7–10 May,* Kuşadasi, *Turkey,* p. 88–91. ISBN: 978-975-8628-33-9.

Altuner, E.M. and Tokuşoğlu, Ö. 2013. The effect of high hydrostatic pressure processing on the extraction, retention and stability of anthocyanins and flavonols contents of berry fruits and berry juices. *International Journal of Food Science and Technology* (IJFST). 48, 1991–1997.

Andersen, Q.M. and Markham, K.R. 2006. *Flavonoids. Chemistry, Biochemistry, and Applications.* Boca Raton, FL: CRC Press, Taylor & Francis.

Baert, L., Uyttendaele, M., Van Coillie, E. and Debevere, J. 2008. The reduction of murine norovirus 1, B-Fragilis HSP40 infecting phage B40-8 and *E. coli* after a mild thermal pasteurization process of raspberry puree. *Food Microbiology,* 25, 871–874.

Barba, F.J., Esteve, M.J. and Frigola, A. 2013. Physicochemical and nutritional characteristics of blueberry juice after high pressure processing. *Food Research International,* 50(2), 545–549.

Barbagallo, R.N., Palmeri, R., Fabiano, S., Rapisarda, P. and Spagna, G. 2007. Characteristic of β-glucosidase from sicilian blood oranges in relation to anthocyanin degradation. *Enzyme and Microbial Technology,* 41(5), 570–575.

Barbosa-Canovas, G.V., Pothakamury, U.R., Palou, E. and Swanson, B.G. 1998. High hydrostatic pressure food processing. In: *Non-Thermal Preservation of Foods.* pp. 9–48, 139–213. New York: Marcel Dekker.

Beecher, G.R. 1999. Flavonoids in foods. In: *Antioxidant Food Supplements in Human Health,* eds. L. Packer, M. Hiramatsu, and T. Yoshikawa. New York: Academic Press.

Bourzeix, M., Weyland, D. and Heredia N. 1986. Étude Des Catéchines Et Des Procyanidols De La Grape De Raisin, Du Vin Et d'Autres Dérives De La Vigne. *Bulletin de l'O.I.V.,* 669–670, 1171–1254.

Cacace, H.E. and Mazza, G. 2003. Mass transfer process during extraction of phenolic compounds form milled berries. *Journal of Food Engineering,* 59 (4), 379–389.

Cano, M.P., Hernandez, A. and De Ancos, B. 1997. High pressure and temperature effects on enzyme inactivation in strawberry and orange products. *Journal of Food Science,* 62, 85–88.

Cao, X., Zhang, Y., Zhang, F., Wang, Y., Yi, J. and Liao, X. 2011. Effects of high hydrostatic pressure on enzymes, phenolic compounds, anthocyanins, polymeric color and color of strawberry pulps. *Journal of the Science of Food and Agriculture,* 91, 877–885.

Chauhan, O.P., Raju, P.S., Ravi, N., Roopa, N. and Bawa, A.S. 2011. Studies on retention of antioxidant activity, phenolics and flavonoids in high pressure processed black grape juice and their modelling. *International Journal of Food Science and Technology,* 46, 2562–2568.

Chen, H., Zuo, Y. and Deng, Y. 2001. Separation and determination of flavonoids and other phenolic compounds in cranberry juice by high performance liquid chromatography. *Journal of Chromatography A,* 913, 387–395.

Connor, A.M., Luby, J.J., Tong, C.B.S., Finn, C.E. and Hancock, J.F. 2002. Variation and heritability estimates for antioxidant activity total phenolic content and anthocyanin content in blueberry progenies. *Journal of the American Society for Horticultural Science,* 1, 82–88.

Corrales, M., Garcia, A.F., Butz, P. and Tauscher, B. 2008a. Extraction of anthocyanins from grape skin assisted by high hydrostatic pressure. *Journal of Food Engineering,* 90, 415–421.

Corrales, M., Toepfl, S., Butz, P., Knorr, D. and Tauscher, B. 2008b. Extraction of Anthocyanins from grape by-products assisted by ultrasonic, high hydrostatic pressure or pulsed electric fields: A comparison. *Innovative Food Science and Emerging Technologies,* 9, 85–91.

Corrales, M., Butz, P. and Tauscher, B. 2008c. Anthocyanin condensation reactions under high hydrostatic pressure. *Food Chemistry*, 110(3), 627–635.

Crozier, A., Jaganath, I.B. and Clifford, M.N. 2006. Phenols, polyphenols and tannins: An overview. In: *Plant Secondary Metabolites: Occurrence, Structure and Role in the Human Diet.*, eds. A. Crozier, M.N. Clifford and H. Ashihara, 1–24. Oxford: Blackwell Publishing, Ltd.

Delfini, C., Conterno, L., Carpi, G. and Amati, A. 1994. Antiseptic effects of high pressure and direct injection of anhydrous sulphur in wine bottles. *Revue des Oenologues et des Techniques Vitivinicoles et Oenologiques*, 73S, 58.

Del Pozo-Insfran, D., Del Follo-Martinez, A., Talcott, S.T. and Brenes, C.H. 2007. Stability of copigmented anthocyanins and ascorbic acid in muscadine grape juice processed by high hydrostatic pressure. *Journal of Food Science*, 72(4), 247–253.

Dixon, R.A. and Paiva, N.L. 1995. Stress induced phenylpropanoid metabolism. *Plant Cell*, 7, 1085–1097.

Dornenburg, H. and Knorr, D. 1993. Cellular permeabilization of cultured plant tissues by high electric field pulses or ultra high pressure for the recovery of secondary metabolites. *Food Biotechnology*, 7, 35–48.

Engmann, N.F., Ma, Y.K., Ying, X. and Qing, Y. 2013. Investigating the effect of high hydrostatic pressure processing on anthocyanins composition of mulberry (*Morus moraceae*) juice. *Czech Journal of Food Science*, 31(1), 72–80.

FDA. 2001. Food and Drug Administration (FDA) survey of imported fresh produce FY 1999 field assignment. Retrieved 3/29, 2012, from http://www.fda.gov/Food/FoodSafety/Product-SpecificInformation/FruitsVegetablesJuices/GuidanceComplianceRegulatoryInformation/ucm118891.htm.

Garcia-Palazon, A., Suthanthangjai, W., Kajda, P. and Zabetakis, I. 2004. The effects of high hydrostatic pressure on β-glucosidase, peroxidase and polyphenoloxidase in red raspberry (*Rubus idaeus*) and Strawberry (Fragaria x ananassa). *Food Chemistry*, 88, 7–10.

Giacarini, G.M.G.C. 2008. *Effect of High Hydrostatic Pressure and Thermal Processing on Cranberry Juice*. M.Sc. Thesis, Graduate School – New Brunswick Rutgers, The State University of New Jersey.

Gu, L., Kelm, M., Hammerstone, J.F., Beecher, G., Cunningham, D., Vannozzi, S. and Prior, R.L. 2002. Fractionation of polymeric procyanidins from lowbush blueberry and quantification of procyanidins in selected foods with an optimized normal-phase HPLC-MS fluorescent detection method. *Journal of Agricultural and Food Chemistry*, 50, 4852–4860.

Güler, A. and Tokuşoğlu, Ö. 2012. Utility of novel food processing techniques on grape and grape derived products. *Advanced Nonthermal Processing in Food Technology: Effects of Quality and Shelf Life of Food and Beverages*. 07–10 May, 2012, Kuşadasi-Turkey. Oral Presentation. *In ANPFT2012 Proceeding Book*, 277–282 pp. ISBN: 978-975-8628-33-9, Celal Bayar University Publishing, Turkey.

Häkkinen, S. 2000. *Flavonols and Phenolic Acids in Berries and Berry Products*. PhD Doctoral dissertation. University of Kuopio, Kuopio.

Häkkinen, S., Heinonen, M., Karenlampi, S., Mykkanen, H., Ruuskanen, J. and Torronen, R. 1999. Screening of selected flavonoids and phenolic acids in 19 berries. *Food Research International*, 32, 345–353.

Hilz, H., Lille, M., Poutanen, K., Schols, H.A. and Voragen, A.G.J. 2006. Combined enzymatic and high-pressure processing affect cell wall polysaccharides in berries. *Journal of Agricultural and Food Chemistry*, 54(4), 1322–1328.

Huang, Y., Ye, M. and Chen, H. 2013. Inactivation of Escherichia coli O157:H7 and Salmonella spp. in Strawberry puree by high hydrostatic pressure with/without subsequent frozen storage. *International Journal of Food Microbiology*, 160, 337–343.

Ju, Z.Y. and Howard, L.R. 2003. Effects of solvent and temperature on pressurized liquid extraction of anthocyanins and total phenolics from dried red grape skin. *Journal of Agricultural and Food Chemistry*, 51, 5207–5213.

Kähkäonen, M.P., Heinamaki, J., Ollilainen, V. and Heinonen, M. 2003. Berry anthocyanins: Isolation, identification and antioxidant activities. *Journal of the Science of Food and Agriculture*, 83, 1403–1411.

Kirca, A. and Cemeroglu, B. 2003. Degradation kinetics of anthocyanins in blood orange juice and concentrate. *Food Chemistry*, 81 (4), 583–587.

Kong, J.M., Chia, L.S., Goh, N.K., Chia, T.F. and Brouillard, R. 2003. Analysis and biological activities of anthocyanins. *Phytochemistry*, 64, 923–933.

Kouniaki, S., Kajda, P. and Zabetakis, I. 2004. The effect of high hydrostatic pressure on anthocyanins and ascorbic acid in blackcurrants (*Ribes nigrum*). *Flavour and Fragrance Journal*, 19, 281–286.

Kühnau, J. 1976. *The Flavonoids. A Class of Semi-Essential Food Components: Their Role in Human Nutrition*. In: *World Review of Nutrition and Dietetics*. Ed. G.H. Bourne. Basel, Switzerland: S. Karger, pp. 117–120.

Lafka, T.I., Sinanoglou, V. and Lazos, E.S. 2007. On the extraction and antioxidant activity of phenolic compounds from winery wastes. *Food Chemistry*, 104, 1206–1214.

Lambert, Y., Demazeau, G., Largeteau, A. and Bouvier, J.M. 1999. Changes in aromatic volatile composition of strawberry after high pressure treatment. *Food Chemistry*, 67, 7–16.

Larrauri, J.A., Ruperez, P. and Saura, C.F. 1997. Effect of drying temperature on the stability of polyphenols and antioxidant activity of red grape pomace peels. *Journal of Agricultural and Food Chemistry*, 45, 1390–1393.

Lou, X., Janssen, H. and Cramers, C.A. 1997. Parameters affecting the accelerated solvent extraction of polymeric samples. *Analytical Chemistry*, 69 (8), 1598–1603.

Mermelstein, N.H. 1998. Beer and wine making. *Food Technology*, 52(4), 84, 86, 88.

Meskin, M.S., Bidlack W.R., Davies, A.J., Lewis, D.S., and Randolph, R.K. 2003. *Phytochemicals: Mechanisms of Action*. Boca Raton, FL: CRC Press.

Moyer, R.A., Hummer, K.E., Finn, C.E., Frei, B. and Wrolstad, R.W. 2002. Anthocyanins, phenolics, and antioxidant capacity in diverse small fruits: Vaccinium, rubus and ribes. *Journal of Agricultural and Food Chemistry*, 50, 519–525.

Mok, C., Song, K.T., Park, Y.S., Lim, S., Ruan, R. and Chen, P. 2006. High hydrostatic pressure pasteurization of red wine. *Journal of Food Science*, 71(8), 265–269.

Morata, A., Benito, S., González, M.C., Palomero, F., Tesfaye, W. and Suárez-Lepe, J.A. 2012. Cold pasteurisation of red wines with high hydrostatic pressure to control dekkera/brettanomyces: Effect on both aromatic and chromatic quality of wine. *European Food Research and Technology*, 235, 147–154.

Niu, M.T., Polish, L.B., Robertson, B.H., Khanna, B.K., Woodruff, B.A., Shapiro, C.N., Miller, M.A., Smith, J.D., Gedrose, J.K., Alter, M.J. and Margolis, H.S. 1992. Multistate outbreak of hepatitis-A associated with frozen strawberries. *Journal of Infectious Diseases*, 166, 518–524.

ODA. 2011. Oregon Department of Agriculture. *Fresh Strawberries From Washington County Farm Implicated In E. coli O157 Outbreak In NW Oregon*. http://oregon.gov/ODA/FSD/strawberries.shtml.

Oey, I., Plancken, I.V., Loey, A.V. and Hendrickx, M. 2008. Does high pressure processing influence nutritional aspects of plant based food systems? *Trends in Food Science and Technology*, 19(6), 300–308.

Patras, A., Brunton, N.P., Pieve, S.D. and Butler, F. 2009. Impact of high pressure processing on total antioxidant activity, phenolic, ascorbic acid, anthocyanin content and colour of strawberry and blackberry purées. *Innovative Food Science and Emerging Technologies*, 10, 308–313.

Prior R.L., Cao G., Martin A., Sofic E., McEwen J., O'Brien C., Lischner N., et al. 1998. Antioxidant capacity as influenced by total phenolic and anthocyanin content, maturity, and variety of vaccinium species. *Journal of Agricultural and Food Chemistry*, 46, 2686–2693.

Puig, A, Vilavella, M, Daoudi, L, Guamis, B. and Minguez, S. 2003. Microbiological and biochemical stabilization of wines using the high pressure technique. *Bulletindde l'OIV*, 76 (869/870), 569–617.

Rajnarayana, K., Reddy, M.S., Chaluvadi, M.R. and Krishna, D.R. 2001. Bioflavonoids classification, pharmacological, biochemical effects and therapeutic potential. *Indian Journal of Pharmacology*, 33, 2–16.

Ransay, C.N. and Upton, P.A., 1989. Hepatitis-A and frozen raspberries. *Lancet*, 1, 43–44.

Rastogi, N.K., Raghavarao, K.S., Balasubramaniam, V.M., Niranjan, K. and Knorr, D. 2007. Opportunities and challenges in high pressure processing of foods. *Critical Reviews in Food Science and Nutrition*, 47(1), 69–112.

Rawson, A., Patras, A., Tiwari, B.K., Noci, F., Koutchma, T. and Brunton, N. 2011. Effect of thermal and non thermal processing technologies on the bioactive content of exotic fruits and their products: Review of recent advances, *Food Research International*, 44, 1875–1887.

Reid, T.M.S. and Robinson, H.G., 1987. Frozen raspberries and Hepatitis-A. *Epidemiology and Infection*, 98, 109–112.

Richard, J.S. 1992. *High Pressure Phase Behaviour of Multicomponent Fluid Mixtures*. Amsterdam: Elsevier.

Sancho, F., Lambert, Y., Demazeau, G., Largeteau, A., Bouvier, J.M. and Narbonne, J.F. 1999. Effect of ultra-high hydrostatic pressure on hydrosoluble vitamins. *Journal of Food Engineering*, 39, 247–253.

San Martin, M.F., Barbosa-Cánovas, G.V. and Swanson Barry, G. 2002. Food processing by high hydrostatic pressure. *Critical Reviews in Food Science and Nutrition*, 42, 627–645.

Sarvikivi, E., Roivainen, M., Maunula, L., Niskanen, T., Korhonen, T., Lappalainen, M., Kuusi, M., 2012. Multiple norovirus outbreaks linked to imported frozen raspberries. *Epidemiology and Infection*, 140, 260–267.

Simpson, M.G. 2010. *Plant Systematics*. Oxford, UK: Elsevier.

Stoner, G.D., Wang, L.S. and Casto, B.C. 2008. Laboratory and clinical studies of cancer chemoprevention by antioxidants in berries. *Carcinogenesis*, 29, 1665–1674.

Stöhr, H. and Hermann, K. 1975. Phenolics in fruits. VI. Phenolics of currants, gooseberries and blueberries and the changes in phenolic acids and catechins during the development of blackcurrants. *Z. Lebensm. Unters. Forsch*, 159, 31–37.

Suthanthangjai, W., Kajda, P. and Zabetakis, I. 2005. The effect of high hydrostatic pressure on the anthocyanins of raspberry (*Rubus idaeus*). *Food Chemistry*, 90(1–2), 193–197.

Takush, D.G. and Osborne, J.P. 2011. Investigating high hydrostatic pressure processing as a tool for studying yeast during red winemaking. *American Journal of Enology and Viticulture*, 62(4), 536–541.

Tiwari, B.K., O'Donnella, C.P. and Cullen, P.J. 2009. Review. Effect of non thermal processing technologies on the anthocyanin content of fruit juices. *Trends in Food Science & Technology*, 20, 137–145.

Tokuşoğlu, Ö. 2001. The Determination of the Major Phenolic Compounds (Flavanols, Flavonols, Tannins and Aroma Properties of Black Teas. PhD Thesis. Department of Food Engineering, Bornova, Izmir, Turkey: Ege University.

Tokuşoğlu, Ö. 2012. ANPFT2012 Proceeding book. International Congree; *Advanced Nonthermal Processing in Food Technology: Effects of Quality and Shelf Life of Food and Beverages*. 07–10 May, 2012, Kuşadasi-Turkey. Ö. Tokuşoğlu. Celal Bayar University Publishing, Turkey, p. 321. ISBN: 978-975-8628-33-9.

Tokuşoğlu, Ö. 2013. High pressure processing and pulsed electrical field strategies on shelf life quality and bioactives in fruit juices and functional soft drinks. In *Nutritional Science 2013. 2nd International Conference and Exhibition on Nutritional Science & Therapy*. July 15–17, 2013, Philadelphia, USA. Lecture Presentation.

Tokuşoğlu, Ö. 2014. Functional and multiPurpose foods. *Lecture Notes*. Celal Bayar University, Manisa. Unpublished.

Tokuşoğlu, Ö., Alpas, H. and Bozoğlu, F. 2010b. High hydrostatic pressure effects on mold flora, citrinin mycotoxin, hydroxytyrosol, oleuropein phenolics and antioxidant activity of black table olives. *Innovative Food Science and Emerging Technologies*, 11(2), 250–258.

Tokuşoğlu, Ö., Bozoğlu, F. and Doona, C.J. 2010a. High Pressure Processing Strategies on Phytochemicals, Bioactives and Antioxidant Activity in Foods. (Division: Nonthermal Processing; Session: 100-3/Novel Bioactives: Approaches for the Search, Evaluation and Processing for Nutraceuticals) July 19, 2010, 9:15 am- 9:35 am; Room N426b,Mc Cormick Place Chicago/ILLINOIS, USA. ORAL PRESENTATION. TechnicalResearchPaper.2010IFT *Annual Meeting + Food Expo. Book of Abstracts* p.115.

Tokuşoğlu, Ö. and Doona, C. 2011. High pressure processing technology on bioactives in fruits & cereals (Chapter 21—Part IV. Functionality, processing, characterization and applications of fruit & cereal bioactives). In: *Fruit and Cereal Bioactives: Sources, Chemistry & Applications*. Ed. Tokuşoğlu Ö. and C. Hall. Boca Raton, Florida, USA: CRC Press, Taylor & Francis Group, 459 pp., ISBN: 9781439806654; ISBN-10: 1439806659.

Tokuşoğlu, Ö. and Hall, C. 2011. Introduction to bioactives in fruits and cereals (Chapter 1 - Part I.Introduction). In: *Fruit and Cereal Bioactives: Sources, Chemistry & Applications*. Ed. Tokuşoğlu Ö. and C. Hall. Florida, USA: CRC Press, Taylor & Francis Group, *Boca Raton*, 459 pp. ISBN: 9781439806654; ISBN-10: 1439806659.

Tokuşoğlu, Ö. and Stoner, G. 2011. Phytochemical bioactives in berry fruits (Chapter 7 - Part II. Chemistry and mechanisms of beneficial bioactives in fruits & cereals). In: *Fruit and Cereal Bioactives: Sources, Chemistry & Applications*. Ed. Tokuşoğlu Ö. and C. Hall. Boca Raton, Florida, USA: CRC Press, Taylor & Francis Group, p. 459ISBN: 9781439806654; ISBN-10: 1439806659.

Tokuşoğlu, Ö. and Yıldırım, Z. 2012. Effects of cooking methods on the anthocyanin levels and antioxidant activity of a local Turkish sweet potato [*Ipomoea batatas* (L.) Lam.] cultivar hatay kırmızı: Boiling, steaming and frying effects. *Turkish Journal of Field Crops*,17(2), 87–90.

Tokuşoğlu, Ö., Swanson, B.G., Powers, J.R., Younce, F. and Güler, A. 2011a. High pressure processed black grape pomaces: Multivariate analysis of major flavan-3-ol phenolics, antioxidant activity and microbiological quality profiles. Nonthermal Processing Section Technical Research Paper. 2011-TRP-2908-IFT2011IFT Annual Meeting + Food Expo, New Orleans, Louisina, USA.

Tokuşoğlu, Ö., Swanson, B.G. Younce, F. and Barbosa-Cánovas, G.V. 2011b. Rheological, microbial quality characteristics and antioxidant activity of high hydrostatic pressure (HHP) processed berry ice-cream mixes: Weibull modeling and response surface methodology approach for treatment pressure and holding time effects. *Nonthermal Processing Division Workshop 2011 – Innovation Food Conference. October 12–14, Osnabrück*, Germany. Oral Presentation.

Tokuşoğlu, Ö., Swanson, B.G., Powers, J.R., Güler, A. and Barbosa-Canovas, G.V. 2012. High pressure processing on black and white grape pomaces. *Advanced Nonthermal Processing in Food Technology: Effects of Quality and Shelf Life of Food and Beverages*. 07–10 May, 2012, Kuşadasi-Turkey. Oral Presentation. *In ANPFT2012 Proceeding Book*, pp. 228–234. ISBN: 978–975-8628-33-9, Celal Bayar University Publishing, Turkey.

Tokuşoğlu, Ö., Ünal M.K. and Yıldırım Z. 2003. HPLC-UV and GC-MS characterization of the flavonol aglycons quercetin, kaempferol and myricetin in tomato pastes and other tomato-based products. *Acta Chromatographica*, 13, 196–207.

Tomas-Barberan, F.A. and Espin, J.C. 2001. Phenolic compounds and related enzymes as determinants of quality in fruits and vegetables. *Journal of the Science of Food and Agriculture*, 81, 853–876.

Vasco, C. 2009. *Phenolic Compounds in Ecuadorian Fruits*. PhD. Thesis, Swedish University of Agricultural Sciences, Uppsala.

Wang, L.-S., Dombkowski, A.A., Rocha, C., Seguin, C., Cukovic, D., Mukundan, A., Henry, C. and Stoner, G.D. 2011. Effects of black raspberries on late events in N-nitrosomethylbenzylamine-induced rat esophageal carcinogenesis. *Molecular Carcinogenesis,* 50(4), 291–300.

Wang, S.Y. 2007. Antioxidant capacity and phenolic content of berry fruits as affected by genotype, preharvest conditions, maturity, and postharvest handling. Chapter 5 in *Berry Fruit: Value-Added Products for Health Promotion,* ed. Y. Zhao. Boca Raton, FL: CRC Press Taylor & Francis.

Wang, S.Y. and Lin H.S. 2000. Antioxidant activity in fruit and leaves of blackberry, raspberry, and strawberry varies with Cultivar and Developmental Stage. *Journal of Agricultural and Food Chemistry*, 48, 140–146.

Xiao, Z.P., Peng, Z.Y., Peng, M.J., Yan, W.B., Ouyang, Y.Z. and Zhu, H.L. 2011. Flavonoids health benefits and their molecular mechanism. *Mini Rev Med Chem.*, 11(2), 169–177.

Zabetakis, I., Leclerc, D. and Kajda, P. 2000. The effect of high hydrostatic pressure on the strawberry anthocyanins. *Journal of Agriculture and Food Chemistry*, 48, 2749–2754.

Zhang, S., Junjie, Z. and Changzhen, W. 2004. Novel high pressure extraction technology. *International Journal of Pharmaceutics*, 78, 471–474.

Zhang, S., Ruizhan, C. and Changzheng, W. 2007. Experiment study on ultrahigh pressure extraction of ginsenosides. *Journal of Food Engineering*, 79, 1–5.

Zhang, S., Xi, J. and Wang, C.Z. 2005. Effect of high hydrostatic pressure on extraction of flavonoids in propolis. *Food Science and Technology International*, 11, 213–216.

Zhao, G.Y., Li, B. and Bai, Y.H. 2009. Studies on optimum process for the extraction of polyphenols from grape skins under ultra-high pressure. *Proceedings of the Sixth International Symposium on Viticulture and Enology*, 287–291.

12 Improving Quality and Shelf-Life of Table Eggs and Olives by High-Pressure Processing

Özlem Tokuşoğlu and Gustavo V. Barbosa-Cánovas

CONTENTS

12.1 TABLE EGG

12.1.1 INTRODUCTION TO TABLE EGG AND THE NECESSITY OF HIGH-PRESSURE PROCESSING

Whole egg (WE) has an excellent nutritional value, containing a high biological value of protein as compared to any dietary protein sources. Egg proteins own all covetable nutritional and functional properties. Liquid egg, homogenized as whole egg or separated into white and yolk, is used as an ingredient and/or as a colorant in a wide variety of processed food products (ICMSF, 1998; Tokuşoğlu, 2013).

Exclusively, liquid whole egg (LWE) contributes physicochemical characteristics to foods, including coagulating, foaming, emulsifying, and gelling (Lee et al., 1999; Ma et al., 1986; Yang and Baldwin, 1995). Due to these important functional properties, LWE can be extensively used as a food ingredient and colorant for many foods such as bakery products, meringues, meat products, chocolate, confectionary products, drinks, infant foods, dressings, noodles, and in the snack food industry. Owing to holding a large quantity of air in the form of fine bubbles, the bubbles of beaten eggs expand in a cake mix and the albumen gives strength to the walls of the air pocket. Egg also contains other nutrients: carbohydrates, vitamins, minerals, phospholipids, and other functional lipids. The phospholipids containing yolk confer stability in emulsions of oils and water and are utilized in the making of mayonnaise; egg phospholipids are also used as an ingredient of dough and ice-cream mix (Ahmed et al., 2003; Dawson and Martinez-Dawson, 1998; Tokuşoğlu, 2013; ICMSF, 1998).

The outer eggshell is made almost entirely of calcium carbonate ($CaCO_3$) and is covered with as many as 17,000 tiny pores. It is a semipermeable membrane that allows air and moisture to pass through its pores. Chalaza parts in opposite directions of the egg serve to keep the yolk centered, and it is stated that the more prominent the chalazae, the fresher the egg (Figure 12.1).

Regrettably, egg and egg-products are also responsible for a large number of foodborne illnesses owing to its anatomy; water, protein, and lipid are the major components of liquid egg that can support microbial growth with inappropriate storage. Microbial contamination of eggs as well as its economic implications for the poultry industry have been reported (Bruce and Drysdal, 1994; EFSA, 2011; Wong and Kitts, 2003).

Traditional thermal treatments used to pasteurize LWE (e.g., 60°C for 3.5 min in the USA, or 64°C for 2.5 min in the UK) ensure food safety by giving 5–9 Log_{10} reductions of the most frequent *Salmonella* serotypes (Alvarez et al., 2006; Mañas et al., 2003), even though some heat-resistant microorganisms can survive the above-mentioned pasteurization requirements and spoil the LWE even under refrigerated conditions (Lee et al., 2001). Pasteurization for LWE is limited to lower pasteurization temperatures and longer holding times because of the coagulation of its proteins at higher temperatures.

Salmonella is primarily the problem in most cases (EFSA, 2011). Recently, *Salmonella* in eggs has emerged as a primary concern for public health agencies in Europe and in the United States (CDC, 2003, 2004; Schroeder et al., 2005). It was reported that pasteurization of egg products became mandatory in the US in 1966 (Cunningham, 1995) and current regulations in the US require that LWE is heated

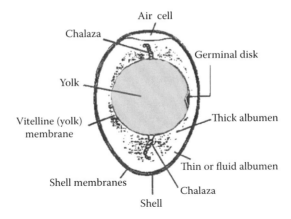

FIGURE 12.1 The parts of egg.

to at least 60°C for a minimum of 3–5 min. The reason may be attributed to the fact that the USDA requires liquid egg to be heated at the above-mentioned temperature and duration time to achieve more than 3.0 log in colony-forming units (CFU)/mL reduction of *Salmonella* (ICMSF, 1998). It is concluded that the functional performance of egg white (EW) is impaired when heated for several minutes above 57°C (Ma et al., 1997). Owing to the incomplete pasteurization at lower temperatures, the foodborne outbreaks comprising *Salmonella enteritidis* have resulted in eggs (Tauxe, 1991; Tood, 1996). The illness risk is greater when the egg is used as a food ingredient in foods rather than when consumed as an individual egg (Todd, 2001).

The thermal pasteurization of LWE in most cases leads to protein denaturation and coagulation, hence affecting the liquid egg consistency. Therefore, pasteurization of LWE is limited to lower pasteurization temperatures and longer holding times owing to the coagulation of its proteins at higher temperatures (Cheftel, 1995; Tauxe, 1991; Tewari et al., 1999; Tood, 1996).

High-pressure processing (HPP) is an industrially tested technology that offers a more natural, environmentally friendly alternative for pasteurization or shelf life extension of a wide range of food products (Barbosa-Cánovas and Juliano, 2007; Welti-Chanes et al., 2005). The great potential of HPP in the food industry has been recently reviewed (Norton and Sun, 2008).

Numerous studies including high-hydrostatic pressure (HHP) technologies have been performed to develop the procedures replacing conventional heat treatment (pasteurization) of liquid egg, which is applied at 60–65°C for 5–10 min (Farr, 1990; San Martín et al., 2002). Using of HHP technology provides the better preservation of native properties of raw foods with similar antimicrobial efficacy as heat treatment. The profitable effects of HHP are demonstrated for many heat-sensitive foods, and liquid foods are treated in their packing material to avoid potential postinfection of the final product (Oey et al., 2008; Seregély et al., 2007).

Previous studies have shown that HHP technology is appropriate for the destruction of various pathogen microorganisms in LWE and egg products (Jankowska et al., 2005; Ponce et al., 1999, 1998a,b). The viscosity of egg product is related to coagulation of specific egg proteins induced by HHP, thereby, the pressure is effective on the rheological product characteristics. For prepotent treatment of LWE, the accomplishing of adequate microbiological condition and the preserving of desirable sensory and functional characteristics of LWE are imperative (Tokuşoğlu, 2013).

12.1.2 MICROBIAL STABILITY, PRESERVATION, AND SHELF LIFE OF TABLE EGGS BY HIGH PRESSURE

HHP processing inactivates microorganisms, denatures proteins, and extends the shelf life of food products, with minor effects on nutritional value and flavor. For the treatment of LWE and shelf stable egg-based products, not only achieving the satisfactory microbiological condition, but also the preservation of its beneficial organoleptic and functional properties are important (Ahmed et al., 2003; Barbosa-Cánovas, 2008; Guamis et al., 2005; Juliano et al., 2006, 2007; Pajan et al., 2006; Ponce et al. 1998a,b). The HHP exposes the foods with pressures in the range of 100–1000 MPa with processing temperatures from below 0° to 100°C, where significant microbial reduction can be achieved (Huang et al., 2006).

It has been shown that the pressure treatments (300–450 MPa) at various temperatures (15°, 20°, or 50°C) for 5–15 min efficiently inactivated *Salmonella enteritidis* inoculated in LWE (Ponce et al., 1999).

Bari et al. (2008) investigated the use of high-pressure pulse treatment to inactivate *Salmonella enteritidis* inoculated in liquid egg. In that study given by Bari et al. (2008), liquid egg was inoculated with *Salmonella enteritidis* (8.0 log CFU/mL) and exposed to hydrostatic pressures (300–400 MPa) and pressure (350 MPa) pulsing at 25°, 40°, and 50°C for up to 40 min to determine the maximum allowable pressure that can inactivate the *Salmonella* with minimal injury. Bari et al. (2008) stated that the SE-2 and SE-3 strains of *Salmonella* were the most sensitive strains at 400 MPa (25°C) pressure treatments for 10 min, and 8.0 and 7.0 $\log_{10}$ CFU/mL reduction were obtained for strains SE-2 and SE-3, respectively. Based on these studies, strains SE-1 and SE-4 were the least sensitive and 5.0 and 4.0 $\log_{10}$ CFU/mL of inhibition were achieved, respectively.

It was shown that the result of HPP treatment of liquid egg inoculated with *Salmonella enteritidis* is shown in Table 12.1 (Bari et al., 2008).

It was reported that at 300 and 350 MPa pressure treatment for *Salmonella enteritidis* in liquid eggs gave the 4.0 and 4.8 $\log_{10}$ CFU/mL of reduction, respectively, whereas 400 MPa pressure treatment gave the 6.0 $\log_{10}$ CFU/mL of reduction at 25°C for up to 40 min (Bari et al., 2008). The effects of HPP temperatures (25°, 40°, and 50°C) on inactivation of *Salmonella enteritidis* in liquid egg were monitored at 350 MPa pressure up to 40 min. It was found that the highest inactivation of *Salmonella* in the liquid egg was observed at 50°C, which resulted in a 6.0 $\log_{10}$ CFU/mL reduction (Figure 12.2).

It was concluded that when the treated liquid eggs were stored at 4°, 25°, and 37°C for 24 h, no *Salmonella* was detected in the samples (Bari et al., 2008).

HHP treatments have been applied to inactivate different microorganisms inoculated in LWE (Guamis et al., 2005). It was shown that treatments at pressure above

TABLE 12.1

Populations of *Salmonella* Strain SE-4 Recovered from Liquid Whole Egg Following High Hydrostatic Pressure Treatment

	Population ($\log_{10}$ CFU/mL)[a]	
Treatment	**Survival**	**Reduction**
Control	9.12 ± 0.12	0.00
300 MPa (30 min)	5.06 ± 0.11	4.06 ± 0.11
350 MPa (30 min)	4.37 ± 0.10	4.75 ± 0.10
400 MPa (10 min)	3.16 ± 0.10	5.96 ± 0.10

Source: Adapted from Bari M.L. et al. 2008. *Food Borne Pathogens and Disease*, 5(3), 175–182.

[a] Mean ± SD ($n = 3$). Populations of *Salmonella* were recovered on tryptose soy agar medium. CFU: colony-forming units.

400 MPa combined with temperature of 50°C were able to reduce the *Salmonella enteritidis* count by 8 $\log_{10}$ units, whereas total bacterial count was also significantly reduced and 10 CFU/mL of reduction was detected after 15 days of storage at 4°C (Guamis et al., 2005).

Ponce et al. (1998b) applied 300–450 MPa/5–15 min at temperatures of −15°, 2°, and 20°C to LWE inoculated with *Listeria innocua* at a pH of 8. *Listeria innocua* inactivation at 400 MPa followed the first-order kinetics for 0–20 min, and exhibited decimal reduction times D of 7.35 min at 2°C while 8.23 min at 20°C. The greatest

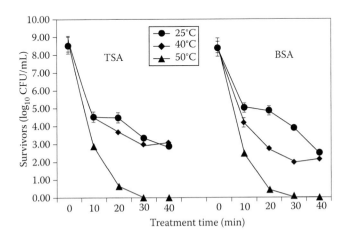

FIGURE 12.2 The effects of HPP temperatures on inactivation of *Salmonella enteritidis* in LWE at 350 MPa/40 min. The values are means ± SD of three experiments with duplicated determinations. (Adapted from Bari M.L. et al. 2008. *Food Borne Pathogens and Disease*, 5(3), 175–182.)

inactivation (>5 log reductions) was obtained at 450 MPa for 15 min at 20°C (Ponce et al., 1998b).

Ponce et al. (1998b) stated that the highest reduction of *Escherichia coli* in LWE was obtained at 50°C and it was reported that *E. coli* in LWE was more resistant to pressure at 20° and −15°C than at 50° and 2°C (Ponce et al., 1998b).

10^7–10^8 CFU/mL inoculation of *Salmonella enteritidis* in LWE were subjected to 350 and 450 MPa at 50°, 20°, 2°, and −15°C for 5, 10, 15 min of treatment times as well as cycles of 5–5 and 5–5–5 min treatments (Ponce et al., 1999). It was concluded that inactivation increased with pressure and exposure time; the greatest inactivation (>8 log cycles) occurred at the severest treatment conditions at 450 MPa/50°C whereas the minimal inactivation (1 log reduction) occurred at the lowest temperature and time conditions (Ponce et al., 1999).

Lee et al. (2003) performed the effects of various pressures on *Listeria seeligeri* and *E. coli* (10^7 and 10^8 CFU/mL, respectively), in LWE at 5°C; *Listeria* reductions were not detected after 250 and 350 MPa treatments for 886 and 200 s of exposure duration, respectively, whereas >2 log reductions of *E. coli* were accomplished (Lee et al., 2003). Yuste et al. (2003) reported that 400 MPa/5 min of treatments resulted in *E. coli* inactivation of 5.5 log cycles in LWE, whereas no *Salmonella typhimurium*, *Yersinia enterocolitica*, and *Listeria monocyctogenes* were detected (Yuste et al., 2003). Isiker et al. (2003) stated that increasing the pressure had a significant effect on *Salmonella enteridis* inactivation in LWE (Isiker et al., 2003).

Dong-Un Lee (2002) stated that kinetic studies on the isothermal HHP inactivation of *E. coli* in LWE were performed at 5°C and 25°C in the pressure range of 250–400 MPa, and the characteristic tailing inactivation curves were described by a first-order biphasic model.

It was prevailed that the degree of *E. coli* inactivation at isothermal pressure condition was independent of applied pressures if the physical characteristics of LWE are considered, that is, between 2.0 and 3.0 log reductions at 5°C, and less than 1.0 log reductions at 25°C in the range of 250–400 MPa, so HHP at 5°C is more favorable than at 25°C (Dong-Un Lee, 2002). It was reported that about 3 log reductions of *E. coli* and over 5 log reductions of *Pseudomonas* and *Paenibacillus* by HHP treatment of LWE at 5°C is regarded to be as effective as conventional thermal pasteurization (Dong-Un Lee, 2002).

Dong-Un Lee (2002) stated that the HHP processing conditions were fixed to either 250 MPa for 886 s at 5°C or 300 MPa for 200s at 5°C, which have been indicated as the optimized HHP processing conditions. It was put forward that the addition of nisin (Figure 12.3) prior to pressure treatments significantly increased the lethal effects of HHP against *Listeria seeligeri*. The individual effects of each nisin and HHP on the *Listeria* reductions were very small, and the increased *Listeria* reductions, up to 5 log cycles were obtained owing to the synergistic action of bactericidal effect of nisin and high pressure effects (Dong-Un Lee, 2002).

The nisin–HHP combination can effectively reduce the microbial loads of Gram-negative *E. coli*. It was concluded that the marginal effects of nisin–HHP synergy on *E. coli* reduction in LWE can be expressed by the membrane structure of Gram-negative *E. coli* or by the protective effects of LWE (Dong-Un Lee, 2002).

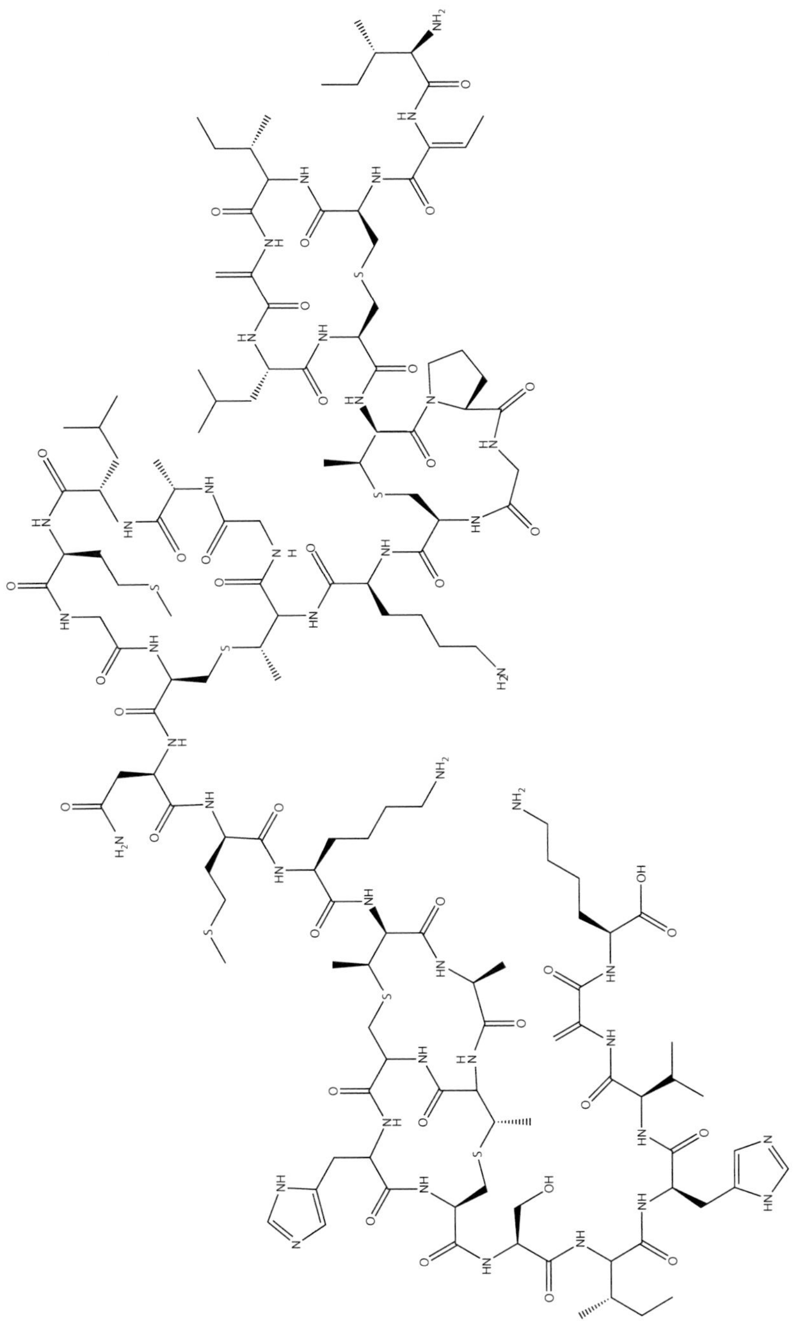

FIGURE 12.3 The chemical formula of nisin.

Juliano (2006) studied the inactivation of *Bacillus stearothermophilus* after different stages in the process: (a) baking an egg mix to form a patty, (b) after preheating, and (c) after high pressure high temperature processing. In the study described by Juliano (2006), *Bacillus stearothermophilus* spore inoculated in the egg mix showed a one log cycle reduction after baking. It was reported that the inactivation of *B. stearothermophilus* (ATCC 7953) spores in egg patties was accelerated after pressure-assisted thermal processing treatment at 700 MPa/105°C and *B. stearothermophilus* spores were inactivated rapidly in egg matrix (4 log reductions in 5 min) when compared to thermal treatment at 121°C (1.5 log reduction in 15 min) (Rajan et al., 2006). Similar result was found by Koutchma et al. (2005) and the inactivation of *B. stearothermophilus* in spore strips located between two egg patties can be reduced by at least 6 log cycles at 688 MPa/105°C in 5 min (Koutchma et al., 2005).

12.1.3 RHEOLOGICAL PROPERTIES OF TABLE EGGS BY HIGH PRESSURE

The egg is a low acid food (higher pH) that necessitates preservation by some means to increase the shelf life. It was found that HHP-treated EW at 600 MPa or more gets fully coagulated to form gels (Bridgeman, 1914).

It is known that the protein can be denatured, coagulated, or gelled and each protein has unique nutritional and physicochemical properties, therefore, processing quality of proteins in foods depends on several factors such as pH, protein type, temperature, applied pressure, and ionic strength (Tokuşoğlu, 2013). HHP has been focused on food proteins and its functional properties, modification, and texture (Ahmed et al., 2003). It is reported that HHP has reduced the alterations of postprocess contamination, and coagulation, creating a better retention of nutritional qualities of eggs as the process is carried out at considerably low temperature in packed form, and also the product could be consumed directly without the heat treatment (Ahmed et al., 2003; Knorr, 1993, 1996). Pressure induces protein denaturation, depending on the protein concentration, pressure level, temperature, and pH (Balny and Masson, 1993).

Eggs contain protein of high biological value as compared to any dietary protein. Egg proteins have all the desirable nutritional and functional properties, so they are widely used in food technology (Hsieh et al., 1993; Lee et al., 1999).

For food quality control, sensory evaluation, food process, equipment design, and also for new product development, the information of rheological properties of foods is necessary. Based on the origin, chemical and nutritional composition, structure behavior, and previous history, the flow behavior of a fluid can be varied from Newtonian to time-dependent non-Newtonian in nature (Rao, 1986). Newtonian and/or time-dependent non-Newtonian flow behavior of the egg was reported (Cornford et al., 1969; Lee et al., 1999) and the rheology of a commercial egg gel white at high temperature using creep and compression measurement was studied (Nagano and Nishinari, 2001).

The denaturation or coagulation or structure breakdown occurs in albumen or EW protein and the role of protein structure on emulsion and gel rheology is important; so, the HHP effects on the rheological properties of the egg is also significant (Ahmed et al., 2003).

Ahmed et al. (2003) stated that whole liquid egg (WLE) and albumen have been denatured at HHP, and it was reported that 100–400 MPa/30 min of HHP

TABLE 12.2

Rate of Thixotropy and Area Under the Curve of Egg Components Obtained from Software

Sample	Pressure (MPa)	Rate of Thixotropy (Pas^{-1})	Area Inside the Curve (s^{-1}Pa)
WLE	0.101	54.85	556.7
	150	51.12	345.6
	200	49.56	329.6
	250	42.52	144.6
	300	514.9	1618
	350	384.7	1253
Albumen	0.101	287.4	790
	150	34.22	111.7
	200	11.91	63.45
	250	7.87	53.87
	300	81.86	423.4
	350	ND	
Yolk	0.101	21,930	84,840
	150	24,620	91,500
	200	25,768	89,680
	250	ND	
	300	30,460	92,657
	350	36,830	99,096

Source: Adapted from Ahmed J. et al. 2003. *Lebensm Wiss und Technol*, 36(5), 517–524.

application affected the rheological characteristics of WLE, albumen, and yolk. In the study given by Ahmed et al. (2003), an advanced controlled stress rheometer was employed to study the rheological properties at a shear rate of 0–200 s^{-1} using a double concentric cylinder for WLE and albumen while parallel plate geometry was used for the yolk with shear rate range of 0–500 s^{-1}. Figure 12.4 shows time dependency of control and pressure treated WLE. It was stated that both WLE and albumen behaved as time-dependent fluids (thixotropic); however, HHP reduced time dependency substantially. It was also determined that albumen individually exhibited more pressure effect compared to WLE and thixotropy of the yolk significantly varied ($p < 0:05$) during HHP (Ahmed et al., 2003). Table 12.2 shows the rate of thixotropy and area under the curve of egg components obtained from software.

The effects of HHP on rheological parameters of WLE, albumen, and yolk as egg components were studied (Tables 12.3 and 12.4). It was shown that for egg albumen, the magnitude of yield stress and consistency coefficient decreased during pressurization; however, the coagulation of protein reversed the trends. It was determined that all egg samples behaved as thixotropic fluid and the structure breakdown of egg protein enhanced with high pressure and was completed at 300 MPa/30 min at 20°C (Ahmed et al., 2003).

A high pressure level at 400–600 MPa can cause enough alterations in the viscosity of egg components so that it can become gel with improved quality characteristics

TABLE 12.3

Effect of Pressure on Rheological Parameters of Liquid Whole Egg (LWE) Using Herschel-Bulkley Model

	Pressure (MPa)	Yield Stress (Pa)	Consistency Coefficient (κ), Pas^n	Flow Behavior Index (n)	Standard Error
Liquid whole egg	0.101 Up	0.536	0.058	0.753	30.11
(LWE) (shear	0.101 Dn	0.171	0.023	0.959	17.07
rate 0–200 s^{-1})	150 Up	0.323	0.044	0.876	17.55
	150 Dn	0.165	0.022	0.961	9.48
	200 Up	0.400	0.054	0.893	15.55
	200 Dn	0.309	0.019	1.079	13.61
	250 Up	0.334	0.324	0.367	19.66
	250 Dn	0.205	0.032	1.094	11.12
	300 Up	0.769	0.786	0.276	16.45
	300 Dn	0.832	0.055	0.962	8.63
	350 Up	1.676	0.887	0.186	18.44
	350 Dn	0.506	0.049	0.903	7.56

Source: Adapted from Ahmed J. et al. 2003. *Lebensm Wiss und Technol,* 36(5), 517–524.

as opposed to the heat-induced gels (Hayashi et al., 1989). It was expressed that pressure-induced gels were softer than untreated samples, more elastic without any cooked taste and flavor, and there was no destruction of vitamins and formation of lysinoalanine (Hayashi et al., 1989).

12.1.4 HIGH PRESSURE EFFECTS ON EGG PHOSVITIN, OVALBUMIN, AND OVOTRANSFERRIN

Egg yolk (EY) phosvitin represents about 7% of the proteins found in EY and is a highly phosphorylated protein of the egg (Abe et al., 1982; Samaraweera et al., 2011). Phosvitin, a highly phosphorylated glycoprotein, represents the major fraction of hen EY phosphoproteins (Anonymous, 2013a) (Figure 12.5). It is known that phosvitin is rich in serine residues and phosphorylated peptides, that is, phosphopeptides, with antioxidant and mineral-binding ability could be a great source of natural functional biopeptides (Jiang and Mine 2000).

Volk et al. (2012) stated that phosvitin structure maintained overall during high-pressure treatment of 600 MPa applied at an initial temperature of 65°C regardless of the pH and treatment duration, confirming the high structural stability of the phosphoprotein. It was reported that treatment of phosvitin with phosphatase increased the degree of dephosphorylation from 24 to 63%, after 2 and 18 h, respectively. It was also found that angiotensin-converting enzyme inhibition and antioxidant activity of dephosphorylated and protease-treated phosvitin were increased by 52 and 39%, respectively, as compared to protease-digested native phosvitin.

TABLE 12.4

Effect of Pressure on Rheological Parameters of Egg Components Albumen and Yolk Using Herschel-Bulkley Model

Egg Components	Pressure (MPa)	Yield Stress (Pa)	Consistency Coefficient (κ), Pas^n	Flow Behavior Index (n)	Standard Error
Albumen (shear	0.101 up	0.768	0.032	0.87	16.77
rate 0–200 s^{-1})	0.101 Dn	0.536	0.015	1.039	13.21
	150 Up	0.096	1.43E-3	0.954	2.57
	150 Dn	0.086	1.28E-3	0.977	2.99
	200 Up	0.053	4.016E-3	0.914	16.13
	200 Dn	0.051	3.00E-3	0.969	15.88
	250 Up	0.033	2.41E-2	0.564	15.22
	250 Dn	0.027	1.83E-3	1.023	11.77
	300 Up	0.220	0.642	0.248	27.33
	300 Dn	0.741	0.016	0.851	19.19
Yolk (Shear Rate	0.101 Up	10.68	21.36	0.473	53.45
0-500 s^{-1})	0.101 Dn	9.529	7.363	0.514	6.85
	150 Up	a			
	150 Dn	10.29	8.13	0.516	5.92
	200 Up	a			
	200 Dn	11.26	8.52	0.509	7.068
	250 Up	a			
	250 Dn				
	300 Up	a			
	300 Dn	21.14	9.38	0.450	6.56
	350 Up				
	350 Dn	28.61	10.09	0.448	7.23

Source: Adapted from Ahmed J. et al. 2003. *Lebensm Wiss und Technol*, 36(5), 517–524.

It was shown that the pressure treatment of EW proteins above 450 MPa resulted in a loss of secondary structure (Hayakawa et al., 1996). In the pressure-induced structural alterations in EW proteins can also be demonstrated by exposing previously buried SH groups and hydrophobic groups (Iametti et al., 1999; Van der Plancken et al., 2004, 2005a,b, 2006). Iametti et al. (1999) stated that the treated albumen had increased viscosity, but retained its foaming and heat-gelling properties (Iametti et al., 1999). It was found that susceptibility of egg albumen proteins to hydrolysis by trypsin increased dramatically after HHP treatment (up to 10 min at 800 MPa).

As it is known, ovalbumin (OVA) (Figure 12.7) is the major protein found in EW, making up 60–65% of the total protein (Huntington and Stein, 2001) and it plays a major role in determining of EW behavior on the application of HHP (Messens et al., 1997). High pressure can result in the structural modification of EW that can be correlated to the enhancement of functional properties (Messens et al., 1997). The

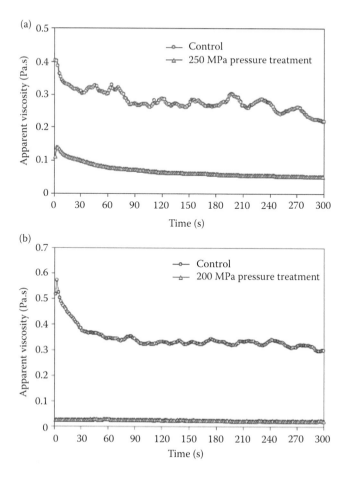

FIGURE 12.4 Time dependency of control and pressure treated WLE at shear rate of $10s^{-1}$, (b) Time dependency of control and pressurized albumen at shear rate of $10s^{-1}$. (Adapted from Ahmed et al. 2003. *Lebensm Wiss und Technol*, 36(5), 517– 524.)

S-form of ovalbumin, the presence of which is an index of egg aging, was not found in any of the pressure-treated samples, which also did not display the evidence for covalent protein aggregation (Iametti et al., 1999).

It was reported that the foaming capacity of EW has been improved owing to the exposure of SH groups that favor foaming stability and capacity (Van der Plancken et al., 2007a; Yang et al., 2010). It was reported that the turbidity, surface hydrophobicity, and exposure of sulfhydryl groups in EW proteins were also increased at pressures over 400 MPa application and a strong increase in surface hydrophobicity was observed between 400 and 700 MPa (Van der Plancken et al., 2005a, 2007b; Yan et al., 2010). Besides, the decrease and increase of total and exposed SH groups, respectively, were enhanced by pressure above 500 MPa (Van der Plancken et al., 2005a). It was reported that the pressure treatment at 410 MPa induced proteins in EY dispersions to aggregate and undergo a sol–gel transition (Aguilar et al., 2007),

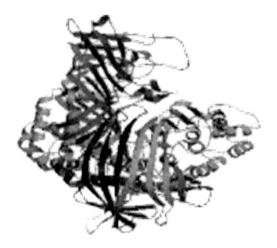

FIGURE 12.5   (**See color insert.**) Phosvitin structure (Adapted from Anonymous, 2013a).

while the treatment at 600 MPa resulted in the modification of emulsifying properties without the effect on the protein solubility of low-density lipoprotein solutions (Speroni et al., 2005).

It is known that ovotransferrin (Figure 12.6), accounting for 12–13% of EW proteins, is a glycosylated protein with an isoelectric point of 6.1. Huopalahti et al. (2007) stated that the ovotransferrin shows 50% homology with mammalian transferrin and lactoferrin, but differs from the other transferrin proteins in its isoelectric point and in the glycosylation pattern. It was also stated that ovotransferrin is also responsible for the ferric ion transfer from the hen oviduct to the developing embryo (Huopalahti et al., 2007), and it was found that ovotransferrin possessed antifungal activity (Valenti et al., 1985), immunomodulatory and antiviral activity (Giansanti et al., 2002, 2005, 2007), and antioxidant and anticancer activities (Ibrahim et al., 2007; Ibrahim and Kiyono, 2009). Ovotransferrin is known for a rich source of bioactive peptides and recently, there is great attention regarding the potential of ovotransferrin as functional food and nutraceutical ingredient (Wu and Acero-Lopez, 2011).

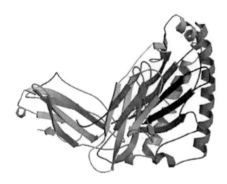

FIGURE 12.6   Ovalbumin structure. (Adapted from Anonymous. 2013b. http://blogs.oregonstate.edu/psquared/2010/04/16/ovalbumin)

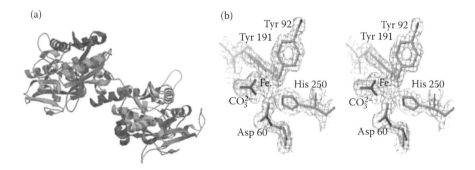

FIGURE 12.7 (a) The chemical structure of ovotransferrin. (Adapted from Kurokawa H., Mikami B., and Hirose M. 1995. *J Mol Biol*, 254(2), 196–207.); (b) The amino acid structure of ovotransferrin (Adapted from Wu J. and Acero-Lopez A. 2012. *Food Research International*, 46(2), 480–487.)

Current research indicated that sonication could affect the exposure of SH groups of ovotransferrin and could release potent antihypertensive peptides (Lei et al., 2011; Majumder and Wu, 2010). After HHP processing, the conformational and physicochemical alterations are important since they affect the functional properties of food proteins and also protein bioactivities.

Acero-Lopez et al. (2012) reported the effect of high pressure treatment on ovotransferrin, and it was determined that HHP treatment caused changes in the ovotransferrin structure depending on the pH of the sample. Acero-Lopez et al. (2012) focused on the determination of a high pressure effect on the structure and physicochemical properties of ovotransferrin concentrate after processing in an acid (pH 3) and in a basic (pH 8) environment. It was found that, a decrease in total sulfhydryl groups and an increase in surface hydrophobicity were observed along with a partial aggregation at pH 8 and pressures higher than 200 MPa. It was also stated, the ovotransferrin adopted a molten globule state at pH 3 and was associated with a significant increase in surface hydrophobicity and reactive sulfhydryl content (Acero-Lopez et al., 2012).

Figure 12.8 shows the alterations in total sulfhydryl groups and in reactive sulfhydryl content (Figure 12.8) (Acero-Lopez et al., 2012). It was stated that ovotransferrin treated at 200 MPa at pH 8 shows a total SH content of 4 μmol SH/g, which is close to the control; whereas further increasing pressure led to considerable decrease in the total SH content to around 2 and 0.9 μmol SH/g at 400 and 700 MPa, respectively (Figure 12.8a).

In the study described by Acero-Lopez et al. (2012), the most evidential alteration in reactive SH content was observed at 600 and 700 MPa where it decreased from 1 μmol SH/g to about 0.2 μmol SH/g (Figure 12.8b). Van der Plancken et al. (2005b) revealed that decreasing of the total SH groups was probably due to rearrangement of cysteine residues and oxidation of SH groups (Van der Plancken et al., 2005b).

In the study reported by Acero-Lopez et al. (2012), a gradual increase in the denaturation peak controlled up to 400 MPa (Figure 12.9) was found. Figure 12.9 shows a differential scanning calorimetry (DSC) thermogram of ovotransferrin samples treated at various pressures between 200 and 700 MPa at pH 8 (Acero-Lopez et al., 2012).

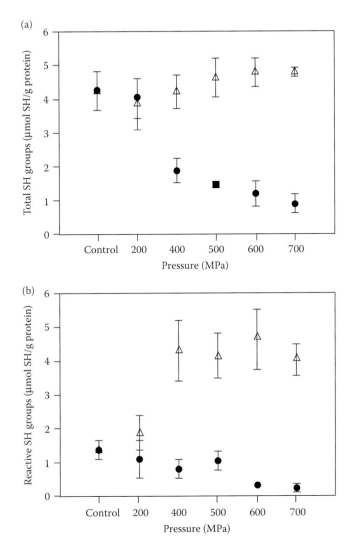

FIGURE 12.8 The changes in total sulfhydryl groups (a) and in reactive sulfhydryl content (b) for samples treated at different pressures and at pH 3 and pH 8. Mean value bars represent standard deviation of triplicate measurement. (Adapted from Acero-Lopez A. et al. 2012. *Food Chemistry*, 135, 2245–2252.)

12.1.5 High Pressure Effects on the Foaming Properties of Egg

Heated liquid eggs coagulate or solidify (as cakes, breads, crackers), whipped EW produces airier and lighter products (meringues, marshmallow, angel cake), and emulsified EY phospholipids and lipoproteins produce special products (mayonnaise, salad dressing, and sauces) (Davis and Reeves, 2002). It is known that food foaming characteristics of egg albumen are quite good. Ferreira et al. (1995) stated

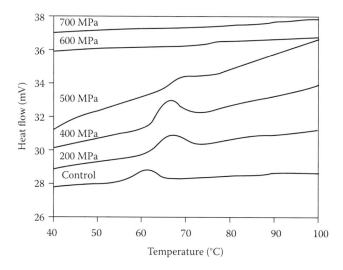

FIGURE 12.9 Differential scanning calorimetry (DSC) thermogram of ovotrans ferrin sample streated at different pressures (200, 400, 500, 600 and 700 MPa) at pH 8 and the control. Samples were heated from 25°C to 110°C at a rate of 10°C/min. (Adapted from Acero-Lopez A. et al. 2012. *Food Chemistry*, 135, 2245–2252.)

that foaming properties are evaluated by foaming capacity (FC) and foam stability (FS). For the determination of FC and FS, the following formulae are used as shown below (Chang and Chen, 2000):

$$FC\ (\%) = (FV/ILV) \times 100\%;\ FS\ (\%) = [(ILV - DV)/ILV] \times 100\%;$$
$$Drainage\ (mL) = LVM - LVS$$

where FV – volume of foam; ILV – volume of the initial liquid phase; DV – volume of drainage; LVM – volume of the liquid phase at $t = 60$ min after foaming was finished; LVS – volume of the liquid phase at $t = 30$ s after foaming.

Lomakina and Míková (2006) reported that various foods are prepared using EW, most of them being based on the foaming properties of EW that are owing to the albumen proteins' ability to encapsulate and retain air (Lomakina and Míková, 2006). Due to the foaming properties of EW, new methods have improved the volume and stability of EW foam (Lomakina and Míková, 2006).

Foaming is affected by water (Baldwin, 1986), temperature, sugar (Stadelman and Cotterill, 1994), EY (Kim and Setser, 1982), oil difference and quantity (Kim and Setser, 1982; Stadelman and Cotterill, 1994), and also by stabilizers and surfactants (Kim and Setser, 1982).

It was expressed that the pasteurization of egg albumen decreased the foaming ability and resulted in the quality reduction and volume of angel cake, this occurred by ovotransferrin denaturation on pasteurization at 53°C. Hatta et al. (1997) stated that metallic ions (Fe, Cu, Al, or other) and salts of phosphoric and citric acids are used to increase the denaturation temperature and for the improvement of the foaming properties of egg albumen after pasteurization.

It was reported that pasteurized EW required a longer whipping to attain a foam comparable in specific gravity to the foam from unpasteurized egg albumen (Stadelman and Cotterill, 1994). Ma (1996) stated the effects of chemical modifications on the physicochemical and cakebaking properties of EW and reported that the overrun and the foam stability of spray-dried EW increased significantly by gamma irradiation processing. Ma et al. (1994) also reported that the time for 50% drainage, an index of the foam stability, increased by irradiation with higher dosages, indicating improvement in the foam stability, whereas it decreased in the overrun at 4 kGy, but there was no alteration in the foam stability for the frozen EW.

It was indicated that the greater increasing foaming power observed in the case of ultrasound high pressure combinations may be explained by the homogenization effect of ultrasound (Knorr et al., 2004). Knorr et al. (2004) expressed the foaming capacity of LWE by ultrasound processing since ultrasound dispersed the protein and fat particles in LWE (Knorr et al., 2004). Table 12.5 shows the combined processing effects on the foaming capacity of LWE (Knorr et al., 2004).

Hoppe (2010) analyzed the foaming properties of EW solutions (10% v/v) with varying levels of pressure and pH. It was found that pressure treatment of 10% EW solutions at pH 9.11 resulted in a homogenous solution with improved foaming capacity over the control (Figure 12.10) (Hoppe, 2010). Figure 12.10 shows the effect of HPP on a foam overrun at pH 9.11 at 0.1 MPa (control), 600 and 800 MPa (Hoppe, 2010). It was shown that increasing pressure resulted in an increase in foam volume and foam overrun increased significantly ($\alpha = 0.05$) at 800 MPa at all-time points (Hoppe, 2010).

Hoppe (2010) expressed that the foaming properties of EW solutions were also highly dependent on pH (Hoppe, 2010). It was found that the greatest foam overrun was achieved at pH 4.5 whereas the foaming ability was significantly decreased at pH 6 (Figure 12.11). In the study described by Hoppe (2010), it was also reported that the increased foam overrun at pH 4.5 could be attributed to major EW proteins,

TABLE 12.5
Combined Processing Effects on the Foaming Capacity of Liquid Whole Egg

Processing	Power (Overrun (%))	Stability (Stability (%))
Control	479	52
HHP	490	56
Nisin–HHP	484	55
Ultrasound–HHP	638	50

Source: Adapted from Knorr D. et al. (2004). *Trends in Food Science & Technology*, 15, 261–266.

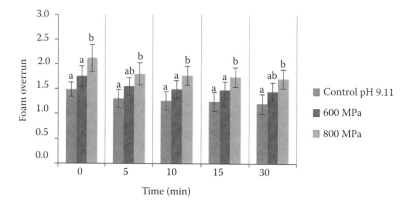

FIGURE 12.10 The effect of HPP (5 min) on 10% egg white solution foam overrun at pH 9.11 and at 0.1 MPa (control), 600 MPa, and 800 MPa. Time intervals represent measurement of foam overrun postfoaming. Error bars are ±1 standard deviation. (Adapted from Hoppe A. 2010. Examination of egg white proteins and effects of high pressure on select physical and functional properties. *MSc Thesis in Food Science and Technology Food Science and Technology Department*. Faculty of the Graduate College at the University of Nebraska, Lincoln.)

ovalbumin and ovomucin, important to foaming properties, which have respective pI of 4.5 and 4.1.

It was found that the foam stability determined the effect of HPP or pH on EW foaming properties (Hoppe, 2010). In the study given by Hoppe (2010), it was also reported that HPP significantly reduced the foam stability, with the exception at

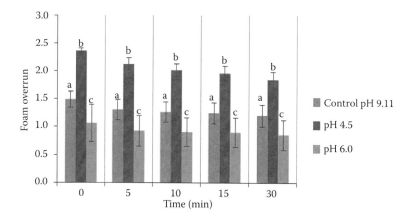

FIGURE 12.11 The pH effect on 10% egg white solution foam overrun. Time intervals represent measurement of foam overrun postfoaming. Error bars are ±1 standard deviation. (Adapted from Hoppe A. 2010. Examination of egg white proteins and effects of high pressure on select physical and functional properties. *MSc Thesis in Food Science and Technology Food Science and Technology Department*. Faculty of the Graduate College at the University of Nebraska, Lincoln.)

800 MPa/0 time point, and this data was in contrast to the study described by Van der Plancken et al. (2007a), where it was found that the HHP treatment increased the overall foam stability.

Figure 12.12 shows the effect of HPP (5 min) on 10% EW solution foam stability at pH 9.11 and at 0.1 MPa (control), 600 and 800 MPa (Figure 12.12) (Hoppe, 2010). In the study described by Hoppe (2010), the increased stability of 800 MPa at the 0 time point was attributed to the increased foam volume and incorporation of liquid in the foam. It was found that the liquid drainage was the greatest over the first 5 min postfoam (800 MPa), as indicated by the slope and drop in stability, as shown in Figure 12.12 (Hoppe, 2010).

12.1.6 High Pressure Effects on Color and Texture Properties of Egg

Singh and Ramaswamy (2010) reported that L*,a*,b* values increased with the increase in pressure intensity for whole egg. In egg yolk, L* remained mostly stable and a* value decreased, whereas the b* value showed a great increase in yellowness. It was also reported that HPP induced an increase in L* (lightness) and a* (redness) values of EW up to 700 MPa, whereas the b* value simultaneously decreased, indicating decreasing yellowness with increasing treatment intensity (Singh and Ramaswamy, 2010).

It was demonstrated that HPP affected the color values of EW, EY, and WLE, respectively, and it was found that the L* value increased linearly at all pressure–time combinations, indicating an increase in brightness of the sample with increasing pressure treatment intensity for EW (Table 12.6) (Singh, 2012).

It was found that the a* values increased with increasing pressure treatment intensity whereas there was a small increase in b* value for EW (Singh, 2012). Table 12.7

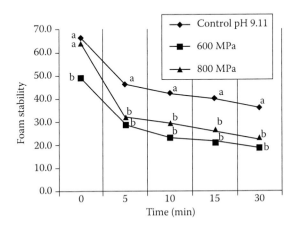

FIGURE 12.12 The HHP (5 min) effect on 10% egg white solution foam stability at pH 9.11 and at 0.1 MPa (control), 600 MPa, and 800 MPa. Time intervals represent measurement of foam stability postfoaming. (Adapted from Hoppe A. 2010. examination of egg white proteins and effects of high pressure on select physical and functional properties. *MSc Thesis in Food Science and Technology Food Science and Technology Department.* Faculty of The Graduate College at the University of Nebraska, Lincoln.)

TABLE 12.6

Hunter L* (Lightness) Values of the Egg White and Yolk Subjected to Pressure Level and Treatment Time

	L* Value			
	Egg White			
		Time		
Pressure	**0**	**5**	**10**	**15**
600	58 ± 0.707	63.5 ± 3.5	66 ± 4.24	80 ± 1.14
700	81 ± 1.41	86	85.5 ± 0.70	91.5 ± 0.70
800	86 ± 3.53	89 ± 4.24	91.5 ± 2.12	96 ± 1.41
900	88 ± 2.82	91.5 ± 3.5	94 ± 1.41	100.5 ± 0.70
	Egg Yolk			
600	57.4 ± 0.84	56.8 ± 0.28	53.85 ± 0.21	61.1 ± 0.14
700	57.05 ± 1.34	55.7 ± 0.49	51.75 ± 0.35	44.8 ± 0.070
800	55.3 ± 0.98	50.6 ± 0.84	50.6 ± 0.56	51.25 ± 0.07
900	57 ± 1.41	53.15 ± 0.21	55.1 ± 0.28	58.5 ± 0.28

Source: Adapted from Singh A. 2012. Evaluation of high pressure processing for improving quality and functionality of egg products. PhD Thesis. Department of Food Science and Agricultural Chemistry, Macdonald Campus, McGill University, Ste. Anne-De-Bellevue, Quebec, Canada.

shows the Hunter a* and b* values of the EW subjected to 600–900 MPa of high pressure for 1–15 min of treatment time (Table 12.7) (Singh, 2012).

Singh (2012) reported that EY containing a high level of xanthophylls (yellow color) showed different color behavior than that of EW, EY color changed from pale yellow to orangish yellow as per visual appearance (Singh, 2012). It was found that the L* value remained constant and the a* value decreased from 9.51 to 2.11, indicating diminution in redness of sample whereas the b* value increased significantly ($p < 0.05$) from 56.5 to 76.5, showing a great deal of increase in yellow color of the EY (Tables 12.6 and 12.7) (Singh, 2012). It is known that the yellow color is desirable from a customer point of view.

It is known that ΔE is the total color difference, and it represents the color variance of foods during processing (Equation 12.1). ΔE is obtained as the combined differences in L*, a*, and b* values and ΔE is calculated using L*, a* and b* values whereas raw egg components acted as reference (Azarpazhooh and Ramaswamy, 2012).

$$\Delta E = \sqrt{(L_0^* - L^*)^2 + (a_0^* - a^*)^2 + (b_0^* - b^*)^2} \tag{12.1}$$

Ahmed et al. (2005) found that ΔE remained constant even after increase in HPP treatment and this situation indicated the stability of pigments (Ahmed et al., 2005). It was concluded that ΔE increased with an increase in pressure level and treatment time (Singh, 2012).

TABLE 12.7
Hunter a* (Redness) and b* (Yellowness) Values of the Egg White and Yolk Subjected to Pressure Level and Treatment Time

	a* Value			
	Egg White			
			Time	
Pressure	0	5	10	15
600	1.2 ± 0.007	2.005 ± 0.021	2.22 ± 0.4	2.26 ± 0.014
700	2.8 ± 0.077	2.91 ± 0.01	3.125 ± 0.035	3.13 ± 0.01
800	1.94 ± 0.65	2.065 ± 0.021	2.005 ± 0.007	2.35 ± 0.07
900	1.9 ± 0.02	2.095 ± 0.021	20.025 ± 0.007	2.25 ± 0.07
	Egg Yolk			
600	9.51 ± 0.014	8.46 ± 0.05	8.405 ± 0.007	61.5 ± 0.07
700	8.63 ± 0.04	5.84 ± 0.06	5.36 ± 0.05	4.15 ± 0.07
800	4.15 ± 0.07	2.95 ± 0.07	2.6 ± 0.28	1.85 ± 0.07
900	3.95 ± 0.07	2.765 ± 0.91	2.36 ± 0.06	2.11 ± 0.014
	b* Value			
	Egg White			
600	1.9	1.61 ± 0.014	1.805 ± 0.0007	2.25 ± 0.07
700	1.5 ± 0.14	1.19 ± 0.014	1.11 ± 0.014	1.25 ± 0.07
800	1.3 ± 0.07	1.11 ± 0.014	0.98 ± 0.0141	0.855 ± 0.021
900	0.99 ± 0.01	1.085 ± 0.021	1.615 ± 0.021	2.214 ± 0.0021
	Egg Yolk			
600	44.19 ± 15.2	56.15 ± 1.20	58.1 ± 0.14	58.1 ± 0.14
700	61.1 ± 0.14	62.5 ± 0.70	62.25 ± 1.06	63.4 ± 1.27
800	64.75 ± 0.35	66.5 ± 3.53	70.4 ± 0.28	70.6 ± 0.28
900	69.3 ± 1.83	70.5 ± 2.12	72.5 ± 2.12	75.5 ± 0.70

Source: Adapted from Singh A. 2012. Evaluation of high pressure processing for improving quality and functionality of egg products. PhD Thesis. Department of Food Science and Agricultural Chemistry, Macdonald Campus, McGill University, Ste. Anne-De-Bellevue, Quebec, Canada.

It is valued that texture is an imperative characteristic of egg and egg-based products that affects consumer perception and overall acceptability. Textural alterations in egg constituents are very sensitive to food processing types and utilized parameters (Hayashi et al., 1989; Kilcast and Lewis, 1990; Tokuşoğlu, 2013). Pons and Fiszman (1996) stated that HPP can be used to modify food proteins in a controlled manner, so as to make egg gels with better quality, uncooked flavor, and better textural properties. HPP not only improved the color but also resulted in a more firmer texture than heat coagulated egg products (Singh and Ramaswamy, 2010).

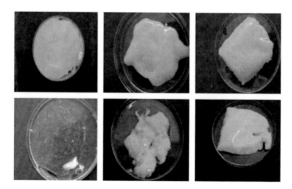

FIGURE 12.13 (**See color insert.**) Texture profiles of egg yolk and egg white. (Control, 400, 600 MPa of HHP treatment). (Adapted from Singh A., Ramaswamy H.S. 2010. Effect of HPP on physicochemical properties of egg components. In: Annual Meeting of Institute of Food Technologists (IFT), Chicago, USA, July 2010.)

With the increase in pressure level and treatment time, texture properties including firmness, springiness, cohesiveness, gumminess, and chewiness improved for EW. For EY, firmness, adhesiveness, gumminess, chewiness, and resilience were enhanced while cohesiveness decreased with an increase in pressure level and treatment severity (Singh and Ramaswamy, 2010) (Figure 12.13). Singh (2012) stated the texture profiles including hardness, adhesiveness, cohesiveness, chewiness, gumminess, and springiness of pressure-treated EW, EY, and WLE samples.

It was reported that the hardness of WLE increased with increasing pressure level and treatment time, but the hardness values were lower than EY and higher than EW (Singh, 2012). With the high pressure application, the form of egg gels was very adhesive and elastic in the study described by Singh (2012) and these data were in accordance with that reported by Hayashi et al. (1989). Similarly, it was found that high-pressure-coagulated EW gels were more adhesive and elastic than thermally-treated gels (Hayashi et al., 1989). According to another study, high pressure processed gels have a softer structure than that of thermal treatments (Carlez et al., 1995). It is known that adhesiveness is related to surface properties. In the study reported by Singh (2012), adhesiveness decreased linearly with an increase in pressure level and treatment time for EW.

Singh (2012) found that increasing the pressure level from 500 to 900 MPa affected the adhesion properties of egg constituents, and WLE followed an increasing trend while the adhesiveness value of EW was two-fold to that of the WLE. It was also found that EY samples were more adhesive physically than EW and WLE. For adhesiveness, egg components followed EY > EW > WLE pattern, where EY demonstrated maximum hardness (Figure 12.14). It was interpreted that the highest increase in EY could be the higher amount of fat level in EY matrix, thus increasing adhesiveness (Singh, 2012; Singh and Ramaswamy, 2010).

It was found that EW showed different behavior than those of EY and WLE because of their high protein content. It was reported that HPP caused EW coagulation and

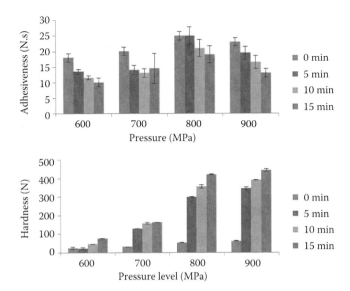

FIGURE 12.14   The effect of high-pressure processing and treatment time on adhesiveness (upper figure), and hardness (bottom figure) of egg yolk (EY), (Adapted from Singh A. 2012. Evaluation of high pressure processing for improving quality and functionality of egg products. PhD Thesis. Department of Food Science and Agricultural Chemistry, Macdonald Campus, McGill University, Ste. Anne-De-Bellevue, Quebec, Canada.)

an increase in the intensity of pressure level and treatment time caused the EW gelation (Singh, 2012).

It was reported that EW turned opaque at 600 MPa/15 min of HHP treatment and was able to form egg gels that can stand by themselves. EY was able to form gels at 700 MPa/15 min while WLE was able to form gels at a very short time processing treatment of 700 MPa/10 min (Singh, 2012).

Singh (2012) reported that the egg constituents changed from liquid state to complete gel with coagulation and gelation of egg constituents by increasing pressure application. It was stated that egg gels were formed at pressure levels greater than 600 MPa with temperatures that required for thermal gel formation (Singh, 2012). It was concluded that HHP led to the formation of full set egg gels with improved physicochemical and functional characteristics and without any cooked flavors.

Hoppe (2010) reported that egg gels formed with heat at 95°C had an average hardness value and over twice the value of HP-induce gels at 800 MPa (Figure 12.15) while the softest gel was observed at 600 MPa (Hoppe, 2010). It was also found that pH reduction decreased the hardness of heat induced gels while it increased the hardness of HHP gels in the study described by Hoppe (2010). Egg gel gumminess was determined and found with similar patterns to gel hardness with heat-induced gels being gummier (Figure 12.15). Figure 12.15 shows the effect of heat and HHP treatment on EW gel hardness and gel gumminess (Hoppe, 2010).

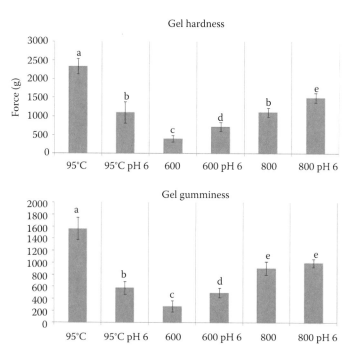

FIGURE 12.15 The effect of heat and HHP treatment on egg white gel hardness and gel gumminess at natural pH (9.11) and pH adjusted with tartaric acid (pH 6.0). Error bars are ±1 standard deviation. (Adapted from Hoppe A. 2010. Examination of Egg White Proteins and Effects of High Pressure on Select Physical and functional Properties. *MSc Thesis in Food Science and Technology Food Science and Technology Department.* Faculty of The Graduate College at the University of Nebraska- Lincoln.)

Monfort et al. (2012) reported the design and evaluation of an HHP-combined process for pasteurization of LWE. The physicochemical and functional properties of non-treated LWE and HHP-treated LWE (300 MPa/3 min at 20°C) followed by heat treatment (52°C/3.5 min or 55°C/2 min) in the presence of 2%, and current heat ultrapasteurization-treated LWE (Table 12.8) (Monfort et al., 2012) were put forward.

In the study reported by Monfort et al. (2012), gels from ultrapasteurized LWE were harder than those prepared with LWE treated by HHP, as reported by Van der Plancken et al. (2005a,b) in EW. It was found that, in the case of gelling properties, gels of treated LWE showed similar values of hardness and water holding capacity (WHC) (Table 12.14). Marco-Moles et al. (2011) rendered that higher hardness values could indicate that protein-based conformational structures may be irreversibly impaired by high-temperature processes. The designed treatment at 52°C/3.5 min by Monfort et al. (2012) resulted in harder gels than the treatment at 55°C/2 min or the non-treated LWE.

Overall, HPP improved the functional and physicochemical properties of EW, EY, and LWE. HPP is an emerging technology with the potential to increase new functional properties to food products.

TABLE 12.8

Physicochemical and Functional Properties of Non-Treated LWE, HHP-Treated LWE (300 MPa/3 min at 20°C) Followed by Heat Treatment (52°C/3.5 min or 55°C/2 min) in the Presence of 2%, and Current Heat Ultrapasteurization-Treated LWE

	Control	TC-HHP-HT		Ultrapasteurization
		HHP + 52°C/3.5 min + 2% TC	HHP + 52°C/3.5 min + 2% TC	71°C/1.5 min
Physicochemical Properties				
pH	7.64 ± 0.06	99.9 ± 0.1	99.7 ± 0.4	102.1 ± 0.7
L*	35.0 ± 0.4	103.8 ± 0.2	108.0 ± 0.2	118.8 ± 1.7
a*	11.4 ± 0.3	114.0 ± 0.2	1143 ± 0.1	66.2 ± 1.2
b*	25.2 ± 0.5	93.6 ± 0.1	98.2 ± 0.1	61.6 ± 1.3
Viscosity (mPa s)	12.7 ± 0.1	156.4 ± 12.4	1323 ± 0.6	239.4 ± 2.9
Soluble protein	0.741 ± 0.024	88.3 ± 0.1	88.3 ± 0.1	84.7 ± 0.7
Functional Properties				
Foaming				
Foaming capacity (%)	504.0 ± 1.6	126.5 ± 1.3	126.4 ± 0.6	31.6 ± 0.8
Foaming stability (min)	4.68 ± 0.37	218.0 ± 28.3	186.2 ± 5.9	16.0 ± 2.5
Emulsifying				
Emulsifying capacity (%)	62.1 ± 0.8	97.5 ± 2.1	95.8 ± 2.5	31.2 ± 0.7
Emulsifying stability (min)	80.6 ± 7.1	95.6 ± 2.5	89.0 ± 1.7	88.4 ± 24.7
Gelling				
WHC (%)	86.6 ± 0.5	97.3 ± 1.4	99.2 ± 1.4	100.3 ± 0.6
Hardness (g)	1041.0 ± 44.5	123.2 ± 2.5	100.7 ± 3.6	126.8 ± 2.0

Source: Adapted from Monfort S. 2012. *Innovative Food Science and Emerging Technologies,* 14, 1–10.

12.2 TABLE OLIVE

12.2.1 INTRODUCTION TO TABLE OLIVE AND THE NECESSITY OF HIGH-PRESSURE PROCESSING

The olive (*Olea europea* L.) is a most widespread and important plant and it probably originates from Mesopotamia. It has been cultivated for many centuries in southern European countries bordering the Mediterranean and in North Africa (Karleskind and Wolff, 1998; O'Brien et al., 2000).

Ninety-eight percent of olive production worldwide is concentrated in the Mediterranean area and also in the United States; California being the biggest area of table olive production. The worldwide production has been estimated to be 1,832,500 tonnes. Table olive fruit is a valuable commodity worldwide that is consumed as whole, stuffed, or sliced, and must be prepared using safe conditions based on the Codex Alimentarius and International Olive Oil Council (IOOC) Directives (Tokuşoğlu, 2010). The olive fruit is a drupe fruit and includes and has a bitter component (oleuropein), a low sugar content (2.6–6%) compared with other drupes (12% or more) and a high oil content (12–30%) depending on the time of year and variety. Table olive is described as 'special fruit' and a considerable part of the olive is processed (IOOC, 2010, Tokuşoğlu, 2010). Table olive is the most popular fermented food in the Mediterranean countries and fermented olive is highly appreciated for its good taste, as well as for its nutritional properties.

The stonefruit table olive is classified as a drupe and is similar to peach and sour cherry fruits; it has an oval shape; it consists of two main parts: pericarp and endocarp (Figure 12.16.). As a detailed classification, the table olive is made up of the epicarp or skin (pellicle), the mesocarp or flesh (pulp), and the endocarp that consists of a woody shell enclosing one or two seeds (Connor and Fereres, 2005; Tokuşoğlu, 2010).

The pericarp comprises 66–85% of the olive fruit weight and the rest 10–27% is endocarp. The epicarp is a protective tissue that accounts for 1–3% of the drupe weight, whereas the seed represents 2–4% of the weight. The mesocarp, the edible portion and the most important part of the olive, comprises 70–80% of the whole fruit (Therios, 2009; Tokuşoğlu, 2010).

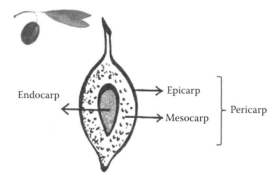

FIGURE 12.16 The olive fruit parts. (Adapted from Tokuşoğlu Ö. 2010. *Special Fruit Olive: Chemistry, Quality Control and Technology*. Seher Publishing. Pub.No: 006–1B; Sidas Medya Ltd. Şti., İzmir. 350 page. ISBN: 978-9944-5660-4-9.)

Pericarp contains 96–98% of the total fat of the table olive, the remaining fat (2–4%) is in the endocarp part. Fresh and ripe fruit comprises water, fat, protein, sugars, pectin compounds, organic acids, tannins, oleuropein, inorganic compounds, and other constituents. Citric, oxalic, malonic, fumaric, tartaric, lactic, acetic, and tricarbalic acids may exist in the flesh part (pulp) of table olive. The chemical composition of the table olive varies by depending on the maturity degree of table olives, environmental factors, and breeding conditions (Tokuşoğlu, 2010).

Table 12.9 shows the chemical composition of black and green table olives and it is seen that the green olive is low in fat and is rich in ash (Table 12.9) (Tokuşoğlu, 2010). High level in ash indicates the high level in minerals. The traditional "Mediterranean diet", in which olive oil is the main dietary fat, is considered to be one of the healthiest because of its strong association with the reduced incidence of cardiovascular diseases and certain cancers (Knoops et al., 2004; Ryan and Robards, 1998; Soler-Rivas et al., 2000). The health benefits of olive oil are mainly ascribed to the presence of a high content of essential fatty acids, especially monounsaturated fatty acid (MUFAs), and functional bioactives including tocopherols, carotenoids, phospholipids, and biologically active phytochemicals containing polyphenols and phenolics (Tokuşoğlu, 2010).

However, table olives including various essential micronutrients, essential fatty acids and biologically active polyphenols, it has been reported that table olives act as a suitable substrate for the production of citrinin mycotoxin (El Adlouni et al., 2006; Heperkan et al., 2006; Tokuşoğlu, 2010, 2013; Tokuşoğlu et al., 2010; Tokuşoğlu and Bozoğlu, 2010).

HHP is a non-thermal technique that is recently receiving a great deal of interest as a technology to destroy pathogenic and spoilage microorganisms in foods (Knorr et al., 1993; San Martín et al., 2002) HHP enables food processing at ambient temperature or even lower temperature; it causes microbial death while virtually eliminating heat damage; it is utilized as chemical preservatives/additives, thereby

TABLE 12.9
Chemical Composition of Black and Green Table Olives

Content	Black Table Olive (%)	Green Table Olive (%)
Water	71.8	75.2
Protein	1.8	1.5
Fat	21.0	13.5
Carbohydrate	2.6	4.0
Raw cellulose	1.5	1.2
Ash	2.8	5.8
Vitamin A (IU)	60.0	30.2
Ca (mg)	8.7	8.7

Source: Adapted from Tokuşoğlu Ö. 2010. *Special Fruit Olive: Chemistry, Quality Control and Technology.* Seher Publishing. Pub.No: 006-1B; Sidas Medya Ltd. Şti., İzmir. 350 page. ISBN: 978-9944-5660-4-9.

leading to improvements in the overall quality of foods; and it can also be utilized to create ingredients with novel functional properties (Rastogi et al., 2007).

HHP treatment as a minimal thermal technology is a valuable tool for microbiologically safe and shelf stable fruit and vegetable production (Guerrero-Beltran et al., 2005). HHP is an innovative, emerging technology with the potential of optimizing the intake of nutrient and nonnutrient phytochemicals in human foods.

HHP technology is already diffused into the markets of Japan, Europe, and the US to preserve liquid and solid foods of high added value (Guerrero-Beltran et al., 2005). The retention of organoleptic attributes and other characteristics of freshness, combined with increased convenience and extended shelf life, will no doubt increase the consumer appeal of foods preserved using HPP. (Bozoğlu et al., 2001; Deliza et al., 2005).

12.2.2 MICROBIOLOGICAL STABILITY AND PRESERVATION OF TABLE OLIVES BY HIGH-PRESSURE PROCESSING

Table olives are rich sources of a wide range of essential micronutrients, essential fatty acids, biologically active phytochemicals containing polyphenols, many of which have purported health benefits (Kountouri et al., 2007; Visioli et al., 1998). However, it has been reported that table olives act as a suitable substrate for the production of citrinin mycotoxin (El Adlouni et al., 2006; Heperkan et al., 2006; Tokuşoğlu et al., 2010; Tokuşoğlu and Bozoğlu, 2010; Tokuşoğlu, 2010).

The mycotoxin citrinin (CIT) (Figure 12.17) is a toxic secondary metabolite, isolated from filamentous fungus *Penicillium citrinum* and also produced by other species of *P. aspergillus* and *Monascus* (Betina, 1989).

The relative toxicity studies showed that citrinin acted in animals as a nephrotoxin that damaged the proximal tubules of the kidney and was implicated as a causative agent in human endemica Balkan nephropathy (Frank, 1992). The LD_{50} (standardized measure for expressing and comparing the toxicity; the dose that kills half (50%) of the animals tested [LD = "lethal dose"]) of CIT has been reported at about 50 mg/kg for oral administration to the rat, 35–58 mg/kg (ip.) to the mouse, and 19 mg/kg (ip.) to the rabbit (Anonymous, 2008).

Black table olives are more favorable and 85% of consumable olives are the black type. Table olives are consumed especially as breakfast foodstuff in all Mediterranean meals and recipes; therefore citrinin is a risk for food safety. The known research on CIT levels in black table olives are limited in the food world.

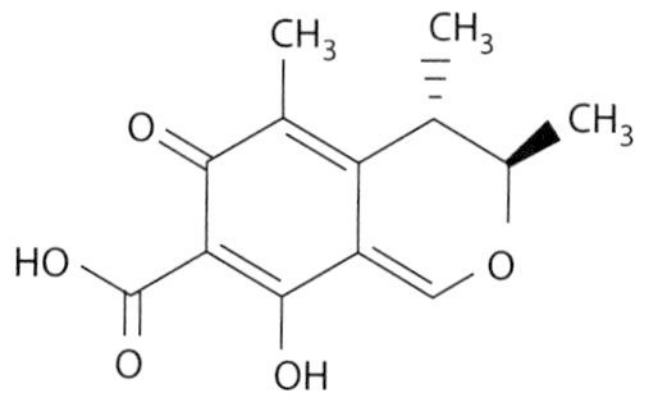

FIGURE 12.17   Mycotoxin citrinin (CIT).

It is reported that 0.45–0.52 µg CIT/kg (avg. 0.5 µg/kg) was found in black table olives collected in growing areas and markets at Morocco (El Adlouni et al., 2006). In one survey, an avg. 0.77 µg CIT/kg was found in 69 black table olives in the Marmara and Aegean regions in Turkey during 2000–2001 via precoated silica gel TLC plates and UV spectrometry (Heperkan et al., 2006). Tokuşoğlu and Bozoğlu (2010) stated that black and green table olives could be a considerable source of CIT in the human diet. In 88 groups of black and green table olives there contained 0.21–7.58 µg kg^{-1} of citrinin and 0.02–0.33 g kg^{-1} of ochratoxin (Tokuşoğlu and Bozoğlu, 2010).

Mold growth was observed in 14 of 30 (47%) samples obtained from the Aegean area control samples, whereas 21 of 33 samples (64%) were obtained from the Marmara Area control samples (Tokuşoğlu and Bozoğlu, 2010). Among the mold species, Penicillium spp. was the dominant flora of black table olives. *Penicillium crustosum, P. roqueforti, P. viridicatum,* and *P. citrinum* were encountered most frequently. *P. crustosum* was determined in 16 of 63 samples of studied olives as major mold species (25%) whereas *P. citrinum* (11%) was found in 4 of 33 Marmara samples and 3 of 30 in Aegean samples (Tokuşoğlu and Bozoğlu, 2010).

Tokuşoğlu et al. (2010) reported the HHP effects on mold flora and citrinin mycotoxin of various selected black table olives.

It was reported that mold flora of olives was reduced, on average ,90% level at 25°C whereas, on average, 100% at 4°C after HHP (Table 12.10) (Figure 12.18). Rose-Bengal Chloramphenicol Agar (RBCA) data of selected black table olives at 25°C (Table 12.10) indicated that total mold counts ranged from 5 to >250 CFU/g in control olive samples whereas they ranged from 0 to 70 CFU/g in HHP-treated black olive samples (Tokuşoğlu et al., 2010).

The HHP treatment reduced the counts on RBCA to 0–70 CFU/g as shown in Table 12.10. It is shown that the HHP process applied on black table olives supressed the mold growth on average 90%. Especially, light U., pastor out-F and washed/selected-F type of olives showed "zero" amount of mold after the HHP process at 25°C of incubation. It is determined that the low temperature (4°C) at which the RBCA agar medium incubated after the HHP process entirely inhibited the mold growth for all groups of olives (Table 12.10) (Tokuşoğlu et al., 2010).

Figure 12.19 shows the HPLC chromatogram of citrinin (CIT) occurrence in controlled "dock out olive-F", in "HHP-treated dock out olive-F", and "standard citrinin" using optimized IAC–HPLC–FD procedure given by Tokuşoğlu et al. (2010).

It was stated that in dock out-F/liquid and pastor out-F/liquid samples provided by controlled company areas, initial citrinin levels were found as 2.26 and 1.71 µg/kg dwt, respectively. It was shown that the HHP process applied for dock out-F/liquid and pastor out-F/liquid resulted in the inhibition of citrinin to 1.19 and 0.58 µg/kg dwt and was reduced to 47% and 66%, respectively. (Tokuşoğlu et al., 2010). Within 8 months of storage, an alteration in the CIT concentration in the stored two types of selected black table olives was determined (Table 12.11). Table 12.11 shows the citrinin (CIT) control and high pressure processed data of the stored two types of black table olives for 8 months (Table 12.11). After the HHP process, the inhibition levels of the CIT in two types of olives was stated by Tokuşoğlu et al. (2010).

TABLE 12.10

HHP Effects on Total Mold Flora of Selected Black Table Olives at 25°C and 4°C

	Vacuum F-XL	Vacuum F-L	F-Tin	Light U Gemlik Type	Light U Aegean Type	Vacuum U Jumbo	Vacuum U Özel	U-Tin	U-Sele	Dock Out-F	Pastör. Out-F	Washed/ Selected-F
						RBCA Data at 25°C (as CFU/g)						
Control												
−1	149	119	159	5	4	170	108	143	155	N250	12	105
−2	19	13	16	0	0	26	9	19	23	22	1	10
HHP												
−1	70	38	57	0	0	48	23	39	41	24	0	0
−2	7	2	0	0	0	5	0	3	4	3	0	0
						RBCA Data at 25°C (as CFU/g)						
Control												
−1	54	41	10	0	0	67	21	40	46	142	0	33
−2	8	4	1	0	0	5	2	3	4	11	0	2
HHP												
−1	0	0	0	0	0	0	0	0	0	0	0	0
−2	0	0	0	0	0	0	0	0	0	0	0	0

Source: Adapted from Tokuşoğlu Ö., Alpas H., and Bozoğlu F.T. 2010. *Innovative Food Science and Emerging Technologies*, 11(2), 250–258.

Note: Data mean two replicate of RBCA analysis.

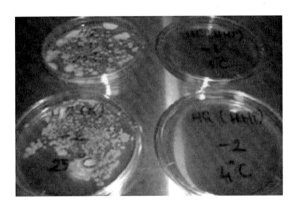

FIGURE 12.18 Incubated control olive pate-HÇ (left side) and HHP olive pate-HÇ (right side) in RBCA during 5–7 days at 25 ± 1°C and at 4 ± 1°C. (From Tokuşoğlu Ö., Alpas H., and Bozoğlu F.T. 2010. *Innovative Food Science and Emerging Technologies*, 11(2), 250–258.)

Figure 12.20 shows the HHP effect on degradation levels of citrinin (in percent) in selective olive pate samples fortificated with various concentrations of citrinin. The citrinin loading study was performed and 1, 1.25, 2.5, 10, 25, and 100 ppb citrinin addition was carried out for selected sample Vacuum F-L without citrinin (Tokuşoğlu et al., 2010). After the HHP processing, citrinin amounts in these olive pates were

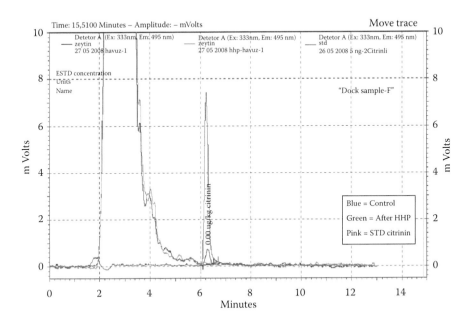

FIGURE 12.19 (**See color insert.**) HPLC chromatogram of citrinin (CIT) occurrence in control "dock-F" and HHP-treated "dock-F" olives and standard citrinin separation. (From Tokuşoğlu Ö., Alpas H., and Bozoğlu F.T. 2010. *Innovative Food Science and Emerging Technologies*, 11(2), 250–258.)

TABLE 12.11

Citrinin (CIT) Control and HHP Data for the 8-Month Stored Two Types of Black Table Olives

	Citrinin (CIT) Mycotoxin (as μg/kg dwt)	
	Control	HHP-treated
Olive Pate (Vacuum F-L)		
Storage 1 month	0.71 ± 0.05	ND
Storage 2 months	0.83 ± 0.02	<0.1 ± 0.01
Storage 4 months	1.82 ± 0.03	0.43 ± 0.02
Storage 8 months	2.35 ± 0.11	0.79 ± 0.06
Olive Pate (Vacuum U. Özel)		
Storage 1 month	0.20 ± 0.01	ND
Storage 2 months	0.25 ± 0.00	<0.1 ± 0.01
Storage 4 months	0.50 ± 0.01	0.18 ± 0.05
Storage 8 months	0.77 ± 0.06	<0.5 ± 0.02

Source: Adapted from Tokuşoğlu Ö., Alpas H., and Bozoğlu F.T. 2010. *Innovative Food Science and Emerging Technologies*, 11(2), 250–258.

reduced as averages 100, 98, 55, 37, 9, and 1.3%, respectively, for citrinin-added pates, as shown in Figure 12.20. 1 ppb of CIT sample contamination and that of less contamination were absolutely inhibited as 100% (Figure 12.20). In fact, olives may contain 1 ppb or less amount of CIT in olive tanks or manufacturing area and it is shown that the HHP process is notably effective on 1 ppb and less-contaminated black olive pates (Tokuşoğlu et al., 2010).

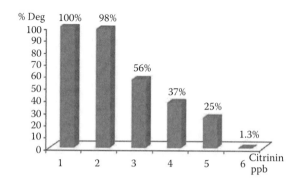

FIGURE 12.20   The HHP effect on degradation levels of citrinin (in percent) in olive pate (Vaccum F-L) fortificated with various concentrations of citrinin. (Adapted from Tokuşoğlu Ö., Alpas H., and Bozoğlu F.T. 2010. *Innovative Food Science and Emerging Technologies*, 11(2), 250–258.)

12.2.3 ENZYME STABILITY AND ACTIVITY OF TABLE OLIVES BY HIGH-PRESSURE PROCESSING

The application of high pressure for food processing has received increased interest from industrial companies and research laboratories. The use of high pressure to process food requires the knowledge of its effect on food components. Enzymes, for example, differ significantly in their baric sensibility (Hendrickx et al., 1998).

Saraiva et al. (2002) stated the effect of high pressure treatments (50 and 110 MPa for 15 and 30 min) on protease and β-galactosidase activities of mature table olives (Portuguese Douro variety). It was determined for the HHP treatment effects on the three cell wall fractions including the soluble, the ionically, and the covalently bound for both protease and β-galactosidase enzymes (Saraiva et al., 2002). For β-galactosidase enzyme, the HHP treatments caused no significant alterations in activity, while proteolytic activity was reduced by more than 90% in all the studied cases. The significantly different baric stabilities for proteolytic activity in all the studied fractions were also found (Saraiva et al., 2002). With the applying HHP studies for soluble fraction (FS), the ionically bound (FI) and covalently bound (FC) to cell wall fractions, the enzymes were quantified.

For the ionically bound (FI) part to cell wall fraction, protease enzyme activity was 0.942 ΔO.D.$_{280\,nm}$/(min.g olive pulp) for nonprocessed table olive while 0.270 ΔO.D.$_{280\,nm}$/(min.g olive pulp) was after 50 MPa (15) pressure treatment (Saraiva et al., 2002). According to the data given by Saraiva et al. (2002), protease enzyme activity (proteolytic activity) was reduced by more than 90% in all cases ($p < 0.05$) while no significant alteration in β-galactosidase activity ($p > 0.05$) was found in all studied mature table olives.

12.2.4 STABILITY OF BIOACTIVE PHENOLICS OF TABLE OLIVES BY HIGH-PRESSURE PROCESSING

Tokuşoğlu et al. (2010) reported the HHP effects on hydroxytyrosol, oleuropein phenolics, and antioxidant activity of black table olives. It was stated that initial hydroxytyrosol levels of studied black table olives were 3937.3–9486.1 mg/kg dwt, whereas these values varied from 4660.1 to 11386.9 mg/kg dwt after the HHP process of olives as shown in Table 12.12 (Tokuşoğlu et al., 2010). It is stated that hyroxytyrosol levels of black table olives were increased at approximately 0.8–2-fold than that of the initials after the HHP process (Tokuşoğlu et al., 2010).

It was determined that the small size of olives contained more hydroxytyrosol and oleuropein phenolics such as in Vacuum F-L. Its phenolic levels were higher than Vacuum F-XL (Table 12.12) while its citrinin level was zero in respect of Vacuum F-XL (Table 12.10). It was assumed that there was a correlation between the phenolics and citrinin concentrations in black table olives ($p < 0.01$) and table olives containing more phenolics absorbed less mycotoxin (Tokuşoğlu et al., 2010).

The initial total phenolics of studied black table olives were 1292.2–1853.7 mg GA/100 g dwt whereas that of the range between 2542.3 and 4895.1 mg GA/100 g dwt after the HHP process, as shown in Table 12.13. It was shown that the total

TABLE 12.12
Major Phenolics Hydroxytyrosol (HYD), Oleuropein (OLE) Profiles of Control and HHP-Treated Black Table Olive Samples

| | Major Phenolics (as mg/kg dwt) | | | |
Olive Sample	Hydroxytyrosol (HYD) Control	Hydroxytyrosol (HYD) HHP-treated	Oleuropein (OLE) Control	Oleuropein (OLE) HHP-Treated
Vacuum F-XL	8200.6 ± 2.93	11386.9 ± 1.87	1036.64 ± 0.56	750.98 ± 1.05
Vacuum F-L	9486.1 ± 2.03	12772.9 ± 0.17	1272.1 ± 0.09	914.0 ± 0.00
F-Tin	5420.6 ± 0.78	6201.1 ± 7.05	2441.2 ± 1.90	1148.8 ± 1.76
Dock out-F	6662.4 ± 0.50	7002.7 ± 3.37	1758.0 ± 0.88	1084.8 ± 2.33
Washed/Selected-F	6820.2 ± 1.23	8015.9 ± 0.05	1802.3 ± 0.44	1021.2 ± 0.56
Pastor out-F	5498.0 ± 1.07	7291.7 ± 1.74	1739.2 ± 2.11	1000.1 ± 0.06
Vacuum U. Özel	7101.6 ± 0.73	8005.4 ± 1.06	1355.6 ± 1.08	957.0 ± 0.18
Light U. Gemlik Type	3937.3 ± 1.37	4660.1 ± 5.08	778.6 ± 0.06	518.5 ± 0.95
Light U. Aegean Type	4522.5 ± 3.86	5240.6 ± 1.77	794.3 ± 0.19	547.2 ± 2.12
Vacuum U. Jumbo	7708.3 ± 4.06	8115.0 ± 0.09	2121.6 ± 0.75	1910.1 ± 1.05
U-Sele	7277.5 ± 3.15	7965.3 ± 0.17	1962.5 ± 1.23	1623.3 ± 0.79
U-Tin	5007.2 ± 0.99	6487.5 ± 1.43	2155.5 ± 3.00	1362.6 ± 3.02

Source: Adapted from Tokuşoğlu Ö., Alpas H., and Bozoğlu F.T. 2010. *Innovative Food Science and Emerging Technologies*, 11(2), 250–258.

TABLE 12.13
Total Phenolics of Control and HHP-treated Black Table Olive Samples

| | Total Phenolics (as mg GA/100 g dwt) | |
	TP Control	TP HHP-treated
Vacuum F-XL	1788.2 ± 6.07	4073.2 ± 1.00
Vacuum F-L	1853.7 ± 3.56	4895.1 ± 3.33
F-Tin	1455.2 ± 2.55	3117.7 ± 0.97
Dock out-F	1677.1 ± 0.99	3551.0 ± 1.54
Washed/Selected-F	1694.6 ± 1.58	3846.2 ± 1.90
Pastor out-F	1416.5 ± 3.00	3677.5 ± 0.11
Vacuum U. Özel	1711.7 ± 2.82	3267.6 ± 2.72
Light U. Gemlik Type	1292.2 ± 0.71	2669.9 ± 1.13
Light U. Aegean Type	1570.8 ± 4.11	3141.5 ± 4.72
Vacuum U. Jumbo	1792.6 ± 1.74	3518.1 ± 0.85
U-Sele	1432.5 ± 0.97	2542.3 ± 1.07
U-Tin	1380.8 ± 5.04	2852.5 ± 2.08

Source: Adapted from Tokuşoğlu Ö., Alpas H., and Bozoğlu F.T. 2010. *Innovative Food Science and Emerging Technologies*, 11(2), 250–258.

phenolics of studied olives increased as 2.1–2.5-fold after the HHP process (as mg gallic acid/100 g) (Table 12.13) (Tokuşoğlu et al., 2010).

Several parameters are known to influence the quantitative phenolic profiles of olive fruits. Among these factors, the ripeness degree, the geographical origin, and the cultivar nature are definitely those that have a pronounced effect on the composition (Tokuşoğlu, 2010; Vinha et al., 2005). HPP is an innovative, emerging technology with the potential of optimizing the intake of nutrient and nonnutrient phytochemicals in human foods. The HHP procedure has provided higher extraction yield, higher extraction selectivity, and efficiency (Tokuşoğlu, 2010; Tokuşoğlu and Bozoğlu, 2009; Tokuşoğlu et al., 2008; Qiu et al., 2006). It was found that there were higher values of hydroxytyrosol and total phenolics after HHP and it was thought that higher phenolic levels were due to the higher extraction yield of olives by the HHP process (Tokuşoğlu et al., 2010).

In the study described by Tokuşoğlu et al. (2010), it is assumed that the increasing bioavailability of these olive phytochemicals after HHP may be more useful for consumers. The increasing amount of phenolics might be attributed to the fact that high pressure can rupture the tissue of olive fruit and thereafter release more phenolics. This phenomenon agreed with the study of Qiu et al. (2006) regarding lycopene in tomatoes. From the viewpoint of the HHP process conditions (pressure, temperature and time) and also extraction time, those will be important for phenolic quantification.

Tables 12.14 and 12.15 show the control and HHP data of the major phenolics and total phenolics for the 8-month stored two types of black table olives. The hydroxytyrosol level of Vacuum F-L was 9002.5 ± 2.15 mg/kg dwt at the 1-month storage while that level declined to 1488.3 ± 0.79 mg/kg dwt and there was found to be a 6-fold decrease for the hydroxytyrosol level of olive fruit. It is shown that after the HHP process, the HYD level of 1-month stored sample increased as 1.33-fold, while the HYD level of 8-month stored sample increased only by 1.08-fold. With 8 months of storage, the HYD level of Vacuum F-L decreased by 6-fold for control whereas it decreased only by 2.4-fold for HHP-treated Vacuum F-L (Table 12.14). It is shown that the HHP process could be an effective preservative, nonthermal method for HYD phenolic of stored olive fruits (Table 12.14). These findings are in accordance with the study given by Qiu et al. (2006). In the study by Qiu et al. (2006), after 16 days of storage, total lycopene content in untreated samples declined more (15.3% loss) than that in HHP-treated (about 3.0% loss) samples.

In the study reported by Tokuşoğlu et al. (2010), the antioxidant activity values varied from 17.238 to 29.344 mmol Fe^{2+}/100 g for control samples whereas 18.579–32.998 mmol Fe^{2+}/100 g for HHP-treated samples. It was determined that the antioxidant activity of HHP processed olives was 1.03–1.12-fold higher. It was thought that the higher antioxidant activity of pressure processed olives might be from total phenolic levels of these processed olives. There was a positive relationship between phenolics and antioxidant activity levels for pressure processed samples (Tokuşoğlu et al. 2010). The highest antioxidant activity that belonged to Vacuum F-L was 29.344 mmol Fe^{2+}/100 g dwt in control and 32.998 mmol $Fe2 + $/100 g dwt in HHP-treated samples. It was shown that the antioxidant activity of olive samples

TABLE 12.14
Control and HHP Data of the Major Phenolics (Hydroxytyrosol and Oleuropein) for the 8 Month Stored Two Types of Black Table Olives

Olive Sample	Major Phenolics (as mg/kg dwt)			
	Hydroxytyrosol (HYD) Control	Hydroxytyrosol (HYD) HHP-Treated	Oleuropein (OLE) Control	Oleuropein (OLE) HHP-Treated
(Vacuum F-L)				
Storage 1 month	8997.5 ± 2.15	11950.2 ± 0.80	1156.3 ± 1.19	887.0 ± 3.07
Storage 2 months	7256.2 ± 5.05	10560.3 ± 4.23	986.5 ± 0.90	498.8 ± 3.05
Storage 4 months	3767.5 ± 3.14	6078.0 ± 1.45	525.1 ± 3.23	378.3 ± 1.90
Storage 8 months	1488.3 ± 0.79	3718.2 ± 3.33	483.5 ± 1.52	275.5 ± 1.16
(Vacuum U. Özel)				
Storage 1 month	6976.6 ± 1.17	7405.5 ± 1.80	1243.2 ± 1.00	824.0 ± 1.83
Storage 2 months	5424.3 ± 0.97	5952.2 ± 0.06	967.5 ± 0.73	556.2 ± 0.98
Storage 4 months	2956.5 ± 1.55	3278.5 ± 0.23	750.7 ± 1.50	345.5 ± 7.04
Storage 8 months	1756.8 ± 2.00	1884.3 ± 4.09	527.3 ± 6.10	215.0 ± 3.51

Source: Adapted from Tokuşoğlu Ö., Alpas H., and Bozoğlu F.T. 2010. *Innovative Food Science and Emerging Technologies*, 11(2), 250–258.

TABLE 12.15
Control and HHP Data of the Total Phenolics for the 8-Month Stored Two Types of Black Table Olives

	Total Phenolics (as mg GA/100 g dwt)	
	TP Control	TP HHP-Treated
(Vacuum F-L)		
Storage 1 month	1727.5 ± 1.73	3765.5 ± 0.90
Storage 2 months	1489.0 ± 0.65	1546.2 ± 1.06
Storage 4 months	640.5 ± 1.88	1055.8 ± 0.82
Storage 8 months	303.6 ± 5.05	446.5 ± 9.20
(Vacuum U. Özel)		
Storage 1 month	1593.5 ± 2.09	2482.9 ± 3.15
Storage 2 months	1115.7 ± 0.13	1407.5 ± 5.02
Storage 4 months	568.5 ± 1.86	955.3 ± 1.25
Storage 8 months	235.7 ± 7.07	287.6 ± 0.76

Source: Adapted from Tokuşoğlu Ö., Alpas H., and Bozoğlu F.T. 2010. *Innovative Food Science and Emerging Technologies*, 11(2), 250–258.

increased by the HHP process (Tokuşoğlu et al., 2010). According to McInerney et al. (2007), the high pressure treatment had differential effects on water-soluble antioxidant activity depending on vegetable type (McInerney et al., 2007). In the study given by McInerney et al. (2007), for HPP-treated broccoli had no effect on antioxidant activity, for carrots there was a modest reduction in antioxidant activity at the 400 MPa pressure while for 600 MPa HHP-treated green beans resulted in a significant increase in lutein availability compared to untreated samples or those processed at the lower pressure level (400 MPa).

McInerney et al. (2007) stated that levels of the most abundant individual carotenoids in carrots, green beans and broccoli were unaffected by HP, regardless of the pressure level that was employed. The total carotenoid concentration in carrots was between 5 and 10-fold greater than that of green beans and broccoli and it was reported that the magnitude of the difference in total carotenoids between the different types of vegetables was markedly greater than differences between treatments (McInerney et al., 2007).

12.2.5 HHP EFFECTS ON SHELF LIFE STABILITY AND PHYSICOCHEMICAL QUALITY PROPERTIES

The feasibility of the use of HHP treatments as preservation methods to extend the shelf life of traditionally dressed Cornezuelo table olives was reported by Pradas et al. (2012). Pradas et al. (2012) found that the initial pH values of HHP-treated olives were slightly higher than the values of the control samples (Figure 12.21). Figure 12.21 shows the pH and acidity changes of HHP-treated olives during 300 days of storage. In the study described by Pradas et al. (2012), 400 MPa/5 min, 400 MPa/10 min, 600 MPa/5 min, 600 MPa/10 min of applications correspond to A, B, C, D letters, respectively (Figure 12.21).

It was found that the initial pH values of treated samples decreased after 60 days of storage. It was stated that the initial free acidity values of HHP-treated samples (expressed as percentage of lactic acid) were slightly lower than the values of control samples (Figure 12.21). It was stated that the room temperature rise to 30°C caused an increase in the acidity from 0.32% (day 186th) to 0.58% (day 280th) while the HHP-treated groups showed constant free acidity values without major alterations during 90 more days (Figure 12.21) (Pradas et al., 2012).

It was reported that the loss of firmness (softening) of olives in brines increased with time for all the samples and treatments after the increasing of ambient temperature and increasing in the acid conditions of the brines (Figure 12.22). 400 MPa/5 min of treatment (treatment A) showed the best properties of firmness and punctuation obtained by the sensory evaluation panel. Pradas et al. (2012) reported that the firmness remained almost constant (no significant differences) for all the samples during the first 186 days of storage.

It is known that during the sterilization application, one of the most important softening processes is believed to be temperature-dependent β-elimination of pectin, that is the degradation of the cell wall (Sajjaanantakul et al., 1989). It is also stated that the lower softening percentage may also be attributed to activation or retention of pectin methylesterase (PME) (Fernandez-Bolanos et al., 2001). Basak and

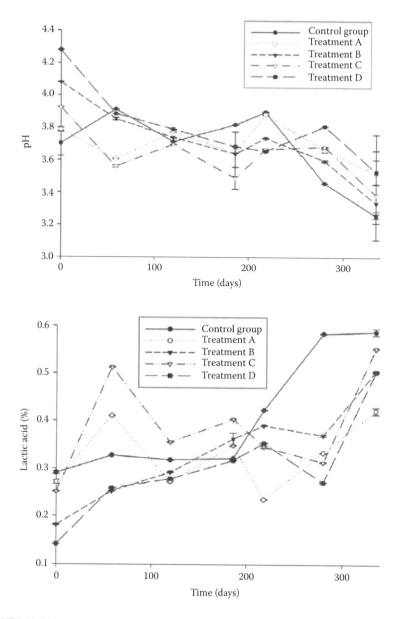

FIGURE 12.21 pH and acidity changes of HHP-treated olives during storage time. (Adapted from Pradas I. et al. 2012. *Innovative Food Science and Emerging Technologies.* 13, 64–68.)

Ramaswamy (1998) stated that the high pressure can enhance the pectin methylesterase (PME) activity and decrease the activity of the polygalacturonase (PG). Owing to the fact that the increase in firmness may be from demethylation reactions with the substrates and the gel-network formed with divalent ions right afterwards. Krebbers et al. (2002) reported the similar data on higher retention of firmness of green beans during storage after HHP treatment (Krebbers et al., 2002).

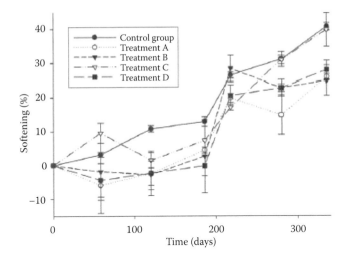

FIGURE 12.22 Softening changes of HHP-treated olives during storage time. (Adapted from Pradas I. et al. 2012. *Innovative Food Science and Emerging Technologies.* 13, 64–68.)

Sánchez et al. (2012) stated the effect of HHP versus thermal pasteurization on the microbiological, sensory aspects, and oxidative stability of olive pate. In the study reported by Sánchez et al. (2012), the effects of HHP as an alternative to thermal pasteurization treatments for the conservation of olive pate and the stability of HHP-treated products during refrigerated storage were stated; the characterization of microbiological, physico-chemical, and sensory aspects were exhibited (Sánchez et al., 2012).

It was stated that high-pressure processed olive pates showed a lower peroxide value than the data obtained by thermal pasteurization (80°C for 20 min) at the starting of storage. As a matter of fact, the peroxide values of HHP processed samples at 30 days of storage showed lower values than the control data. Also, there was an increase in the oxidative stability for high-pressure processed olive pates (Table 12.16). Table 12.16 shows the peroxide values and oxidative stability of heat-treated olive pates and HHP-treated olive pates under four different conditions (Sánchez et al., 2012).

It was reported that the HP-treated samples showed a further reduction in the presence of microorganisms including mesophiles, psychrotrophs, yeasts, greater clarity, and less browning regarding the colorimetric coordinates in comparison with those treated by thermal pasteurization (Sánchez et al., 2012).

The higher sensory acceptance in appearance, flavor, and texture of HHP-treated olive pates (Table 12.17) in comparison with those treated by thermal pasteurization at 80°C/20 min (Sánchez et al., 2012) was found.

Sánchez et al. (2012) found that the study of product shelf life in refrigeration would indicate the feasibility to implement HP technology to obtain foods with a similar shelf life to foods obtained through the traditional thermal pasteurization treatment, but with better sensory quality.

TABLE 12.16

Comparison of Peroxide Value and Oxidative Stability of Heat Treated, and High Pressure Processed Olive Pates

	Control	Heat Treatment	High-Pressure Processing (450 MPa)		High-Pressure Processing (600 MPa)	
		(80°C/20 min)	5 min	10 min	5 min	10 min
			1 Day of Storage			
Peroxide Value (meq. O$_2$/kg fat)	30.6^c ± 0.1	26.9^{bc} ± 7.3	22.4^b ± 3.0	15.9^a ± 0.6	25.9^{bc} ± 4.7	22.8^b ± 1.4
Oxidative Stability (Rancimat,h)	62.4^{ab} ± 3.8	64.0^a ± 10.9	63.5^{ab} ± 5.0	73.1^b ± 2.2	62.6^{ab} ± 5.3	63.4^{ab} ± 9.7
			30 Day of Storage			
Peroxide Value (meq. O$_2$/kg fat)	64.5^d ± 0.2	25.9^{ab} ± 6.4	41.9^c ±1.2	20.9^a ±7.7	45.2^c ±13.0	33.0^b ± 7.9
Oxidative Stability (Rancimat,h)	32.9^a ± 1.0	62.4^c ± 8.0	54.0^c ± 0.6	67.0^d ± 5.5	56.5^{bc} ±13.0	57.5^{bc} ± 2.4

Source: Compiled from Sánchez P.J. et al. 2012. *Grasas y Aceites,* 63 (1),100–108.

As a conclusion, HPP improved the functional and physicochemical quality properties, and sensory characteristics of table olives and olive-based products. HHP provided the microbial stability, stability of enzymatic activities, and also possible mycotoxin inhibition of table olives and pates by pressure effects. Also, HPP enhanced the bioactive phenolic properties of table olives and pates and, as an emerging technology, HHP could be used in the olive industry as nonthermal processing.

TABLE 12.17

Sensory Evaluation of Heat Treated and High Pressure Processed Olive Pates at 1 Day of Storage ($p < 0.05$)

	Heat Treatment	High-Pressure Processing (450 MPa)		High-Pressure Processing (600 MPa)	
	(80°C/20 min)	5 min	10 min	5 min	10 min
Appearance	5.7 ± 2.6	7.3 ± 2.0	7.3 ± 2.2	7.5 ± 1.8	7.3 ± 1.8
Flavor	6.1 ± 1.9	7.3 ± 1.5	7.2 ± 1.5	6.8 ± 1.7	7.3 ± 2.3
Texture	5.5 ± 2.5	6.7 ± 1.6	6.5 ± 2.0	6.7 ± 1.8	7.2 ± 1.8

Source: Compiled from Sánchez P.J. et al. 2012. *Grasas y Aceites,* 63 (1),100–108.

REFERENCES

Abe Y., Itoh T., and Adachi S. 1982. Fractionation and characterization of hen's egg yolk phosvitin. *J Food Sci*, 47, 1903–1907.

Acero-Lopez A., Ullah A., Offengenden M., Jung S., and Wua J. 2012. Effect of high pressure treatment on ovotransferrin. *Food Chem*, 135, 2245–2252.

Aguilar J.M., Cordobes F., and Jerez A. 2007. Influence of high pressure processing on the linear viscoelastic properties of egg yolk dispersions. *Rheol Acta*, 46, 731–740.

Ahmed J., Ramaswamya H.S., Allia I., and Ngadi M. 2003. Effect of high pressure on rheological characteristics of liquid egg. *Lebensm Wiss und Technol*, 36(5), 517–524.

Ahmed J., Ramaswamy H.S., and Hiremath N. 2005. The effect of high pressure treatment on rheological characteristics and colour of mango pulp. *Int J Food Sci Technol*, 40(8), 885–895.

Alvarez I., Niemira B.A., Fan X.T., and Sommers C.H. 2006. Inactivation of salmonella serovars in liquid whole egg by heat following irradiation treatment. *Journal of Food Protection*, 69(9), 2066–2074.

Anonymous 2008. http://www.micotoxinas.com.br/citrininfacts.htm. Leatherhead Food Research Association, Randalls Road, Leatherhead, Surrey KT227RY, England.

Anonymous. 2013a. http://www.uscnk.com/directory/Phosvitin(PV)-1679.htm

Anonymous. 2013b. http://blogs.oregonstate.edu/psquared/2010/04/16/ovalbumin

Azarpazhooh E. and Ramaswamy H.S. 2012. Modeling and optimization of microwave osmotic dehydration of apple cylinders under continuous-flow spray mode processing conditions. *Food and Bioprocess Technology*, 5(5), 1486–1501.

Balny C. and Masson P. 1993. Effects of high pressure on proteins. *Food Rev Int*, 9, 611–628.

Barbosa-Cánovas G.V. 2008. Shelf stable egg-based products processed by ultra high pressure technology. *Final Technical Report. OMB No. 0704-0188.* Combat Ration Network for Technology Implementation (Coranet II). Short Term Project (STP) # 2015. DLA Contracting Services Office.

Barbosa-Cánovas G.V. and Juliano P. 2007. Food sterilization by combining high pressure and thermal energy. In: Gutiérrez-López G.F., Barbosa-Cánovas G.V., Welti-Chanes J.S., and Parada-Arias, E. (Eds.), *Food Engineering: An Integrated Approach*. Springer, New York.

Barbosa-Cánovas G.V., Koutchma T.N., Balasubramaniam V.M., Sadler G.D., Clark S., and Juliano P. 2004. High pressure processing of eggbased products. *Annual Meeting of Institute of Food Technologists*, Las Vegas, NA, July 12–16 (Abstract number 86–4).

Baldwin R.E. 1986. Functional properties of eggs in foods. Chap 16. In: Stadelman N.W.I., Cotterill O.J. (Eds.), *Egg Science and Technology*. 3rd edition, AVI Publishing Co., New York, p. 346.

Bari M.L., Ukuku D.O., Mon M., Kawamoto S., and Yamamoto K. 2008. Effect of hydrostatic pressure pulsing on the inactivation of *Salmonella* enteritidis in liquid whole egg. *Food Borne Pathogens Dis*, 5(3), 175–182.

Basak S. and Ramaswamy H.S. 1998. Effect of high pressure processing on the texture of selected fruits and vegetables. *J Texture Stud*, 29(5), 587–601.

Betina V. 1989. *Mycotoxins: Chemical, Biological and Environmental Aspects*. New York, USA: Elsevier.

Bianchi G. 2003. Lipids and phenols in table olives. *Eur J Lipid Sci Technol* 105: 229–242.

Bianco A. and Uccella N. 2000. Biophenolic components of olives. *Food Res Int*, 33, 475–485.

Bozoğlu F., Deák T., and Ray B. 2001. Novel processes and control technologies in the food industry. In: Bozoğlu F., (Ed.), *NATO Science for Peace and Security Series*. IOS Press, Springer-Verlag GmbH, Berlin, Heidelberg. p. 246.

Bridgeman P. 1914. The coagulation of albumen by pressure. *J Biol Chem*, 19(4), 511–512.

Bruce J. and Drysdale L. 1994. Trans-shell transmission. In: Board, R.G. and Fuller, R. (Eds.), *Microbiology of the Avian Egg*. Chapman and Hall, London, pp. 63–91.

Carlez A., Veciana-Nogues T., and Cheftel J.C. 1995. Changes in colour and myoglobin of minced beef meat due to high pressure processing. *Lebensm Wiss und Technol*, 28(5), 528–538.

CDC. 2003. Centers for disease control and prevention. Outbreaks of *Salmonella* serotype enteritidis infection associated with eating shell eggs—United States, 1999–2001. *Morb Mort Wkly Rep*, 51, 1149–1152.

CDC. 2004. Centers for disease control and prevention. *Salmonella* serotype typhirnurium outbreak associated with commercially processed egg salad—Oregon, 2003. *Mort Mort Wkly Rep*, 53, 1132–1134.

Chang Y.I. and Chen T.C. 2000. Functional and gel characteristics of liquid whole egg as affected by pH alteration. *J Food Eng*, 45, 237–241.

Cheftel J.C. 1995. Review: High pressure, microbial inactivation and food preservation. *Food Sci Technol*, 1, 75–90.

Connor D.J. and Fereres E. 2005. The physiology of adaptation and yield expression in olive. *Horticult Rev*, 31, 155–229.

Cornford S.J., Parkinson T.L., and Robb J. 1969. Rheological characteristics of processed whole egg. *J Food Technol*, 4, 353–361.

Cunningham F.E. 1995. Egg product pasteurization. In: Stadelman W.J. and Cotterill O.J. (Eds.), *Egg Sci Technol*, Food Products Press, New York, NY, pp. 289–315.

Davis C. and Reeves R. 2002. High value opportunities from the chicken egg. RIRDC Publication, No. 02/094.

Dawson P.L. and Martinez-Dawson R. 1998. Using response surface analysis to optimize the quality of ultrapasteurized liquid whole egg. *Poultry Sci*, 77(3), 468–474.

Deliza R., Rosenthal A., Abadio F.B.D., Silva C.H.O., and Castillo C. 2005. Application of high pressure technology in the fruit juice processing: Benefits percieved by consumers. *J Food Eng*, 67, 241–246.

Dong-Un Lee. 2002. Application of combined nonthermal treatments for the processing of liquid whole egg. PhD Thesis, von der Fakultät III—Prozesswissenschaften der Technischen Universität Berlin zur Erlangung des akademischen Grades. Berlin, Germany.

EFSA. 2011. European Food Safety Authority, European Centre for Disease Prevention and Control. The European Union Summary Report on Trends and Sources of Zoonoses, Zoonotic.

El Adlouni C., Tozlovanu M., Naman F., Faid M., and Pfohl-Leszkowicz A. 2006. Preliminary data on the presence of mycotoxins (Ochratoxin A, Citrinin, and Aflatoxin B1) in black table olives "Greek Style" of Moroccan origin. *Nahrung/Food*, 50(6), 507–512.

Farr D. 1990. High pressure technology in the food industry. *Trends Food Sci Technol*, 1(1), 14–16.

Fernandez-Bolanos J., Heredia A., Saldana C., Rodriguez R., Guillen R., and Jimenez A. 2001. Effect of dressings "(alinos)" on olive texture: Cellulase, polygalacturonase and glycosidase activities of garlic and lemon present in brines. *Eur Food Res Technol*, 212(4), 465–468.

Ferreira M., Benringer R., and Jost R. 1995. Instrumental method for characterizing protein foams. *J Food Sci*, 60, 90–93.

Frank H.K. 1992. Citrinin. *Zeitschrift fur Ernahrungswissenschaft*, 31, 164–177.

Giansanti F., Giardi M.F., Massucci M.T., Botti D., and Antonini G. 2007. Ovotransferrin expression and release by chicken cell lines infected with Marek's disease virus. *Biochem Cell Biol*, 85, 150–155.

Giansanti F., Massucci M.T., Giardi M.F., Nozza F., Pulsinelli E., Nicolini C., Botti D., and Antonini G. 2005. Antiviral activity of ovotransferrin derived peptides. *Biochem Biophys Res Commun*, 331(1), 69–73.

Giansanti F., Rossi P., Massucci M.T., Botti D., Antonini G., Valenti P., Seganti L. 2002. Antiviral activity of ovotransferrin discloses an evolutionary strategy for the defensive activities of lactoferrin. *Biochem Cell Biol*, 80(1), 125–130.

Guamis B., Pla R., Trujillo A.J., Capellas M., Gervilla R., Saldo J., and Yuste J. 2005. High pressure processing of milk and dairy and egg products, In: Barbosa-Cánovas G.V., Tapia M.S., and Cano M.P. (Eds.) *Novel Food Processing Technologies*, Marcel Dekker/ CRC Press, Boca Raton, pp. 394–412.

Guerrero-Beltran J.A., Barbosa-Canovas G., and Swanson B.G. 2005. High hydrostatic pressure processing of fruit and vegetable products. *Food Rev Int*, 21(4), 411–425.

Hatta H., Hagi T., and Hirano K. 1997. Chemical and physicochemical properties of hen eggs and their application in foods. In: Yamamoto T., Junea L.R., Hatta H., and Kim M. (Eds.) *Hen Eggs. Their Basic and Applied Science*. CRC Press, Boca Raton, pp. 117–134.

Hayakawa I., Linko Y.Y., and Linko P. 1996. Mechanism of high pressure denaturation of proteins. *LWT-Food Sci Technol*, 29(8), 756–762.

Hayashi R., Kawamura Y., Nakasa T., and Okinaka O. 1989. Application of high pressure to food processing: Pressurization of egg white and yolk, and properties of gels formed. *Agric Biol Chem*, 53(11), 2935–2939.

Hendrickx M., Ludikhuyze L., Van den Broeck I., and Weemaes C. 1998. Effects of high pressure on enzymes related to food quality. *Trends Food Sci Technol*, 9, 197.

Heperkan D., Meriç B.E., Şişmanoğlu G., Dalkiliç G., and Güler Karbancioglu F. 2006. Mycobiota, mycotoxigenic fungi, and citrinin production in black olives. In: Hocking, A.D., Pitt, J.I., Samson R.A. and Thrane, U. (Eds.), *Advances in Mycology. Advances in Experimental Medicine & Biology*. Vol. 571, pp. 203–210.

Hoppe A. 2010. Examination of egg white proteins and effects of high pressure on select physical and functional properties. *MSc Thesis in Food Science and Technology Food Science and Technology Department*. Faculty of The Graduate College at the University of Nebraska, Lincoln.

Hsieh Y.L., Regenstein J.M., and Rao M.A. 1993. Gel point of whey and egg proteins using dynamic rheological data. *J Food Sci*, 58, 116–119.

Huang E., Mittal G.S., and Griffiths M.W. 2006. Inactivation of *Salmonella* enteritidis in liquid whole egg using combination treatments of pulsed electric field, high pressure and ultrasound. *Biosyst Eng*, 94 (3), 403–413.

Huntington J.A. and Stein P.E. 2001. Structure and properties of ovalbumin. *J Chromatogr B*, 756 (1–2), 189–198.

Huopalahti R., López-Fandiño R., Anton M., and Schade R. 2007. *Bioactive Egg Compounds*. Springer, Berlin, Germany. ISBN: 978-3-540-37883-9 (Print) 978-3-540-37885-3 (Online)

Iametti S., Donnizzelli E., Pittia P., Rovere P.P., Squarcina N., and Bonomi F. 1999. Characterization of high pressure treated egg albumen. *J Agric Food Chem*, 47, 3611–3616.

Ibrahim H.R., Hoq M.I., and Aoki T. 2007. Ovotransferrin possesses SOD-like superoxide anion scavenging activity that is promoted by copper and manganese binding. *Int J Biol Macromol*, 41, 631–640.

Ibrahim H.R. and Kiyono T. 2009. Novel anticancer activity of the autocleaved ovotransferrin against human colon and breast cancer cells. *J Agric Food Chem*, 57, 11383–11390.

ICMSF. 1998. International commission on microbiological specification for foods. In: *Microorganisms in Foods*. Microbial ecology of food commodities. Blackie Academy and professional, New York, p. 615.

IOOC. 2010. *About Olives*. International Olive Oil Council, Madrid, Spain. http://www.internationaloliveoil.org/estaticos/view/77-about-olives

Isiker G., Gurakan C.G., and Bayindirli A. 2003. Combined effect of high hydrostatic pressure treatment and hydrogen peroxide on *Salmonella enteritidis* in Liquid Whole Egg. *European Food Research and Technology*, 217, 244–248.

Jankowska A., Reps A., Proszek A., and Krasowska M. 2005. Effect of high pressure on microflora and sensory characteristics of yoghurt. *Pol J Food Nutr Sci*, 55(1), 79–84.

Jiang B. and Mine Y. 2000. Preparation of novel functional oligophosphopeptides from hen egg yolk phosvitin. *J Agric Food Chem*, 4, 990–994.

Juliano P. 2006. High pressure thermal sterilization of egg products. PhD Thesis. Washington State University, Department of Biological Systems Engineering, WA, USA.

Juliano P., Clark S., Koutchma T.N., Ouattara M., Mathews J.W., Dunne C.P., and Barbosa-Cánovas G.V. 2007. Consumer and trained panel evaluation of high pressure thermally treated scrambled egg patties. *J Food Qual*, 30(1), 57–80.

Juliano P., Li B., Clark S., Mathews J.W., Dunne C.P., and Barbosa-Cánovas G.V. 2006. Descriptive analysis of precooked egg products after high pressure processing combined with low and high temperatures. *J Food Qual*, 29(5), 505–530.

Karleskind A. and Wolff JP. 1998. Oils and fats manual, A. Comprehensive treatise properties-production-applications. 1(11), 225–233, rue Lavoisier F-75384 Paris Cedex 08.

Kilcast D. and Lewis D. 1990. Structure and texture – their importance in food quality. *Nutr Bull*, 15(2), 103–113.

Kim K. and Setser C.S. 1982. Foaming properties of fresh and commercially dried eggs in the presence of stabilizers and surfactants. *Poultry Sci*, 61, 2194–2199.

Knoops, K.T., de Groot, L.C., and Kromhout, D. 2004. Mediterranean diet, lifestyle factors, and 10-year mortality in elderly European men and women. *J Am Med Assoc*, 292, 1433–1439.

Knorr D., Zenker M., Heinz V., and Lee D.U. 2004. Application and potential of ultrasonics in food processing. *Trends Food Sci Technol*, 15, 261–266.

Knorr D. 1993. Effects of high-hydrostatic-pressure processes on food safety and quality. *Food Technol*, 47(6), 156–161.

Knorr, D. 1996. Advantages, opportunities and challenges of high hydrostatic pressure application to food systems. *Prog Biotechnol*, 13, 279–287.

Koutchma T., Guo B., Patazca E., and Parisi B. 2005. High pressure–high temperature inactivation of clostridium sporogenes spores: From kinetics to process verification. *J Food Process Eng*, 28(6), 610–629.

Kountouri A.M., Mylona A., Kaliora A.C., and Andrikopoulos N.K. 2007. Bioavailability of the phenolic compounds of the fruits (drupes) of olea europaea (olives): Impact on Plasma Antioxidant Status in humans. *Phytomedicine*, 14, 659–667.

Krebbers B., Matser A.M., Koets M., and Van den Berg R.W. 2002. Quality and storage stability of high pressure preserved green beans. *J Food Eng*, 54(1), 27–33.

Kurokawa H., Mikami B., and Hirose M. 1995. Crystal structure of diferric hen ovotransferrin at 2.4 A resolution. *J Mol Biol*, 254(2), 196–207.

Lee D., Heinz V., and Knorr D. 1999. Evaluation of processing criteria for the high pressure treatment of liquid whole egg: Rheological study. *Lebensm Wiss und Technol*, 32, 299–304.

Lee D.U., Heinz V., and Knorr D. 2001. Biphasic inactivation kinetics of *Escherichia coli* in liquid whole egg by high hydrostatic pressure treatments. *Biotechnol Progr*, 17(6), 1020–1025.

Lee D.U., Heinz V., and Knorr D. 2003. Effects of combination treatments of nisin and high intensity utrasound with high pressure on the microbial inactivation in liquid whole egg. *Innov Food Sci Emer Technol* 4, 387–393.

Lei B., Majumder K., Shen S., and Wu J. 2011. Effect of sonication on thermolysin hydrolysis of ovotransferrin. *Food Chem*, 124, 808–815.

Lomakina K. and Míková K. 2006. A study of the factors affecting the foaming properties of egg white—A review. *Czech J Food Sci*, 24(3), 110–118.

Ma C.Y. 1996. Effects of gamma irradiation on physicochemical and functional properties of eggs and egg products. *Radiat Phys Chem*, 48(3), 375.

Ma L., Chang R.J., and Barbosa-Canovas G.V. 1997. Inactivation of *E. coli* in liquid whole eggs using pulsed electric fieldstechnology. In: Barbosa Canovas G.V., Lombardo S.,

Narsimhan G. and Okos M. (Eds.), *New Frontiers in Food Engineering*, Proceedings of Fifth Conference of Food Engineering, American Institute of Chemical Engineers, New York, NY, pp. 216–221.

Ma C.Y., Poste L.M., and Holme J. 1986. Effects of chemical modifications on the physico-chemical and cakebaking properties of egg white. *Can Inst Food Sci Technol J*, 19, 17–22.

McInerney J.K., Seccafien C.A., Stewart C.M., and Bird A.R. 2007. Effects of high pressure processing on antioxidant activity and total carotenoid content and availability in vegetables. *Innov Food Sci Emerg Technol*, 8, 543 – 548.

Majumder K. and Wu J. 2010. A new approach for identification of novel antihypertensive peptides from egg proteins by QSAR and bioinformatics. *Food Res. Int.*, 43, 1371–1378.

Mañas P., Pagan R., Raso J., and Condon S. 2003. Predicting thermal inactivation in media of different pH of *Salmonella* grown at different temperatures. *Int J Food Microbiol*, 87(1–2), 45–53.

Messens W., VanCamp J. and Huyghebaert A. 1997. The use of high pressure to modify the functionality of food proteins. *Trends Food Sci Technol*, 8(4), 107–112.

Marco-Moles R., Rojas-Grau M.A., Hernando I., Perez-Munuera I., Soliva-Fortuny R., and Martin Belloso O. 2011. Physical and structural changes in liquid whole egg treated with high-intensity pulsed electric fields. *J Food Sci*, 76(2), 257–264.

Monfort S., Ramos S., Meneses N., Knorr D., Raso J., and Álvarez I. 2012. Design and evaluation of a high hydrostatic pressure combined process for pasteurization of liquid whole egg. *Innov Food Sci Emerg Technol*, 14, 1–10.

Nagano T. and Nishinari K. 2001. Rheological studies on commercial egg white using creep and compression measurements. *Food Hydrocolloids*, 15(4), 415–421.

Norton T. and Sun D.W. 2008. Recent advances in the use of high pressure as an effective processing technique in the food industry. *Food Bioprocess Technol*, 1, 2–34.

Oey I., Lille M., Loey A.V., and Hendrickx M. 2008. Effect of high pressure processing on colour, texture and flavour of fruit and vegetable based food products: A review. *Trends Food Sci Technol*, 19(6), 320–328.

O'Brien R.D., Farr W.E., and Wan P.J. 2000. Introduction to fats and oils technology. American Oil Chemists' Society (AOCS), Champaign, p. 618. ISBN: 1893997138, 9781893997134.

Pajan S., Pandrangi S., Balasubramaniam V.M., and Yousef A.E. 2006. Inactivation of *Bacillus stearothermophilus* spores in egg patties by pressure assisted thermal processing. *LWT – Food Sci Technol*, 39(8), 844–851.

Pradas I., del Pino B., Peña F., Ortiz V., Moreno-Rojas J.M., Fernández-Hernández A., and García-Mesa J.A. 2012. The use of high hydrostatic pressure (HHP) treatments for table olives preservation. *Innov Food Sci Emerg Technol,* 13, 64–68.

Ponce E, Pla R, Sendra E, Guamis B, and Mor-Mur M. 1999. Destruction of *Salmonella enteritidis* inoculated in liquid whole egg by high hydrostatic pressure: Comparative study in selective and nonselective media. *Food Microbiol*, 16(4), 357–365.

Ponce E, Pla R, Sendra E, Guamis B, and Mor-Mur M. 1998a. Combined effect of nisin and high hydrostatic pressure on destruction of *Listeria innocua* and *Escherichia coli* in liquid whole egg. *Int J Food Microbiol*, 43(1–2), 15–19.

Ponce E., Pla R., Capellas M., Guamis B., and Mor-Mur M. 1998b. Inactivation of *Escherichia coli* inoculated in liquid whole egg by high hydrostatic pressure. *Food Microbiol*, 15, 265–272.

Pons M. and Fiszman S. 1996. Instrumental texture profile analysis with articular reference to gelled systems. *J Texture Stud*, 27(6), 597–624.

Pradas I., del Pino B., Peña F., Ortiz V., Moreno-Rojas J.M., Fernández-Hernández A., García-Mesa J.A. 2012. The use of high hydrostatic pressure (HHP) treatments for table olives preservation. *Innov Food Sci Emerg Technol*, 13, 64–68.

Qiu W., Jiang H., Wang H., and Gao Y. 2006. Effect of high hydrostatic pressure on lycopene stability. *Food Chem.*, 97, 516–523.

Rajan S., Pandrangi S., Balasubramaniam V.M., and Yousef A.E. 2006. Inactivation of Bacillus stearothermophilus spores in egg patties by pressure-assisted thermal processing. *LWT-Food Sci Technol*, 39, 844–851.

Rao M.A. 1986. Rheological properties of fluid foods. In: Rao, M.A. and H. Rizvi, S.S. (Eds.), *Engineering Properties of Foods*. Marcel Dekker, New York, pp. 1–47.

Rastogi N.K., Raghavarao K.S.M.S., Balasubramaniam V.M., Niranjan K., Knorr D. 2007. Opportunities and challenges in high pressure processing of foods. *Crit Rev Food Sci Nutr*, 47, 69–112.

Ryan D. and Robards K. 1998. Phenolic compounds in olives. *Analyst*, 123, 31R–44R.

Samaraweera H., Zhang W.G., Lee E.J., and Ahn D.U. 2011. Egg yolk phosvitin and functional phosphopeptides: Review. *J Food Sci*, 7, 143–150.

Sajjaanantakul T., Van Buren J.P., and Downing D.L. 1989. Effect of methyl ester content on heat degradation of chelator soluble carrot pectin. *J Food Sci*, 54(5), 1272–1277.

San Martín M.F., Barbosa-Cánovas G.V. and Swanson B.G. (2002). Food processing by high hydrostatic pressure. *Crit Rev Food Sci Nutr*, 42(6), 627–645.

Sánchez P.J., De Miguel C., Ramírez M.R., Delgado J., Franco M.N., Martín D. 2012. Efecto De Las Altas Presiones Hidrostáticas Respecto a La Pasteurización Térmica En Los Aspectos Microbiológicos, Sensoriales y Estabilidad Oxidativa de Un Paté de Aceituna (Effect of High Hydrostatic Pressure Versus Thermal Pasteurization on the Microbiological, Sensory Aspects and Oxidative Stability of Olive Pate. *Grasas y Aceites*, 63(1), 100–108.

Saraiva J., Vitorino R., b, Nunes C., Coimbra M.A. 2002. Effect of high pressure treatments on protease and β-Galactosidase activities of table olives. *High Press Res*, 22, 669–672.

Schroeder C.M., Naugle A.L., Schlosser W.D., Hogue A.T., Angulo F.J., Rose J.S., Ebel E.D., Disney W.T., Holt K.G., and Goldman D.P. 2005. Estimate of illnesses from *Salmonella* enteritidis in eggs, United States, 2000. *Emerg Infect Dis*, 11, 113–115.

Seregély Z., Dalmadi I., and Farkas J. 2007. The effect of the high hydrostatic pressure technology on livestock products monitored by NIR spectroscopy and chemosensor array. *Int J. High Press Res*, 27(1), 23–26.

Singh A. 2012. Evaluation of high pressure processing for improving quality and functionality of egg products. PhD Thesis. Department of Food Science and Agricultural Chemistry, Macdonald Campus, McGill University, Ste. Anne-De-Bellevue, Quebec, Canada.

Singh A. and Ramaswamy H.S. 2010. Effect of HPP on physicochemical properties of egg components. In: *Annual Meeting of Institute of Food Technologists (IFT)*, Chicago, USA, July 2010.

Soler-Rivas C., Epsin J.C., and Wichers H.J. 2000. Oleuropein and related compounds. *J Sci Food Agric*, 80, 1013–1023.

Speroni F., Puppo M.C., Chapleau N., De Lamballerie M., Castellani O., Añon M.C., and Anton M. 2005. High pressure induced physicochemical and functional modifications of low density lipoproteins from hen egg yolk. *J Agric Food Chem*, 53, 5719–5725.

Stadelman W.I. and Cotterill O.J. 1994. *Foaming. In: Egg Science & Technology*. Food Product Press, Haworth Press Inc., Binghamton.

Tauxe R.T. 1991. *Salmonella*: A postmortem pathogen. *J Food Protect*, 54, 563–568.

Tewari G., Jayas D.S., and Holley R.A. 1999. High pressure processing of foods: An overview. *Science des Aliments*, 19, 619–661.

Todd E.C.D. 1996. Worldwide surveillance of foodborne disease: The need to improve. *J Food Protect*, 59, 82–92.

Todd E.C.D. 2001. Epidemiology and globalization of food borne disease. In: Labbi R.G. and Garcia S. (Eds.) *Guide to Foodborne Pathogens*. Wiley-Interscience, New York, pp. 1–22.

Therios I. 2009. Structure and composition of the olive fruit. In: CABI (Ed.), *Olives. Series-Crop Production Science in Horticulture*, Oxford University Press, Oxford, UK, pp. 25.

Tokuşoğlu Ö. 2010. *Special Fruit Olive: Chemistry, Quality Control and Technology* Book. Seher Publishing. Pub.No: 006–1B; Sidas Medya Ltd. Şti., İzmir. 350 page. ISBN: 978-9944-5660-4-9.

Tokuşoğlu Ö. 2013. Table egg quality and technology (Yemeklik Yumurta Kalite ve Teknolojisi) In Turkish. *Graduate Course Notes*, Celal Bayar University, Manisa, Turkey.

Tokuşoğlu Ö., Alpas H., and Bozoğlu F. 2008. Effect of high hydrostatic pressure (HHP) on mycotoxin citrinin, major phenolics oleuropein, hydroxytyrosol, and total antioxidant activity in black table olives. 135–26 Technical Research Paper. 2008 IFT Annual Meeting + Food Expo. Book of Abstracts, p. 183, June 28–July 1, New Orleans, LA, USA.

Tokuşoğlu Ö., Alpas H., and Bozoğlu F.T. 2010. High hydrostatic pressure effects on mold flora, citrinin mycotoxin, hydroxytyrosol, oleuropein phenolics and antioxidant activity of black table olives. *Innov Food Sci Emerg Technol*, 11(2), 250–258.

Tokuşoğlu Ö. and Bozoğlu F. 2009. High hydrostatic pressure (HHP) effects on detailed chemical and microbiological aspects of table olives. 156–37 Technical Research Paper. 2009. IFT Annual Meeting + Food Expo. June 6–9, Anaheim/Orange County, CAL, USA.

Tokuşoğlu Ö. and Bozoğlu F.T. 2010. Citrinin risk in black and green table olives: Simultaneous determination with ochratoxin-A by optimized extraction and IAC-HPLC-FD. *Italian Journal of Food Science*, 22 (3), 284–291.

Valenti P., Visca P., Antonini G., and Orsi N. 1985. Antifungal activity of ovotransferrin towards genus Candida. *Mycopathologia*, 89, 169–175.

Van der Plancken I., Delattre M., Indrawati A., Van Loey A., and Hendrickx M.E.G. 2004. Kinetics study on the changes in the susceptibility of egg white proteins to enzymatic hydrolysis induced by heat and high hydrostatic pressure pretreatment. *J Agric Food Chem*, 52, 5621–5626.

Van der Plancken, I., Van Loey, A., and Hendrickx, M.E. G. 2005a. Combined effect of high pressure and temperature on selected properties of egg white proteins. *Innov Food Sci Emerg Technol*, 6, 11–20.

Van der Plancken I., Van Loey A., and Hendrickx M.E.G. 2005b. Changes in sulfhydryl content of egg white proteins due to heat and pressure treatment. *J Agric Food Chem*, 53, 5726–5733.

Van der Plancken I., Van Loey A., and Hendrickx M.E.G. 2006. Effect of heat treatment on the physicochemical properties of egg white proteins: A kinetic study. *J Food Eng*, 75, 316–326.

Van der Plancken I., Van Loey A., and Hendrickx M.E. 2007a. Foaming properties of egg white proteins affected by heat or high pressure treatment. *J Food Eng*, 78(4), 1410–1426.

Van der Plancken, I., Van Loey, A., and Hendrickx, M.E. 2007b. Kinetic study on the combined effect of high pressure and temperature on the physico-chemical properties of egg white proteins. *J Food Eng*, 78, 206–216.

Vinha A.F., Ferreres F., Silva B.M., Valentão P., Gonçalves A., Pereira J.A., Beatriz Oliveira M., Seabra R.M., and Andrade P.B. 2005. Phenolic profiles of portuguese olive fruits (*Olea europaea* L.): Influences of cultivar and geographical origin. *Food Chem*, 89(4), 561–568.

Visioli F., Bellomo G., and Galli C. 1998. Free radicals scavenging properties of olive oil polyphenols. *Biochem Biophys Res Commun*, 247, 60–64.

Volk S.P., Ahn D.U., Zeece M., and Jung S. 2012. Effects of high pressure processing and enzymatic dephosphorylation on phosvitin properties. *J Sci Food Agric*, 92, 3095–3098.

Welti-Chanes J., López-Malo A., Palou E., Bermúdez D., Guerrero-Beltrán J.A., and Barbosa-Cánovas G.V. 2005. Fundamentals and applications of high pressure processing of foods, In: Barbosa-Cánovas G.V., Tapia M.S., Cano M.P. (Eds.), *Novel Food Process Technol*, CRC Press, New York, pp. 157–182.

Wong P.Y. and Kitts D. 2003. Physicochemical and functional properties of shell eggs following electron beam irradiation. *J Sci Food Agric*, 83, 44–52.

Wu J. and Acero-Lopez A. 2012. Ovotransferrin: Structure, bioactivities, and preparation. *Food Res Int*, 46(2), 480–487.

Yan W., Qiao L., Gu X., Li J., Xu R., Wang M., Reuhs B., and Yang Y. 2010. Effect of high pressure treatment on the physicochemical and functional properties of egg yolk. *Eur Food Res Technol*, 231(3), 371–377.

Yang S.C. and Baldwin R.E. 1995. *Functional properties of eggs in foods*. In: Stadelman W.J. and Cotterill O. (Eds.), *Egg Sci Technol,* 4th edition. The Haworth Press, New York, pp. 405–463.

Yang R.X., Li W.Z., Zhu C.Q., and Zhang Q. 2010. Effects of ultra high hydrostatic pressure on foaming and physical–chemistry properties of egg white. *J Biomed Sci Eng*, 2, 617–620.

Yuste J., Capellas M., Pla R., Llorens S., Fung D.Y.C., and Mor-Mur M. 2003. Use of conventional media and thin agar layer method for recovery of foodborne pathogens from pressuretreated poultry products. *J Food Sci*, 68, 2321–2324.

13 Applications of High Pressure as a Nonthermal Fermentation Control Technique

Toru Shigematsu

CONTENTS

13.1 INTRODUCTION

High hydrostatic pressure (HP) processing is a nonthermal technology for microbial inactivation, which can inactivate microorganisms while preventing alterations in the flavor and nutrient contents of foods (Gänzle et al. 2001; Patterson et al. 2007). Since the 1990s, the development and commercial introduction of this technology has encouraged researchers to focus mainly on the inactivation of food-borne pathogens and spoilage organisms typically using HP treatment in the range of 200–800 MPa (Michiels et al. 2008). Recently, it has been proposed that HP processing may be potentially suitable not only for microbial inactivation but also for alteration of characteristics of agricultural products. HP treatment in the range of approximately 100–400 MPa can induce a transformation of agricultural products into an alternative form, where membrane systems are damaged (Ueno et al. 2009c) but certain enzymes are still active. HP-induced transformation, or "Hi-Pit (Ueno et al. 2009a)", of agricultural products has been successfully applied for the generation of functional compounds in agricultural products. High accumulation of γ-aminobutyric acid (GABA) and free amino acids in water-soaked brown rice (Kinefuchi et al. 1999; Shigematsu et al. 2010a, d) and soybeans (Ueno et al. 2010, 2013) were enhanced by HP treatment. Ueno et al. also reported that HP treatment caused the formation of green–blue compounds in turnip roots (Ueno et al. 2009b).

There are a number of fermented foods that are made from the conversion of agricultural products by microorganisms and require fermentation control technologies to prevent overfermentation. The conventional techniques for fermentation control are the application of heat and/or salt. If HP treatment in the range of 100–400 MPa could be applied for fermentation control rather than heat and/or salt, a wide range of fermented foods could retain more of their health-giving compounds and nutritional properties. It is necessary to develop techniques generating HP-sensitive (piezosensitive) microorganisms that are effectively sensitive in the range below 400 MPa, with sufficient fermentation abilities and ideally without a genetic engineering technique.

13.2 PRESSURE INACTIVATION OF MICROORGANISMS

Since the first scientific report by Roger in 1895 describing HP inactivation of *Escherichia coli* and *Staphylococcus aureus* (cited by Zobell 1970), a number of researches concerning HP effect on microbial cells have been investigated. In general, growth of microorganisms is inhibited under more than approximately 50 MPa at ambient temperature, which was reported in a bacterial species *E. coli* (Aertsen and Michiels 2008), as well as a eukaryotic species *Saccharomyces cerevisiae* (Abe 2008; Kawarai et al. 2006). At pressures exceeding 100 MPa, microbial species, such as *E. coli* as well as *S. cerevisiae*, increasingly loses viability within minutes (Fernandes 2008). Therefore, lethal pressures of more than 100 MPa have been applied for sterilization and microbial control in food materials.

Pressure inactivation behaviors of microorganisms appear as the loss of viable cell ratio depending on the pressuring time. Especially in a short pressuring time, the inactive kinetics in a number of microbial species and/or strains follows the first-order kinetics, which is similar with heat (thermal) inactivation.

HP treatment over approximately 100 MPa on microbial cells leads to phase transition of lipid bilayer, which causes damage in the cellular membrane systems (Ueno et al. 2009a,b). By transmission electron microscopic observation of *S. cerevisiae* cells, the nuclear membrane demonstrated damage by HP treatment over 100 MPa (Osumi 1990). The pressure inactivation behavior of *E. coli* was reportedly varied by concentration and the kind of coexisting materials such as salt and osmotic solutes (Ueno et al. 2011). Some auxotrophic phenotypes, such as Ade⁻, Trp⁻, Tyr⁻, Phe⁻, Met⁻, and Thr⁻ mutants of *S. cerevisiae*, showed piezosensitive phenotypes (Abe and Minegishi 2008; Iwahashi et al. 1993). The function of amino acid transporters located in the cellular membrane of *S. cerevisiae* was impaired by pressure (Abe and Horikoshi 2000; Abe 2008). Using *S. cerevisiae* deletion library, ergosterol was shown to be required for growth under HP at 25 MPa and a low temperature of 15°C (Abe and Minegishi 2008). Increased concentration of trehalose, which is a protectant of the cellular membrane, showed increased piezotolerance (Iwahashi et al. 2000). On the basis of these researches, the damage of structure and function of cellular and organelle membranes by pressure would be the main cause for the microbial inactivation by pressure. The damage of the protein complex, which forms the cytoskeleton, by pressure and the damage of the degradation system function of misfolding proteins were also some of the main causes for pressure inactivation of microorganisms (Kobori et al. 1995). Microbial cells, which were subjected to heat treatment to allow expression of heat-shock proteins, showed increased piezotolerance (Iwahashi et al. 1991). Genes encoding molecular chaperones, such as Hsp104 and Hsc70, have been shown to contribute to piezotolerance at 180 and 140 MPa, respectively, in the research on their deletion mutants (Iwahashi et al. 1997, 2001).

Despite these important studies, knowledge to generate piezotolerant and/or piezosensitive strains for applicative use is limited to date. For further application of HP processing as sterilization and fermentation control, it is important to obtain more knowledge on the mechanisms for HP inactivation of microorganisms.

13.3 A HIGH-THROUGHPUT METHOD FOR PRESSURE MICROBIAL INACTIVATION ANALYSIS

For further knowledge on the mechanisms for microbial HP inactivation, studies on the HP inactivation of microorganisms under a number of pressure conditions with coexisting materials are required. Therefore, we constructed a new method for analyzing the HP inactivation kinetics of microorganisms using microplates as cultivation and pressure vessels (Hasegawa et al. 2012).

The schematic diagram of the new method was shown in Figure 13.1. Cell suspensions of *E. coli* or *S. cerevisiae* were introduced into each well of eight-well microplate strips. A silicon capmat was carefully put on the microplate strip to exclude air from the wells. The microplate strip with the capmat was soaked with distilled water, vacuum packed in a polyethylene bag, and subjected to the HP treatment. After HP treatment, each cell suspension was transferred into a well of 96-well microplate, in which culture medium was prepared, and was subjected to liquid cultivation using the Bio Microplate Reader and growth was automatically monitored as optical density. The enumeration of viable cells of the cell suspension after HP treatment was carried out according to the growth delay method (Takano and Tsuchido 1982) in the liquid cultivation in microplates. In a batch-wise cultivation of a microorganism, the growth delays with decreases in the initial cell concentration. The $t_{\Delta 0.5}$ value, which is the cultivation time when the absorbance increased by 0.5 from the initial absorbance, was determined by the cultivation. This value was negatively linear and proportional to the logarithmic concentration of the diluted cell suspension, which was used as the inoculum for the microplate cultivation. Thus, we could evaluate the viability loss of microorganisms using microplate cultivation.

Using the new method, we evaluated the HP inactivation kinetics of *E. coli* and *S. cerevisiae*. The results indicated apparent reproducibility of the HP inactivation behaviors analyzed by the microplate method and the conventional method. We used microplates as the pressure vessels and cultivation vessels for the development of a new high-throughput microbial pressure inactivation kinetics analysis system (HT-PIKAS). This new system would facilitate an accelerated accumulation of useful knowledge on the HP-dependent inactivation of microorganisms. The

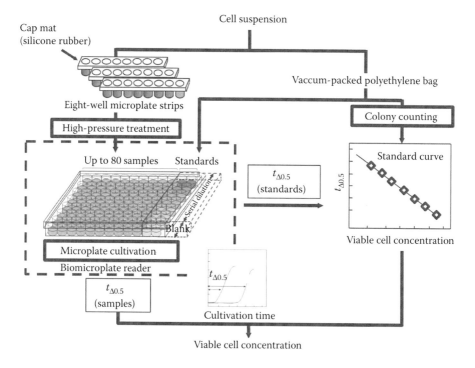

FIGURE 13.1   Schematic diagram of the HT-PIKAS. (Adapted from Hasegawa, T. et al. 2012. *J. Biosci. Bioeng.* 113: 788–91.)

accumulation of such knowledge could reveal the mechanism of the HP inactivation of microorganisms and could lead to better application of HP technology as sterilization and fermentation control.

13.4 RELATIONSHIP BETWEEN ENVIRONMENTAL STRESS RESPONSE AND PIEZOTOLERANCE IN *E. COLI*

As described above, a number of studies have revealed that the sensitivities of microorganisms to HP depend on the species and strains, as well as the environmental conditions such as temperature, pH, and the kind and concentrations of food components. Salt content of foodstuffs influences the microbial environment. High concentrations of salts in foods decrease water activity and create a hyperosmotic environment, which draws water from microorganisms by osmosis (Record et al. 1998). In a hypertonic solution, the cell loses water and shrinks. Conversely, in a hypotonic solution, there may be swelling and the cell may actually burst if its integrity is not maintained by its rigid cell wall. If external osmolality increases, cell collapse can be prevented by the cell's homeostatic mechanisms, which counterbalance increased external pressure by transporting compatible solutes into the cell

(Poolman et al. 2004). These osmotic and/or ionic stress events depend on the molar concentrations of salt solutions (Barron et al. 1986).

Since microbial HP stress responses may depend on environmental conditions, we focused on HP inactivation under varying salt stress imposed by different salts. Three salts (KCl, NaCl, and LiCl) were compared for effect on microbial physiology (Ueno et al. 2011). The cells of *E. coli* strain K12, suspended in each salt solution at 0.07–0.29 mol L^{-1}, were subjected to HP treatment at 250–400 MPa (20°C). After HP treatment, the survival ratios were calculated based on colony counting. Under all conditions tested, the inactivation behaviors could be approximated by first-order kinetics. This result suggested that inactivation of *E. coli* could be explained by one-hit hypothesis based on the target theory. If inactivation of microbial cells follows first-order kinetics, inactivation must then result from inactivation of a single sensitive molecule or site per cell (Moats 1971). The inactivation rate constants (k [s^{-1}]), were calculated based on the equation for the first-order kinetics. The k values of cells suspended in NaCl, KCl, and LiCl varied not only by the kind of salts but also by their concentration. (The result for NaCl solution was shown in Figure 13.2a.) These results indicated cationic ions effects on the pressure inactivation of *E. coli*. Inactivation of *E. coli* may be caused by effects on cell membrane structure and its functional properties; these effects are likely exerted by a combination of cationic and osmotic stress while cells are under pressure.

In each salt solution, the logarithmic k value linearly increased with the increase in the pressure levels. Such pressure dependencies of microbial inactivation rate constant have been demonstrated by several reports (Palou et al. 1997; Shigematsu et al. 2010b). These results indicated that the HP inactivation reaction follows the basic equation for pressure dependency of reaction velocity (Laidler 1951):

$$k = k_0 \cdot e^{-(\Delta V^* P / RT)}, \tag{13.1}$$

where k_0 is a preexponential factor [s^{-1}], P is pressure [MPa], ΔV^* is the activation volume [cm^3 mol^{-1}], R is the gas constant [J K^{-1} mol^{-1}], and T is the absolute

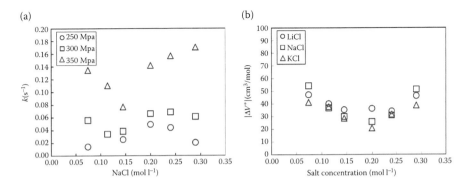

FIGURE 13.2 Effect of NaCl concentration on the inactivation rate constant k [s^{-1}] of *E. coli* cells (a) at 250 (circles), 300 (squares), and 350 (triangles) MPa at ambient temperature. The activation volume of pressure inactivation of *E. coli* (b) in LiCl (circles), NaCl (triangles), and KCl (boxes) during HP treatment (250–400 MPa, up to 300 s, ambient temperature).

temperature [K]. The activation volume is assumed by the volume change of the sensitive site between the initial and the transition state during the microbial inactivation reaction.

From the behavior of the activation volume (Figure 13.2b), we discussed the effect of salts on the pressure inactivation kinetics. At concentrations below 0.145 mol L^{-1} and above 0.240 mol L^{-1}, pressure inactivation is not salt specific. On the other hand, in the intermediate salt concentration range of 0.145–0.240 mol L^{-1}, inactivation kinetics in the presence of the Na$^+$ and K$^+$ significantly differed from those in the presence of Li$^+$, indicating salt specific. In this concentration range, the effect of salt stress and osmotic stress significantly differed from those in concentrations below 0.145 mol L^{-1} or above 0.240 mol L^{-1}. Inactivation of *E. coli* may be caused by effects on the cell membrane structure and its functional properties; these effects are likely exerted by a combination of salt stress and osmotic stress while cells are under pressure.

Recently, libraries comprising single-gene knockout mutations have been developed for many microbial species. Such single-gene knockout mutants can be used to analyze genes of unknown function, and can be the foundation for experimental databases that can in turn be used in the system biological approaches. A library of single-gene knockout mutants can also be used to identify genes responsible for a certain phenotype. A complete set of single-gene knockout mutations of all nonessential genes that comprises mutations in 3985 different genes was created with the *E. coli* strain BW25113, which is a derivative of strain K12; this library was named as the Keio collection (Baba et al. 2006). In addition, we have developed a HT-PIKAS based on a 96-well microplate format (Hasegawa et al. 2012) that enabled us to analyze pressure inactivation of microorganisms with very high efficiency. Therefore, we used single-gene knockout strains from the Keio collection to examine the effects of individual genes involved in osmoregulation or the cationic stress response on HP-induced inactivation of *E. coli* (Hasegawa et al. 2013).

We examined the HP inactivation kinetics of *E. coli* strain BW25113 (wild type) at 350–450 MPa with 0–0.35 mol L^{-1} NaCl. The behavior of inactivation rate constants of strain BW25113 was shown to be essentially similar to that of strain K12. Then, we selected five mutant strains that each lacked one gene reportedly involved in osmoregulation ($\Delta mzrA$, $\Delta envZ$, $\Delta ompR$, $\Delta ompC$, and $\Delta ompF$) and four mutant strains that each lacked one gene reportedly involved in the cationic stress response ($\Delta mscS$, $\Delta mscL$, $\Delta chaA$, and $\Delta nhaA$) and assessed kinetics of HP-induced inactivation in each strain at 400 MPa in solutions of 0–0.35 mol L^{-1} NaCl. The nine strains were found to fall into three different categories with regard to the effects of the mutation on k values (Figure 13.3).

Group 1 comprised five mutants—$\Delta envZ$, $\Delta ompR$, $\Delta ompC$, $\Delta ompF$, and $\Delta chaA$— whose k values were equivalent to those of BW25113 at NaCl concentrations of 0–0.35 mol L^{-1} (Figure 13.3a). Group 2 comprised three mutants—$\Delta mzrA$, $\Delta mscS$, and $\Delta nhaA$—whose k values were larger than those of BW25113 at NaCl concentrations of 0.145–0.25 mol L^{-1} (Figure 13.3b). Group 3 comprised only the $\Delta mscL$ mutant, whose k values were larger than those of BW25113 at low NaCl concentrations below 0.1 mol L^{-1} (Figure 13.3c).

The genes *mzrA*, *envZ*, *ompR*, *ompC,* and *ompF* are reportedly involved in osmoregulation (Gerken and Misra 2010; Matsubara and Mizuno 1999; Tokishita et al.

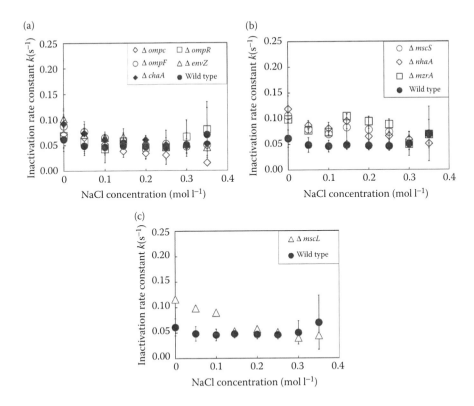

FIGURE 13.3 The effects of salt concentration on kinetics of HP-induced inactivation of *E. coli* (mutants and wild type). (Adapted from Hasegawa, T. et al. 2013. *High Press. Res.: An International J.* 33(2), 292–298.) (a) The *k* values of the five mutants (Δ*envZ*, Δ*ompR*, Δ*ompF*, Δ*ompC*, and Δ*chaA*), which were equivalent with those of the wild-type strain. (b) The *k* values of the three mutants (Δ*mscS*, Δ*nhaA*, and Δ*mzrA*), which were larger than those of the wild-type strain at NaCl concentrations of 0.145–0.25 M. (c) The *k* values of Δ*mscL* mutant, which were larger than those in the wild-type strain at low NaCl concentrations below 0.1 M.

1992). The *k* values of Δ*envZ*, Δ*ompR*, Δ*ompC*, and Δ*ompF* mutants were equivalent to those of BW25113 (Figure 13.3a). This result indicated that *envZ*, *ompR*, *ompC*, and *ompF* genes did not contribute to piezotolerance of strain BW25113 in the presence of NaCl at 400 MPa. In contrast, the *k* values of Δ*mzrA* mutant were larger than those of strain BW25113 at NaCl concentrations over a wide range, especially of 0.145–0.25 mol L^{-1} (Figure 13.3b). From this result, *mzrA* gene would contribute to increased piezotolerance of strain BW25113 at the investigated NaCl concentration. MzrA has been known as an upstream member of the EnvZ/OmpR signal transduction system, which interacts with EnvZ and leads to a high level of OmpR phosphorylation (Gerken and Misra 2010). Our results suggested the unknown function of MzrA, which contributes to piezotolerance in response to NaCl concentrations, possibly independent from the EnvZ/OmpR system.

The genes *mscS* and *mscL* encode mechanosensitive channels and play major roles in cell survival against hypotonic stress (Li et al. 2002). The k values of $\Delta mscS$ mutant were larger than those of strain BW25113 at NaCl concentrations over a wide range, especially of 0.145–0.25 mol L^{-1} (Figure 13.3b). The k value of the $\Delta mscL$ mutant was larger than those of strain BW25113 at a low NaCl concentration below 0.1 mol L^{-1} (Figure 13.3c). These results indicated that *mscS* and *mscL* genes would contribute to an increased piezotolerance of strain BW25113 at the NaCl concentrations of 0.145–0.25 and of 0–0.1 mol L^{-1}, respectively.

The genes *nhaA* and *chaA* encode Na$^+$/H$^+$ antiporters, which regulate internal pH and Na$^+$ concentrations to maintain homeostasis in the cell (Herz et al. 2010; Maes et al. 2012; Radchenko et al. 2006). The k values of $\Delta nhaA$ mutant were larger than those of strain BW25113 at NaCl concentrations over a wide range, especially of 0.145–0.25 mol L^{-1} (Figure 13.3b). In contrast, the k values of $\Delta chaA$ mutant were equivalent to those of strain BW25113 at NaCl concentrations of 0.0–0.35 mol L^{-1} (Figure 13.3a). These results indicate that the *nhaA* gene would contribute to an increased piezotolerance of strain BW25113, but that the *chaA* gene would not contribute to piezotolerance at the NaCl concentration. NhaA is reportedly the main antiporter in *E. coli* and is indispensable to the adaptation of high salt concentrations, survival of Li$^+$ toxicity, and growth under alkaline pH. ChaA is a Na$^+$ (Ca$^+$)/H$^+$ antiporter and has a key role in the K$^+$ extrusion system. ChaA has a role in K$^+$ efflux against a K$^+$ concentration gradient, and thereby, in maintaining cellular homeostasis. The contribution of *chaA* gene for cationic stress response would possibly be smaller than that of the *nhaA* gene at the investigated NaCl concentrations.

We used the k values during the HP treatment to assess the osmoregulation, cationic stress, and stress response of *E. coli*. On the basis of the results from the nine mutants that each lacked an osmoregulatory gene or a cationic stress response gene, we concluded that four of the nine genes were suggested to be involved in piezotolerance via osmoregulation and/or cationic stress response. The relationship between piezotolerance and genes involved in osmoregulation and/or the cationic stress response was thus demonstrated. Further analyses on the mechanisms for the HP inactivation behavior could contribute to the application of pressure as a sterilization technology for fermentation control.

13.5 PIEZOTOLERANT AND PIEZOSENSITIVE MUTANTS OF *S. CEREVISIAE*

Nonthermal food-processing technology using HP remains attractive as it can inactivate microorganisms without degrading vitamins, flavor, or texture. Fermented foods require fermentation controls to prevent taste deterioration and a reduction in shelf life due to overfermentation. Applying HP to inactivate the microorganisms involved in fermentation might be a promising approach to maintaining the quality of fermented foods. *S. cerevisiae* is a yeast species that is commonly used in the preparation of fermented foods and beverages such as bread and sake. A number of studies have been carried out to analyze the piezotolerance and piezosensitive mechanisms as described above and have suggested that piezotolerance–piezosensitivity can be genetically controlled. We applied random mutagenesis on *S. cerevisiae* haploid strains KA31a and KA31α by UV (ultraviolet) irradiation. From strain KA31a, three mutant strains were generated: a piezosensitive strain a924E1 compared with parent strain KA31a, a piezotolerant strain a2568D8, and strain a1210H12, which showed variability in piezotolerance (Nanba et al. 2013; Shigematsu et al. 2010b). From strain KA31α, a piezosensitive strain α1013E6 was generated (Shigematsu et al. 2010b). The pressure inactivation behaviors, based on the colony counting using a Compact dry Nissui YM dry sheet medium culture plates, of the parent strains and three mutant strains, a924E1, a2568D8, and α1013E6 could be approximated by the first-order kinetics (Figure 13.4). This result suggested that inactivation of the strains could be explained by one-hit hypothesis based on the target theory. The inactivation rate constants (k) at each pressure level were calculated and shown in Table 13.1. At all pressure levels tested, strains a924E1 and α1013E6 showed higher k values compared with their parent strains KA31a and KA31α, respectively. In contrast, strain a2568D8 showed lower k values, compared with the parent strain KA31a, at all pressure levels tested. These results indicated the piezosensitivity of strains a924E1 and α1013E6, and piezotolerance of strain a2568D8.

Interestingly, another mutant strain a1210H12 did not follow first-order kinetics (Figure 13.4). Inactivation curves could be approximated based on the multihit hypothesis based on the target theory (Uden et al. 1968):

$$N = N_0 \cdot [1 - (1 - e^{-kt})^m], \tag{13.2}$$

where N is the viable cell count [CFU mL^{-1}] at time t [s], N_0 is the viable cell count [CFU mL^{-1}] at time 0 s, k is the inactivation rate constant [s^{-1}], and m is the number of the sensitive sites per cell. At a pressure level of 225 MPa, the survival ratios (N/N_0) of strain a1210H12 showed larger treatment below 240 s than those of the parent strain KA31a. However, the survival ratio of strain a1210H12 showed smaller treatment than the parent strain at 300 s. This variable piezotolerant–piezosensitive phenotype of strain a1210H12, compared with the parent strain, was also observed by HP treatment at 250 MPa. This strain was thus recognized as a variable piezotolerant mutant.

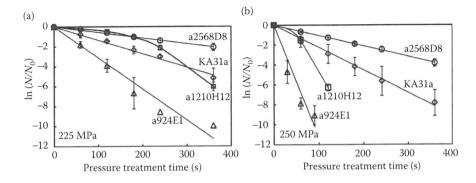

FIGURE 13.4 Survival curves of *S. cerevisiae* strains at 225 and 250 MPa as determined by colony counting using a Compact dry Nissui YM dry sheet medium culture plates. Strains KA31a (diamonds), a924E1 (triangles), a2568D8 (circles), and a1210H12 (squares).

TABLE 13.1

Kinetic Parameters of HP Inactivation of the Barosensitive Mutants (Based on Colony Counting)

	Strain	Inactivation Rate Constant (k) (s⁻¹)					Activation Volume (ΔV^*) (cm³ mol⁻¹)	Reference
		150 MPa	175 MPa	200 MPa	225 MPa	250 MPa		
KA31a	Wild type	ND	0.0019 ± 0.0002	0.0035 ± 0.0004	0.0089 ± 0.0004	0.0269 ± 0.0009	−86.5 ± 3.1	Laboratory stock
a924E1	Barosensitive mutant from KA31a	0.0019	0.0058 ± 0.0006	0.0065 ± 0.0006	0.0286 ± 0.0005	0.0847 ± 0.0055	−89.6 ± 0.4	Shigematsu et al. (2010b)
a2568D8	Barotolerant mutant from KA31a	ND	ND	ND	0.0057 ± 0.0005	0.0107 ± 0.0009	ND	Nanba et al. (2013)
a1210H12	Variably barotolerant mutant from KA31a	ND	ND	0.0058 ± 0.0008	0.0294 ± 0.0013	0.0828 ± 0.0046	−131.0 ± 4.2	Nanba et al. (2013)
KA31α	Wild type	ND	0.0023 ± 0.0008	0.0031 ± 0.0001	ND	0.0117 ± 0.0006	−54.6 ± 1.1	Laboratory stock
α1013E6	Barosensitive mutant from KA31α	ND	0.0037	0.0047 ± 0.0002	0.0119 ± 0.0006	0.0191 ± 0.0008	−57.0 ± 1.9	Shigematsu et al. (2010b)

Note: Means ± standard deviations are shown from at least three experiments with exceptions: one experiment for 150 MPa of strain a924E1and for 175 MPa of strain a1013E6. ND, not determined.

The auxotrophic properties of these mutants were identical to those of the parent strains: His⁻, Leu⁻, Trp⁻, and Ura⁻. This result suggested that the mutation causing the piezosensitive or piezotolerant phenotypes was not directly related to genes for amino acid or uracil metabolisms. There have been reports describing the relationship between amino acid metabolisms and piezotolerance. Iwahashi et al. (1993) reported that mutations from Ade⁻ to Ade⁺ and from Trp⁻ to Trp⁺ give piezotolerance under HP treatment at 180 MPa. Abe and Horikoshi (2000) reported that a mutation of Trp⁻ to Trp⁺ released the arrested growth of *S. cerevisiae* under the pressure of 25 MPa. In our case, mutations would occur on the genes not for amino acid metabolism but for other functions.

Pressure dependencies of the pressure inactivation rate constant (k) values of the mutant strains, as well as their parent strains, were analyzed. For each strain, the logarithmic k value was linearly proportional with pressure levels, as described in *E. coli* (Ueno et al. 2011). These results indicated that the HP inactivation reaction of *S. cerevisiae* strains also follows the basic equation for pressure dependency of reaction velocity (Laidler 1951). The activation volume (ΔV^*) values of the strains were calculated (Table 13.1). The ΔV^* values of strains KA31a and a924E1 were essentially the same, at -86.5 ± 3.1 and -89.6 ± 0.4 cm³ mol⁻¹, respectively. The ΔV^* values of strains KA31α and α1013E6 were also essentially the same, at -54.6 ± 1.1 and -57.0 ± 1.9 cm³ mol⁻¹, respectively. Interestingly, the ΔV^* values of each mutant–parent strain pair (a924E1 compared with KA31a and α1013E6 compared with KA31α) appeared to be equivalent. This result suggested that the mutation did not drastically change the sensitive sites or molecules for HP inactivation in volume variations between the transition state and their initial state. However, the ΔV^* value of strain a1210H12 showed a remarkably smaller value, being -131.0 ± 4.2 cm³ mol⁻¹, than the parent strain KA31a. The mutation in strain a1210H12 might occur in the gene responsible for the sensitive site or molecule.

Ethanol fermentation properties of the piezosensitive mutant strains a924E1 and α1013E6 were evaluated by batch cultivation at 30°C for 36 h using 15% yeast extract-peptone-dextrose (YPD) medium (Shigematsu et al. 2010b). The growth, glucose consumption, and ethanol production of strain a924E1 were equivalent with those of strain KA31a. The calculated ethanol yields of strains KA31a and a924E1 were 85.5% and 95.3%, respectively. The growth, glucose consumption, and ethanol production of strain α1013E6 were also equivalent with those of strain KA31α. The ethanol yields of strains KA31α and α1013E6 were 90.5% and 89.7%, respectively. These results suggested that the mutations in the piezosensitive strains did not occur in the gene responsible for the fermentation abilities.

The thermosensitivity of the piezosensitive mutant strain a924E1 was analyzed (Shigematsu et al. 2010c). The survival rates of strain a924E1 during thermal treatment at 50°C, 53°C, 55°C, and 58°C at 0.1 MPa were determined and compared with the parent strain KA31a. The logarithmic survival rates for each strain were negatively proportional to the heating time under all temperatures tested. This result indicates that the inactivation of the piezosensitive mutant, as well as the parent strain, followed first-order kinetics within 600 s of thermal treatment. The first-order inactivation rate constant (k) for each temperature was determined (Table 13.2) from the linear-fitted curves. The k values of the piezosensitive mutant strain a924E1

TABLE 13.2
Kinetic Parameters of Thermal Inactivation of the Barosensitive Mutant Strain KA31a (Based on Colony Counting)

Strain		Inactivation Rate Constant (k) (s^{-1})				Activation Energy (E) (kJ mol^{-1})
		50°C	53°C	55°C	58°C	
KA31a	Wild type	0.0034 ± 0.0004	0.0125 ± 0.0009	0.0284 ± 0.0057	0.0596 ± 0.0037	321.2 ± 15.5
a924E1	Barosensitive mutant from KA31a	0.0109 ± 0.0014	0.0324 ± 0.0013	0.0849 ± 0.0117	0.1849 ± 0.0368	320.3 ± 14.7

Note: Means ± standard deviations were shown from three experiments.

were larger than those of the parent strain in the range of 5058°C, indicating that the mutant was significantly more thermosensitive compared to the parent strain, at least in this temperature range. These results indicate that the mutation introduced to strain KA31a enhanced thermosensitivity as well as piezosensitivity. HP has been demonstrated to cause decreased viscosity of water, resulting in the destruction of hydrogen bonding, as does an increase in temperature (Bett and Cappi 1965). The effect of HP and high temperature is thus recognized to be analogous for organisms. The contribution of the accumulation of trehalose and a heat-shock protein to piezotolerance as well as thermotolerance (Iwahashi et al. 1997) would support this analogous effect of temperature and HP. The mutations are likely to be located in a gene or genes highly contributing to both piezotolerance and thermotolerance. The temperature dependency of the inactivation rate constant of strains KA31a and a924E1 was analyzed. The logarithmic k value was negatively proportional to the $1/T$ [K^{-1}] value. This finding indicates that the thermal inactivation of the piezosensitive mutant and the parent strain could be described as chemical reactions based on the transition state theory. According to Arrhenius equation, the activation energies (E) for each strain were, thus, determined. The E values of strains KA31a and a924E1 were essentially equivalent, being 321.2 ± 15.5 and 320.3 ± 14.7 kJ mol^{-1}, respectively. Since the activation volume (ΔV^*) of strains KA31a and a924E1 was also essentially equivalent (Table 13.1), the mechanism for HP and thermal inactivation reaction of strain a924E1 would be basically the same as that of strain KA31a.

The inactivation behaviors of the mutant strains were analyzed at 200–250 MPa of pressure by using the HT-PIKAS, in which the viable cell count after HP treatment is based on a growth delay in the subsequent cultivation in liquid medium (Nanba et al. 2013). For strains KA31a, a924E1, and a2568D8, the inactivation behavior could be approximated by first-order kinetics based on the single-hit death theory as the same manner with the results obtained by colony counting. Strain a1210H12 could be approximated by multihit death theory in the same manner as the results obtained by colony counting. The relative piezotolerant–piezosensitive properties compared with the parent strain were also identical to the results obtained by colony counting. The inactivation rate constants k values were calculated at the pressure levels of 200–275 MPa. The resulting k values (Table 13.3) were different from those obtained by colony counting (Table 13.1). This discrepancy occurred by the different methods for a viable cell count. The survival ratios determined by HT-PIKAS were lower than those determined by colony counting on 2% YPD agar medium at any time after HP treatment of strains KA31a, a924E1, and a2568D8 (Figure 13.5a through c). The difference in the survival ratios of the three strains increased with increasing HP treatment time. However, the survival ratio did not differ in strain a1210H12 (Figure 13.5d).

Since HT-PIKAS is a method to count viable cells based on a growth delay caused by a reduction in the number of viable cells, we speculated that HP damage on cells would cause lag time for growth. If the suspension after HP treatment contained "cells with a growth delay caused by pressure-induced damage," the viable cell concentration would be underestimated by HT-PIKAS compared with colony counting because cells that had not been affected by HP would continue to grow normally and faster than those with a growth delay. Therefore, lag time was calculated as

TABLE 13.3
Kinetic Parameters of HP Inactivation of the Mutant Strains (Based on HT-PIKAS)

Strain		Inactivation Rate Constant (k) (s^{-1})				Activation Volume (ΔV^*) (cm^3 mol^{-1})
		200 MPa	225 MPa	250 MPa	275 MPa	
KA31a	Wild type	0.0074 ± 0.0008	0.0121 ± 0.0013	0.0255 ± 0.0003	0.0371 ± 0.0051	−54.8 ± 8.1
a924E1	Barosensitive mutant from KA31a	0.0135 ± 0.0021	0.0238 ± 0.0087	0.1134 ± 0.0095	ND	−105.1 ± 7.7
a2568D8	Barotolerant mutant from KA31a	ND	0.0084 ± 0.0005	0.0123 ± 0.0017	0.0184 ± 0.0017	−38.5 ± 6.6
a1210H12	Variably barotolerant mutant from KA31a	0.0030 ± 0.0005	0.0104 ± 0.0012	0.0311 ± 0.0010	ND	−118.2 ± 7.3

Note: Data are shown as means ± standard deviations from three experiments. ND, not determined.

the difference in growth between viable cells determined by colony counting on 2% YPD agar medium and growth curves generated using HT-PIKAS. Lag times were detected in all strains. Lag times in strains KA31a, a924E1, and a2568D8 increased depending on the length of HP treatment (Figure 13.5a through c). The results suggested that the proportions of growth-delayed cells increased depending on the length of HP treatment. Lag time was the longest in strain a924E1, following KA31a and a2568D8, whereas it did not obviously change in strain a1210H12 during HP treatment (Figure 13.5d). These findings suggested that HP does not damage strain a1210H12 and cause a growth delay, or that this strain cannot recover from the damage that causes such a growth delay. The existence of the growth-delayed cells caused by HP treatment was confirmed by analysis on the distribution of colony size during cultivation after HP treatment at 225 MPa. The distribution of colony size of strains KA31a, a924E1, and a2568D8 shifted to smaller sizes according to the length of HP treatment. An increase in the numbers of smaller colonies induced by HP indicated the generation of growth-delayed cells. On the other hand, the size of strain a1210H12 colonies did not change. This result indicated that HP did not induce a growth delay in strain a1210H12. The proportions of growth-delayed cells were

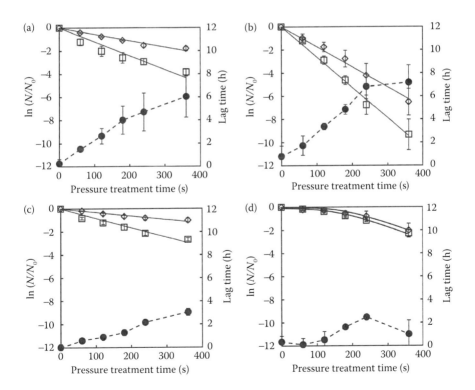

FIGURE 13.5 Survival curves of *S. cerevisiae* strains KA31a (a), a924E1 (b), a2568D8 (c), and a1210H12 (d) at 225 MPa at ambient temperature. Survival ratios determined by colony counting on 2% YPD agar medium (diamonds) and by HT-PIKAS (squares) and lag time caused by growth-delayed cells (circles).

calculated by subtracting the concentration of viable cells in 2% YPD liquid medium determined by HT-PIKAS from that of cells on 2% YPD agar medium determined by colony counting. The proportions of viable cells with no effect on growth in 2% YPD liquid medium were determined by HT-PIKAS. The calculated numbers of viable cells were divided by those of viable HP-untreated cells and multiplied by 100 to give the proportions.

The proportions of growth-delayed KA31a and a2568D8 cells were 20% and 34%, respectively, which were higher, and 12% in strain a924E1, which was lower than that of the wild-type strain KA31a. A growth delay was undetectable in strain a1210H12. Therefore, the proportion of growth-delayed cells is an important indicator of piezosensitivity and piezotolerance.

The piezosensitivity of strain a924E1 would be caused by a mutagenesis-induced decreased ability to recover from HP damage compared with strain KA31a. On the other hand, the piezotolerance of strain a2568D8 would be caused by a mutagenesis-induced increased ability to recover from HP damage compared with strain KA31a. Thus, piezotolerance and piezosensitivity are apparently related to an increase or decrease in the ability to recover from HP damage. The variable piezotolerance of strain a1210H12 would be caused by a deficiency in the ability to recover from HP damage conferred by mutagenesis, compared with strain KA31a.

Although the molecular mechanism is unclear and requires further investigation, the function of energy metabolisms in each strain may play an important role for the recovery ability from HP damage. Using 2,3,5-triphenyl-2H-tetrazolium chloride (TTC) staining method, the respiration abilities of the strains were assayed. The variable piezotolerant strain a1210H12 and piezosensitive strain a924E1 showed deficient less respiration abilities, whereas the piezotolerant strain a2568D8 showed sufficient respiration ability that was equivalent with the wild-type strain KA31a. Therefore, these results suggested that the ability to recover from HP damage could be affected by respiration ability, possibly based on the function of mitochondria in the strains. The ability to recover from damage induced by HP that is sufficient to cause a growth delay would be important for piezosensitivity and piezotolerance. Information about mutated genes would allow control of the piezosensitivity and piezotolerance in yeast strains. HP could be used to control fermentation if piezosensitive genes are inserted into industrial strains and if molecular breeding is successful.

13.6 CONCLUSIONS

Application of HP processing as fermentation control is potentially attractive, because HP can inactivate microorganisms while preventing alteration in the flavor and nutrient contents of foods. For this purpose, it is important to develop techniques generating piezosensitive microorganisms. Although techniques to generate piezotolerant and/or piezosensitive strains for applicative use are limited to date, useful knowledge on genetic and environmental factors affecting piezotolerant and/or piezosensitive microorganisms based on the pressure inactivation mechanisms has been accumulated. We indicated that the interactions between cationic stress and cellular response systems affect piezotolerance of E. coli based on a newly developed high-throughput experimental system. We also provided a finding that UV mutagenesis could alter the piezotolerance of *S. cerevisiae* strain without lacking fermentation and auxotrophic properties. Another important finding was that cellular changes in piezotolerance caused by mutations are remarkably affected by the ability to recover from cellular damage by HP that results in a growth delay. Further studies on the mechanisms of microbial pressure inactivation would lead to developing a technique for molecular breeding of piezosensitive microorganisms, which enable a novel use of HP processing as fermentation control.

REFERENCES

Abe, F. 2008. Effect of growth-permissive pressures on the physiology of *Saccharomyces cerevisiae*. In: *High-Pressure Microbiology*, eds., C. Michiels, D. H. Bartlett, and A. Aertsen, 167–79.Washington, DC: ASM Press.

Abe, F. and Horikoshi, K. 2000. Tryptophan permease gene *TAT2* confers high-pressure growth in *Saccharomyces cerevisiae*. *Mol. Cell. Biol.* 20:8093–102.

Abe, F. and Minegishi, H. 2008. Global screening of genes essential for growth in high pressure and cold environments: Searching for basic adaptive strategies using a yeast deletion library. *Genetics* 178: 851–72.

Aertsen, A. and Michiels, C. W. 2008. Cellular impact of sublethal pressures on *Escherichia coli*. In: *High-Pressure Microbiology*, eds., C. Michiels, D. H. Bartlett, and A. Aertsen, 87–100. Washington, DC: ASM Press.

Baba, T., Ara, T., Hasegawa, M., Takai, Y., Okumura, Y., Baba, M., Datsenko, K. A., Tomita, M., Wanner, B. L. and Mori, H. 2006. Construction of *Escherichia coli* K-12 in-frame, single-gene knockout mutants: The Keio collection. *Mol. Syst. Biol.* 2: 2006.0008.

Barron, A., May, G., Bremer, E. and Villarejo, M. 1986. Regulation of envelope protein composition during adaptation to osmotic stress in *Escherichia coli*. *J. Bacteriol.* 167: 433–38.

Bett, K. E. and Cappi, J. B. 1965. Effect of pressure on the viscosity of water. *Nature* 207: 620–21.

Fernandes, P. M. B. 2008. *Saccharomyces cerevisiae* response to high hydrostatic pressure. In: *High-Pressure Microbiology*, eds., C. Michiels, D. H. Bartlett, and A. Aertsen, 145–66. Washington, DC: ASM Press.

Gänzle, M. G., Ulmer, H. M. and Vogel, R. F. 2001. High pressure inactivation of *Lactobacillus plantarum* in a model beer system. *J. Food Sci.* 66: 1174–81.

Gerken, H. and Misra, R. 2010. MzrA–EnvZ interactions in the periplasm influence the EnvZ/OmpR two-component regulon. *J. Bacteriol.* 192: 6271–78.

Hasegawa, T., Hayashi, M., Nomura, K., Hayashi, M., Kido, M., Ohmori, T., Fukuda, M., Iguchi, A., Ueno, S., Shigematsu, T., Hirayama, M. and Fujii, T. 2012. High-throughput method for a kinetics analysis of the high-pressure inactivation of microorganisms using microplates. *J. Biosci. Bioeng.* 113: 788–91.

Hasegawa, T., Nakamura, T., Hayashi, M., Kido, M., Hirayama, M., Yamaguchi, T., Iguchi, A., Ueno, S., Shigematsu, T. and Fujii, T. 2013. Influence of osmotic and cationic stresses on high-pressure inactivation of *Escherichia coli*. *High Press. Res.: An International J.* 33: 292–98.

Herz, K., Rimon, A., Olkhova, E., Kozachkov, L. and Padan, E. 2010. Transmembrane segment II of NhaA Na⁺/H⁺ antiporter lines the cation passage, and Asp^{65} is critical for pH activation of the antiporter. *J. Biol. Chem.* 285: 2211–20.

Iwahashi, H., Fujii, S., Obuchi, K., Kaul, S. C., Sato, A. and Komatsu, Y. 1993. Hydrostatic pressure is like high temperature and oxidative stress in the damage it causes to yeast. *FEMS Microbiol. Lett.* 108: 53–7.

Iwahashi, H., Kaul, S. C., Obuchi, K. and Komatsu, Y. 1991. Induction of barotolerance by heat shock treatment. *FEMS Microbiol. Lett.* 80: 325–28.

Iwahashi, H., Nwaka, S. and Obuchi, K. 2000. Evidence for the contribution of neutral trehalose in barotolerance of *Saccharomyces cerevisiae. Appl. Environ. Microbiol.* 66: 5182–85.

Iwahashi, H., Nwaka, S. and Obuchi, K. 2001. Contribution of Hsc70 to barotolerance in the yeast *Saccharomyces cerevisiae. Extremophiles* 5: 417–21.

Iwahashi, H., Obuchi, K., Fujii, S. and Komatsu, Y. 1997. Barotolerance is dependent on both trehalose and heat shock protein 104 but is essentially different from thermotolerance in *Saccharomyces cerevisiae. Lett. Appl. Microbiol.* 25: 43–47.

Kawarai, T., Arai, S., Furukawa, S., Ogihara, H. and Yamasaki, M. 2006. High-hydrostatic-pressure treatment impairs actin cables and budding in *Saccharomyces cerevisiae. J. Biosci. Bioeng.* 101: 515–18.

Kinefuchi, M., Sekiya, M., Yamazaki, A. and Yamamoto, K. 1999. Accumulation of GABA in brown rice by high pressure treatment. *Nippon Shokuhin Kagaku Kougaku Kaishi* 46: 323–28. (in Japanese)

Kobori, H., Sato, M., Tameike, A., Hamada, K., Shimada, S. and Osumi, M. 1995. Ultrastructural effects of pressure stress to the nucleus in *Saccharomyces cerevisiae*: A study by immunoelectron microscopy using frozen thin sections. *FEMS Microbiol. Lett.* 132: 253–58.

Laidler, K. H. 1951. The influence of pressure on the rates of biological reactions. *Arch. Biochem.* 30: 226–36.

Li, Y., Moe, P. C., Chandrasekaran, S., Booth, I. R. and Blount, P. 2002. Ionic regulation of MscK, a mechanosensitive channel from *Escherichia coli. EMBO J.* 21: 5323–30.

Maes, M., Rimon, A., Kozachkov-Magrisso, L., Friedler, A. and Padan, E. 2012. Revealing the ligand binding site of NhaA Na$^+$/H$^+$ antiporter and its pH dependence. *J. Biol. Chem.* 287: 38150–57.

Matsubara, M. and Mizuno, T. 1999. EnvZ-independent phosphotransfer signaling pathway of the OmpR-mediated osmoregulatory expression of OmpC and OmpF in *Escherichia coli. Biosci. Biotechnol. Biochem.* 63: 408–14.

Michiels, C., Barlett, D. H. and Aertsen, A. 2008. *High-Pressure Microbiology.* Washington, DC: ASM Press.

Moats, W. A. 1971. Kinetics of thermal death of bacteria. *J. Bacteriol.* 105: 165–71.

Nanba, M., Nomura, K., Nasuhara, Y., Hayashi, M., Kido, M., Hayashi, M., Iguchi, A., Shigematsu, T., Hirayama, M., Ueno, S. and Fujii, T. 2013. Importance of cell damage causing growth delay for high-pressure inactivation of *Saccharomyces cerevisiae. High Press. Res.* 33: 299–307.

Osumi, M. 1990. Effects of hydrostatic pressure to ultrastructure of yeast cells. In: *Pressure-Processed Food-Research and Development*, ed. R. Hayashi, 157–64. Kyoto: Sanei Shuppan. (in Japanese)

Palou, E., López-Malo, A., Barbosa-Cánovas, G. V., Welti-Chanes, J. and Swanson, B. G. 1997. Kinetic analysis of *Zygosaccharomyces bailii* inactivation by high hydrostatic pressure. *Lebensum-Wiss Technol.* 30: 703–8.

Patterson, M. F., Linton, M. and Doona, C. J. 2007. Introduction to high pressure processing of foods. In: *High Pressure Processing of Foods*, eds., C. J. Doona, F. E. Feeherry, 1–14. Oxford: Blackwell Publishing.

Poolman, B., Spitzer, J. J. and Wood, J. M. 2004. Bacterial osmosensing: Roles of membrane structure and electrostatics in lipid–protein and protein–protein interactions. *Biochim. Biophys. Acta.* 1666: 88–104.

Radchenko, M. V., Tanaka, K., Waditee, R., Oshimi, S., Matsuzaki, Y., Fukuhara, M., Kobayashi, H., Takabe, T. and Nakamura, T. 2006. Potassium/proton antiport system of *Escherichia coli. J. Biol. Chem.* 281: 19822–29.

Record, M. T., Courtenay, E. S., Cayley, D. S. and Guttman, H. J. 1998. Responses of *E. coli* to osmotic stress: Large changes in amounts of cytoplasmic solutes and water. *Trends Biochem. Sci.* 23: 143–48.

Shigematsu, T., Hayashi, M., Nakajima, K., Uno, Y., Sakano, A., Murakami, M., Narahara, Y., Ueno, S. and Fujii, T. 2010a. Effects of high hydrostatic pressure on distribution dynamics of free amino acids in water soaked brown rice grain. *J. Phys.: Conf. Ser.* 215: 0121271.

Shigematsu, T., Murakami, M., Nakajima, K., Uno, Y., Sakano, A., Narahara, Y., Hayashi, M., Ueno, S. and Fujii, T. 2010d. Bioconversion of glutamic acid to γ-aminobutyric acid (GABA) in brown rice grains induced by high pressure treatment. *Japan J. Food Eng.* 11: 189–99.

Shigematsu, T., Nasuhara, Y., Nagai, G., Nomura, K., Ikarashi, K., Hirayama, M., Hayashi, M., Ueno, S. and Fujii, T. 2010b. Isolation and characterization of barosensitive mutants of *Saccharomyces cerevisiae* obtained by UV mutagenesis. *J. Food Sci.* 75: M509–14.

Shigematsu, T., Nomura, K., Nasuhara, Y., Ikarashi, K., Nagai, G., Hirayama, M., Hayashi, M., Ueno, S. and Fujii, T. 2010c. Thermosensitivity of a barosensitive *Saccharomyces cerevisiae* mutant obtained by UV mutagenesis. *High Press. Res.* 30: 524–29.

Takano, M. and Tsuchido, T. 1982. Availability of growth delay analysis for the evaluation of total injury of stressed bacterial populations. *J. Ferment. Technol.* 60: 189–98.

Tokishita, S., Kojima, A. and Mizuno, T. 1992. Transmembrane signal transduction and osmo-regulation in *Escherichia coli*: Functional importance of the transmembrane regions of membrane-located protein kinase, EnvZ. *J. Biochem.* 111: 707–13.

Uden, N. V., Abranches, P. and Silva, C. C. 1968. Temperature functions of thermal death in yeasts and their relation to the maximum temperature for growth. *Arch. Microbiol.* 61: 381–93.

Ueno, S., Hayashi, M., Shigematsu, T. and Fujii, T. 2009b. Formation of green–blue compounds in *Brassica rapa* root by high pressure processing and subsequent storage. *Biosci. Biotechnol. Biochem.* 73: 943–45.

Ueno, S., Izumi, T. and Fujii, T. 2009c. Estimation of damage to cells of Japanese radish induced by high pressure with drying rate as index. *Biosci. Biotechnol. Biochem.* 73:1699–703.

Ueno, S., Katayama, T., Watanabe, T., Nakajima, K., Hayashi, M., Shigematsu, T. and Fujii, T. 2013. Enzymatic production of γ-aminobutyric acid in soybeans using high hydrostatic pressure and precursor feeding. *Biosci. Biotechnol. Biochem.* 77: 706–13.

Ueno, S., Shigematsu, T., Hasegawa, T., Higashi, J., Anzai, M., Hayashi, M. and Fujii, T. 2011. Kinetic analysis of *E. coli* inactivation by high hydrostatic pressure with salts. *J. Food Sci.* 76: M47–53.

Ueno, S., Shigematsu, T., Kuga, K., Saito, M., Hayashi, M. and Fujii, T. 2009a. High-pressure induced transformation of onion. *Japan J. Food Eng.* 10: 37–43. (in Japanese)

Ueno, S., Shigematsu, T., Watanabe, T., Nakajima, K., Murakami, M., Hayashi, M. and Fujii, T. 2010. Generation of free amino acids and γ-aminobutyric acid in water-soaked soybean by high-hydrostatic pressure processing. *J. Agric. Food Chem.* 58: 1208–13.

Zobell, C. E. 1970. Pressure effects on morphology and life processes of bacteria. . In: *High Pressure Effects on Cellular Processes*, ed. A. M. Zimmerman, 85–130. New York: Academic Press.

14 Food Allergies
High-Pressure Processing Effects on Food Allergens and Allergenicity

Özlem Tokuşoğlu and Faruk T. Bozoğlu

CONTENTS

14.1   INTRODUCTION TO FOOD SENSITIVITY AND FOOD ALLERGY

Food sensitivity is an adverse reaction to a food which other people can safely eat, and includes food allergies, food intolerances, microbial toxications, and chemical sensitivities, whereas food allergy is an abnormal response to a food triggered by body's immune system. Foodborne allergic reactions can sometimes cause serious illness and death (Bozoğlu and Tokuşoğlu, 2014).

Different types of food sensitivities may arise from a wide variety of reasons making it complex, most of the time confusing, and is not an easily defined area of study. It was stated that diagnosis could also be difficult since symptoms may be postponed for up to two days after the food consumption. As a whole, food sensitivities are the result of toxic responses to food. Food sensitivities are divided into two classes, including allergic responses and food intolerances (Bozoğlu, 2011; Anonymous, 2013a; Bozoğlu and Tokuşoğlu, 2014).

As a matter of fact, only up to 3% of adults and 6–8% of children have clinically proven true allergic reactions to food. For those with food allergies, sensitivities, or intolerances, avoiding specific foods and ingredients is an important health challenge. It was stated that food sensitivities are more common and have a wider and more varied impact on our health than previously realized. Food sensitivities are equated with food allergies, but also include food intolerances which, unlike allergies, are toxic reactions to foods that do not involve the immune system and are often more difficult to diagnose (Bozoğlu, 2011; Bozoğlu and Tokuşoğlu, 2014).

Various symptoms of food sensitivities include vomiting, diarrhea, blood in the stool, eczema, urticaria (hives), skin rashes, wheezing, and runny noses associated with an allergic reaction to specific foods or beverages. Food sensitivities may also cause fatigue, gas, bloating, mood swings, nervousness, migraines, and eating disorders (Bozoğlu, 2011; Bozoğlu and Tokuşoğlu, 2014).

Based on clinical data, the sensitivity to food can also increase the severity of the symptoms of rheumatoid arthritis, asthma, and other diseases normally not considered food related. It is known that tree nuts and peanuts are the leading causes of deadly allergic reactions called anaphylaxis. It is stated that another group of symptoms such as bloating, muscle and joint aches, pains, and tiredness that a consumer reports when he/she eats certain foods are often collectively known as food intolerance. This group of symptoms is less defined and poorly understood, and consequently, generally much harder to diagnose than the classical allergy. Another kind of food intolerance is the adverse reaction to specific compounds (food additives) that are added to food for enhancing taste, providing color stability, or protecting against the microorganism growth as a preservative. The consumption of large amounts of food additives can produce symptoms that mimic the entire range of allergic symptoms (Bozoğlu and Tokuşoğlu, 2014).

On the other side, food intolerance is an adverse reaction to some sort of food or food ingredient that occurs every time the food is eaten, generally if larger quantities are consumed (Johansson et al., 2001). Food intolerance is not the same as a food allergy, because the immune system is not activated. Neither is it the same as food poisoning, which is caused by toxic substances that would cause symptoms in anyone who ate the food. It is stated that food intolerance symptoms vary and can be mistaken for those of a food allergy. The general symptoms of food intolerance can be nausea, stomach pain, gas, cramps, bloating, vomiting, heartburn, diarrhea, headaches, irritability, or nervousness. Food intolerances are more likely to originate in the gastrointestinal system and are usually caused by an inability to digest or absorb certain foods, or components of those foods. Food intolerance occurs when the body is unable to deal with a certain type of foodstuff (Bozoğlu, 2011; Bozoğlu and Tokuşoğlu, 2014).

Food poisoning is the result of eating food that is contaminated with the toxins produced in the food by pathogen microorganisms. Thereby, the food ingestion with microbial toxins can produce symptoms resembling food allergies (Bozoğlu and Tokuşoğlu, 2014).

14.2   FOOD ALLERGY CONCEPT

Food allergy is a reaction of the body's immune system to a certain food or beverage. In this context, food allergy is a very specific reaction involving the immune system of the body. At this point, distinguishing food allergy from other food sensitivities is the most important. Whereas food allergies are rare, food intolerances, which are the other classification of food sensitivities, are more prevalent (Anonymous, 2013c; Bozoğlu and Tokuşoğlu, 2014).

There are generally several types of allergens in each food, and it is known that allergens are generally proteins. So far, it is not exactly understood why some foods can cause allergies and others do not, but a theory is that some proteins in foods mimic very closely the proteins present in viruses and bacteria (Anonymous, 2013b,c).

Some consumers are genetically predisposed and their immune system is not able to differentiate the food protein from the virus or bacteria, thereby attacks occur. Some proteins or protein fragments are resistant to digestion and those that are not broken down in the digestive process are tagged by the Immunoglobulin E (IgE). These tags trick the immune system into thinking that the protein is harmful. It is stated that the IgE acts like a tag, sticking to molecules in food or pollen called allergens. If an allergic consumer eats a beleaguered food, IgE attaches to the allergens, setting off an allergic reaction in the body (Bozoğlu and Tokuşoğlu, 2014). Another recognized effect that IgE induces is the histamine, which can lead to the alterations. It can be seen in our bodies as symptoms, like nettle rash or wheezing, and these reactions can be ranged from mild to severe. It is stated that an antibody only binds one specific antigen and when the antibody binds/sticks to the dangerous molecule, it acts like a red flag identifying the molecule as something potentially damaging and that it should be removed. Macrophage cells are known as scavenger cells in the immune system; these are particularly designed for removing damaging molecules from the body. The macrophages consume the molecule, taking it out of circulation and exterminating it; afterward the antibody binds to the dangerous molecule (Anonymous, 2013a; Bozoğlu and Tokuşoğlu, 2014).

Severe allergic reactions to foods are usually rapid, emerging within an hour or sometimes even seconds of consumption, also in some situations, they may be postponed and appear up to 4 h after consuming.

Skin rashes (nettle rash) called urticaria or hives can emerge that generally are short lived and disappear within several days, whereas longer lasting rashes or chronic skin reactions such as scaly patches or atopic dermatitis can appear. Also, an itchy nose and eyes, sneezing, and a runny nose may appear, while asthmatic symptoms, such as wheezing, breathlessness, and coughing may occur. The above-mentioned symptoms are not seen so often with food allergies. When contacting a food, itching and swelling around the lips and mouth may occur and also nausea, cramping pains, bloating, vomiting or diarrhea may arise (Bozoğlu and Tokuşoğlu, 2014).

14.3 A GLANCE AT MAJOR ALLERGIC FOODS AND THEIR CHARACTERISTIC SYMPTOMS

Several specific foods are responsible for the majority of food allergies, even though any food can stimulate an immune response in allergic individuals. It is known that peanuts are the leading cause of severe allergic reactions, followed by shellfish, fish, tree nuts, and eggs. Peanuts, tree nuts including almonds, brazil nuts, cashews, hazelnuts (filberts), macadamia nuts, pecans, pine nuts (pignolias), pistachio nuts, walnuts, sesame seeds, milk, eggs, fish including shellfish and crustaceans, soy, glüten, fava beans, garlic and onion, mustard are some of the most known allergic foods (Anonymous, 2013b; Bozoğlu and Tokuşoğlu, 2014).

14.3.1 PEANUTS AS ALLERGIC FOOD

Peanut allergy symptoms can range from a minor irritation to a life-threatening reaction (anaphylaxis) and is common, especially in children. Many peanut-sensitive patients have severe reactions, such as fatal and near-fatal. Even tiny amounts of peanuts can cause a serious reaction for some people suffering from peanut allergy (Skripak, 2008).

It is stated that the symptoms of peanut allergy are related to the action of Immunoglobulin E (IgE) or other anaphylatoxins, which act to release histamine. It is reported that the only real way for treating a nut allergy is to avoid peanuts and tree nuts. Avoiding nuts mean more than just not eating them; it is very difficult because peanuts are commonly used as an adulterant in the preparation of foods (Høst et al., 2008). It is reported that new processes to make allergen-free peanuts in testing showed a 100% deactivation of peanut allergens in whole roasted kernels, and human serums from severely allergic individuals showed no reaction when exposed to the processed peanuts (Bozoğlu and Tokuşoğlu, 2014; Li, 2007).

14.3.2 GLUTEN AS ALLERGIC FOOD

Kaukinen et al. (2000) reported that a gluten allergy, like any other food allergy, is when the body's immune system reacts against gluten protein, resulting in a number of food allergy symptoms (Kaukinen et al., 2000).

Gluten allergy could appear early in life, and then disappear as the child grows older, or it could appear later in life, either vanishing some years later, or hanging around for the rest of a person's life. Its symptoms may appear as dermatitis, but could also be present as difficulty in breathing during exercise. Also, gastrointestinal symptoms may occur in both cases. In the case of idiopathic gluten sensitivity, all known symptoms are restrained to the nervous system. Gluten allergy symptoms may be similar to the symptoms for celiac disease and specific symptoms, along with the severity of each one, varies from one person to the next (Nelson, 2002). People with gluten allergies need not worry about the tiny amount of gluten in their diet as long as they feel good. In other respects, if people have celiac disease, they must eliminate all gluten, even if they feel good. With idiopathic gluten sensitivity, the antibodies that correlate with disease are anti-gliadin antibodies. Whether these antibodies are pathogenic or are simply indicators of circulating gliadin is unknown (Garsed and Scott, 2007; Størsrud et al., 2003; Workman et al., 1984).

Upper repository tract problems, chronic fatigue syndrome, mouth ulcers, anaemia, osteoporosis, weight loss, short stature as the natural height position in children, diarrhea, abdominal bloating, diverticulitis, depression, attention and behavioral problems for children and adults, skin problems, asthma, irritability are known symptoms of gluten allergy. It is stated that the gluten source needs to be declared when a food contains gluten protein or modified gluten protein from barley, oats, rye, triticale, or wheat. It is reported that gluten-free oats can provide a valuable source of fiber, vitamin B, iron, zinc, and complex carbohydrates (Bozoğlu and Tokuşoğlu, 2014; Garsed and Scott, 2007; Størsrud et al., 2003; Workman et al., 1984).

14.3.3 Soy as Allergic Food

It is known that soy is one of the so-called "big eight" allergens, and along with milk allergies and egg allergies, it is one of the three allergies children are most likely to outgrow.

Cordle (2004) stated that most soy allergies are fairly mild and may cause hives (red and sometimes itchy bumps on your skin), nausea, or rhinitis (stuffy nose). Soy allergic people need to avoid very minor sources of soy protein such as soy oil or soy lecithin. Severe reactions are most likely in people who also have peanut allergies and asthma. Soy rarely causes severe reactions, including breathing difficulty and anaphylaxis. Many people with soy allergy can tolerate small to moderate amounts of soy protein (Cordle, 2004).

It is stated that research has been done on soy allergies. By using a "gene silencing" technique, researchers were able to "knock out" a gene that makes a protein called P34, which is thought to trigger most allergic reactions to soy. It is concluded that tests on blood from people allergic to soy showed no antibody response to the plant with the knocked-out gene, indicating that the allergen could not be detected. It is reported that some fermented soy foods such as tempeh, shoyu, and miso cause less allergy than whole soybeans, because the fermentation process partly breaks down the proteins (Cordle, 2004).

14.3.4 Egg as Allergic Food

It is revealed that egg allergy usually first appears when children are very young and most outgrow it by the time they are 5 years old. It is reported that when something made with eggs enters the digestive system of a person with an egg allergy, the immune system responds by creating specific antibodies (IgE) to that food that trigger the release of certain chemicals in the body, one of which is histamine. Egg yolk allergies may be somewhat more prevalent in adults. Some people can be allergic to proteins in both the egg white and the egg yolk, whereas a person who reacts only to a protein in the egg yolk may be able to easily tolerate egg whites. Moreover, a few people who are allergic to eggs can develop an allergy to chicken or other poultry meats. The respiratory system, gastrointestinal tract, skin, and the cardiovascular system can be affected by egg allergy and wheezing, nausea, headache, stomach ache, or itchy hives can occur as allergy symptoms (Venter and Skypala, 2009).

It is stated that ovomucoid is the immunodominant protein fraction in egg white and that the use of commercially purified ovalbumin has led to an overestimation of the dominance of ovalbumin as a major egg allergen and antigen in human beings (Bernhisel-Broadbent et al., 1994). Research on lysozyme and ovotransferrin suggests that, one of the causes of the allergy is from the chelating capacity of the proteins with metals, especially those of heavy metals (Li, 2005; Moreau et al., 1995).

14.3.5 Milk as Allergic Food

It is known that cow's milk is the usual cause of milk allergy; milk from sheep, goats and buffalo can also cause a reaction. It is stated that cow's milk allergy is one of the most common food allergies in children. Some children who are allergic to cow's milk are allergic to soy milk too. Almost all infants are fussy at times. However, some are excessively fussy owing to having an allergy to the protein in cow's milk, which is the basis for most commercial baby formulas. It is also reported that a person of any age can have a milk allergy, but it is more common among infants (about 2–3% of babies), though most outgrow it (Høst, 2002).

It is revealed that a milk allergy usually occurs a few minutes to a few hours after milk is consumed. It is reported that signs and symptoms of milk allergy range from mild to severe and can include wheezing, vomiting, hives, or digestive problems. Rapid-onset reactions come on suddenly with symptoms that can include irritability, vomiting, wheezing, swelling, hives, other itchy bumps on the skin, and bloody diarrhea. In scarce situations, a potentially severe allergic reaction (anaphylaxis) can occur and affect the baby's skin, stomach, breathing, and blood pressure (Brill, 2008).

Lactose intolerance means the body cannot easily digest lactose, a type of natural sugar found in milk and dairy products. This is not the same thing as a food allergy to milk. People who have lactose intolerance cannot digest any milk products. When lactose moves through the large intestine (colon) without being properly digested, it can cause uncomfortable symptoms such as gas, belly pain, and bloating (Montalto et al., 2006; Bozoğlu & Tokuşoğlu,2014).

Milk allergy is a food allergy, that is, an adverse immune reaction to a food protein that is normally harmless to the non-allergic individual whereas milk protein intolerance (MPI) is different than the milk allergy. Milk protein intolerance produces a range of symptoms very similar to milk allergy symptoms, but can also include blood and/or mucus in the stool. MPI is a delayed reaction to a food protein that is normally harmless to the non-allergic, non-intolerant individual. MPI produces a non-IgE antibody and is not detected by allergy blood tests (Bozoğlu, 2011).

14.3.6 SHELLFISH AS ALLERGIC FOOD

It is mostly known that shellfish allergy is the most common food allergy among adults. About 2% of adults have a shellfish allergy, and 0.1% of children have a shellfish allergy (CFIA, 2009; NIAID, 2005). It is reported that shellfish allergy is more likely to develop in adulthood than in early childhood, unlike other food allergies. Shellfish allergies tend to be severe, lifelong food allergies. It is stated that allergies to shellfish are in two classes of foods: mollusks including clams, mussels, and oysters and crustaceans including shrimp, lobster, and crabs (CFIA, 2009).

Hives, itching, eczema, swelling of the lips, face, tongue, throat, or other parts of the body, wheezing, nasal congestion, trouble in breathing, abdominal pain, diarrhea, nausea or vomiting, dizziness, lightheadedness or fainting, or tingling in the mouth are the most common shellfish allergy symptoms. It is stated that shrimp or crab allergy symptoms starts within 30 min after ingestion. A severe allergic reaction to shellfish called anaphylaxis is rare, but can be life-threatening if it interferes with breathing (CFIA, 2009).

It is stated that there is a high rate of allergic cross reactivity between the above-mentioned two shellfish groups. Therefore, many people who are allergic to any one shellfish are advised to avoid all shellfish. It is stated that shrimp is the most allergenic shellfish. The shrimp protein that most commonly causes shellfish allergies is the tropomyosin. People who are allergic to one type of crustacean, such as shrimp, are generally allergic to all other crustaceans (mollusks such as clams or oysters) (CFIA, 2009).

14.3.7 FAVA BEANS AS ALLERGIC FOOD

It is reported that people with G6PD deficiency are at risk of hemolytic anemia in states of oxidative stress. The name favism is sometimes used to refer to the enzyme deficiency as a whole, although this is misleading as not all people with G6PD deficiency will manifest a physically observable reaction to consumption of broad beans. Broad beans, for example, fava beans, contain high levels of vicine, divicine, convicine, and isouramil, all of which are oxidants. Some studies claim that it may be a common disorder in American and African Blacks. It is stated that males are more affected than females. Infants and young children are more affected than adults (Frank, 2005). The symptoms result from red cell haemolysis and include nausea, abdominal pain, fever, chills, pallor (a pale color of skin due to reduced amount of oxyhaemoglobin in skin or mucous membrane), shortness of breath, and fatigue. Renal failure occurs in the more severe cases. It is stated that symptoms appear within 24 h following ingestion of the bean and continue for 2 days. There is a 10% mortality if not treated properly and recovery is usually spontaneous. G6PD deficiency is a genetic disorder and favism can be determined with a simple blood test, which will check a patient's enzyme levels to determine whether or not the patient has favism and how severe the condition is (Frank, 2005).

Recently, fava beans are used as flavoring or filling in some packaged products and are becoming more widely used in many products. It is stated that consumer should read all the package labels when purchasing commercially produced products even if the person knows the product, as recipes are often altered (Bozoğlu and Tokuşoğlu, 2014).

14.3.8 Mustard as Recent Allergic Food

Mustard is a condiment prepared from mustard seeds. It is widely used in numerous kinds of seasonings and sauces as well as in other industrial preparations and can often arise as a masked allergen, leading to serious allergic reactions. European data indicates mustard as the third or fourth most common food allergen in France and mustard is recognized as an allergen by the International Union of Immunological Societies (Figueroa et al., 2005).

It is reported that about 1.1% of children with food allergies are allergic to mustard. Symptoms can include difficulty breathing, a rash or hives, or itchy skin. The skin can become swollen, too, as in other allergic reactions. In some severe situations, a reaction to mustard can result in painful, sensitive rashes on the skin of the back or abdomen, stomach cramps, and vomiting. In scarce situations, an allergic reaction to mustard can prompt an anaphylactic response. Anaphylaxis can include dizziness, shortness of breath, difficulty breathing, swollen tongue, rapid heartbeat, dramatic change in blood pressure and fainting (Figueroa et al., 2005). The results from characterization studies of allergenic proteins indicate that proteins in mustard are resistant to degradation by heat and digestive enzymes. It is stated that the thermostable allergenic proteins in mustard have the potential to be hidden within certain ingredients, preparations, and mixtures in processed and prepackaged foods, with exemptions from individual component declaration (Figueroa et al., 2005).

14.3.9 Onion/Garlic as Recent Allergic Foods

It is reported that garlic and/or onion is a food allergen. A 12 kDa protein band to young garlic, garlic, and onion extracts was detected in a garlic-allergic individual. Reactions to these allergens among allergic individuals range from mild to severe (Kao et al., 2004). It is stated that trouble in breathing, speaking, or swallowing, a drop in blood pressure, rapid heart beat, and/or loss of consciousne, flushed face, hives or a rash, or red and itchy skin, swelling of the eyes, face, lips, throat, and tongue, anxiousness, distress, faintness, paleness, and/or weakness, cramps, diarrhea, and/or vomiting are common symptoms of onion/garlic allergy. Trace amounts of onion/garlic as food ingredients can potentially be found in a wide range of food products including snack foods, health foods, baked goods, seasonings, and many other foods. It is stated that the potential for severe allergic reactions to occur as a result of hidden sources of garlic and/or onion in pre-packaged foods is considered minimal (Kao et al., 2004).

14.4 LABELING PRECAUTION FOR MINIMIZING OF FOOD ALLERGY RISKS

The Food and Drug Administration require that most prepackaged foods carry a label and that the ingredients appear on labels in decreasing order of proportion. Despite this, certain ingredients utilized in food products are currently exempt from declaration in the list of ingredients, for example, components of margarine, seasoning, and flour. Moreover, these regulations do not require constituents, including main ingredients and the ingredients of certain foods and products as flavoring, seasoning, spices, and vinegar, to be listed on food labels (Bozoğlu, 2011).

Precautionary labeling, however, must be truthful and must not be used in lieu of adherence to legal requirements. It is stated that when an allergen is likely to be present in a product, the use of precautionary labeling is not acceptable and the presence of the allergenic ingredient should be accurately declared on the label. National food regulatory agencies must use the Codex Alimentarius Commission advices on food additives and other chemicals and ingredients in food to build on the Codex list and develop their own lists of priority foods, which should be targeted for mandatory labeling on foods available for sale in the country or region under their oversight (Bozoğlu, 2011).

14.5 CONVENTIONAL PROCESSING EFFECTS ON FOOD ALLERGENICITY

Food allergy is an IgE-mediated abnormal response to normally tolerated food proteins. Food allergy symptoms vary from mild localized symptoms to severe anaphylaxis that, at times, may be fatal. It is stated that the causes for an individual to become intolerant toward a specific food protein are unclear (Sathe et al., 2005). It is reported that the amount of protein required, the threshold, to elicit an allergic response in a sensitized person varies considerably from patient to patient and protein to protein. Besides, thresholds for many allergenic proteins remain unknown (Sathe et al., 2005).

Due to the ubiquitous presence of allergens in the food supply, reducing/elimination strategies of allergens in foods are important. Using food processing to reduce/eliminate allergenicity have been performed (Sathe et al., 2005). It has been shown that food processing has important effects on the structural and allergenic properties of food allergens (Mills et al., 2009).

It was reported that some food processing methods are mechanical such as separation, isolation, and/or purification while others are thermal, like pasteurization, cooking, and roasting including diverse chemical reactions such as Maillard, or there are biochemical methods like enzymatic treatment of food (Sathe et al., 2005). Thermal processing of foods includes moist heat and dry heat whereas nonthermal processing includes high-pressure processing, γ-irradiation, and multiple techniques (Sathe et al., 2005). In previous sample studies, Besler et al. (2001) reported that allergenic elements from bovine milk are by combining heat and ezymatic treatments with ultrafiltration, and Watanabe et al. (1990) reported that

TABLE 14.1

Impact of Food Processing on Different Types of Food Allergens

Type of Processing Behavior	Effect of Thermal Processing	Types of Food Allergen
Processing-labile allergens	Protein unfolding, modification by Maillard adducts in sugar-rich foods, modification by polyphenols	Bet v 1 homolgoues from fuirts such as Mal d 1, Pru av 1
Partially-denatured allergens	Partial unfolding of proteins, aggregation to form networks as emulsifiers around lipid or gelled systems. Maillard modifications may potentiate allergenicity	Cupins allergens, such as Ara h 1 from peanut. Lipocalins such as β-Lg and α-lactalbumin from milk;
Allergens able to refold	Proteins unfold to a limited extent during heating but can re-fold on cooling. Maillard modification may potentiate allergenicity	Prolamin superfamily members belonging to the ns LTP, and 2S albumin sub-families such as Mal d 3; tropomyosins and parvalbumins
Mobile rheomorphic proteins	Proteins do not adopt a rigid conformation but are very mobile and consequently do not denature following thermal treatment	Caseins, seed storage prolamins of wheat (gluten), ovomucoid

Source: Adapted from Clare Mills E.N. et al. 2009. *Mol Nutr Food Res* 53(8), 963–969.

allergenic components from rice are by an enzymatic treatment. It is known that the hypoallergenic milk and rice are now produced commercially and can be supplied to allergic patients.

Table 14.1 shows the thermal food processing effects on various food allergens.

It is stated that cupins allergens, such as Ara h 1 from peanut and Lipocalins such as β-Lg, and α-lactalbumin from milk may partially denature by thermal processing. The effects of thermal processing in these allergens are partial unfolding of proteins and aggregation to form networks as emulsifiers around lipid or gelled systems. Also, Maillard modifications may potentiate allergenicity by thermal processing (Table 14.1) (Clare Mills et al., 2009).

Clare Mills et al. (2009) stated that another aspect of allergen management in foods is the allergenic risk assessment process, which forms part of the regulatory framework of novel foods and processes (Clare Mills et al., 2009).

14.6 HIGH-PRESSURE PROCESSING EFFECTS ON FOOD ALLERGENICITY

Although thermal treatments, enzymatic treatments, and other conventional methods have generally been used for eliminating food allergenicity, some treatments result in a degradation of the processing food characteristics, as well as a deterioration in the flavor and taste; for instance, the development of bitterness or an unpleasant odor. Besides, the ezymatic treatment applications to foods give a high level of protein; this situation is not practicable, especially for meats (Yamamoto et al., 2010).

High-pressure (HP) processing treatments are novel-processing techniques that have the potential to alleviate the need for thermal processing of foods. High-pressure (400–700 MPa) processing is combined with temperatures around room temperature (5–40°C). It is stated that treatments offer an alternative to high-temperature pasteurization, or chemical preservation and fresh-like properties of foods are preserved (Sasagawa and Yamazaki, 2002).

It is known that the current recommended strategy for allergy sufferers is avoidance of allergen foods and also the recommended strategy for manufacturers is the necessity of labeling regarding potential changes in food manufacturing and/or information of ingredient/additive used in food preparation.

It was reported that novel-processing techniques including HP processing focused on enzyme research including proteins. Currently, limited studies have been performed on the HP processing effects on the structure of known allergens and the elimination of allergen compounds in foods.

14.6.1 Meat (Beef) Allergens By HP Processing

Meat is an important food due to its high nutritional value and functional properties. Hovever, meat allergy is considered to be a scarce pathological problem, and the current reports have demonstrated that meat allergy can be a serious problem, particularly in children. It is reported that bovine serum albumin (BSA) and bovine gamma globulin (BGG) are the major beef allergens (Han et al., 2000). Beef allergy patients also show allergic symptoms against bovine milk owing to BSA and BGG present in both beef and milk as allergens (Martelli et al., 2002).

The HP processing is considered to be a useful food-processing technique and such a treatment affects the protein structures resulting in alterations in the food characteristics. Recently, the high pressure effects on the antigenicity of beef extracts (Han et al., 2002), BSA (beef allergen) (Nogami et al., 2006), and ovomucoid (chicken egg white allergen) (Odani et al., 2007) were reported; it was concluded that the structural changes in the allergens induced by pressure may lead to the reduction or eliminating of antigenicity.

Yamamoto et al. (2010) stated that the effects of a HP treatment (100–600 MPa/5–7 °C/5 min) on the IgE-specific binding activity and structural changes to bovine gamma globulin (BGG), a beef allergen, were investigated and then the allergenicity of pressure-treated BGG was evaluated. It was found that the IgE-specific binding activity and allergenicity of BGG were decreased by the HP treatment. In the study reported by Yamamoto et al. (2010), no significant alteration in secondary protein structure for pressurized BGG (Figure 14.1), but an alteration in the tertiary structure was detected (Yamamoto et al., 2010). It was stated that the decreased IgE-specific binding activity and allergenicity of pressurized BGG were probably owing to changes in the tertiary structure caused by high pressurization. It was concuded that a high-pressure treatment would be an effective food processing technique to reduce the allergenicity of BGG (Yamamoto et al., 2010).

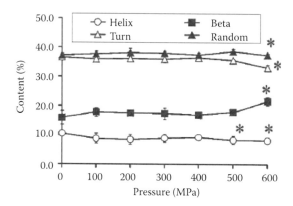

FIGURE 14.1 The effect of high-pressure treatment on the secondary structure of BGG. Each value is the mean ± standard deviation ($n = 4$) ($p < 0.05$). (Adapted from Yamamoto S. et al. 2010. *Biosci Biotechnol Biochem* 74(3), 525–30.)

Yamamoto et al. (2010) reported that the IgE-specific binding activities of pressurized and unpressurized BGG in the sera from patients allergic to BGG, as assessed by dot-plotting, are shown in Figure 14.2 (Yamamoto et al., 2010). It was shown that the IgE-specific binding activities of BGG decreased significantly at 100–400 MPa and 600 MPa in patient No. 5 (Figure 14.2a), at up to 300 MPa in patient No. 64 (Figure 14.2b), and at up to 200 MPa in patient No. 82 (Figure 14.2c). These data showed that HHP treatment affected the reduction of IgE-specific binding activity (Yamamoto et al., 2010).

It was stated that the impact of high pressure on allergenicity in high-pressure-treated meat products is studied by Han et al. (2002) and Hajos et al. (2004). It is concluded that conformational changes of proteins may alter antigenicity and/or immunological crossreactivity by changing the binding abilities of their epitopes, so that HHP might either reduce allergenicity or reveal new antigenic sites (Hajos et. al., 2004; Han et al., 2002).

Han et al. (2002) stated that the sera of bovine gamma globulin (BGG) positive beef allergic patients were used to investigate the changes in IgE-specific binding activity with regard to beef extract altered by heat or HP treatment. It was reported that in inhibition-ELISA, the sample treated at 60°C did not show any significant changes in the antigenicity of BGG, but the sample treated at 100°C showed a decrease of the antigenicity (Han et al. 2002). In the study described by Han et al. (2002), in the case of the treatment with heating at 100°C, heat coagulation occurred in the beef extract. It was found the persistent antigenicity, even after the treatment at 100°C in inhibition-ELISA, remained principally in the heat-coagulated fraction, which indicated the importance of the method of handling the heat coagulation in heat treatment. Han et al. (2002) also reported that HP treatments (200–600 MPa) of beef extract did not show any significant changes in binding with the sera

Hajos et al. (2004) stated the high pressure (600 MPa/20 min/at ambient temperature) induced conformational alterations in pork batter proteins with of some

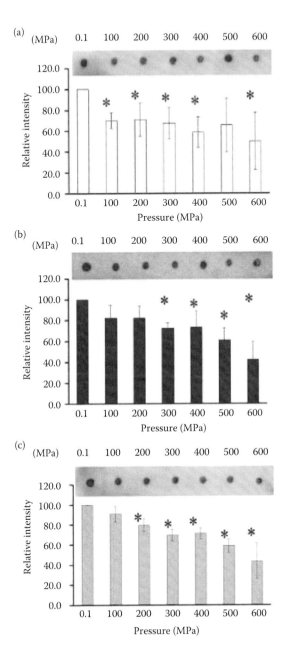

FIGURE 14.2 Dot-blotting of the IgE-binding activities to BGG in Sera from patients with beef allergy after a high-pressure treatment. Relative intensity is expressed as a percentage of the value at 0.1 MPa. A, B, C show different patients. Each value is the mean ± standard deviation ($n = 4$) ($p < 0.05$). (Adapted from Yamamoto S. et al. 2010. *Biosci Biotechnol Biochem* 74(3), 525–30.)

of the epitope structures (Hajos et al., 2004). It was stated that some protein groups lost their immunoreactivity in HP-treated samples, others kept the same level of immunoreactivity, and pressure created one group of slightly immunoreactive species (Hajos et al., 2004).

The few existing studies indicate that high pressure does not enhance the migration of compounds from the packaging material and displays only a small effect on protein allergenicity in meat. Simonin et al. (2012) stated that high pressure decreases biogenic amine levels in meats and it seems to be a safe technology for the treatment of meat products (Simonin et al., 2012).

14.6.2 HAZELNUT ALLERGENS BY HP PROCESSING

Hazelnut is an edible nut with health benefits. It is reported that hazelnut is as a causative agent of allergic reactions. It is shown that allergies to hazelnuts, tend to be of a more severe nature, causing life-threatening and sometimes fatal reactions. Indeed, symptoms upon hazelnut ingestion are often confined to the mouth and throat, but are severe systemically. Type I food allergy is defined as an IgE-mediated response to a protein (or proteins) in a food source (López et al., 2012; Calhoun and Schofield, 2010; Genuis, 2010). It was indicated that post-translational modifications (PTMs) like phosphorylation and glycosylation can play a relevant role in allergenicity related to edible nuts including hazelnuts (Dupont et al., 2010).

In previous studies, it has been suggested that modifications in allergenicity are caused by thermal treatments. Hansen et al. (2003) indicated that roasting of hazelnuts reduced the allergenicity. Recently, López et al. (2012) evaluated the effects of autoclaving (AC) and high hydrostatic pressure (HHP) processing effects on the allergenic characteristics of proteins from hazelnuts.

López et al. (2012) aimed to combine functional research protein-allergen analysis via western blotting and structural analysis of allergen proteins from the hazelnuts to observe what is occurring in relation to the specific recognition of the allergenic proteins from hazelnuts by IgE using sera from patients allergic to hazelnuts with clinical significance.

In the study described by López et al. (2012), hazelnut samples were subjected to AC and HHP processing and the specific IgE-reactivity was studied in 15 allergic clinic patients via western blotting analyses. It was reported that a series of homology-based-bioinformatic 3D models (Cora 1, Cora 8, Cora 9, and Cora 11) were generated for the antigens included in the study to analyze the complexity of their protein structure (López et al., 2012). It was concluded that a severe reduction *in vitro* in allergenicity to hazelnut after AC processing was observed in the allergic clinic patients studied. The specific-IgE binding of some of the described immunoreactive hazelnut protein bands: Cora 1 ~18 KDa, Cora 8 ~9 KDa, Cora 9 ~35–40 KDa and Cora 11 ~47–48 KDa decreases (López et al., 2012). It was also stated that a relevant glycosylation was assigned and visualized via structural analysis of proteins (3D modeling) for the first time in the protein-allergen Cora 11 showing a new role that could open a door for allergenicity-unravellings (López et al., 2012).

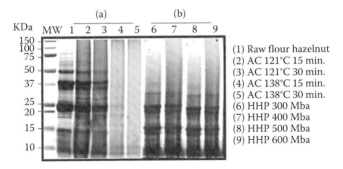

(1) Raw flour hazelnut
(2) AC 121°C 15 min.
(3) AC 121°C 30 min.
(4) AC 138°C 15 min.
(5) AC 138°C 30 min.
(6) HHP 300 Mba
(7) HHP 400 Mba
(8) HHP 500 Mba
(9) HHP 600 Mba

FIGURE 14.3 Hazelnut flour samples analyzed by SDS-PAGE 4–20%: The wild-raw proteins and the raw-processed proteins by autoclaving and high pressure with different treatments are illustrated. (Adapted from López E. et al. 2012. *J Clin Bioinf* 2(1), 12. doi: 10.1186/2043-9113-2-12.)

In the study reported by López et al. (2012), it was obtained that protein patterns from the AC hazelnut processed samples via SDS-PAGE analysis were different compared to the protein-pattern of the raw hazelnut (without AC treatment) (Figure 14.3).

López et al. (2012) also stated that hazelnut allergenicity studies *in vivo* via Prick-Prick and other means using AC processing are crucial to verify the data observed via *in vitro* analyses. It was reported that the glycosylation studies provided the mechanisms of the structures that contribute to hazelnut allergenicity, which thus, in turn, help alleviate food allergens (López et al., 2012).

14.6.3 EGG PROTEIN ALLERGENS BY HP PROCESSING

It is known that egg proteins are responsible for one of the most common forms of food allergy, especially in children, and one of the major allergens is ovalbumin (OVA).

López-Expósito et al. (2008) examined the potential of high pressure to enhance the enzymatic hydrolysis of OVA and modify its immunoreactivity, hence the protein was proteolyzed with pepsin under HP conditions (400 MPa). In the study described by López-Expósito et al. (2008), characterization of the hydrolysates and peptide identification was performed by reversed phase high-performance liquid chromatography-tandem mass spectrometry (RP-HPLC-MS/MS). López-Expósito et al. (2008) stated that the antigenicity (binding to IgG) and binding to IgE, using the sera of patients with specific IgE to OVA were also determined. It was shown that upon treatment with pepsin at 400 MPa, all of the intact protein was removed in minutes, leading to the production of hydrolysates with lower antigenicity than those produced in hours at atmospheric pressure (López-Expósito et al., 2008). It was found that the exposure of new target residues only partially facilitated the removal of allergenic epitopes. The peptides disappeared at later stages of proteolysis, although reactivity toward IgG and IgE was not completely abolished (López-Expósito et al., 2008).

López-Expósito et al. (2008) stated that some fragments identified in the hydrolysates including Leu124-Phe134, Ile178-Ala187, Leu242-Leu252, Gly251-Ile259, Lys322-Gly343, Phe358-Phe366, and Phe378-Pro385 carried previously identified IgE-binding epitopes due to some of the peptides found, such as Phe358-Phe366, probably contain only one binding site for IgE (López-Expósito et al., 2008).

14.6.4 Peanut Protein Allergens By HP Processing

Johnson et al. (2010) examined the effects of HP/temperature treatment and pulsed electric field treatment on native peanut Ara h 2, 6 and apple Mal d 3 and Mal d 1b prepared by heterologous expression. It was stated that alterations in secondary structure and the aggregation state of treated proteins were characterized by circular dichroism spectroscopy and gel-filtration chromatography. It was concluded that HP/temperature at 20°C did not change the structure of the Ara h 2, 6 or Mal d 3 and resulted in only minor changes in structure of Mal d 1b. Ara h 2, 6 was stable to HPP at 80°C, whereas changes in circular dichroism spectra were observed for both apple allergens (Johnson et al., 2010). In the study reported by Johnson et al. (2010), these alterations were attributable to aggregation and adiabatic heating during HPP. An ELISA assay of temperature-treated Mal d3 showed the antibody reactivity correlated well with the loss of structure (Johnson et al., 2010). Johnson et al. (2010) concluding that novel-processing techniques had little effect on purified allergen structure. It is also stated that further studies will demonstrate if these stability properties are retained in food matrices (Johnson et al., 2010).

14.6.5 Milk Protein Allergens By HP Processing

It was stated that cow's milk protein allergy (CMPA) was the most prevalent allergy for infants or young children, with an incidence of about 2% to 7.5% in population-based studies in different countries (Fiocchi et al., 2010). It was reported that β-lactoglobulin (β-LG), α-lactalbumin (α-LA), and caseins are the main allergens in cow's milk; other proteins, such as bovine serum albumin (BSA) and even lactoferrin (present in trace amounts) are also potential allergens (Fritsche, 2003; Sharma et al., 2001). It is known that the differences between cow's milk protein and human milk protein composition may be one of the reasons that induce cow's milk allergy in infants (Fox and McSweeney, 1998).

Chicón et al. (2006a, 2006b) stated that β-lactoglobulin can be efficiently hydrolyzed by various enzymes under high pressure (Chicón et al. 2006a, 2006b).

It was shown that the hydrolysates obtained via the enzymatic treatment of β-LG under high pressure exhibited reduced antigenicity and IgE binding (Bonomi et al. 2003; Peñas et al. 2006). It was reported that HP treatment (100–300 MPa) enhanced dairy whey protein hydrolysis and reduced the residual antigenicity of the hydrolysates, depending upon the choice of enzymes including trypsin, chymotrypsin, and pepsin (Peñas et al., 2006).

Bu et al. (2013) stated that a combination of different technologies including heat treatment, glycation reaction, HP processing, enzymatic hydrolysis, lactic acid fermentation may be crucial for reducing cow's milk allergy. It was also stated that the conformation of proteins was effected by high pressure (HP) processing and HP made enzymatic digestion easier; the composed structural changes in milk proteins by HHP influenced the allergenic potential of milk proteins (Bu et al., 2013).

López-Expósito et al. (2012) stated that the major milk allergen β-lactoglobulin (β-LG) exhibited an enhanced susceptibility to proteolysis under HHP. It was stated that the method used by López-Expósito et al. (2012) was efficient to produce hypoallergenic hydrolysates. It was evaluated *in vivo* allergenicity of three β-LG hydrolysates produced under atmospheric pressure or HP conditions. A complete removal of intact β-LG was provided but differed *in vitro* IgE-binding properties that could be traced to the peptide pattern. López-Expósito et al. (2012) reported that the ability to trigger systemic anaphylaxis was assessed using C3H/HeJ mice orally sensitized to β-LG. In the study described by López-Expósito et al. (2012), mast cell degranulation *in vivo* was assessed by passive cutaneous anaphylaxis. It was demonstrated that the peptides presented in the hydrolysates had lost their ability to cross-link two human IgE antibodies to induce mast cell degranulation, thus indicating that most of the peptides formed retain just one relevant IgE-binding epitope (López-Expósito et al., 2012). López-Expósito et al. (2012) concluded that the orally sensitized mouse model was a useful tool to address the *in vivo* allergenicity of novel milk formulas and demonstrated the safety of hydrolysates produced under HP conditions (López-Expósito et al., 2012).

14.6.6 Apple and Celeriac Tissue Protein Allergens By HP Processing

It is known that apple is one of the most common allergenic foods in Europe and contains two predominant allergens, Mal d1 and Mal d3. Apple Mal d3 is a non-specific lipid transfer protein, a member of the prolamin structural family (Jenkins et al., 2005) and is a problematic allergen in Southern European populations (Fernandez-Rivas, 2006). It is stated that another allergen, Mal d 1, is a member of the PR-10 structural family. It is allergenic due to cross-reactivity with another member of the same structural family, Bet v 1 from Birch pollen. It is also revealed that Mal d1 is a major allergen in Northern European populations and is implicated in pollen-associated fruit allergy (Fernandez-Rivas, 2006).

It is known that the celery root or celeriac is widely consumed in central Europe whereas UK consumers mainly prefer celery stem. In many central Europe areas, the celery root or celeriac is consumed in processed form (such as celery salt) that is made from a combination of salt and either celery seeds or extract of celeriac (Husband et al., 2011).

It was shown that HP processing at 600 MPa at 20°C significantly reduced the polyclonal reactivity of Api g1. It was reported that *in vivo* studies involving food challenges in celeriac allergic patients have shown that the Api g1 as PR 10 allergen in celeriac was very much reduced in its allergenicity by thermal processing (110°C/15 min) (Ballmer-Weber et al., 2002).

In the study descibed by Husband et al. (2011), the impact of thermal and HP processing on the immunoreactivity of the allergens Mal d1, Mal d3 and Api g 1 has been investigated in apple and celeriac tissue. It was stated that the extracted proteins were assessed using SDS-PAGE and Western blot. The obtained data showed that Mal d1 was subjected to loss of immunoreactivity as soon as the apple tissue was disrupted although it was remarkably resistant to both thermal and HP processing. It was found that the major allergen Mal d3 in apple was found to be resistant to chemical modification and thermal processing in apple, which is in contrast to the behavior in solution. It was also found that Api g1 was susceptible to thermal processing and the pressure and temperature combination significantly decreased the allergic potential of Api g1 in celeriac (Husband et al., 2011).

Pectin was found to protect Mal d3 from thermal denaturation in solution and is a possible candidate for the protective effect of the fruit (Husband et al., 2011). It was concluded that the combination of HP and thermal processing is an effective method to reduce the allergenicity of both apple and celeriac (Husband et al., 2011).

14.6.7   RICE PROTEIN ALLERGENS BY HP PROCESSING

Rice is a cereal grain belonging to the family of *Gramineae* and it is known that cereals such as wheat, barley, rye, oats, maize, and rice are reported to elicit allergic reactions (Anand, 1978).

Rice is the main and most important food taken every day in Eastern Asia. The prevalence of IgE-mediated rice allergy is about 10% in atopic subjects in Japan, whereas the frequency of rice allergic reactions is much lower in Europe and in the US (Besler et al., 2001).

Izumi et al. (1999) stated that the 16-kDa rice allergen, RA17, belonging to the alpha-amylase/trypsin inhibitor family was isolated from rice seed and structurally characterized by identifying cystine-containing peptides and predicting the secondary structure and hydrophobic regions (Izumi et al., 1999).

It is known that rice allergy is more prominent in adults than in children, and the symptoms associated with rice allergy are atopic dermatitis, eczema, or asthma, and also anaphylactic reactions for severe cases (Besler et al., 2001).

Kato et al. (2000) stated that protein is released from rice grains during HP treatment. In the study described by Kato et al. (2000), polished rice grains were immersed in distilled water and pressurized at 100–400 MPa, so a considerable amount of proteins (0.2–0.5 mg per gram of grains) were released. The released proteins were identified as 16 kDa albumin, alpha-globulin, and 33 kDa globulin, which were known as major rice allergens after sodium dodecyl sulfate-polyacrylamide gel electrophoresis and immunoblot analyses (Kato et al., 2000). A partial morphological alteration was detected in endosperm cells while no apparent structural changes in protein bodies was found by scanning electron microscopic observation of rice grains pressurized at 300 MPa (Kato et al., 2000).

It was concluded that these allergenic proteins content decreased by pressurization and almost completely disappeared from rice grains by the pressurization in the presence of proteolytic enzyme (Kato et al., 2000). The data obtained by Kato et al. (2000) suggested that partial destruction of endosperm cells caused by pressurization enhanced permeability of a surrounding solution into rice grains and the part of the proteins was solubilized and subsequently released into a surrounding solution (Kato et al., 2000).

As a conclusion, HHP processing improved the reducing of allergenic structure and allergenicity of some foods. Recently, limited studies have been performed on HHP effects on the structure of known allergens and the elimination of allergen compounds in foods. Further studies are needed for some allergenic proteins in various food matrices.

REFERENCES

Anand B.S., Piris J., Truelove S.C. 1978. The role of various cereals in coeliac disease. *Q J Med* 47(185), 101–110.

Anonymous 2013a. Food sensitivities. http://www.whfoods.com/genpage.php?tname = faq& dbid = 30

Anonymous 2013b. Food allergy: Causes, symptoms, treatment, tests and diagnosis. http:// www.foodreactions.co.uk/allergy

Anonymous 2013c. Food allergy-general facts. http://www.ifr.ac.uk/protall/infosheet.htm

Ballmer-Weber B.K., Hoffmann A., Wuthrich B., Luttkopf D. et al. 2002. Influence of food processing on the allergenicity of celery: DBPCFC with celery spice and cooked celery in patients with celery allergy. *Allergy* 57, 228–235.

Bernhisel-Broadbent J., Dintzis H.M., Dintzis R.Z., Sampson H.A. 1994. Allergenicity and antigenicity of chicken egg ovomucoid (Gal d III) compared with ovalbumin (Gal d I) in children with egg allergy and in mice. *J Allergy Clin Immunol* 93(6), 1047–1059.

Besler M., Steinhart H., Paschke A. 2001. *J Chromatogr B* 756, 207–228.

Besler M., Tanabe S., Urisu A. 2001. Rice (*Oryza sativa*). In *Food Allergens*. Internet symposium on food allergens 3(Suppl. 2): 1–17. http://www.food-allergens.de

Bonomi F., Fiocchi A., Frøkiaer H., Gaiaschi A., Iametti S., Poiesi C., Rasmussen P., Restani P., Rovere P. 2003. Reduction of immunoreactivity of bovine beta-lactoglobulin upon combined physical and proteolytic treatment. *J Dairy Res* 70, 51–59.

Bozoğlu F.T. and Tokuşoğlu Ö. 2014. Food sensitivity and allergy. In *Food Toxicology: Toxicological Compounds in Foods and Beverages* by Tokuşoğlu Ö. Sidas Publishing, Turkey (In Turkish). In Press.

Bozoğlu F.T. 2011. Food allergies, intolerances and food-borne intoxications. Chapter 9. In *Strategies for Achieving Food Security in Central Asia*. NATO Science for Peace and Security Series-C. Environmental Security. Editors: Alpas H., Smith M., Kulmyrzaev A. Springer, Dordrecht, The Netherlands. ISBN 978-94-007-2504-1.

Brill H. 2008. Approach to milk protein allergy in infants. *Can Fam Physician* 54(9), 1258–1264.

Broadley K.J., Anwar M.A., Herbert A.A. et al. 2009. Effects of dietary amines on the gut and its vasculature. *Br J Nutr* 101(11), 1645–1652.

Bu G., Luo Y., Chen F., Liu K., Zhu T. 2013. Milk processing as a tool to reduce cow's milk allergenicity: A mini-review. *Dairy Sci Technol* 93, 211–223.

Calhoun K.H., Schofield M.L. 2010. IgE mediated food allergy. *Curr Opin Otolaryngol Head Neck Surg* 18(3), 182–186.

Carroccio A., Montalto G., Cavera G., Notarbatolo A. 1998. Lactose intolerance and self reported milk intolerance: Relationship with lactose maldigestion and nutrient intake. Lactase Deficiency Study Group. *J Am Coll Nutr* 17(6), 631–636.

CFIA. 2009. Seafood (Fish, Crustaceans and Shellfish)—One of the nine most common food allergens. Canadian Food Inspection Agency. 2009-06-12. http://www.inspection.gc.ca/ english/fssa/labeti/allerg/fispoie.shtml

Chicón R., Belloque J., Recio I., López-Fandiño R. 2006a. Influence of high hydrostatic pressure on the proteolysis pattern of beta-lactoglobulin a treated with trypsin. *J Dairy Res* 73, 121–128.

Chicón R., López-Fandiño R., Quirós A., Belloque J. 2006b. Changes of chymotrypsin hydrolysis of betalactoblobulin induced by high hydrostatic pressure. *J Agric Food Chem* 54, 2333–2341.

Clare Mills E.N., Sancho A.I., Rigby N., Jenkins J.A., Mackie A.R. 2009. Impact of food processing on the structural and allergenic properties of food allergens. *Mol Nutr Food Res* 53(8), 963–969.

Cordle C.T. 2004. Soy protein allergy: Incidence and relative severity. *J Nutr* 134(5), 1213–1219.

Dupont D., Mandalari G., Mollé D., Jardin J., Rolet-Répécaud O., Duboz G., Léonil J., Mills C.E., Mackie A.R. 2010. Food processing increases casein resistance to simulated infant digestion. *Mol Nutr Food Res* 54(11), 1677–1689.

Fernandez-Rivas M. 2006. Cross-reacting allergens at the molecular scaled a north-south comparison. *Revue Francaise D Allergologie et D Immunol Clin* 2006, 46, 167–169.

Figueroa J., Blanco C., Dumpiérrez A.G., Almeida L., Ortega N., Castillo R., Navarro L., Pérez E., Gallego M.D., Carrillo T. 2005. Mustard allergy confirmed by double-blind placebo-controlled food challenges: Clinical features and cross-reactivity with mugwort pollen and plant-derived foods. *Allergy* 60(1), 48–55.

Fiocchi A., Brozek J., Schünemann H., Bahna S.L., von Berg A., Beyer K., Bozzola M., Bradsher J., Compalati E., Ebisawa M., Guzman M.A., Li H., Heine R.G., Keith P., Lack G., Landi M., Martelli A., Rancé F., Sampson H., Stein A., Terracciano L., Vieths S. 2010. World allergy organization (WAO) diagnosis and rationale for action against cow's milk allergy (DRACMA) guidelines. *Pediatr Allergy Immunol* 21, 1–125.

Fox P.F., McSweeney P.L.H. 1998. *Dairy chemistry and biochemistry*. Blackie Academic and Professional, London.

Frank J.E. 2005. Diagnosis and management of G6PD deficiency. *Am Fam Physician* 72(7), 1277–1282.

Fritsche R. 2003. Role for technology in dairy allergy. *Aust J Dairy Technol* 58, 89–91.

Garsed K., Scott B.B. 2007 Can oats be taken in a gluten-free diet? A systematic review. *Scand J Gastroenterol* 42(2), 171–178.

Genuis S.J. 2010. Sensitivity related illness: The escalating pandemic of allergy, food intolerance and chemical sensitivity. *Sci Total Environ* 408(24), 6047–6061.

Hajós G, Polgár M, Farkas J. 2004. High pressure effects on IgE immunoreactivity of proteins in a sausage batter. *Innov Food Sci Emerg Technol* 5(4), 443–449.

Han G.D., Matsuno M, Ikeuchi Y., Suzuki A. 2002. Effects of heat and high-pressure treatments on antigenicity of beef extract. *Biosci Biotechnol Biochem* 66(1), 202–205.

Han G.D., Matsuno M., Ito G., Ikeucht Y., Suzuki A. 2000. Meat allergy: Investigation of potential allergenic proteins in beef. *Biosci Biotechnol Biochem* 64, 1887–1895.

Hansen K.S., Ballmer-Weber B.K., Lüttkopf D., Skov P.S., Wüthrich B., Bindslev-Jensen C., Vieths S., Poulsen L.K. 2003. Roasted hazelnuts allergenic activity evaluated by double-blind, placebo controlled food challenge. *Allergy* 58(2), 132–138.

Høst A., Halken S., Muraro A. et al. 2008. Dietary prevention of allergic diseases in infants and small children. *Pediatr Allergy Immunol* 19(1), 1–4.

Høst A. 2002. Frequency of cow's milk allergy in childhood. *Ann Allergy Asthma Immunol* 89(6 Suppl 1), 33–37.

Husband F.A., Aldick T., Vander Plancken I., Grauwet T., Hendrickx M., Skypala I., Mackie A.R. 2011. High-pressure treatment reduces the immunoreactivity of the major allergens in apple and celeriac. *Mol Nutr Food Res* 55(7), 1087–95.

Izumi H, Sugiyama M, Matsuda T, Nakamura R. 1999. Structural characterization of the 16-kDa allergen, RA17, in rice seeds. Prediction of the secondary structure and identification of intramolecular disulfide bridges. *Biosci Biotechnol Biochem* 63(12), 2059–2063.

Jenkins J.A., Griffiths-Jones S., Shewry P.R., Breiteneder H., Mills E.N.C. 2005. Structural relatedness of plant food allergens with specific reference to cross reactive allergens: An in silico analysis. *J Allergy Clin Immunol* 2005, 115, 163–170.

Johnson P.E., Vander Plancken I., Balasa A., Husband F.A., Grauwet T., Hendrickx M., Knorr D., Mills E.N., Mackie A.R. 2010. High pressure, thermal and pulsed electric-field-induced structural changes in selected food allergens. *Mol Nutr Food Res* 54(12), 1701–1710.

Johansson S.G., Hourihane J.O., Bousquet J. et al. 2001. A revised nomenclature for allergy. An EAACI position statement from the EAACI nomenclature task force. *Allergy* 56(9), 813–824.

Kato T., Katayama E., Matsubara S., Omi Y., Matsuda T. 2000. *J Agric Food Chem* 48(8), 3124–3129.

Kao S.H., Hsu C.H., Su S.N. et al. 2004. Identification and immunologic characterization of an allergen, alliin lyase from garlic (Allium sativum). *J Allergy Clin Immunol* 113(1), 161–168.

Kaukinen K., Turjanmaa K., Mäki M. et al. 2000. Intolerance to cereals is not specific for coeliac disease. *Scand J Gastroenterol* 35(9), 942–946.

Li X. 2007. Traditional Chinese herbal remedies for asthma and food allergy. *J Allergy Clin Immunol* 120, 25.

Li S.J. 2005. Structural details at active site of hen egg white lysozyme with di- and trivalent metal ions. *Biopolymers* 81(2), 74–80.

López-Expósito I, Chicón R, Belloque J, Recio I, Alonso E, López-Fandiño R. 2008. Changes in the ovalbumin proteolysis profile by high pressure and its effect on IgG and IgE binding. *J Agric Food Chem* 6(24), 11809–11816.

López E., Cuadrado C., Burbano C., Jiménez M.A., Rodríguez J., Crespo J.F. 2012. Effects of autoclaving and high pressure on allergenicity of hazelnut proteins. *J Clin Bioinfor* 2(1), 12. doi: 10.1186/2043-9113-2-12.

López-Expósito I., Chicón R., Belloque J., López-Fandiño R., Berin M.C. 2012. *In vivo* methods for testing allergenicity show that high hydrostatic pressure hydrolysates of β-lactoglobulin are immunologically inert. *J Dairy Sci* 95(2), 541–548.

Martelli A., Chiara A.D., Corvo M., Restani P., Fiocchi A. 2002. *Ann Allergy Asthma Immunol* 89, 38–43.

Mills E.N., Sancho A.I., Rigby N.M., Jenkins J.A., Mackie A.R. 2009. Impact of food processing on the structural and allergenic properties of food allergens. *Mol Nutr Food Res* 53(8), 963–969.

Montalto M., Curigliano V., Santoro L. et al. 2006. Management and treatment of lactose malabsorption. *World J Gastroenterol* 12(2), 187–191.

Moreau S.A., Abalo C., Molle D., Le Graet Y., Brule G. et al. 1995. Hen egg white lysozymemetal ion interactions: Investigation by electrospray ionization mass spectrometry. *J Agric Food Chem* 43(4), 883–889.

Nelson D.A. 2002. Gluten-sensitive enteropath (celiac disease): More common than you think. *Am Fam Physician* 66(12), 2259–2266.

NIAID. 2005. National Institutes of Health, NIAID Allergy Statistics. http://www.niaid.nih. Gov/factsheets/allergystat.htm

Nogami N., Matsuno M., Hara T., Joh T., Nishiumi T., Suzuki A. 2006. *Rev High Pressure Sci Technol* 16, 11–16.

Odani S., Kanda Y., Hara T., Matsuno M., Suzuki A. 2007. *Proc 4th Int Conf High Pressure Biosci Biotech* 1, 252–258.

Peñas E., Préstamo G., Baeza M.L., Martínez-Molero M.I., Gomez R. 2006. Effects of combined high pressure and enzymatic treatments on the hydrolysis and immunoreactivity of dairy whey proteins. *Int Dairy J* 16, 831–839.

Sasagawa A., Yamazaki A. 2002. In *Trends in High-Pressure Bioscience and Biotechnology* Hayashi, R. (ed.), Elsevier, Amsterdam 2002, pp. 375–384.

Sathe S.K., Teuber S.S., Roux K.H. 2005. Effects of food processing on the stability of food allergens. *Biotechnol Adv* 23(6), 423–429.

Sharma S., Kumar P., Betzel C., Singh T.P. 2001. Structure and function of proteins involved in milk allergies. *J Chromatogr B* 756, 183–187.

Simonin H., Duranton F., de Lamballerie M. 2012. New insights into the high-pressure processing of meat and meat products. *Comprehensive Rev Food Sci Food Safety* 11, 285–306.

Skripak J.M. 2008. Educational clinical case series: Peanut and tree nut allergy in childhood. *Pediatr Allergy Immunol* 19, 368.

Størsrud S., Hulthén L.R., Lenner R.A. 2003. Beneficial effects of oats in the gluten-free diet of adults with special reference to nutrient status, symptoms and subjective experiences. *Br J Nutr* 90(1), 101–107.

Venter C., Skypala I. 2009. Food hypersensitivity: Diagnosing and managing food allergies and intolerance. Wiley-Blackwell, Chichester, pp. 129–131.

Watanabe M., Miyakawa J., Ikezawa Z., Suzuki Y., Hirano T., Yoshizawa T., Arai S. 1990. *J Food Sci* 55, 781–783.

Workman E.M., Alun Jones V., Wilson A.J., Hunter J.O. 1984. Diet in the management of Crohn's disease. *Hum Nutr Appl Nutr* 38(6), 469–473.

Yamamoto S., Mikami N., Matsuno M., Hara T., Odani S., Suzuki A., Nishiumi T. 2010. Effects of a high-pressure treatment on bovine gamma globulin and its reduction in allergenicity. *Biosci Biotechnol Biochem* 74(3), 525–30.

Part III

Improving Food Quality with Pulsed Electric Field Technologies

15 Effects of Pulsed Electric Field Processing on Microbial Quality, Enzymatic, and Physical Properties of Milk

Kambiz Shamsi

CONTENTS

15.1 INTRODUCTION

Milk is a sterile and pathogen-free food when still in the cow's udder. The contamination occurs as soon as it comes in contact with utensils, milking apparatus, handling trucks, and processing systems, making it a perishable product. High-temperature short-time (HTST) pasteurization has been effectively used for decades as a method of choice to extend the shelf life of milk through inactivating the spoilage bacteria. Various time and temperature protocols are used to reduce the number of pathogens while incurring less damage to organoleptic and nutritional properties of milk. The most common organoleptic change in pasteurized milk is the generation of cooked flavor that distinguishes fresh milk from processed milk. This has given way to the emerging of novel nonthermal technologies such as pulsed electric field (PEF) and high hydrostatic pressure (HHP) as a potential alternative to traditional thermal pasteurization of milk with the advantage of minimizing sensory and nutritional damage, thus providing fresh-like products. However, more investigation is required to understand the nature of PEF effects on milk properties to achieve a controlled rate of enzymatic and higher microbial inactivation to make PEF technology applicable in the dairy industry.

During the past two decades, there have been numerous studies on the effects of PEF treatment on bovine milk and dairy products. However, there is a gap in knowledge of the effects of PEF treatment on native milk microflora and enzymes as well as on the functionality of milk fat and proteins.

In this chapter, the effects of PEF processing on microbial quality, enzymatic, and physical properties of milk will be discussed.

15.2   A BRIEF HISTORY OF PEF TECHNOLOGY

The application of electricity for processing milk was introduced in the early 1900s, using a process known as the electro-pure method. However, this method was, in fact, a thermal process as the milk was heated up by ohmic resistance. The treatment voltages applied to milk ranged from 220 to 4200 V and only those researchers who had applied high voltages reported the ability of the process to kill the bacteria "below their thermal death point" (Beattie 1915; Beattie and Lewis 1916, 1925).

In the late 1940s, electric fields were used by Flaumenbaun to increase the permeability of fruits in order to facilitate the subsequent extraction of juice, which is currently an important application of PEF technology (Heinz and Knorr 2001). In 1960, Doevenspeck in Germany patented a PEF equipment piece and during the same period, Sale and Hamilton (1967) and Hamilton and Sale (1967) published a short series of articles on microbial inactivation by PEF.

In the field of genetic engineering, a method was developed by Zimmermann et al. (1974) to promote *in-vitro* cell-to-cell fusion using PEF that increased the permeability in localized zones of the membrane, currently known as the reversible electrical breakdown, electro-permeabilization, or electroporation. Hülsheger et al. (1981 and 1983) and Hülsheger and Niemann (1980) published a series of papers discussing the sensitivity of various bacteria to PEF and also mathematically simulated the effect of the electric field intensity and treatment time on microbial kills.

Up to the late 1980s, many researchers continued to develop PEF applications for preservation of foods and several patents were filed. By 1992, PEF was recognized as a nonthermal preservation technology, a novel method that was able to provide consumers with fresh-like foods and valuable sensory properties (Grahl et al. 1992; Jayaram et al. 1992).

The PEF process by some researchers is considered to be more energy efficient than thermal pasteurization; however, it is still a matter of controversy. The microbial inactivation is achieved at ambient or moderately elevated temperatures by the application of short bursts of high-intensity electric fields to liquid foods flowing between two electrodes. Liquid foods (e.g., milk or fruit juices), due to their chemical composition and physical properties and in the presence of electrical charge carriers, may have different conductivities resulting in various flux of electrical current through the food (Barbosa-Canovas et al. 2000; Zhang et al. 1995, 1996).

Currently, there are various research groups in Australia, Belgium, Canada, China, France, Germany, Iceland, Japan, the Netherlands, New Zealand, Scotland,

Spain, Sweden, Switzerland, Taiwan, the United Kingdom, and the United States working on different stages of PEF applications. In 1995, the CoolPure[R] PEF process developed by PurePulse Technologies (4241, Ponderosa Ave., San Diego, CA, 92123, USA) was approved by the FDA (Food and Drug Administration) for treatment of pumpable foods. It was the first regulatory effort to introduce PEF process in the food industry (FDA 2000).

15.3   INACTIVATING MICROFLORA AND ALKALINE PHOSPHATASE IN BOVINE MILK BY PEF

The study on the effects of PEF on inactivation of alkaline phosphatase (AlP) in milk to date has been limited to only a few researchers who have determined the AlP activity in spiked simulated milk ultrafiltrate (SMUF) or milk after treatment with exponentially decaying pulses with the aim of observing the effects of field intensity on the structure of enzyme and inactivation level. The indigenous AlP is technologically significant in the dairy industry since it is used as an indicator of milk pasteurization adequacy. However, no study to date has investigated the possibility of using AlP as an indicator of PEF treatment adequacy.

In a recent study by Shamsi (2010), the effects of PEF treatment in combination with mild heat on microflora and native enzymes in raw bovine milk as well as its effects on rheological and textural properties of rennet-induced gels of milk were investigated using an OSU-4 PEF unit with four treatment chambers and a PEF unit in the Technical University of Berlin, Germany with two treatment chambers capable of generating bipolar and monopolar square pulses with maximum field intensity of 38 and 50 kV cm^{-1}, respectively.

In the first part of the study, the effects of PEF treatments at field intensities of 25–37 kV cm^{-1}, final outlet temperatures of 30°C and 60°C and treatment time of 19.2 μs on the inactivation of AlP, total plate count (TPC), *Pseudomonas,* and *Enterobacteriaceae* counts were determined in skim milk. For the outlet temperatures of 30°C and 60°C, the inlet temperatures were maintained at 20°C and 50°C, respectively. Table 15.1 indicates the experimental conditions of PEF.

At 30°C, PEF treatments of 28, 31, 34, and 37 kV cm^{-1} inactivated AlP by 24%, 25%, 31%, and 42% and <1 log reduction was observed in TPC and *Pseudomonas* count, while the *Enterobacteriaceae* count was reduced by at least 2.1 logs below the detection limit of 1 colony-forming units (CFUs) mL^{-1}. At 60°C, PEF treatments of 25, 29, 31, and 35 kV cm^{-1} resulted in 29%, 42%, 56%, and 67% inactivation of AlP and up to 2.4 logs reduction in TPC, while the *Pseudomonas* and *Enterobacteriaceae* counts were reduced by at least 5.9 and 2.1 logs, respectively, to below the detection limit of 1 CFU mL^{-1}. A combined effect was observed between the field intensity and heat in the inactivation of both AlP enzyme and the natural microflora in skim milk. Enzymes are stabilized by weak noncovalent forces, such as hydrogen bonds and hydrophobic interactions, and the application of high electric field pulses may have affected the three-dimensional structure of the globular protein in AlP (Ho et al. 1997). Castro (1994) related the PEF inactivation of AlP to degradation of the enzyme's secondary structure and alteration of the entire globular configuration of AlP.

TABLE 15.1
PEF Treatment Conditions of Skim Milk

Outlet T(°C)	Input Voltage(kV)	Field Intensity (kV cm^{-1})	Current(A)	Energy Input (kJ L^{-1})
	6.5	28	20	65
	7.5	31	24	87
30	8.5	34	28	112
	9.5	37	36	139
	6.5	25	24	65
	7.5	29	28	87
60	8.5	31	36	112
	9.5	35	44	139

Note: The pulse frequency and flow rate were maintained at 200 Hz and 60 mL min^{-1} giving a fixed treatment time of 19.2 μs. The electrical conductivity of skim milk at 30°C and 60°C was 4.53 and 5.11 mS cm^{-1}, respectively.

The efficacy of microbial inactivation by PEF is affected by a number of factors including field intensity, pulse characteristics, temperature, and the presence of different types of microflora. Raw milk contains a relatively wide range of microorganisms including Gram-positive and Gram-negative bacteria, yeasts and molds, and a range of bacterial spores. Typically, some of the Gram-positive thermoduric bacteria and spores will survive pasteurization and PEF processes, and be enumerated in TPC of the treated product.

There was a significant difference ($p < 0.05$) between the inactivation levels of TPC at 30°C (Figure 15.1a) and 60°C (Figure 15.1b); however, this difference was mostly due to the effect of heat rather than field intensity. Therefore, the resistance to PEF treatment at 30°C may be due to the survival of Gram-positive bacteria. The *Pseudomonas* population decreased with an increase in the field intensity and treatment temperature. A combined treatment of 60°C and 35 kV cm^{-1} resulted in >5.9 logs reduction. The additional 1–3.5 logs reduction on the native *Pseudomonas* number in raw skim milk resulted from the PEF treatment at 60°C compared with that at 30°C that demonstrated the enhanced microbial kill of PEF at mild heating conditions. When the PEF was switched off, a slight inactivation was achieved (only 0.2 log) by heating the milk to 55°C, indicating the efficiency of PEF in inactivation process.

Enterobacteriaceae was more inactivated by PEF and heat than TPC and *Pseudomonas* since it was totally inactivated at both 30°C and 60°C (Figure 15.1). However, compared to *Pseudomonas* (with initial count of 5.9 logs CFU mL^{-1}), the initial number of *Enterobacteriaceae* was nearly 4 logs lower that contributed to its total elimination.

Regression analysis of the experimental data (Figure 15.2) obtained from the Hülsheger kinetic model showed a good correlation ($R^2 = 0.897$ at 30°C, $R^2 = 0.0.9701$ at 60°C) between the natural log of the remaining fraction of AlP and

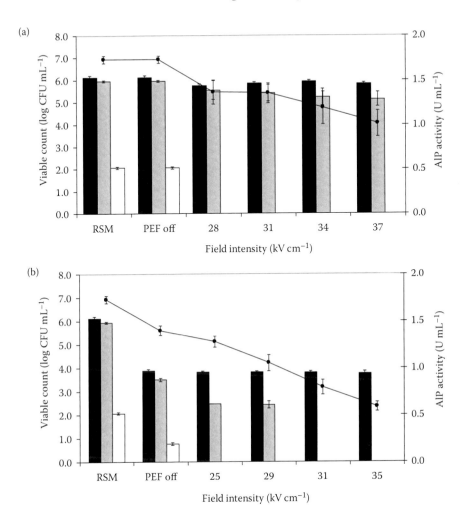

FIGURE 15.1 Effects of pulsed electric field with final temperature of 30°C (a) and 60°C (b) on microbial flora and native AlP activity in raw skim milk at different field intensities (kV cm⁻¹). PEF conditions: frequency of 200 Hz, monopolar pulses of 2 μs with total treatment time of 19.6 μs; RSM: raw skim milk, ■ total plate count, ▨ *Pseudomonas*, □ *Enterobacteriaceae*, and ● alkaline phosphatase.

field intensity at both temperatures. The high R^2 values indicated that the first-order inactivation model is valid for describing the inactivation of AlP by PEF within the range of applied field intensities. The proportionality constant (b_E) was 0.0351 for 30°C and 0.0808 for 60°C, respectively, which suggests that increasing the temperature has an enhancing effect on enzyme inactivation, as demonstrated by a 2.3× increase in the slope of line (b_E value) when the PEF treatment temperature increased from 30°C to 60°C.

As comparisons, raw skim milk was thermally pasteurized by HTST and low-temperature long-time (LTLT) methods, and up to 98% of the native AlP was inactivated

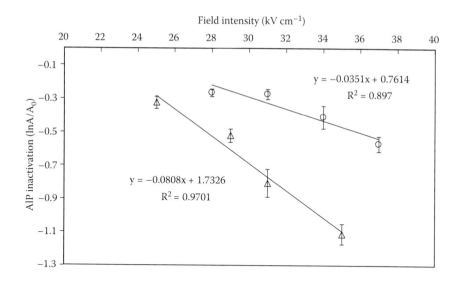

FIGURE 15.2   The correlation between various PEF field intensities and AlP inactivation in skim milk at 30°C (O) and 60°C (Δ).

in both pasteurization methods. These results indicated that under these experiment conditions, AlP cannot be used as an indicator of PEF treatment adequacy as the maximum inactivation achieved by PEF treatment at 60°C was only 67% versus 98% inactivation achieved with pasteurization. On the basis of the IDF definition of pasteurized milk, the phosphatase test of an HTST-treated milk should be negative, that is, no yellow color should be released into the assay (IDF 1984).

15.4   COMBINED EFFECTS OF FIELD INTENSITY AND HEAT ON MILK ENZYMES INACTIVATION

The question still remains as to what extent heat affects the inactivation of enzymes and microorganisms. Although PEF is considered a nonthermal process in which the electric field itself is assumed as the cause of enzymatic or microbial inactivation, the heat generated by applied field intensity has an impact on the efficiency of the process. It is important to note that PEF is referred to as a nonthermal technology due to the alternative origin of the inactivating parameters (i.e., field intensity or treatment time), and not for the complete absence of thermal effects. In fact, after the electric field intensity and treatment time, processing temperature is one of the most relevant processing parameters in PEF technology. Thus, the exposure time of the product to heat should be kept to a minimum before and after the liquid enters the treatment chamber as well as inside the treatment chamber to minimize the heat damage to the product. Figure 15.3 shows a schematic diagram of the PEF unit used in this study.

The partial conversion of electric pulses into heat causes an increase in milk temperature that if high enough can contribute to thermal inactivation of enzymes. The thermal enhancement of PEF treatments can be considered as an additive

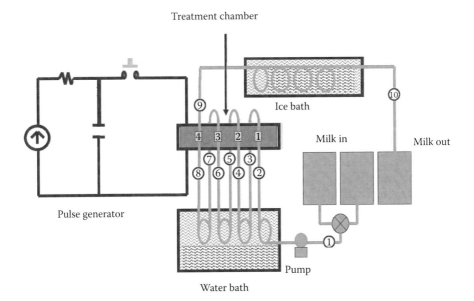

FIGURE 15.3 Schematic diagram of the modified OSU-4 PEF unit with numbers 1–10 representing the 10 in-line thermocouples referred to as T1 to T10. From T1 to T9, the exposure time of milk to heat was 100 s before it enters the ice bath. (From Sui, Q. et al. 2008. Differentiation of temperature from pulsed electric field effects on lactoperoxidase activity using thermal inactivation kinetics. Poster presented at the *4th Innovative Foods Centre Conference*, Brisbane, Australia, September 17–18.)

within the boundaries of moderately low and high temperatures (30–60°C for few seconds). After this limit is surpassed, it is difficult to differentiate between the thermal and nonthermal effects of the process and the thermal effect may dominate the process.

Many indigenous enzymes in milk including AlP, lipase, xanthine oxidase (XO), and plasmin are technologically significant because of their roles in tracing the thermal history of milk and its shelf life. Microbial lipase and XO in milk play a positive role in cheese ripening while they can also render the milk rancid and unpalatable since both of them survive HTST pasteurization. Plasmin survival in ultra high temperature (UHT) milk accelerates the proteolysis resulting in age gelation (Fox and McSweeney 1998; Girotti et al. 1999). Therefore, from an industrial point of view, it is important to investigate how PEF affects these enzymes in bovine milk under various treatment conditions. However, what complicates the investigation is the simultaneous and synergistic effects of various parameters in PEF such as field intensity, treatment time, total specific pulsing energy, and heat on the inactivation process.

Table 15.2 shows the thermal profile of skim milk treated at various field intensities (29 and 35 kV cm^{-1}) and temperatures (35°C, 45°C, 55°C, 65°C, and 75°C). At field intensity of 29 kV cm^{-1} and with temperatures set to 35°C, 45°C, 55°C, 65°C, and 75°C, AlP was inactivated by 21%, 24%, 46%, 94%, and 97%. Under the same conditions, the inactivation levels of total lipase, XO, and plasmin were 22%, 33%,

TABLE 15.2
Thermal Profile of Skim Milk during PEF Treatment

Field Intensity (kV cm⁻¹)	Temperature Profile (°C)				
	Target[a]	35	45	55	65	75
	Water bath[b]	38	48	58	68	77
	T1	13	14	14	15	15
	T2	32	39	44	55	64
	T3	34	40	45	56	65
0 (non-PEF control)	T4	39	47	54	65	69
	T5	39	47	54	64	71
	T6	40	49	57	66	73
	T7	37	45	53	63	74
	T8	38	47	56	65	75
	T9	38	46	54	65	75
	T10	29	32	34	30	32
	Target	35	45	55	65	75
	Water bath	25	38	49	60	72
	T1	12	12	12	12	13
	T2	20	29	37	45	54
	T3	31	38	47	53	62
	T4	26	37	47	57	65
29	T5	35	45	54	62	67
	T6	27	39	49	59	69
	T7	34	45	56	64	73
	T8	25	37	46	56	74
	T9	35	45	54	64	75
	T10	20	25	27	30	34
	Target	35	45	55	65	75
	Water bath	25	38	49	60	72
	T1	12	12	12	13	13
	T2	20	29	37	47	55
	T3	31	38	46	56	62
	T4	26	37	47	56	67
35	T5	35	45	54	64	70
	T6	27	40	50	60	72
	T7	34	45	55	65	73
	T8	25	37	46	63	74
	T9	33	44	55	65	75
	T10	20	24	25	32	37

[a] Target temperature refers to the final temperature achieved after the fourth treatment chamber that is measured by thermocouple T9.

[b] Water bath temperatures were set to those temperatures to achieve the final treatment temperatures. The thermocouples sensitivity was from ±0.1°C to ±0.5°C; milk temperature fluctuated by ±1°C depending on the flow rate (60 mL min⁻¹ and pulse frequency of 200 Hz) and location of thermocouples (Figure 15.3).

44%, 78%, and 89%; 20%, 23%, 46%, 74%, and 88%; 17%, 19%, 31%, 53%, and 61%, respectively. When the field intensity was increased to 35 kV cm^{-1}, the inactivation levels of AlP, total lipase, XO, and plasmin further increased to 24%, 46%, 53%, 96%, and 99%; 33%, 44%, 56%, 89%, and 99%; 23%, 38%, 49%, 78%, and 89%; and 17%, 22%, 42%, 53%, and 64%, respectively (Table 15.3).

The factorial statistics presented in Table 15.4 demonstrated that once the PEF treatment temperature reached 65°C and beyond, the enzyme inactivation was mostly dominated by heat rather than field intensity. However, the PEF effects cannot be ignored because there was a significant difference ($p < 0.05$) between the activity levels of enzymes in PEF-treated milk and the control (non-PEF-treated) samples at 65°C and 75°C. Compared to the other three enzymes, plasmin showed resistance to both heat and PEF effect. AlP was inactivated by >90% at temperature 65°C at both field intensities. The sensitivity of the enzymes to PEF could be related to their thermostability and size as the molecular weight of plasmin (81 Da) is much less than AlP (170–190 kDa), lipase (40–400 kDa), and XO (300 kDa) (Fox and McSweeney 1998) that makes it more resistant to heat and PEF treatment.

Figure 15.4 shows the correlation between the natural log of relative enzymes activity at 29 or 35 kV cm^{-1} and different treatment temperatures. The proportionality

---

TABLE 15.3

Activity Levels of AlP, Total Lipase, XO, and Plasmin at Different Temperatures with and without PEF Treatment

| | | Residual Enzyme Activity | | | |
Field Intensity (kV cm^{-1})	T (°C)	AlP* (U mL^{-1})	Total Lipase (mU mL^{-1})	XO** (mU mL^{-1})	Plasmin (AMC U mL^{-1})
	35[a]	2.50[b] ± 0.02	0.09[b] ± 0.002	82[b] ± 0.2	0.36[b] ± 0.02
	45	2.40[b] ± 0.03	0.08[b] ± 0.002	79[b] ± 1.1	0.36[b] ± 0.05
0 (non-PEF control)	55	1.87[c] ± 0.04	0.07[bc] ± 0.002	67[c] ± 1.1	0.34[b] ± 0.02
	65	1.27[d] ± 0.05	0.05[c] ± 0.002	48[d] ± 0.2	0.31[b] ± 0.01
	75	0.15[e] ± 0.01	0.02[d] ± 0.002	22[e] ± 0.6	0.26[c] ± 0.01
	35	1.98[c] ± 0.05	0.07[c] ± 0.003	66[c] ± 0.7	0.30[c] ± 0.01
	45	1.90[c] ± 0.12	0.06[c] ± 0.003	63[c] ± 0.5	0.29[c] ± 0.02
29	55	1.35[d] ± 0.03	0.05[cd] ± 0.003	44[d] ± 0.9	0.25[d] ± 0.03
	65	0.14[d] ± 0.05	0.02[e] ± 0.003	21[e] ± 0.5	0.17[e] ± 0.04
	75	0.07[e] ± 0.07	0.01[e] ± 0.003	10[f] ± 0.8	0.14[f] ± 0.01
	35	1.90[c] ± 0.04	0.06[c] ± 0.004	63[c] ± 1.4	0.30[c] ± 0.02
	45	1.35[d] ± 0.02	0.05[c] ± 0.004	51[d] ± 3.1	0.28[c] ± 0.03
35	55	1.17[d] ± 0.07	0.04[cd] ± 0.004	42[e] ± 1.8	0.21[d] ± 0.02
	65	0.10[e] ± 0.07	0.01[e] ± 0.003	18[f] ± 0.9	0.17[e] ± 0.04
	75	0.02[e] ± 0.04	0.00[f] ± 0.002	9.0[f] ± 1.0	0.13[f] ± 0.04

* Alkaline phosphatase, **xanthine oxidase.

[a] Initial activity of all enzymes was measured at 35°C since no inactivation occurs at this temperature. The different letters in each column show a significant difference ($p < 0.05$).

TABLE 15.4

Statistical Matrix Based on Factorial Design Indicating the Interaction of Field Intensity and Temperature on Enzyme Inactivation

Field Intensity (kV cm⁻¹)	T (°C)	AlP*	Total Lipase	XO**	Plasmin
	35	0	0	0	0
	45	0	0	0	0
PEF off (E = 0)	55	×	×	×	×
	65	××	××	××	××
	75	×××	×××	×××	×××
	35	+	+	+	+
	45	×+	×+	×+	×+
29	55	×+	×+	×+	×+
	65	××+	××+	××+	××+
	75	×××+	×××+	×××+	×××+
	35	++	++	++	++
	45	×++	×++	××++	×++
35	55	×++	×++	××++	×++
	65	××++	××++	×××++	××++
	75	×××++	×××++	×××++	×××++

Note: (×) Heat effect; (+) PEF effect; *alkaline phosphatase; and **xanthine oxidase; the residence time of milk in the highest targeted temperature after fourth treatment chamber was 5 s but the total exposure time of milk to a range of temperatures below target temperature was 100 s.

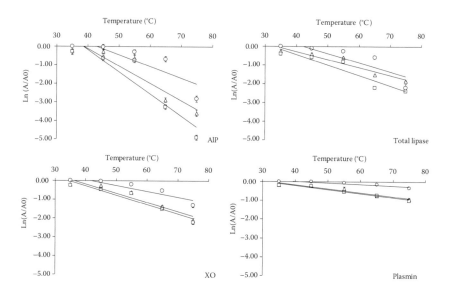

FIGURE 15.4 Correlation between two field intensities and various temperatures for enzyme inactivation in skim milk. The total specific pulsing energy was 163 kJ L⁻¹ for all PEF treatments. ○ PEF off; ▵ 29 kV cm⁻¹; □ 35 kV cm⁻¹.

constant (k_T) for temperature represents the slope of the inactivation curve, with a larger k_T value indicating a greater inactivation efficacy with the increase in the PEF treatment temperature (Table 15.5). The proportionality constant for all enzymes and all temperatures significantly increased ($p < 0.05$) from control to PEF-treated samples that indicate the additive effects of PEF treatment and heat on enzyme inactivation. The k_T of AlP, total lipase, XO, and plasmin, varied significantly with the highest k_T for AlP and lowest for plasmin. The slope of the curve for AlP and XO had a sudden fall from 65°C onward while XO and plasmin had a milder slope that indicates the dominant effect of heat on AlP and total lipase while XO and plasmin were more resistant to both heat and PEF treatment. However, although the enzyme inactivation was dominated by heat at 65–75°C, the role of PEF treatment was also evident on enzyme inactivation since the curves of all enzymes had a steeper slope than those of non-PEF-treated control samples (PEF off, $E = 0$). The inactivation curves in all temperatures were positioned close together, which showed that the difference in field intensities used (29 or 35 kV cm^{-1}) did not result in a significant difference ($p > 0.05$) in enzyme inactivation. The linear relationship between enzyme inactivation, field intensity, and treatment temperature showed a regression coefficient of $R^2 \geq 0.71$, which indicates that the first-order inactivation model was adequate for describing the enzyme inactivation in raw skim milk within the experimental conditions tested (Min et al. 2003; Zar 1999).

The mechanism of enzyme inactivation by PEF has been attributed to the effects of electrical pulses on the secondary and tertiary structures of the enzymes (Castro et al. 2001; Zhong et al. 2007). The effects of electric fields on proteins include the association or dissociation of functional groups, movements of charged chains, and changes in alignment of helices. The factors that mainly influence PEF enzymatic

TABLE 15.5
Proportionality Constant and Regression Coefficients of Enzyme Inactivation by Two Field Intensities as a Function of Treatment Temperatures

Enzyme	Field Intensity (kV cm^{-1})	Regression Coefficients	k_T
	0	$R^2 = 0.71$	0.0616
AlP	29	$R^2 = 0.87$	0.088
	35	$R^2 = 0.84$	0.1231
	0	$R^2 = 0.72$	0.0486
Total lipase	29	$R^2 = 0.87$	0.0644
	35	$R^2 = 0.89$	0.0571
	0	$R^2 = 0.83$	0.0313
XO	29	$R^2 = 0.91$	0.0554
	35	$R^2 = 0.89$	0.0467
	0	$R^2 = 0.85$	0.008
Plasmin	29	$R^2 = 0.96$	0.0205
	35	$R^2 = 0.93$	0.0193

inactivation are (1) electric parameters (e.g., electric field intensity, total treatment time, and pulse width); (2) enzymatic structures (e.g., secondary and tertiary structures); (3) PEF treatment temperatures; and (4) treatment media (Min et al. 2007). It has also been reported that the larger and more complex an enzyme molecule, the more susceptible it is to heat (Yang et al. 2004).

These results indicated that PEF treatment temperatures exceeding 60°C would result in the dominance of heat effect on enzyme inactivation.

15.5 EFFECTS OF PEF ON RHEOLOGICAL AND TEXTURAL PROPERTIES OF RENNET-INDUCED GEL OF BOVINE MILK

Milk gelation is a critical step in cheese and yogurt production processes. The overall visual appearance as well as rheological properties in milk gel's microstructure are important as they contribute to the sensory properties of these products. Milk is coagulated through cleaving of κ-casein by the rennet enzyme and subsequent aggregation of casein micelles (Walstra 1990; Zoon et al. 1988). Heating milk above 70°C even for normal HTST pasteurization purpose hinders rennet action on κ-casein through denaturing whey proteins to varying degrees. The denatured whey proteins interact with κ-casein and adversely affect the coagulation process and gel strength (Beaulieu et al. 1999; Qi et al. 1995).

The effects of PEF on microorganisms and enzymes in liquid foods, that is, milk or juices, are well reported (Barbosa-Canovas and Sepulveda 2005; Calderon-Miranda et al. 1999; Craven et al. 2008; Shamsi 2010; Shamsi and Sherkat 2009). However, there is still a need to investigate the effects of PEF on protein components of foods, such as milk proteins.

Table 15.6 shows PEF treatment conditions for skim and whole milk. Treatment of whole milk even at 50 kV cm^{-1} did not have a significant effect on casein micelles size; however, in the high-heat-treated milk (97°C for 10 min) samples, the micelles size increased (Figure 15.5). The mechanism of PEF impact on casein micelles still needs further investigation, but Floury et al. (2006) suggested an apparent charge modification of micelles after exposure to intense electrical field, resulting in a reduction in hydrodynamic volume of casein micelles. The increased casein micelle size after heat treatment of milk is reported to be due to aggregation of heat-denatured whey proteins at the surface of casein micelles via the β-lg/κ-casein complex formation (Beaulieu et al. 1999; Singh and Latham 1993).

However, PEF treatment did not affect fat globule size but a marginal reduction in fat globule size was observed in HTST and high-heat-treated milk samples with a drop from ca. 3.2 μm to 2.7 that increases the percentage distribution of this size by 2% (Figure 15.6). Fat globules in milk are surrounded by a thin layer of surface-active membrane material consisting of a complex mixture of proteins, glycoproteins, triglycerides, cholesterol, and enzymes. Exposing milk to heat above 70°C causes the denaturation of the cryoglobulins, resulting in size reduction of fat globules (Fox and McSweeney 1998; Je Lee and Sherbon 2002; Mather 2000).

In terms of elasticity and viscosity, the gels made from milks treated at 40, 45, and 50 kV cm^{-1} and 60°C were significantly less elastic ($p > 0.05$) than gels made from

TABLE 15.6
PEF Treatment Conditions of Milk

Treatment Conditions	Skim Milk							Whole Milk						
Field intensity (kV cm⁻¹)	50	45	40	50	45	40	off	50	45	40	50	45	40	off
Pulse frequency (Hz)	28	34	41	16	18	23	0	16	17	22	28	34	43	0
Treatment time (μs)	19	23	27	11	12	15	0	10.5	11	13.5	18	23	29	0
Number of pulses	6	8	9	4	4	5	0	4	4	5	6	8	9	0
Inlet T (°C)	10	10	10	10	10	10	10	10	10	10	10	10	10	10
Outlet T (°C)	60	60	60	30	30	30	60	30	30	30	60	60	60	60
Water bath T (°C)	6	6	6	6	6	6	67	6	6	6	6	6	6	67
Total energy (kJ L⁻¹)	203	203	203	94	94	94	0	87	87	87	193	193	193	0
Peak voltage (kV)	30	27	24	30	27	24	0	30	27	24	30	27	24	0
Peak current (A)	94	86	76	76	68	61	0	70	62	56	86	80	71	0

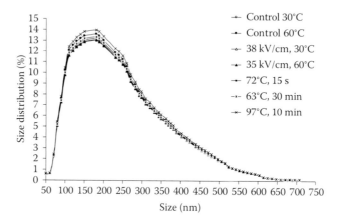

FIGURE 15.5 Size distribution of casein micelles measured by Zetasizer. The standard deviation of particle size for duplicate samples of different treatments was from ±5 to ±8 nm.

HTST-pasteurized milk samples (Table 15.7). Inversely, gels made from PEF-treated milk at 30°C and 40°C, 45 and 50 kV cm^{-1} had significantly ($p < 0.05$) higher G' and G'' compared to HTST-pasteurized samples. LTLT pasteurization (63°C/30 min) of milk resulted in the weakest gel compared to all other gels (Figure 15.7). Long exposure of milk to heat had an adverse effect on gel structure due to denaturation of whey proteins and their aggregation with casein micelle that hinders the coagulation process resulting in a soft gel (Dalgleish 1990; Omar 1985).

In view of the observed similarities in physical attributes of gels made from PEF- and heat-treated milk samples, it is postulated that PEF treatment may cause similar changes in milk proteins as the heat treatment. Therefore, the formation of lower elasticity and viscosity in gels was attributed to the possible aggregation of the denatured whey proteins with κ-casein that hinders the coagulating action of the rennet (Shalabi and Wheelock 1976; Van Hooydonk et al. 1987). Raising the temperature

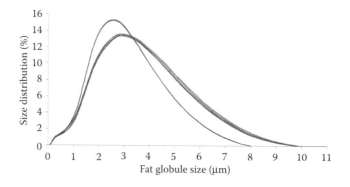

FIGURE 15.6 The average size of fat globule particle measured by Zetasizer. Peak A shows fat globule size of high-heat-treated and HTST milk. The standard deviation of duplicate samples for different treatments was from ±5 to ±8 μm.

TABLE 15.7

Rheological and Textural Properties and Rennet Coagulation Time of Gels from PEF-Treated and Control Milk Samples

T (°C)	Attributes	Control (°C)	PEF Treatment 40 (kV cm⁻¹)	45 (kV cm⁻¹)	50 (kV cm⁻¹)	Pasteurization 72 °C/15 s	63 °C/30 min
30	G' (pa)	52[a] ± 3	37[b] ± 3	31[c] ± 3	27[c] ± 3	24[c] ± 1	12[e] ± 1
	G'' (pa)	14[a] ± 2	12[a] ± 1	11[a] ± 1	10[b] ± 1	9[c] ± 0.51	4[d] ± 1
	Firmness (g)	99[a] ± 4	83[b] ± 2	80[b] ± 3	77[b] ± 6	42[d] ± 4	36[e] ± 3
60	G' (pa)	46[a] ± 3	21[d] ± 2	14[e] ± 2	11[c] ± 1		
	G'' (pa)	13[a] ± 2	9[c] ± 1	8[c] ± 0.5	7[c] ± 0.2		
	Firmness (g)	94[a] ± 2	71[b] ± 4	66[c] ± 6	61[c] ± 5		

during PEF treatment to 60°C had an additive effect on further lowering the G' and G'' of milk gel.

Perez and Pilosof (2004) reported aggregation and denaturation of β-lactoglobulin after PEF treatment at 12.5 kV cm⁻¹, using 3–10 exponential decaying pulses of 2–2.3 ms, as shown by differential scanning calorimetry and sodium dodecyl sulfate-polyacrylamide gel electrophoresis (SDS-PAGE) analyses. However, a study by a different group showed that PEF treatment of skim milk at 29 kV cm⁻¹ using 200 exponentially decaying pulses did not cause marked unfolding and aggregation of β-lactoglobulin (Barsotti et al. 2002).

The results of the texture analyzer were consistent with the rheology findings (Figure 15.8). The gel firmness showed an inverse relationship with increasing field intensity (40, 45, and 50 kV cm⁻¹) and treatment temperature (30°C and 60°C). According to Floury et al. (2006), only PEF treatments above 40 kV cm⁻¹ could modify the casein micelles behavior. The decrease in firmness of gels made from PEF- or heat-treated milks could be (similar to rheological changes) due to hindering the coagulation process since the denatured whey proteins aggregate with casein micelles and hamper the action of the rennet enzyme on cleaving κ-casein (Fox and McSweeney 1998) that could result in a less-firm gel compared to the gels made from controlled 30°C and 60°C milks. Although PEF treatment lowered the gel firmness (due to the combined effects of heat and PEF treatment), the gels were still firmer than those made from pasteurized milk samples, which indicated that the PEF impact on gel structure was less pronounced compared with that of pasteurization. LTLT treatment of milk resulted in the least-firm gel due to a longer exposure time of milk to heat.

Scanning electron microscopy (SEM) of gels provided clear micrographs for observation of changes to gel matrix before and after thermal and PEF treatments (Figure 15.9). The gels made from 30°C and 60°C control milks had a dense structure (Figure 15.9a and b). By increasing the field intensity from 40 to 45 and 50 kV cm⁻¹ and temperature from 30°C to 60°C, the pore size became larger. The pores are, in fact, the sublimated gel matrix once filled with trapped whey during gel formation.

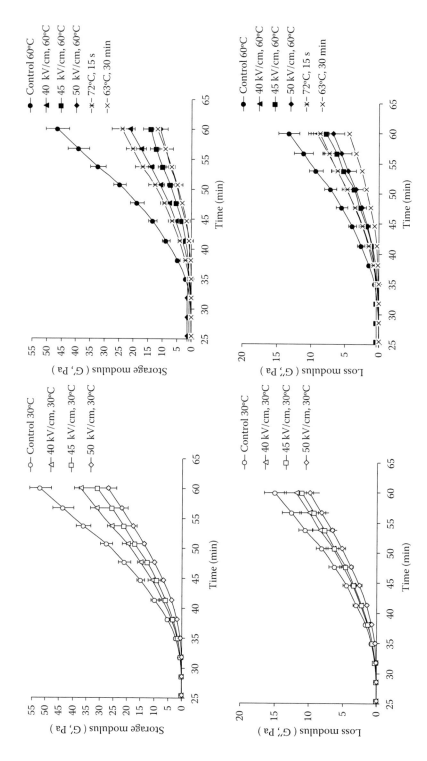

FIGURE 15.7 Storage modulus (G′) and loss modulus (G″) of gels made from thermal and PEF-treated milk samples at 30°C and 60°C.

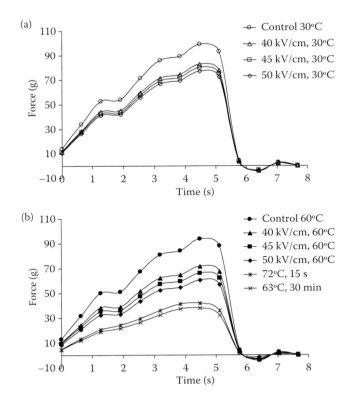

FIGURE 15.8 Firmness of gels made from PEF-treated and untreated skim milk samples. The standard deviations of duplicate samples for different treatments were ±5 to ±9.

Large pores are the result of weaker interactions between the casein micelles making up the gel network, resulting in a lower G', G'' and firmness and a gel that is easier to fracture. They also indicate that heat (or as in this study, PEF treatment alone or combined with heat) has considerably rearranged the gel matrix (Lucey et al. 2001) compared to gels made from control milks since the size of the pores in the former gels has become much larger. The size of the pores indicates the water-holding capacity of the gel matrix that increases by the extent of whey protein denaturation. Compared to PEF effects, the impact of heat on the gel structure was much more pronounced as the micrographs indicated much larger pores in the structure of gels made from HTST- and LTLT-pasteurized milks. Figure 15.9 (i and j) shows that the effects of higher field intensities (>45 kV cm^{-1}) at 60°C on gel matrix are similar to thermal pasteurization. On the basis of the results of the textural analyses, PEF treatment could result in less damage to the gel structure, thus resulting in a firmer gel compared to gels made from pasteurized milks.

15.6　CONCLUDING REMARKS

PEF is potentially an alternative to thermal pasteurization and is capable of providing microbiologically safe products with longer shelf life, fresh-like quality, and

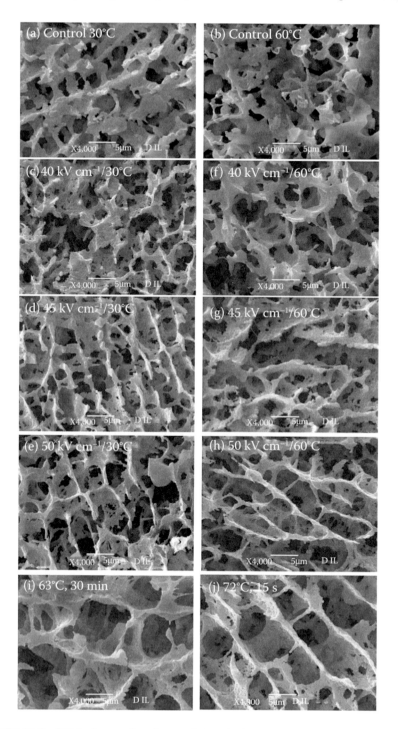

FIGURE 15.9 SEM micrographs of rennet-induced milk gels made from PEF-treated and untreated skim milk.

nutritional properties. However, to date, a number of contradictory results regarding the inactivation of enzymes and microorganisms have been reported that arise from the lack of a standard procedure for treating various products. The versatility in the number and type of PEF units and treatment chambers that are often custom designed by the research groups have added to the variability of the reported results. The relatively high cost of the installations including pulse generators and discharge switches, control of heat generation within the system, and concerns with the survival of spores are among the hurdles in the way of adopting PEF technology as a processing method of choice for milk and dairy products. As a guide, the cost of a bipolar 400 L h^{-1} pilot PEF system is approximately US\$ 250,000, while the commercial PEF system with higher capabilities and process volume is US\$ 500,000 (Gaudreau et al. 2008). Nonetheless, research is ongoing to make the PEF treatment industrially and commercially viable.

There are a number of technical hurdles to overcome in achieving successful application of PEF technology at the industrial scale such as designing treatment chambers with maximum throughput and minimum ohmic heating, (which can adversely affect heat-sensitive products) as well as building highly specific pulse generators and switches capable of handling high voltages and developing corrosion-resistant electrodes (Fox et al. 2007). Considering these challenges, more research is necessary to make PEF applicable in dairy industry and to achieve a higher rate of enzymatic and microbial inactivation in milk.

REFERENCES

Barbosa-Canovas, G.V., Pierson, M.D., Zhang, Q.H. and Schaffner, D.W. 2000. Pulsed electric fields. Special supplement: Kinetics of microbial inactivation for alternative food processing technologies. *Journal of Food Science* 65: 65–79.

Barbosa-Canovas, G.V. and Sepulveda, D. 2005. Present status and the future of PEF technology. In: *Novel Food Processing Technologies,* eds., G.V. Barbosa-Canovas, M.S. Tapia, and M.P. Cano. Boca Raton, Florida: CRC Press, 1–44.

Barsotti, L., Dumay, E., Mu, T.H., Fernandez Diaz, M.D. and Cheftel, J.C. 2002. Effects of high voltage electric pulses on protein-based food constituents and structures. *Trends in Food Science and Technology* 12: 136–144.

Beattie, J.M. 1915. Report on the electrical treatment of milk to the city of Liverpool. Liverpool: C, Tinling and Co.

Beattie, J.M. and Lewis, F.C. 1925. The electric current (apart from the heat generated). A bacteriological agent in the sterilization of milk and other fluids. *Journal of Hygiene* 24: 123–137.

Beaulieu, M., Pouliot, Y. and Pouliot, M. 1999. Thermal aggregation of whey proteins in model solutions as affected by casein/whey protein ratios. *Journal of Food Science* 64 (5): 776–780.

Calderon-Miranda, M.L., Barbosa-Canovas, G.V. and Swanson, B.G. 1999. Inactivation of *Listeria innocua* in skim milk by pulsed electric fields and nisin. *International Journal of Food Microbiology* 51 (1): 19–30.

Castro, A. 1994. Pulsed electric field modification of activity and denaturation of alkaline phosphatase. PhD dissertation. Washington State University, Pullman, WA.

Castro, A.J., Swanson, B.G., Barbosa-Canovas, G.V. and Zhang, Q.H. 2001. Pulsed electric field modification of milk alkaline phosphatase. In: *Pulsed Electric Fields in Food Processing—Fundamental Aspects and Applications. Food Preservation Technology*

Series, eds., G.V. Barbosa-Canovas, and Q.H. Zhang, Pullman: Technomic Publishing Co. Lancaster (PA), 65–82.

Craven, H.M., Swiergon, P., Ng, S., Midgely, J., Versteeg, C., Coventry, M.J. and Wan, J. 2008. Evaluation of pulsed electric field and minimal heat treatments for microbial inactivation of pseudomonads in milk and enhancement of milk shelf-life. *Innovative Food Science and Emerging Technologies* 9 (2): 211–217.

Dalgleish, D.G. 1990. Denaturation and aggregation of serum proteins and caseins in heated milk. *Journal of Agriculture and Food Chemistry* 38: 1995–1999.

Floury, J., Grosset, N., Leconte, N., Pasco, M., Madec, M. and Jeantet, R. 2006. Continuous raw skim milk processing by pulsed electric field at non-lethal temperature: Effect on microbial inactivation and functional properties. *Lait* 86: 43–57.

Food and Drug Administration. 2000. Kinetics of microbial inactivation for alternative food processing technologies. http://www.cfsan.fda.gov/ ~ comm/ift-arc.html, retrieved on 05.19.2008.

Fox, M.B., Esveld, D.C. and Boom, R.M. 2007. Conceptual design of a mass parallelized PEF microreactor. *Trends in Food Science and Technology* 18: 484–491.

Fox, P.F. and McSweeney, P.L.H. 1998. *Dairy Chemistry and Biochemistry*. 1st edition. 317. London: Blackie Academic and Professional, London.

Gaudreau, M., Hawkey, T., Petry, J. and Kempkes, M. 2008. Pulsed electric field processing for food and waste streams. *Food Australia* 60 (7): 323–325.

Girotti, S., Lodi, S., Ferri, E., Lasi, G., Fini, F., Ghini, S. and Budini, R. 1999. Chemiluminescent determination of xanthine oxidase activity in milk. *Journal of Dairy Research* 66: 441–448.

Grahl, T., Sitzmann, W. and Märkl, H. 1992. Killing of microorganisms in fluid media by high-voltage pulses. In: *DECHEMA Biotechnology Conference Series* 5B, Karlsruhe, Germany, pp. 675–678.

Hamilton, W.A. and Sale, A.J.F. 1967. Effects of high electric fields on microorganisms II. Mechanism of action of the lethal effect. *Biochimica et Biophysica Acta* 148: 789–800.

Heinz, V. and Knorr, D. 2001. Effects of high pressure on spores. In: *Ultra High Pressure of Foods*, ed., Kluwer NY: Academic/Plenum New York, 77–113.

Ho, S.Y., Mittal, G.S. and Cross, J.D. 1997. Effects of high field electric pulses on the activity of selected enzymes. *Journal of Food Engineering* 31 (1): 69–84.

Hülsheger, H., Potel, J. and Niemann, E.G. 1983. Electric field effects on bacteria and yeast cells. *Radiation and Environmental Biophysics* 22 (2): 149–162.

Hülsheger, H., Potel, J. and Niemann, E.G. 1981. Killing of bacteria with electric pulses of high fields strength. *Radiation and Environmental Biophysics* 20 (1): 53–65.

Hülsheger, H. and Niemann, E.G. 1980. Lethal effects of high-voltage pulses on *E. coli* K12. *Radiation and Environmental Biophysics* 18: 281–288.

IDF, 1984. Definitions of heat treatment as applied to milk and fluid products. Document1, Brussels, pp. 65.

Jayaram, S., Castle, G.S.P. and Margaritis, A. 1992. Kinetics of sterilization of *Lactobacillus brevis* cells by the application of high voltage pulses. *Biotechnology and Bioengineering* 40 (11): 1412–1420.

Je Lee, S. and Sherbon, J.W. 2002. Chemical changes in bovine milk fat globule membrane caused by heat treatment and homogenization of whole milk. *Journal of Dairy Research* 69 (4): 555–567.

Lucey, J.A., Tamehana, M., Singh, H. and Munro, P.A. 2001. Effect of heat treatment on the physical properties of milk gels made with both rennet and acid. *International Dairy Journal* 11 (4–7): 559–565.

Mather, I.H. 2000. A review and proposed nomenclature for major proteins of the milk-fat globule membrane. *Journal of Dairy Science* 83: 203–247.

Min, S., Min, S.K. and Zhang, Q.H. 2003. Inactivation kinetics of tomato juice lipoxygenase by pulsed electric fields. *Journal of Food Science* 68 (6): 1995–2001.

Min, S., Evrendilek, G.A. and Zhang, H.Q. 2007. Pulsed electric fields: Processing system, microbial and enzyme inhibition, and shelf life extension of foods. *IEEE Transactions on Plasma Science* 35 (1): 59–73.

Omar, M.M. 1985. Size distribution of casein micelles during milk coagulation. *Die Nahrung* 29 (2): 119–124.

Perez, O.E. and Pilosof, A.M.R. 2004. Pulsed electric fields effects on the molecular structure and gelation of β-lactoglobulin concentrate and egg white. *Food Research International* 37:110–120.

Qi, X.L., Brownlow, S., Holt, C. and Sellers, P. 1995. Thermal denaturation of b-lactoglobulin: Effect of protein concentration at pH 6.75 and 8.05. *Biochimica et Biophysica Acta* 1248: 43–49.

Sale, A.J.H. and Hamilton, W.A. 1967. Effects of high electric fields on microorganisms I. Killing of bacteria and yeast. *Biochimica et Biophysica Acta* 163: 34–43.

Shalabi, S.I. and Wheelock, J.V. 1976. The role of α-lactalbumin in the primary phase of chymosin action on heated casein micelles. *Journal of Dairy Research* 43: 331–335.

Shamsi, K. 2010. *Effects of Pulsed Electric Field Processing on Milk Properties.* Germany: VDM Publishing House Ltd, ISBN 9783639222098.

Shamsi, K. and Sherkat, F. 2009. Application of pulsed electric field in nonthermal processing of milk. A review paper. *Asian Journal of Food and Agro-Industry* 2 (03): 216–244.

Singh, H. and Latham, J.M. 1993. Heat stability of milk: Aggregation and dissociation of protein at ultra-high temperatures. *International Dairy Journal* 3: 225–237.

Sui, Q., Roginski, H., Williams, R., Knoerzer, K., Buckow, R., Versteeg, C. and Wan, J. 2008. Differentiation of temperature from pulsed electric field effects on lactoperoxidase activity using thermal inactivation kinetics. Poster presented at the *4th Innovative Foods Centre Conference*, Brisbane, Australia, September 17–18.

Van Hooydonk, A.C.M., De Koster, P.G. and Boerrigter, I.J. 1987. The renneting properties of heated milk. *Netherlands Milk and Dairy Journal* 41: 3–18.

Walstra, P. 1990. On the stability of casein micelles. *Journal of Dairy Science* 73: 1965–1979.

Yang, R.J., Li, Q. and Zhang, Q.H. 2004. Effects of pulsed electric fields in the activity of enzymes in aqueous solution. *Journal of Food Science* 69 (4): 241–248.

Zar, J.H. 1999. *Biostatistical Analysis.* NJ: Prentice-Hall International, Upper Saddle River, NJ, 324–359.

Zhang, Q.H., Barbosa-Canovas, G.V. and Swanson, B.G. 1995. Engineering aspects of pulsed electric fields pasteurization. *Journal of Food Engineering* 25: 261–281.

Zhang, Q.H., Qin, B.L., Barbosa-Canovas, G.V., Swanson, B.G. and Pedrow, P.D. 1996. Batch mode food treatment using pulsed electric fields. U.S. Patent 5,549,041.

Zhong, K., Wu, J., Wang, Z., Chen, F., Liao, X., Hu, X. and Zhang, Z. 2007. Inactivation kinetics and secondary structural change of PEF-treated POD and PPO. *Food Chemistry* 100: 115–123.

Zimmermann, U., Pilwat, G. and Riemann, F. 1974. Dielectric breakdown of cell membranes. *Biophysical Journal* 14: 881–899.

Zoon, P., van Vliet, T. and Walstra, P. 1988. Rheological properties of rennet-induced skim milk gels. *Netherlands Milk Dairy Journal* 42: 249–269.

16 Modification of Cheese Quality Using Pulsed Electric Fields

L. Juan Yu, Malek Amiali, and Michael O. Ngadi

CONTENTS

16.1 INTRODUCTION

16.1.1 HISTORY OF CHEESE

Most of the milk consumed by humans is obtained from dairy cows, although milk from goats and sheep are also available. In Canada, and in many industrialized countries, raw milk (RM) is processed before it is consumed. During normal processing, the fat content of the milk may be adjusted, various vitamins could be added, and pathogenic bacteria inactivated. In addition to being consumed as a beverage, milk is also processed to make butter, cream, yogurt, and cheese. It is believed that cheese was discovered by accident by a traveling Arabic merchant who had brought some milk along for his journeys (Ashford et al., 2012). The pouch in which he placed the milk was made of a sheep's stomach lining. As he traveled, the milk turned into liquid whey and solid curd or cheese components as a result of activity of the enzyme from the pouch's lining. The process of cheese making was extended during the Roman Empire. Cheese arrived in America around 1620 when the Mayflower landed in the continent and it quickly became a staple component of the American diet. In 2011, cheese production in the United States was about 11 billion pounds (projected from data in Cheese Reporter, 2012). Other major cheese producers include the European Union and Brazil.

16.2 INTRODUCTION TO MILK AND CHEESE

Bovine milk, the milk mostly used to make cheese, is an important source of lipids, proteins, amino acids, vitamins, and minerals. It contains immunoglobulins, hormones, growth factors, cytokines, nucleotides, peptides, polyamines, enzymes, and other bioactive peptides. The lipids in milk are emulsified in globules coated with membranes. The proteins are in colloidal dispersions as micelles. The casein micelles occur as colloidal complexes of protein and salts, primarily calcium. The later and other minerals such as phosphate, citrate, and pH influence the casein-to-fat ratio, total solids, lactose, mineral content, consequently moisture levels, and extent of acid development in the finished cheese. The composition of milk varies with the cow's stage of lactation, age, breed, nutrition, energy balance, and health status of the udder. Accordingly, the concentration of protein may be about double in colostrum compared to mature milk (Haug et al., 2007). Milk contains about 87% water and it is a good source of water in the diet. Water is extremely important in human metabolism. It maintains body temperature regulation through sweating. In general, lack of water (dehydration) results in fatigue, mental impairment, cramping, decreased athletic performance, and sometimes severe dehydration that can be life threatening (Milk Facts, 2012). The energy in milk comes from its protein, carbohydrate, and fat contents. Carbohydrates, the primary source of energy for activity, are present in milk in the form of lactose (4.9%) that is a disaccharide made up of glucose and galactose bonded together. Glucose is the only form of energy that can be used by the brain. Excess glucose is stored in the form of glycogen in the muscles and liver for later use. Carbohydrates are important in hormonal regulation in the body. Lack of adequate levels of glucose in the blood and carbohydrate stores leads to muscle fatigue and lack of concentration. Lactose needs to be broken down by the enzyme lactase in the small intestine before it can be used by the body. Some people are lactose intolerant or have problems digesting and absorbing lactose due to decreased activity of lactase in their small intestine.

Whole milk contains about 3.4% fat (USDA Database, 2012). Fats are structural components of cell membranes and hormones. They are a concentrated energy source and are the main energy source used by the body during low-intensity activities and prolonged exercise. The linoleic (18:2) and linolenic (18:3) fatty acids cannot be produced by the body and must come from diet. These fatty acids are used to synthesize the longer-chain fatty acids such as arachidonic acid (AA, 20:4o-6), docopentaenoic acid (DPA, 22:4o-6), eicosapentaenoic acid (EPA, 20:5o-3), and docohexaenoic acid (DHA, 22:6o-3). These longer-chain fatty acids are essential for the synthesis of hormones such as prostaglandins, thromboxanes, and leukotrienes that are involved in muscle contraction, blood clotting, and immune response. Milk fat is composed of approximately 65% saturated, 29% monounsaturated, and 6% polyunsaturated fatty acids. The protein content of milk is approximately 3.3% and it contains all the essential amino acids. Milk protein consists of about 82% casein and 18% whey (serum) proteins. In most cheeses, the casein is coagulated to form curd, and the whey is drained leaving only a small amount of whey proteins in the cheese.

During cheese making, the 6-casein is cleaved between specific amino acids and results in a unique protein fragment that is drained with the whey. This fragment, called milk glycomacropeptide, does not have any phenylalanine and can be used as a source of protein for people with phenylketonuria, the inability to digest proteins that contain phenylalanine (Milk Facts, 2012). Whey proteins have become popular ingredients in foods as an additional source of protein or for functional benefits. They are used as a protein source in high-protein beverages and energy bars targeted to athletes. Whey proteins also contain immunoglobulins that are important in the immune responses of the body. It contains branched-chain amino acids (leucine, isoleucine, and valine) and has been proposed to have some benefits to athletes for muscle recovery and for preventing mental fatigue. There are at least 1000 named cheese varieties, which may be classified into three major groups based on the method used to coagulate the milk (Fox et al., 2000):

(1) Rennet-coagulated (Cheddar) cheeses, which represent about 75% of the total production and include most major international cheese varieties; (2) acid-coagulated cheeses (e.g., Cottage); and (3) heat- and acid-coagulated cheeses (e.g., Ricotta). Cheddar-style cheese is by far one of the most important cheese varieties. In Canada, Cheddar cheese amounts to about 42.9% of the total cheese production (Agriculture and Agri-Food Canada, 2008).

16.3   CHEESE QUALITY AND CONTROL

In general, composition, microbial load, and indigenous enzymes of milk are the major factors that influence different characteristics and quality of all kinds of cheeses. Several authors have reported that cheese quality is largely dependent on fat and protein components (Abd El Gawad and Ahmed, 2011; Brito et al., 2002; Guo et al., 2004; Van den Berg, 1994). Van Boekel (1993) observed the transfer of the components from milk to cheese and figured out the amount of each single-protein fraction that is retained in the curd and that is lost in the whey. Brito et al. (2002) and Lolkema (1993) established relationships between milk components (fat and casein) or cheese composition (moisture, protein, and fat) for varieties of cheese, such as Cheddar and Gouda. The casein fraction of milk protein is the dominant factor affecting curd firmness (CF), syneresis rate, moisture retention, and ultimately affecting cheese quality and yield (Abd El Gawad and Ahmed, 2011; Lawrence, 1993). Therefore, casein content, along with that of fat, is included in all compositions of cow cheese.

The other important factors, namely microbiological and enzymatic, must be considered in assessing cheese quality. Milk may be contaminated with pathogenic microorganisms. The principal pathogens of concern of milk are *Listeria monocytogenes*, *Escherichia coli* O157:H7.

Shigella erwinia, Campylobacter, Staphylococcus, Salmonella app., and *Mycobacterium paratuberculosis.* Some of these bacteria may die under the typically hostile conditions in well-made cheeses (low pH; 5.3, high salt content; 5–10% salt in moisture, S/M, and bacteriocins). Therefore, public health authorities of many countries require that all RMs intended for cheese making must be pasteurized (Fox and Cogan, 2004). However, many cheeses are made from RM in some countries such as France, Germany, and Southern countries. Therefore, good-quality cheese is produced from good-quality RM when subsequently handled under hygienic conditions.

The final cheese composition can be manipulated by controlling the composition of the starting milk to meet the legal definition of a target-specific variety and to improve yields. The use of standardized milk avoids the manufacture of cheese containing excess fat and minimizes fat and casein losses into the whey (Lucey and Kelly, 1994; Scott, 1998). There are three main methods for standardizing milk for cheesemaking. These are namely the addition of skimmed milk powder, the addition of liquid skim milk, and the removal of cream. In the first two cases, the quality and temperature history of the skimmed milk powder or liquid are important, particularly in the manufacture of high-quality Cheddar cheese (Banks et al., 1984). The removal of fat by gravity or centrifugation creaming is practiced in the manufacture of Parmigiano Reggiano (Fox and Cogan, 2004). The pH of milk is a critical factor in cheesemaking. The addition of 1–2% starter culture to milk reduces the pH of the milk immediately by about 0.1 unit. However, the widely used starter concentrates (direct-to-vat starters; DVS), have no immediate, direct acidifying effect. The starters are typically added 30–60 min before adding rennet. During this period, the starter began to grow and produce acid, a process referred to as "ripening." This serves a number of functions such as allowing the bacteria to be highly active during cheesemaking, and favoring rennet action and gel formation (Fox and Cogan, 2004).

16.4 PRETREATMENT OF MILK DESTINED FOR CHEESE MANUFACTURE

16.4.1 Traditional Thermal Pasteurization

Pasteurization of cheese milk became increasingly popular from the beginning of the twentieth century, primarily as a result of public health concerns. Under pasteurization conditions, several important changes occur in milk apart from microbial reduction. These changes are closely related to the final quality of the cheese.

16.4.2 Whey Protein Denaturation and Its Interactions with Casein

Heat treatment during commercial processing results in a series of physicochemical changes in milk constituents. These changes include denaturation of whey proteins, interactions between the denatured whey proteins and the casein micelles, and the transfer of soluble minerals to the colloidal state. Casein micelles are very stable at processing temperatures although they undergo some association–dissociation reactions at severe heating temperatures (Fox, 1981; Singh and Kanawjia, 1988; Singh and Creamer, 1992). For instance, heating milk up to 90°C is known to cause only minor changes in the size of casein micelles, but severe treatment at higher temperatures increases the micelle size. When milk is heated above 65°C, whey proteins are denatured by the unfolding of their polypeptides, thus exposing the side-chain groups that are originally buried within the native structure. The unfolded proteins then interact with casein micelles or simply aggregate with themselves, involving thiol–disulfide interchange reactions, hydrophobic interactions, and ionic linkages (Corredig and Dalgleish, 1999; Oldfield et al., 1998, 2000; Vasbinder et al., 2003). The interaction of denatured whey proteins with the casein micelles is largely between β-lactoglobulin and κ-casein and involves both disulfide and hydropho-bic interactions (Sandra and Dalgleish, 2007; Singh and Fox, 1987; Smits and van Brouwershaven, 1980).

16.4.3   RENNET COAGULATION PROPERTIES OF HEAT-PASTEURIZED CHEESE MILKS

The interaction of whey proteins with the casein micelles has both positive and negative implications in cheese manufacture. From a positive perspective, dena-tured whey proteins could be incorporated into cheese curd, resulting in a higher yield from a given quantity of milk. In a negative sense, the interactions of whey protein with casein micelles interfere with the rennet coagulation process, resulting in long coagulation times and weak curd structures (Singh and Waungana, 2001). The reduction in gel firmness is caused by the disruption of the continuity of the gel network caused by attachment of denatured whey proteins to casein micelles. The denatured whey proteins may sterically hinder the close approach between casein micelles, resulting in a weaker, looser network due to reduced cross-linking (Vasbinder et al., 2003).

It is generally agreed that the secondary phase of rennet coagulation is more adversely affected by heating than the enzymatic phase. Denatured whey proteins on the surface of casein micelles sterically hinder the aggregation of rennet-altered micelles, resulting in prolonged rennet coagulation time (Singh and Kanawjia, 1988). Other factors, such as the formation of heat-induced colloidal calcium phosphate (CCP), may also affect the rate of aggregation of renneted micelles in heated milk (Darling and Dickson, 1979; Schmidt and Poll, 1986).

16.4.4   PROTEOLYSIS PROPERTIES OF HEAT-PASTEURIZED CHEESE MILKS

Limited information is available on the effects of heat treatment of cheese milk on the proteolysis during cheese ripening. Benfeldt et al. (1997) reported that cheeses made from heated milk show decreased proteolysis of caseins, which resulted in slower formation of small peptides and free amino acids (FAA). Also, examination of various cheeses by scanning electron microscopy (SEM) showed that as the inten-sity of heat treatment increases, the protein matrices of cheese become coarser and less homogeneous in appearance and contain numerous small holes or cracks (Singh and Waungana, 2001). Thus, cheeses made from heat-treated milk differ from RM cheese in body, texture, and flavor profiles.

16.5 PULSED ELECTRIC FIELD PROCESSING OF MILK AND CHEESE QUALITY

16.5.1 INTRODUCTION TO PEF

Pulsed electric field processing (PEF) applies short bursts of high-voltage electricity for microbial inactivation and causes no or minimum effect on food-quality attributes. Briefly, the foods being treated by PEF are placed between two electrodes, usually at room temperature. The applied high voltage results in an electric field that causes microbial inactivation. The applied high voltage is usually in the order of 20–80 kV for microseconds. The common types of electrical field waveform applied include exponentially decaying and square wave (Amiali and Ngadi, 2012; Barbosa-Cánovas et al., 1999; Knorr et al., 1994; Ngadi et al., 2009; Zhang et al., 1995).

The principles of PEF processing have been explained by several theories including the transmembrane potential theory, electromechanical compression theory, and the osmotic imbalance theory. One of the most accepted theories is associated with the electroporation of cell membranes. It is generally believed that electric fields induce structural and functional changes in the membranes of microbial cells based on generation of pores in the cell membrane, consequently leading to microbial destruction and inactivation (Barbosa-Cánovas et al., 1999; Tsong, 1991). Excellent reviews of PEF applications in food processing are available in the literature (Amiali et al., 2010; Huang et al., 2012; Jeyamkondan et al., 1999; Sobrino-Lopez and Martin-Belloso, 2010).

Compared with thermal processing, PEF processing has many advantages. It can preserve the original sensory and nutritional characteristics of foods due to the relatively short processing time and low processing temperatures. Energy savings for PEF processing are also important compared with conventional thermal processing. Moreover, it is environmentally friendly with no waste generated. Owing to these advantages, PEF processing has been widely used in food product preservation including milk and milk products. A summary of literature data on inactivation of different microorganisms in milk is presented in Table 16.1.

16.5.2 EFFECT OF PEF TREATMENT ON MILK QUALITY·

In recent years, with increased demand for high-quality milk and milk products, more researchers have focused on studies of changes in organoleptic and physicochemical characteristics in milk and milk products following treatment with PEF (Bendicho et al., 1999; Dunn, 1996; Evrendilek et al., 2001; Li et al., 2003; Michalac et al., 1999; Sepulveda et al., 2005; Shamsi et al., 2008; Shin et al., 2007; Qin et al., 1995; Yeom et al., 2001).

Dunn (1996) reported that milk treated with PEF (E = 20–80 kV/cm) suffered less flavor degradation when compared to RM. The author proposed the possibility of manufacturing dairy products such as cheese, butter, and ice cream using PEF-treated milk although limited information was given in the report. Qin et al. (1995) carried out a study on shelf life, physicochemical properties, and sensory attributes of milk with 2% milk fat, treated with 40 kV/cm electric field and 6–7 pulses. No

TABLE 16.1

Inactivation of Selected Microorganisms in Milk by PEF[a]

Source	Milk Type[b]	Microorganisms	Maximum Log Reduction[c]	Treatment Vessel[c]	Process Conditions[d]
Dunn et al. (1987)	PM	E. coli	3	B, parallel plate	43°C, 43 kV/cm, 23 pulses
Grahl et al. (1996)	UHT milk (1.5% fat)	E. coli	4	B, 25 mL, d = 0.5 cm	<45°C, 22.4 kV/cm, 20 pulses
Zhang et al. (1994b)	Skim milk	E. coli	3	B, d = 0.95 cm, v = 25.7 mL	<25°C, 25 kV/cm, 20 pulses, exponential
Zhang et al. (1995b)	Modified SMUF	E. coli	9	B, parallel plate, 14 mL, d = 0.51 cm	7°C, 20°C, and 33°C, 70 kV/cm, 2 μs, 80 pulses
Qin et al. (1994)	SMUF	E. coli	1.5	B, 80 J/pulse, parallel plate	<30°C, 40 kV/cm, eight pulses, oscillatory decay
Qin et al. (1994)	SMUF	E. coli	3	B, 60 J/pulse, parallel plate	<30°C, 40 kV/cm, four pulses, bipolar
Qin et al. (1994)	Skim milk	E. coli	2.5	B, parallel plate, 14 mL	<30°C, 50 kV/cm, 62 pulses, 2 μs, square wave
Qin et al. (1994)	Skim milk	E. coli	3.5	C, parallel plate	<30°C, 50 kV/cm, 48 pulses, 2 μs, square wave
Qin et al. (1995)	SMUF	E. coli	3.6	C, parallel plate, 8 cm³, d = 0.51 cm	<30°C, 50 kV/cm, 62 pulses, 2 μs, square wave
Qin et al. (1995)	SMUF	E. coli	7	C, coaxial, 29 mL, d = 0.6 cm	<30°C, 25 kV/cm, ±300 pulses, 2 μs, square wave
Qin et al. (1998)	SMUF	E. coli	6	C, coaxial, 29 mL, d = 0.6 cm	<40°C, 60 kV/cm, 50 pulses, 2 μs, exponential decay
Pothakamury et al. (1995)	SMUF	E. coli	4	B, parallel plate, 12.5 mL, d = 0.5 cm	<30°C: 16 kV/cm; 200–300 μs, exponential decay
Pothakamury et al. (1996)	SMUF	E. coli	4.5	B, parallel plate	7°C, 20°C: 36 kV/cm, 60 pulses
Vega-Mercado et al. (1996)	Modified SMUF	E. coli	2.56	B, parallel plate, 12.5 mL, d = 0.5 cm	15°C, 20°C, and 40°C, 55 kV/cm, 2 μs, eight pulses

continued

TABLE 16.1 (continued)
Inactivation of Selected Microorganisms in Milk by PEF[a]

Source	Milk Type[b]	Microorganisms	Maximum Log Reduction	Treatment Vessel[c]	Process Conditions[d]
Martin et al. (1997)	Skim milk diluted with water (1:2:3)	E. coli	Nearly 3	B, parallel plate, 13.8 mL, d = 0.51 cm	15°C, 45 kV/cm, 6 μs, 64 pulses
Martin et al. (1997)	Skim milk	E. coli	2	C, parallel plate, 45 mL/s, v = 8 mL	15°C, 25 kV/cm; 1.8 μs
Dutreux et al. (2000)	Fat-free milk	E. coli	4.0	C, coaxial, 29 mL, flow rate 0.5 L/min, d = 0.6 cm	Inlet 17°C, outlet 37°C, 41 kV/cm, 2.5 μs, up to 63 pulses
Manas et al. (2001)	Sterile dairy cream (33% fat)	E. coli	5.0	B, parallel electrode, cylindrical chamber, d = 0.5 cm, 5.7 mL	<30°C; 34 kV/cm; 7 μs, 1.1 Hz, 64 pulses, exponential decay
Evrendilek et al. (2005)	Skim milk	E. coli O157:H7	1.96	C, 1 mL/s, cofield	24 kV/cm, 2.8 μs, 700 Hz, bipolar pulses, square wave, t = 141 μs
Alkhafaji et al. (2006)	SMUF	E. coli	6.6	C, Multipass treatment chambers	17°C, 43.4 kV/cm, 1.7, 880 μs, bipolar square wave, up to 200 Hz
Dunn et al. (1987)	Milk	Salmonella dublin	4	B, parallel plate	63°C, 36.7 kV/cm, 36 μs, 40 pulses
Sensoy et al. (1997)	Skim milk	S. dublin	3.0	C, cofield	10–50°C, 15–40 kV/cm, 12–127 μs
Floury et al. (2005)	Skim milk	Salmonella enteritidi	1.0	C, coaxial, d = 2 mm	<50°C, 35 kV/cm, 500 ns, 64-pulses square wave
Qin et al. (1994)	SMUF	B. subtilis	4.5	B, parallel plate, 100 μL, d = 0.1 cm	16 kV/cm, monopolar, 180 μs, 13 pulses
Qin et al. (1994)	SMUF	B. subtilis	5.5	B, parallel plate, 100 μL, d = 0.1 cm	16 kV/cm, bipolar, 180 μs, 13 pulses

Reference	Product	Microorganism	Log reduction	Treatment chamber	PEF conditions
Michalac et al. (2003)	Raw skim milk and UHT skim milk	*B. subtilis*	3.0	Cofield, d = 0.19 cm, diameter = 0.23	52–22°C, 35 kV/cm, bipolar square wave, 3 µs, 500 Hz, 64 pulses
Pothakamury et al. (1995b)	SMUF	*Lactobacillus delbrueckii* ATCC 11842	4–5	B, 1 mL, d = 0.1 cm	< 30°C, 16 kV/cm, 200–300 µs, 40-pulses exponential decay, t = 10,000 µs
Pothakamury (1995b)	SMUF	*B. subtilis* spores	4–5	B, parallel plate, 1 mL, d = 0.1 cm	< 30°C, 16 kV/cm, 200–300 µs, 50 pulses, exponential decay, t = 12,500 µs
Pothakamury (1995a)	SMUF	*S. aureus* ATCC 6538	3	B, parallel plate, 1 mL, d = 0.1 cm	< 30°C, 16 kV/cm, 200–300 µs, 60 pulses, exponential decay
Everendilek et al. (2004)	Skim milk	*S. aureus*	3.0	Cofield chamber, d = 0.6 cm, diameter = 2.3 mm	40°C, 250 Hz, 35 kV/cm, 3.7 µs, 250 pulses, bipolar square wave
Sobrino-lopez et al. (2006)	Skim milk	*S. aureus*	4.5	C, cofield, d = 0.29, v = 0.012 cm^3	25°C, 100 Hz, 35 kV/cm, 8 µs, 150 pulses, bipolar square wave
Sobrino-lopez et al. (2006)	Skim milk	*S. aureus*	6.3	C, cofield, d = 0.29, v = 0.012 cm^3	25°C, 75 Hz, 35 kV/cm, 6 µs, 200 pulses, bipolar square wave, pH 6.8, lysozyme (300 IU/mL) and nisin (1IU/mL)
Reina et al. (1998)	PM (whole and skim)	*L. monocytogenes*	3.0–4.0	C, cofield flow, 20 mL,	10–50°C, 0.07 L/s, 30 kV/cm,1.5 µs, 1700 Hz, bipolar pulses, t = 600 µs
Calderon-Miranda et al. (1999)	Raw skim milk	*Listeria innocua*	2.5	C, 29 mL, d = 0.6 cm	22–34°C, 0.5 L/min, 2 µs, 3.5 Hz, 32 pulses, 50 kV/cm, exponential decay
Fernandez et al. (2001)	Pasteurized skim milk (0.2% milk fat)	*L. innocua*	2.6	C, coaxial, 29 mL, d = 0.63	15–28°C, 0.5 L/min, 100 pulses, 50 kV/cm, 2 µs, 3.5 Hz, exponential decay
Fernandez et al. (2001)	Pasteurized skim milk (0.2% milk fat)	*Pseudomonas fluorescens*	2.7	C, coaxial, 29 mL, d = 0.63	15–28°C, 0.5 L/min, 30 pulses, 50 kV/cm, 2 µs, 4.0 Hz, exponential decay
Michalac et al. (2003)	Raw skim milk and UHT skim milk	*P. fluorescens*	2.5	Cofield, d = 0.19 cm, diameter = 0.23	52–22°C, 35 kV/cm, bipolar square wave, 3 µs, 500 Hz, 64 pulses

continued

TABLE 16.1 (continued)

Inactivation of Selected Microorganisms in Milk by PEF[a]

Source	Milk Type[b]	Microorganisms	Maximum Log Reduction	Treatment Vessel[c]	Process Conditions[d]
Shamsi et al. (2008)	Raw skim milk	*Pseudomonas*	5.9	Cofield chamber, flow rate 60 mL/min, d = 0.6 cm, d = 2.3 mm	60°C, 200 Hz, 35 kV/cm, 2 µs, 19.6 µs, bipolar square wave
Dutreux et al. (2000)	Fat-free milk	*L. innocua*	3.9	C, coaxial, 29 mL, flow rate 0.5 L/min, d = 0.6 cm	Inlet 17°C, outlet 37°C, 41 kV/cm, 2.5 µs, up to 63 pulses
Pichart et al. (2002)	Whole milk, skim milk, and dairy cream	*L. innocua*	6	B, cylindrical vessel, d = 39 mm, 5 mm gap, v = 5.6 mL	21.5–42°C, 17–46 kV/cm, 545 pulses, 1.1 or 100 Hz, 1.51 µs, exponential decay
Fleischman et al. (2004)	Water and skim milk	*L. monocytogenes*	Zero	B, cylindrical chamber, v = 20 mL	35°C, 3.25 µs, 20 kV/cm, 10 pulses
Fernandez et al. (2006)	Skim milk	*L. innocua*	3	C, coaxial, flow rate 8.33 mL/min, d = 0.69 cm	<30°C, 38.9 kV/cm, 3 µs, 100 pulses

Source: Adapted from Yu L.J., Ngadi M.O. Raghavan G.S.V. 2009. *Journal of Food Engineering* 95(1): 115–118.

[a] Part of the information was adapted from Barbosa-Cánovas et al. 1999.

[b] UHT: ultra-high-temperature sterilization; SMUF: simulated milk ultrafiltrate.

[c] B: batch; C: continuous; and d: gap.

[d] Temperature, peak electric field, pulse width, number of pulse and shape, and total treatment time (t).

physicochemical or sensory changes were observed after the treatment, in comparison with a sample treated with thermal pasteurization. Bendicho et al. (1999) studied the destruction of riboflavin, thiamine (water soluble), and tocopherol (liposoluble) in milk by treatment with PEF (E = 16–33 kV/cm; N = 100 pulses). The vitamin concentrations before and after treatment were determined by high-performance liquid chromatography (HPLC). The authors observed no destruction of vitamins by treatment with pulses. Michalac et al. (1999) studied the variation in color, pH, proteins, moisture, and particle size of ultra-high-temperature (UHT) skim milk subjected to treatment with PEF (E = 35 kV/cm; W = 3 μs and time = 90 μs). The authors observed no differences in the parameters studied (color, pH, proteins, moisture, and particle size) before and after treatment.

Yeom et al. (2001) studied a commercial, plain, low-fat yogurt mixed with strawberry jelly and strawberry syrup. The authors noted changes in physical (pH, color, and Brix) and sensory attributes during storage at 4°C after treatment with both PEF (E = 30 kV/cm, t = 32 μs) and heat (T = 65°C, 30 s). Sensory evaluation indicated that there were no changes between the control and treated samples. There was also no variation in the color, pH, and Brix. Evrendilek et al. (2001) studied color, pH, Brix, and conductivity at 4°C, 22°C, and 37°C in chocolate milk using treatment with PEF (E = 35 kV/cm; W = 1.4 μs; and time = 45 μs), and PEF + heat (112°C and 105°C, 33 s). The authors compared the results with a control sample not treated by either PEF or heat. Measurement of the color parameters (L, a, and b) at 4°C revealed that the treatments of PEF at 105°C and at 112°C did not cause changes in color. Some studies have suggested that changes do indeed occur in milk as a result of PEF treatment. Xiang et al. (2011a) studied the influence of PEF treatment on flow behavior and viscosity of reconstituted skimmed milk. Electric field intensities of 15, 18, and 20 kV/cm with 20, 40, and 60 pulses were applied. It was reported that the PEF treatments increased the apparent viscosity of the skimmed milk samples as shown in Figure 16.1. A similar result was obtained for soy milk (Figure 16.2). Increasing PEF intensity increased the consistency indices of skimmed milk (Table 16.2). An

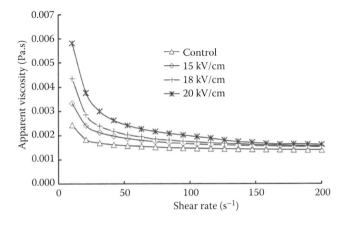

FIGURE 16.1 Apparent viscosity of reconstituted skim milk after treatment with 40 pulses of electric field at 0, 15, 18, and 20 kV/cm.

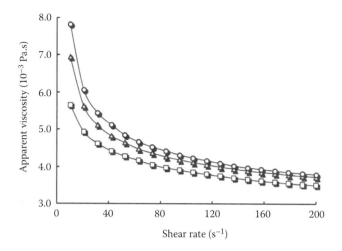

FIGURE 16.2   Apparent viscosity of soy milk after treatment with 75 pulses of electric field at 0, control (□); 20 (Δ); and 22 kV/cm (○). (From Xiang et al., 2011a. *International Journal of Food Sciences and Nutrition* 62(8): 787–793.)

increase in viscosity of the skimmed milk results from changes in the milk globule and micelle size. The increase in viscosity probably reflected intermolecular interactions as a result of the attractions between adjacent denatured molecules (milk protein, milk globule), which may have led to the formation of weak transient networks. The interactions subsequently caused an increase in the effective volume of the milk globules. As the protein aggregation became larger, they occupied more space and thus contributed to an increase in the apparent viscosity of the fluid system (Adapa et al., 1997).

Sepulveda et al. (2005) treated HTST (high-temperature short-time-) pasteurized milk (PM) with an electric field of 35 kV/cm and 2.3 μs of pulse width, at a temperature of 65°C for <10 s. PEF treatments were applied either immediately after thermal pasteurization to produce an extended-shelf-life product, or 8 days after thermal pasteurization to simulate processing after bulk shipping. Application of PEF immediately after HTST pasteurization extended the shelf life of milk from 45 to 60 d, while

---

TABLE 16.2

Electric Field Intensity and Number of Pulses Effect on Apparent Viscosity of Skim Milk

Electric Field Intensity	Consistency	Number of Pulses	Consistency
(kV/cm)	Index	(n)	Index
15	1.473[a]	20	1.474[a]
18	1.500[b]	40	1.501[b]
20	1.531[c]	60	1.529[c]

Note: Values are means of triplicate determinations. Means with different letters (a, b, c) in each column are significantly different (P 0.05).

PEF processing after 8 days caused a shelf-life extension of 78 d; both were proving to be successful strategies for extending the shelf life of milk.

Li et al. (2003) investigated the effects of PEFs and thermal processing on the stability of bovine immunoglobulin G (IgG) in enriched soymilk. PEF at 41 kV/cm for 54 μs caused a 5.3-log reduction of natural microbial flora, with no significant change in bovine IgG activity. Analysis using circular dichroism spectrometry revealed no detectable changes in the secondary structure or the thermal stability of the secondary structure of IgG after the PEF treatment (Li et al., 2005). However, in an experiment investigating the effect of temperature on the stability of IgG during PEF treatment (30 kV/cm, 54 μs), up to 20% of IgG was inactivated when the temperature was increased to 41°C (Li et al., 2003). The effect of PEFs on structural modification and surface hydrophobicity of whey protein isolate (WPI) has been studied using intrinsic and extrinsic fluorescence spectroscopy (Xiang et al., 2011b). The tryptophan residues of proteins were selected as intrinsic fluorescent probes whereas hydrophobicity probe, 1-anilinonaphthalene-8-sulfonic acid (ANS) were used as extrinsic probes to monitor structural modification of whey protein with PEF treatments. The emission spectrum for the control sample showed the maximum intrinsic fluorescence intensity at the emission wavelength of 333 nm. There were red shifts of 2 nm (from 333 to 335 nm) and 4 nm (from 333 to 337 nm) when samples were treated with 30 pulses at the electric fields of 16 and 20 kV/cm, respectively. Similarly, extrinsic fluorescence emission spectra of the control and treated samples showed blue shifts in the range of 2–4 nm. The red shifts in the emission spectra are indicative that microenvironments of tryptophan residues were modified during the process. The blue shifts suggested that the PEF-treated WPI bound more ANS in the less polar environment than did the native WPI, indicating partial denaturation of WPI fractions.

Overall, studies show that PEF as a nonthermal process results in little or no effect on the quality of milk. There could be limited structural modification of milk proteins, in which case, there might be an opportunity to control PEF intensity to achieve the desired quality and functionality of milk products such as cheese.

16.5.3 Effect of PEF Treatment on Cheese Quality

Knowledge of PEF effects on major cheese-making steps, such as coagulation and ripening, are crucial to developing higher quality cheeses made from PEF-PM. However, very little work has been reported so far about PEF effects on cheese-making properties (Sepulveda-Ahumada et al., 2000; Yu et al., 2009,2012). Sepulveda-Ahumada et al. (2000) compared the textural properties and sensory attributes of Cheddar cheese made with heat-treated milk, PEF-treated milk (E = 35 kV/cm, N = 30 pulses), and untreated milk. In the hardness and springiness study, the cheeses made from milk pasteurized by any method were harder than those made from untreated milk. In the sensory evaluation, the panelists also found differences between the cheeses made from untreated milk and milk treated by PEF or heat. Regardless of the differences, the authors still considered using PEF-treated milk to obtain cheese as a feasible option to improve product quality.

The effect of PEF treatment on rennet coagulation properties of milk has been studied by Yu et al. (2009). PEF treatment was achieved in a continuous chamber

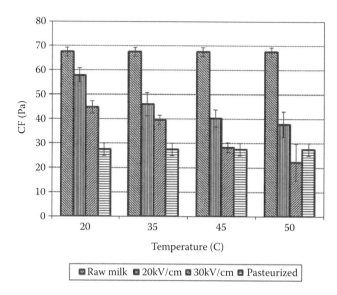

FIGURE 16.3 The effect of electric field intensity and temperature on CF. The applied pulse width is 2 μs, pulse frequency is 2 Hz, and pulse number is 120. Electric field intensity is 20 and 30 kV/cm. Temperature is 20°C, 35°C, 45°C, and 50°C. (Adapted from Yu L.J., Ngadi M.O., Raghavan G.S.V. 2009. *Journal of Food Engineering* 95(1): 115–118.)

using a bipolar square waveform. The maximum 30 kV/cm electric field intensity (E), 120 pulses, and up to 50°C treatment temperatures were applied. A dynamic rheological test was conducted to analyze milk coagulation properties. Results indicated that, in most cases, PEF-treated milk showed better rennetability compared with thermally PM. Figure 16.3 shows the effect of PEF treatment on CF when milk samples were treated using 20 and 30 kV/cm electric field intensity, 120 pulses at different temperatures. At 20 kV/cm, increasing treatment temperature from 20°C to 45°C significantly decreased CF of PEF-treated milk. However, additional increase in temperature from 45°C to 50°C only resulted in a slight increase in CF. The CF values obtained for samples treated using the lower electric field intensity of 20 kV/cm and lower temperature of 20°C were close to the value obtained for RM. However, the CF values obtained at both 30 and 20 kV/cm (except those treated at the higher electric field of 30 kV/cm and higher temperature of 50°C) were significantly higher than the CF values measured for heat-treated samples. The result indicated that treating milk with PEF at 20 kV/cm and 20°C, did not impact much changes in terms of milk coagulation properties. Applying an electric field of 30 kV/cm with the higher temperature of 50°C may affect coagulation properties similar to thermal treatment. Thus, for the purpose of cheese production, the treatment temperature during PEF treatment with higher electrical intensity (30 kV/cm) and longer pulses (120 pulse numbers) should not exceed 50°C to obtain desirable curd formation. Yu et al. (2012) investigated the effect of PEF treatment on the proteolysis process of cheese by using cheese curd slurries under the same PEF treatment

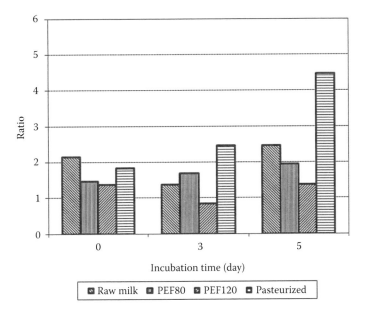

FIGURE 16.4 The ratio of hydrophobic to hydrophilic peptides in WSF of cheese curd slurries made from (1) RM; (2) PEF-treated milk at 80 pulses (PEF80); (3) PEF-treated milk at 120 pulses (PEF120); and (4) PM at the incubation time of 0, 3, and 5 days. (Adapted from Yu L.J., Ngadi M.O., Raghavan G.S.V. 2012. *Food and Bioprocess Technology* 5(1): 47–54.)

conditions. Cheese curds were made from RM, PM, and PEF-treated milk, and their proteolysis processes were compared using curd slurry incubated at 30°C for 5 days. The profiles of water-soluble peptides were measured using a reversed-phase high-performance liquid chromatography (RP-HPLC) system. It was found that both hydrophilic and hydrophobic peptides increased with increase in incubation time of the cheese slurry. However, there were differences in the ratio of hydrophobic to hydrophilic peptides. Heat-PM recorded the highest hydrophobic to hydrophilic peptides compared with those of raw and PEF-treated milk, indicating potential differences in flavor of Cheddar cheese made from these milks (Figure 16.4). The concentration of FAAs was measured by Cd–ninhydrin method. Results indicated that PEF-treated milk has intermediate proteolysis profiles between RM and PM in terms of peptide and FAA concentration. Figure 16.5 shows the concentration of FAAs in the water soluble fraction (WSF) extracted from cheese curd slurries made from RM, PM, and PEF-treated milks at 120 pulses (PEF120) and 80 pulses (PEF80). RM- and PEF80- treated milk had higher FAA than heat-PM, particularly with longer incubation time. The result suggests that increase in PEF pulses may have led to a more severe modification of the enzyme (Giner et al., 2001; Perez and Pilosof, 2004). The reduced enzyme activity may lead to less production of peptides and FAA. Thus, PEF as a nonthermal process could be used for the treatment of RM in mild temperature to produce high-quality cheeses with minimum sacrifice of the natural characteristics of the products.

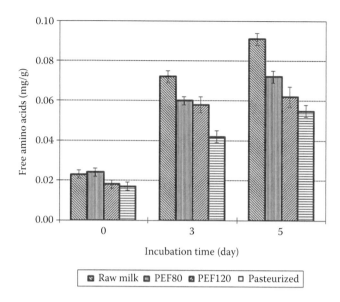

FIGURE 16.5   Concentration of FAAs in WSF of cheese curd slurries made from (1) RM; (2) PEF treated milk at 80 pulses (PEF80); (3) PEF treated milk at 120 pulses (PEF120); (4) PM at the incubation time of 0, 3 and 5 days (Adapted from Yu L.J., Ngadi M.O., Raghavan G.S.V. 2012. *Food and Bioprocess Technology* 5(1): 47–54.)

16.6   CONCLUSION

PEF treatment is a promising pasteurization approach for milk and cheese process-ing. As a nonthermal process, PEF can be employed for the treatment of RM in mild temperature to achieve adequate safety and shelf life while preserving the heat-sensitive nutrients and bioactive compounds. PEF may induce slight modifications in milk proteins and enzymes involved in cheese production. These will have implica-tions on the texture and flavor profiles of cheese. Studies have shown that cheeses made from PEF-treated milk have texture and flavor closer to RM and are therefore superior in terms of quality when compared to heat-PM.

16.7   FUTURE TRENDS

Further investigation is required to study the effect of PEF on the microstructure of casein and the effect of PEF on the interaction of casein and β-lactoglobulin, as well as with other ingredients to explain the effect of coagulation and gel characteristics. A scale-up of cheese production using PEF-treated milk is needed.

REFERENCES

Abd El-Gawad M.A.M., Ahmed N.S. 2011. Cheese yield as affected by some parameters. Review. *Acta Scientiarum Polonorum Technologia Alimentaria* 10(2): 131–153.

Adapa S., Schmidt K.A., Toledo R. 1997. Functional properties of skim milk processed with continuous high pressure throttling. *Journal of Dairy Science* 80(9): 1941–1948.

Agriculture and Agri-Food Canada, 2008. Canada Brand International, http://www.marque-canadabrand.agr.gc.ca/toolbox/dairy_e.htm. Last accessed, December 2012.

Amiali M., Ngadi M.O., Muthukumaran A., Raghavan G.S.V. 2010. Physiochemical property changes and safety issues of foods during pulsed electric field processing. In: *Physicochemical Aspects of Food Engineering and Processing* (edited by Sakamon D.), Chapter 6: pp. 117–218. CRC Press, Taylor & Francis Group, Boca Raton, FL.

Amiali M., Ngadi M.O. 2012. Microbial decontamination of food by pulsed electric fields. In: *Microbial Decontamination in Food Industry: Novel Methods and Applications* (edited by Demirci A. and Ngadi M.O.) Chapter 14: pp. 407–449. Woodhead Publishing Ltd, U.K.

Ashford E., Brooks S., Desart K., Knight S., Purdie J. 2012. Cheese: The website. http://shannak.myweb.uga.edu/. Last accessed, December 2012.

Banks W., Clapperton J.L., Girdler A.K., Steele W. 1984. Effect of inclusion of different forms of dietary fatty acid on the yield and composition of cow's milk. *Journal of Dairy Research* 51:387–395.

Barbosa-Cánovas G.V., Góngora-Nieto M.M., Pothakamury U.R., Swanson B.G. 1999. *Preservation of Foods with Pulsed Electric Fields*, pp. 4–47, 108–180. Academic Press, San Diego, USA.

Bendicho S., Espachs A., Stevens D., Arántegui J., Martín O. 1999. Effect of high intensity pulsed electric fields on vitamins of milk. *European Conference of Emerging Food Science and Technology*. Tampere, Finland, pp. 108.

Benfeldt C., Sorensen J., Ellegard K.H., Petersen T.E. 1997. Heat treatment of cheese milk: Effect on plasmin activity and proteolysis during cheese ripening. *International Dairy Journal* 7: 723–731.

Brito C., Niklitschek L., Molina L.H., Molina I. 2002. Evaluation of mathematical equations to predict the theoretical yield of Chilean Gouda cheese. *International Journal of Dairy Technology* 55: 32–39.

Cheese Reporter 2012. Vol. 136, No. 47. www.cheesereporter.com/May%2018,%202012.pdf. Last accessed, December 2012.

Corredig M., Dalgleish D.G. 1999. The mechanisms of the heat induced interaction of whey proteins with casein micelles in milk. *International Dairy Journal* 9: 233–236.

Darling D.F., Dickson J. 1979. Electrophoretic mobility of casein micelles. *Journal of Dairy Research* 46: 441–451.

Dunn J. 1996. Pulsed light and pulsed electric field for foods and eggs. *Poultry Science* 75(9): 1133–1136.

Evrendilek G.A., Dantzer W.R., Streaker C.B., Ratanatriwong P., Zhang Q.H. 2001. Shelf-life evaluations of liquid foods treated by pilot plant pulsed electric field system. *Journal of Food Processing and Preservation* 25: 283–297.

Fox P.F. 1981. Heat-induced changes in milk preceding coagulation. *Journal of Dairy Science* 64: 2127–2137.

Fox P.F., Guinee T.P., Cogan T.M., McSweeney P.L.H. 2000. *Fundamentals of Cheese Science*, Aspen Publishers, Maryland, USA.

Fox P.F., Cogan T.M. 2004. Factors that affect the quality of cheese. In: *Cheese Chemistry, Physics and Microbiology*, edited by Fox P.F., Paul McSweeney L.H., Timothy M.C., Timothy P.G. pp. 583–608. ISBN: 978-0-12-263652-3, London, Elsiever.

Giner J., Gimeno V., Barbosa-Canovas G.V., Martin O. 2001. Effects of pulsed electric fields processing on apple and pear polyphenoloxidases. *Food Science and Technology International* 7(4): 339–345.

Guo M., Park Y.W., Dixon P.H., Gilmore J.A., Kindstedt P.S. 2004. Relationship between the yield of cheese (Chevre) and chemical composition of goat milk. *Small Ruminant Research* 52:103–107.

Haug A., Hostmark A.T., Harstad O.M. 2007. Bovine milk in human nutrition—A review. *Lipids Health Disease* 6:25–41.

Huang K., Tian H., Gai L., Wang J. 2012. A review of kinetic models for inactivating microorganisms and enzymes by pulsed electric field processing. *Journal of Food Engineering* 111(2): 191–207.

Jeyamkondan S., Jayas D.S., Holley R.A. 1999. Pulsed electric field processing of foods: A review. *Journal of Food Protection* 62(9): 1088–1096.

Knorr D., Geulen M., Grahl T., Sitzmann W. 1994. Food application of high electric field pulses. *Trends in Food Science and Technology* 5: 71–75.

Lawrence R.C. 1993. Relationship between milk protein genotypes and cheese yield capacity. In: *Factors Affecting the Yield of Cheese*, ed. D.B. Emmons. International Dairy Federation. Brussels, 121–127.

Li S.Q., Zhang Q.H., Lee Y.Z., Pham T.V. 2003. Effects of pulsed electric fields and thermal processing on the stability of bovine immunoglobulin G (IgG) in enriched soymilk. *Journal of Food Science* 68(4): 1201–1207.

Li S.Q., Bomser J.A., Zhang Q.H. 2005. Effects of pulsed electric fields and heat treatment on stability and secondary structure of bovine immunoglobulin G. *Journal of Agricultural and Food Chemistry* 53(3): 663–670.

Lolkema H. 1993. Cheese yield used as an instrument for process control—Experience in Friesland, the Netherlands. In: *Factors Affecting the Yield of Cheese*, ed. D.B. Emmons. International Dairy Federation. Brussels, 156–197.

Lucey J., Kelly J. 1994. Cheese yield. *Journal of the Society of Dairy Technology* 47(1): 1–14.

Michalac S.L., Alvarez V.B., Zhang Q.H. 1999. *Microbial Reduction in Skim Milk Using PEF Technology*. IFT *Meeting*, Chicago, IL, USA.

Milk Facts, 2012. http://www.milkfacts.info/Nutrition%20Facts/Nutritional%20Components. htm. Last accessed, December 2012.

Ngadi M.O., Yu L.J., Amiali M., Ortega-Rivas E. 2009. Food quality and safety issues during pulsed electric field processing. In: E. Ortega-Rivas (ed.), *Processing Effects on Safety and Quality of Foods*, Chapter 16: pp. 446–472. CRC press, Taylor and Francis Group, Boca Raton, FL.

Oldfield D.J., Singh H., Taylor M.W. 1998. Association of β-lactoglobulin and α-lactalbumin with the casein micelles in skim milk heated in an ultra high temperature plant. *International Dairy Journal* 8: 765–770.

Oldfield D.J., Singh H., Taylor M.W., Pearce K.N. 2000. Heat induced interactions of β-lactoglobulin and α-lactalbumin with the casein micelle in pH adjusted skim milk. *International Dairy Journal* 10: 509–518.

Perez O.E., Pilosof A.M.R. 2004. Pulsed electric fields effects on the molecular structure and gelation of b-lactoglobulin concentrate and egg white. *Food Research International* 37: 102–110.

Qin B., Pothakamury U.R., Vega H., Martin O., Barbosa-Cánovas G.V., Swanson B.G. 1995. Food pasteurization using high intensity pulsed electric fields. *Journal of Food Technology* 49(12): 55–60.

Sandra S., Dalgleish D.G. 2007. The effect of ultra high-pressure homogenization (UHPH) on rennet coagulation properties of unheated and heated fresh skimmed milk. *International Dairy Journal* 17: 1043–1052.

Schmidt D.G., Poll J.K. 1986. Electrokinetic measurements on heated and unheated casein micelle systems. *Netherlands Milk and Dairy Journal* 40: 269–280.

Scott R. 1998. *Cheese Making Practice*. Aspen Publishers, Gaithersburg.

Sepulveda D.R., Gongora-Nieto M.M., Guerrero J.A., Barbosa-Cánovas G.V. 2005. Production of extended-shelf life milk by processing pasteurized milk with pulsed electric fields. *Journal of Food Engineering* 67: 81–86.

Sepulveda-Ahumada D.R., Ortega-Revas E., Barbosa-Canovas G.V. 2000. Quality aspects of Cheddar cheese obtained with milk pasteurized by pulsed electric fields. *Food and Bioproducts Processing* 78(2): 65–71.

Shamsi K., Versteeg C., Sherkat F., Wan J. 2008. Alkaline phosphatase and microbial inactivation by pulsed electric field in bovine milk. *Innovative Food Science and Emerging Technologies* 9: 217–223.

Shin J.K., Jung K.J., Pyun Y.R., Chung M.S. 2007. Application of pulsed electric fields with square wave pulse to milk inoculated with *E. coli*, *P. fluorescens*, and *B. stearothermophilus*. *Food Science and Biotechnology* 16(6): 1082–1084.

Singh H., Fox P.F. 1987. Heat stability of milk: Role of β-lactoglobulin in the pH dependent dissociation of micellar k-casein. *Journal of Dairy Research* 54: 509–521.

Singh S., Kanawjia S.K. 1988. Development of manufacturing technique for Paneer from cow milk. *Indian Journal of Dairy Science* 41: 322–325.

Singh H., Creamer L.K. 1992. Heat stability of milk. In: *Advances in Dairy Chemistry*, Fox P.F., vol. 1: *Proteins* (2nd ed.), pp. 621–656. Elsevier Science Publishers Ltd, London, UK.

Singh H., Waungana A. 2001. Influence of heat treatment of milk on cheese making properties. *International Dairy Journal* 11:543–551.

Smits P., van Brouwershaven J.H. 1980. Heat-induced association of β-lactoglobulin and casein micelles. *Journal of Dairy Research* 47: 313–325.

Sobrino-Lopez A., Martin-Belloso O. 2010. Review: Potential of high-intensity pulsed electric field technology for milk processing. *Food Engineering Reviews* 2(1): 17–27.

Tsong T.Y. 1991. Electroporation of cell membranes. *Biophysical Journal* 60: 297–306.

USDA Database, 2012. http://ndb.nal.usda.gov/. Last accessed, December 2012.

Van Boekel M.A.J.S. 1993. Transfer of milk components to cheese: Scientific considerations. In: *Cheese Yield and Factors Affecting Its Control*. IDF Seminar. Cork (Ireland), 19–28.

Van den Berg M.G. 1994. *The Transformation of Casein in Milk into the Paracasein Structure of Cheese and Its Relation to Non-Casein Milk Components*. IDF, International Dairy Federation. Brussels, 35–47.

Vasbinder A.J., Alting A.C., de Kruif C.G. 2003. Quantification of heat-induced casein–whey protein interactions in milk and its relation to gelation kinetics. *Colloids and Surfaces B* 31: 115–123.

Xiang B.Y., Simpson M.V., Ngadi M.O., Simpson B.K. 2011a. Effect of pulsed electric field on the rheological and colour properties of soy milk. *International Journal of Food Sciences and Nutrition* 62(8): 787–793.

Xiang B.Y., Ngadi M.O., Ochoa-Martinez L.A., Simpson M.V. 2011b. Pulsed electric field-induced structural modification of whey protein isolate. *Food and Bioprocess Technology* 4(8): 1341–1348.

Yeom H.W., Evrendilek G.A., Jin Z.T., Zhang Q.H. 2001. *Processing of Yogurt Based Product with Pulsed Electric Fields. IFT Meeting*, New Orleans, Louisiana, USA.

Yu L.J., Ngadi M.O., Raghavan G.S.V. 2009. Effect of temperature and pulsed electric field treatment on rennet coagulation properties of milk. *Journal of Food Engineering* 95(1): 115–118.

Yu L.J., Ngadi M.O., Raghavan G.S.V. 2012. Proteolysis of cheese slurry made from pulsed electric field-treated milk. *Food and Bioprocess Technology* 5(1): 47–54.

Zhang Q.H., Barbosa-Cánovas G.V., Swanson B.G. 1995. Engineering aspects of pulsed electric field pasteurization. *Journal of Food Engineering* 25(2): 261–281.

17 Quality, Safety, and Shelf-Life Improvement in Fruit Juices by Pulsed Electric Fields

Özlem Tokuşoğlu, Isabella Odriozola-Serrano, and Olga Martín-Belloso

CONTENTS

17.1 INTRODUCTION TO THERMAL PASTEURIZATION AND PULSED ELECTRICAL FIELD FOR FRUIT JUICES

Fruit juices are popular beverages worldwide and are perceived as nutritious food products due to their health-promoting constituents, including phenolic antioxidants. Even though, the consumption of minimally processed fruit juices has increased in recent years, distribution and marketing are limited owing to their shorter shelf life (Song et al., 2007). It is stated that various microorganisms, in particular, acid-tolerant bacteria and fungi (yeasts and molds) can cause spoilage and produce undesirable organoleptic alterations in juices. Pathogen microorganisms in products can also cause human illness (Tournas et al., 2006).

However, heat processing is the most widely used fruit juice pasteurization technique and juice pasteurization is based on a 5-log reduction of the most resistant microorganisms of public health significance. Recently, nonthermal technologies have been extensively implemented. It is known that traditional thermal pasteurization can be classified into low-temperature/long-time (LTLT) and high-temperature/short-time (HTST) processes. It is stated that LTLT pasteurization involves heating a food at about 63°C for not <30 min, while, for fruit juices, HTST pasteurization is applied at temperatures around 72°C with holding times of 15 s and above (FDA, 2001). Both LTLT and HTST methods may degrade the taste, color, flavor, and nutritional quality of foods (Charles-Rodríguez et al., 2007).

However, higher temperature was not only more efficient for bacterial growth controlling, but also resulted in a higher reduction of phenolics in fruit juices. It is reported that thermal pasteurization can cause the degradation of phenolic compounds in most fruit juices (Chen et al., 2013). It is revealed that thermal processing may cause complex physical and chemical reactions in juices. The releasing of phenolic compounds from their bonded forms, the degradation of polyphenols, and the breakdown and transformation of phenolics can occur (Nagy et al., 1989).

Table 17.1 shows the effect of thermal pasteurization on the microbial inactivation and phenolic degradation of fruit juices (Table 17.1) (Chen et al., 2013). Noci et al. (2008) reported that the phenolic degradation (48%) was detected in apple juices pasteurized at 94°C compared with those pasteurized at 72°C. Odriozola-Serrano et al. (2008) stated that holding the pasteurization at 90°C/1 min led to a higher degradation of total phenolics than holding at 90°C/30 s for strawberry juice. It was concluded that the processing duration was an important factor for phenolic degradation in the juices (Odriozola-Serrano et al., 2008). It was also stated that pasteurization

TABLE 17.1
Effect of Thermal Pasteurization on the Microbial Inactivation and Phenolic Degradation of Fruit Juices

Types of Juices	Conditions	Microbial Inactivation ($\log_{10}$)	Phenolic Degradation (%)	References
Longan juices	100°C/1 min	7.0	42	Zhang et al. (2010)
Apple juices	72°C/26 s	6.0	40	Noci et al. (2008)
Apple juices	94°C/26 s	6.7	48	Noci et al. (2008)
Apple/cranberry juices	72°C/26 s	N/A	NSS	Caminiti et al. (2011)
Orange juices	90°C/1 min	2.5	26	Elez-Martinez et al. (2006)
Strawberry juices	90°C/30 s	N/A	51[a]	Odriozola-Serrano et al. (2008)
Strawberry juices	90°C/60 s	N/A	55[a]	Odriozola-Serrano et al. (2008)
Tomato juices	98°C/40 s	N/A	12	Pérez-Conesa et al. (2009)
Tomato juices	108°C/40 s	N/A	22	Pérez-Conesa et al. (2009)
Tomato juices	128°C/40 s	N/A	24	Pérez-Conesa et al. (2009)
Fruit smoothie	72°C/15 s	3.5	N/A	Walkling-Ribeiro et al. (2008a,b)
Fruit smoothie	72°C/15 s	E.coli 6.3	N/A	Walkling-Ribeiro et al. (2008a)

Source: Adapted from Chen Y., Yu L.J., and Rupasinghe H.P.V. 2013. Journal of Scientific Food Agriculture, 93, 981–986.

Note: N/A, not reported; NSS, no significant change after processing ($p < 0.05$).

[a] Phenolic reduction after 56 days of storage at 4°C.

at HTST showed much better performance on single juices than on more complex/ viscous multiple fruit juices (Chen et al., 2013).

It is reported that nonthermal techniques provide the required overall stability and microbial safety, but with minimal impact on sensory and nutritive properties of the products (Ross et al., 2003). Various nonthermal processing technologies have been developed for food pasteurization, including high-voltage pulsed electric field (PEF) (Ngadi et al., 2009), ultraviolet (UV) exposure (Noci et al., 2008), membrane filtration (Zárate-Rodríguez et al., 2001), and high hydrostatic pressure (HHP) (Ramaswamy et al., 2005). The above-mentioned technologies have the advantage of not only maintaining the "fresh-like" characteristics of food but also saving time and energy.

PEFs can cause electroporation of cell membranes that, depending on the field intensity, may induce irreversible cell damage (Hamilton and Sale, 1967). It is stated that PEF can be applied as an alternative method for cell disintegration. Biological tissues exposed to high electric field pulses develop pores in the cell membrane and these actions result in increased membrane permeability and loss of the cell content (Knorr et al., 2001; Vorobiev and Lebovka, 2008).

It is stated that the novel nonthermal technology PEF for pasteurization or sterilization can inactivate microorganisms and enzymes with minor increasing in temperature, providing fresh-like products with improved flavor and color properties as well as highly preserved nutritive value (Aguilar-Rosas et al., 2007).

PEF could be regarded as a suitable treatment and continuous processing technology for the inactivation of vegetative cells such as *Escherichia coli* and yeasts, for fruit juice preservation (Lado and Yousef, 2002). Owing to the reports on outbreaks that have identified pathogenic microorganisms such as *E. coli* O157:H7 in fruit juices (CDC, 1995), PEF could be used as an effective preservation method for fruit juices. Table 17.2 shows the effects of PEF pasteurization on microbial inactivation and phenolic degradation of fruit juices (Table 17.2) (Chen et al., 2013). It is stated that the PEF treatment efficacy on microbial inactivation can be affected by various factors including electric field strength, pulse length and shape, treatment time, treatment temperature, product composition, and specific energy input (Figure 17.1) (Chen et al., 2013; Heinz et al., 2002).

Significant relations between electric field strength and microbial inactivation efficiency were found (Alkhafaji and Farid, 2007; Huang and Wang, 2009; Toepfl et al., 2007).

It was considered that Gram-positive bacteria were generally more resistant to PEF than Gram-negative bacteria due to their different cell structure (Hülsheger et al., 1983). It was also stated that PEF was more effective on juices with simple composition than on more complex multiple-juice products (Rodrigo et al., 2003). Rodrigo et al. (2003) revealed that blending orange juice with an increasing proportion of carrot juice reduced the effectiveness of PEF on *E. coli*, could be attributed to the more complex matrix of the multicomponent juice (Rodrigo et al., 2003).

It is stated that PEFs of moderate intensity (0.3–3 kV/cm) have emerged as a useful technique for improving solid–liquid expression from plant foods (Vorobiev and Lebovka, 2008).

The increase in juice yield (Grimi et al., 2009; Schillinget al., 2007), the reduction of processing time, and the energy requirement in comparison to enzymatic and

TABLE 17.2

PEF Pasteurization Effects on Microbial Inactivation and Phenolic Degradation of Fruit Juices

Types of Juices	Conditions	Microbial Inactivation ($\log_{10}$)	Phenolic Degradation (%)	References
Longan juices	32 kV cm^{-1}, 90 s	2.0	18	Zhang et al. (2010)
Apple juices	40 kV cm^{-1}, 100 μs	5.4	7	Noci et al. (2008)
Apple juices	35 kV cm^{-1}, 4800 μs	L.brevis 6.3	14.5	Aguilar-Rosas et al. (2007)
Apple juices	35 kV cm^{-1}, 4800 μs	S.cerevisiae 4.2	14.5	Aguilar-Rosas et al. (2007)
Orange juices	35 kV cm^{-1}, 1000 μs	2.5	15	Elez-Martinez et al. (2006)
Orange juices	40 kV cm^{-1}, 150 μs	S.aureus 5.5	N/A	Walkling-Ribeiro et al. (2008)
Strawberry juices	35 kV cm^{-1}, 1700 μs	N/A	49[a]	Odriozola-Serrano et al. (2008)
Watermelon juices	35 kV cm^{-1}, 50 μs	N/A	NSS	Oms-Oliu et al. (2009)
Fruit smoothie	34 kV cm^{-1}, 150 μs	E.coli 5.4	N/A	Walkling-Ribeiro et al. (2008)

Source: Adapted from Chen Y., Yu L.J., and Rupasinghe H.P.V. 2013. *Journal of Scientific Food Agriculture*, 93, 981–986.

Note: N/A, not reported; NSS, no significant change after processing ($p < 0.05$).

[a] Phenolic reduction after 56 days of storage at 4°C.

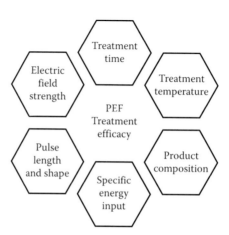

FIGURE 17.1 Various factors affecting the PEF treatment efficacy on microbial inactivation.

thermal treatments of mashes (Toepfl et al., 2006) are important qualities of PEF. Also, the enhancing of released secondary plant metabolites (Fincan et al., 2004; Lebovka et al., 2003) and the possibility to recovering of native pectins from apple pomace (Schilling et al., 2008) can be performed through cell disintegration by PEF.

It was revealed that the enhancement of the juice yield after PEF application was accompanied by a noticeable decrease of absorbance (Praporscic et al., 2007) and improvement in juice cloudiness (Grimi et al., 2009, 2010).

17.2 EFFECT OF PEF PROCESSING ON ASCORBIC ACID (VITAMIN C)

A number of studies have proven the effectiveness of PEF technologies in achieving higher ascorbic acid (vitamin C) content in comparison with heat treatments. Vitamin C retention in a heat-treated (90°C, 60 s) tomato juice was 79.2%, whereas in a PEF-processed juice (35 kV/cm for 1500 µs in bipolar 4-µs pulses at 100 Hz), a 86.5% retention was attained just after processing (Odriozola-Serrano et al., 2008a). Consistently, vitamin C retention reported in a heat-treated strawberry juice (94%) was significantly lower than in a PEF-treated juice (98%) (Odriozola et al., 2008b). Most of the differences in vitamin results between PEF and heat treatments can be explained through the temperatures reached during processing. The concentration of vitamin C in thermally and PEF-processed juices gradually decreased with storage time. However, it has been demonstrated that vitamin C is better retained in PEF-treated juices than in thermally processed juices after 56 days of storage at 4°C (Morales-de la Peña et al., 2010a; Odriozola-Serrano et al., 2008a,b) (Figure 17.2).

Vitamin C juices in PEF juices also depend on PEF-processing factors and, thus, the lower the electric field strength, the treatment time, the pulse frequency, or the pulse width, the higher the vitamin C retention in orange (Elez-Martínez and Martín-Belloso, 2007), tomato (Odriozola-Serrano et al., 2007, 2008b), and strawberry juices (Odriozola-Serrano et al., 2008c, 2009a). Ascorbic acid is an unstable compound, which under less-desirable conditions, decomposes easily; hence, the milder the treatment, the better the vitamin C retention in juices. However, differences in vitamin C pressure stability during storage could be explained by the initial oxygen content and possible endogenous prooxidative enzyme activity. Bioavailability is defined as the proportion of the nutrients, bioactive compounds, or phytochemicals that are digested, absorbed, and metabolized throughout the normal pathway (Sánchez-Moreno et al., 2009). Drinking two glasses of PEF-treated or HP (high-pressure)-treated orange juice (500 mL/day) containing approximately 180 mg of vitamin C was associated with a significant increase in plasma vitamin C concentration and a decrease in plasma levels of 8-epiPGF2α (biomarkers of lipid peroxidation) (Sánchez-Moreno et al., 2003, 2004).

17.3 EFFECT OF PEF PROCESSING ON CAROTENOIDS

Recent studies have suggested that carotenoids content increases significantly after PEF processing compared to the untreated juice. Odriozola-Serrano et al. (2007) observed an enhancement of up to 46.2% in the lycopene-relative concentration of

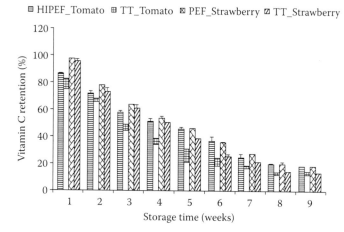

FIGURE 17.2 The effects of PEF and heat treatments on vitamin C of fruit juices throughout storage at 4°C. (Adapted from Odriozola-Serrano I., Soliva-Fortuny R., and Martín-Belloso O. 2008. *European Food Research Technology*, 228, 239–248; Odriozola-Serrano I., Soliva-Fortuny R., and Martín-Belloso O. 2008a. *Innovative Food Science and Emerging Technologies*, 9, 272–279.)

tomato juices after applying different PEF treatments (35 kV/cm for 1000 μs). It has been hypothesized that thermal treatments may lead to an increase in some individual carotenoids, owing to greater stability, inactivation of oxidative and hydrolytic enzymes, and unaccounted moisture loss, which concentrates the sample (Rodríguez-Amaya, 1997). Nguyen and Schwartz (1999) suggested that homogenization and heat treatment disrupt cell membranes and protein–carotenoids complex, making carotenoids more accessible for extraction and probably more bioaccessible. Results obtained by Odriozola-Serrano et al. (2009) offer some controversy, provided that changes in the relative amounts of carotenoids are not consistent for similar compounds. Although the reason for these results is not known, it could be speculated that carotenoid conversions could be triggered by the PEF treatments. This would explain the increase in lycopene at the expense of its precursors. However, to confirm this hypothesis, it should be demonstrated that an effect of PEF on enzymes involved in lycopene synthesis in tomato juice, namely phytoene synthase, which catalyzes the conversion of geranyl–geranyl diphosphate (GGPP) into phytoene, and phytoene desaturase, which mediates the conversion of phytoene into phytofluene, β-carotene, and lycopene (Fraser et al., 1994).

Recent studies also show a higher stability of carotenoids throughout storage in PEF-treated products compared to heat-treated equivalents. PEF-processed tomato juices at 35 kV/cm for 1500 μs maintained higher contents of carotenoids (lycopene, neurosporene, and γ-carotene) through refrigerated storage than heat-processed juices at 90°C for 30 s (Odriozola-Serrano et al., 2009b). The major cause of carotenoid losses in vegetable products is the oxidation of the highly unsaturated carotenoid structure.

Research efforts have been made to obtain fruit and vegetable juices by HP processing without the quality and nutritional damage caused by heat treatments (Sánchez-Moreno et al., 2009). Individual carotenoids with antioxidant activity (β-cryptoxanthin, zeaxanthin, and lutein) appeared to be resistant to a HP treatment of 400 MPa at 40°C for 1 min, thus resulting into a better preservation of the antioxidant activity of the juices with respect to the thermally pasteurized juices.

17.4 EFFECT OF PEF PROCESSING ON PHENOLIC COMPOUNDS

Flavonoids are the most common and widely distributed group of plant phenolics. Among them, flavones, flavonols, flavanols, flavanones, anthocyanins, and isoflavones are particularly common in fruits. Regarding the main flavanones identified in orange juice, HP treatments (400 MPa/40°C/1 min) increased the content of naringenin by 20% and the content of hesperetin by 40% in comparison with an untreated orange juice (Sánchez-Moreno et al., 2005). These results are in accordance with those obtained by other authors showing higher extraction of phenolic compounds due to HP processing. PEF processing (35 kV/cm for 1500 μs with 4-μs bipolar pulses at 100 Hz) and thermal treatments (90°C, 30 s and 90°C, 60 s) did not affect phenolic content of tomato juices. Both PEF- and heat-treated tomato juices undergo a substantial loss of phenolic acids (chlorogenic and ferulic) and flavonols (quercetin and kaempferol) during 56 days of storage at 4°C. Caffeic acid content was slightly enhanced over time, regardless of the kind of processing, whereas PEF- and heat-treated tomato juices underwent a substantial depletion of p-coumaric acid during storage.

The increase of caffeic acid in tomato juices after 28 days of storage could be directly associated with residual hydroxylase activities, which convert coumaric acid into caffeic acid (Odriozol-Serrano et al., 2009). In strawberry juices, p-hydroxybenzoic content was enhanced slightly, but significantly after PEF processing (35 kV/cm for 1700 μs in bipolar 4-μs pulses at 100 Hz) compared to the untreated juice, whereas ellagic acid was substantially reduced when the heat treatment was conducted at 90°C for 60 s. No significant differences in flavonol (kaempferol, quercetin, and myricetin) contents were obtained between fresh and treated strawberry juices; thus, these phenolic compounds were not affected by processing (Odriozola-Serrano et al., 2008b). The degradation of phenolic compounds during storage has been mainly related to the residual activity (RA) of polyphenol oxidase (PPO) and polyphenol peroxidase (POD) (Odriozola-Serrano et al., 2009).

PEF seems to be a good technology to obtain fruit juice–soymilk beverage with a high content of isoflavones and fresh-like characteristics. In a blend fruit juice–soymilk beverage, PEF treatment (35 kV/cm with 4-μs bipolar pulses at 200 Hz for 800 or 1400 μs) did not cause significant changes in the total isoflavone content and, during the storage period, the total isoflavone content tended to increase throughout the time. Genistein, daidzein, and daidzin content increased, while genistin showed a slight decrease, irrespective of the treatment applied (Morales et al., 2010b). These authors suggested that the concentration of some isoflavones in the fruit juice–soymilk beverage might increase during storage from the flavonoids (mainly naringenin from orange) present in the fruits used for the elaboration of the beverage. Nevertheless, there is a need for more in-depth research to provide biochemical evidence of the observed changes.

Anthocyanins are a widespread group of plant phenolic compounds that have been regarded as a natural alternative to replace synthetic food colorants. The content of individual anthocyanins significantly depended on the high-intensity pulsed electric fields (HIPEF) treatment time and electric field strength applied during HIPEF processing of the strawberry juices. Anthocyanins were not affected by HIPEF processing when strawberry juice was treated at 22 kV/cm. At electric field strengths from 27 to 32 kV/cm, it was observed that the lower the treatment time and electric field strength, the greater the anthocyanin retention. By the contrary, strawberry juices subjected to the most intensive treatment (37 kV/cm) exhibited the highest anthocyanins content (110–151%) when the longest HIPEF treatment was conducted (Figure 17.3). It has been reported that proanthocyanins are converted into anthocyanins after processing in acidic water-free conditions (Saint-Cricq de Gaulejac et al., 1999). Thus, intensive HIPEF treatments might stimulate the transformation of proanthocyanins into anthocyanins. Further investigations are still needed to explain the mechanisms that mediate this conversion.

17.5   RECENT STUDIES ON PEF-TREATED APPLE JUICES

PEF effects on pH, soluble solids, acidity, PPO and POD activities, polyphenolic contents, and color of apple juice were extensively studied by various researchers (Aguilar-Rosas et al., 2007; Akdemir-Evrendilek et al., 2000; Charles-Rodríguez et al., 2007; Noci et al., 2008; Ortega-Rivas et al., 1998; Riener et al., 2008; Schilling et al., 2007, 2008). It was reported that PEF treatments did not affect the physicochemical properties (such as pH, acidity, soluble solids, color, and ascorbic acid) of apple juices (Aguilar-Rosas et al., 2007; Akdemir Evrendilek et al., 2000; Charles-Rodríguez et al., 2007; Noci et al., 2008; Ortega-Rivas et al., 1998; Schilling et al., 2007).

Riener et al. (2008) determined the highest level of decreasing PPO and POD enzyme activities, as 71% and 68%, respectively, by using a combination of preheating to 50°C, and a PEF treatment time of 100 μs at 40 kV/cm; the kinetic data for the inactivation of these enzymes were described as a first-order model (Riener et al., 2008).

It was found that the polyphenol levels and antioxidant capacities of the apple juices extracted from PEF-treated apple mash (1–5 kV/cm, n = 30 pulses) did not significantly differ from the controls (Schilling et al., 2007), whereas PEF (35 kV/cm, and a frequency of 1200 pulses per second) caused a 14.49% reduction in phenolics, and a significant decrease in volatile compounds in apple juice (Aguilar-Rosas et al., 2007).

The effects of electric field strength (as 0–35 kV/cm) and pulse rise time (PRT) (as 2 and 0.2 μs) on the quality properties including enzymatic activity, vitamin C, total phenols, color, antioxidant capacities, and rheological characteristics of apple juice were investigated by Bi et al. (2013). It was found that with increasing the electric field strength and PRT, the RA of PPO and POD decreased, and an almost complete inactivation of both enzymes was achieved at 35 kV/cm and 2 μs-PRT. In

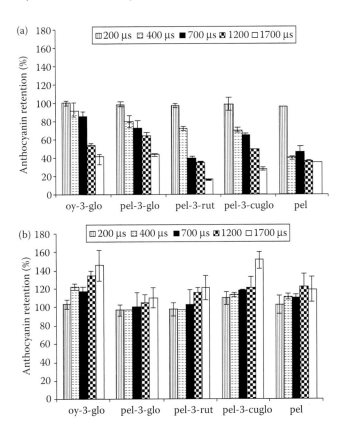

FIGURE 17.3 The effect of treatment time and electric field strength on the individual anthocyanins of strawberry juice. Treatments were performed at 232 Hz and square bipolar pulses of 1 μs. Electric field strengths: (a) 32 and (b) 37 kV/cm. Anthocyanins: cy-3-glc (cyanidin-3-glucoside), pel-3-glc (pelargonidin-3-glucoside), pel-3-rut (pelargonidin-3-rutinoside), pel-3-suglc (pelargonidin-3-succinylglucoside), and pel (aglycon pelargonidin).

the study described by Bi et al. (2013), the vitamin C content in apple juice significantly decreased during PEF treatment; the largest loss was 36.6% at 30 kV/cm and 2 μs-PRT whereas the total phenols were not affected by PEF with 2 μs-PRT, but decreased significantly by PEF with 0.2 μs-PRT (Bi et al., 2013). As antioxidant activity values, 1,1-diphenyl-2-picrylhydrazyl (DPPH) level was not affected by PEF, whereas the ferric reducing ability of plasma (FRAP) (FRAP assay is as a measure of "antioxidant power") and oxygen radical absorbance capacity (ORAC) values increased with increasing the electric field strength and decreasing the PRT.

Figure 17.4 shows the formation of free OH• radicals, determined by the concentration of 2,5-dihydroxybenzoic acid (2,5-DHBA) (Bi et al., 2013). A2,5-DHBA concentration was positively correlated to electric field strength (R^2 was 0.998 for 2 μs-PRT and 0.997 for 0.2 μs-PRT) using a linear model (Figure 17.4). The concentration of 2,5-DHBA was greater after PEF processing with 0.2 μs-PRT, suggesting that more free OH• radicals were generated by treatment with shorter PRT (Bi et al., 2013).

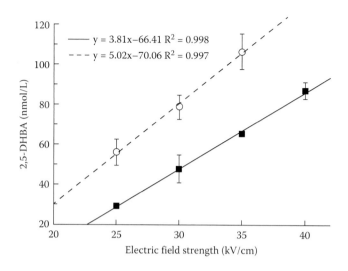

FIGURE 17.4   Formation of OH• in apple juice treated by pulsed electric field. (■) 2 μs PRT and (○) 0.2 μs PRT. (Adapted from Bi X. et al. 2013. *Innovative Food Science and Emerging Technologies*, 17, 85–92.)

In PEF-treated apple juices, the lightness (L^*) and the yellowness (b^*) values were significantly higher than the controlled juices. Table 17.3 shows the color parameters of apple juice after PEF treatment and Table 17.4 shows the vitamin C content, total phenols, and antioxidant capacity (Tables 17.3 and 17.4) (Bi et al., 2013). In the study reported by Odriozola-Serrano et al. (2009), vitamin C loss in PEF-treated watermelon juice was 54.2% at 35 kV/cm and it was found that vitamin C level of

TABLE 17.3
Color Parameters of Apple Juice After Pulsed Electric Field Treatment

Treatment Parameters		L	a	b	ΔE
Control		44.16 ± 0.06^a	-2.75 ± 0.01^a	12.84 ± 0.03^a	0.00
PRT = 2 μs	0 kV/cm	44.38 ± 0.08^a	-2.70 ± 0.07^a	13.16 ± 0.23^a	0.39
	25 kV/cm	44.84 ± 0.14^a	-2.91 ± 0.01^a	13.03 ± 0.09^a	0.72
	30 kV/cm	49.47 ± 0.62^b	-2.97 ± 0.04^a	15.61 ± 0.03^b	5.99
	35 kV/cm	51.86 ± 1.22^c	-2.97 ± 0.10^a	16.12 ± 0.68^b	8.36
PRT = 0.2 μs	0 kV/cm	43.85 ± 0.27^a	-2.65 ± 0.24^a	12.73 ± 0.18^a	0.42
	25 kV/cm	45.14 ± 0.49^a	-2.72 ± 0.30^a	13.43 ± 0.57^a	1.28
	30 kV/cm	49.23 ± 0.85^b	-2.81 ± 0.10^a	15.21 ± 0.53^b	5.59
	35 kV/cm	53.97 ± 0.56^d	-2.54 ± 0.06^a	17.50 ± 0.28^c	10.85

Source: Adapted from Bi X. et al. 2013. *Innovative Food Science and Emerging Technologies*, 17, 85–92.
Note: PRT, pulse rise time; values were mean ± SD of three measurements, n = 4; different letters represent a significant difference within the same column ($p < 0.05$).

TABLE 17.4

Vitamin C Content, Total Phenols, and Antioxidant Capacity in Apple Juice After Pulsed Electrical Field

Treatment Parameters		Vc (µg/g)	Total Phenols (µg GAE/g)	DPPH (mmol TE/L)	FRAP (mmol TE/L)	ORAC (mmol TE/L)
Control		119.9 ± 5.2^a	704.6 ± 13.8^{abc}	1.03 ± 0.04^a	2.70 ± 0.07^{ab}	51.24 ± 1.34^{ab}
PRT = 2 µs	0 kV/cm	110.8 ± 1.8^a	647.0 ± 10.9^d	1.02 ± 0.04^a	2.33 ± 0.03^c	40.65 ± 1.62^c
	25 kV/cm	79.5 ± 3.8^b	663.3 ± 19.0^{cd}	1.00 ± 0.06^a	2.55 ± 0.05^b	42.89 ± 1.65^c
	30 kV/cm	76.0 ± 0.3^b	672.7 ± 14.8^{bcd}	0.99 ± 0.06^a	2.73 ± 0.02^{ab}	46.50 ± 0.78^{bc}
	35 kV/cm	78.3 ± 1.7^b	713.3 ± 19.9^{ab}	1.01 ± 0.06^a	2.81 ± 0.04^a	54.65 ± 1.58^{ad}
PRT = 0.2 µs	0 kV/cm	111.3 ± 1.8^a	644.5 ± 6.5^d	1.03 ± 0.01^a	2.34 ± 0.02^c	41.91 ± 2.04^c
	25 kV/cm	83.5 ± 5.0^b	650.4 ± 6.4^d	1.02 ± 0.01^a	2.84 ± 0.08^a	54.45 ± 3.09^{ad}
	30 kV/cm	79.8 ± 4.6^b	657.8 ± 15.6^d	1.02 ± 0.01^a	2.88 ± 0.04^a	60.07 ± 2.39^{de}
	35 kV/cm	86.3 ± 4.1^b	668.3 ± 14.9^{cd}	1.02 ± 0.02^a	3.91 ± 0.09^d	63.35 ± 2.83^e

Source: Adapted from Bi X. et al. 2013. *Innovative Food Science and Emerging Technologies*, 17, 85–92.
Note: PRT, pulse rise time; values were mean ± SD of three measurements, n = 4; different letters represent a significant difference within the same column ($p < 0.05$).

PEF-processed tomato juice decreased drastically after being treated compared to untreated juice and the more sever the PEF treatment was, the lower the vitamin C retention (Odriozola-Serrano et al., 2009b).

Bi et al. (2013) determined that the apparent viscosity and consistency index (K) of apple juice decreased while the flow behavior index (n) increased with increasing the electric field strength, and apple juice treated at 2 µs-PRT had

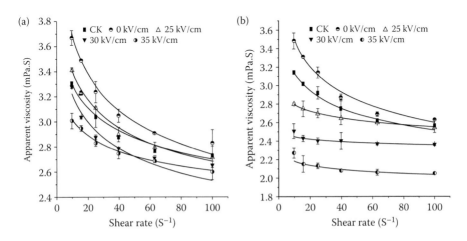

FIGURE 17.5 The effect of PEF on the rheological behavior of apple juice. (a) 2 µs PRT and (b) 0.2 µs PRT. (Adapted from Bi X. et al. 2013. *Innovative Food Science and Emerging Technologies*, 17, 85–92.)

TABLE 17.5

Rheological Parameters of Apple Juice After Pulsed Electric Field Treatment

Treatment Parameters		K(mPa.S^n)	n	R^2
Control		4.10	0.91	0.99
PRT= 2 μs	0 kV/cm	4.82	0.88	0.99
	25 kV/cm	4.32	0.90	0.98
	30 kV/cm	3.96	0.91	0.97
	35 kV/cm	3.50	0.94	0.99
Control		3.91	0.91	0.99
PRT = 0.2 μs	0 kV/cm	4.73	0.87	0.99
	25 kV/cm	3.07	0.96	0.99
	30 kV/cm	2.62	0.98	0.87
	35 kV/cm	2.45	0.94	0.71

Source: Adapted from Bi X. et al. 2013. *Innovative Food Science and Emerging Technologies*, 17, 85–92.

Note: PRT, pulse rise time; K, consistency index; n: flow behaviour index; R^2, regression coefficient by power law model; values were mean ± SD of three measurements, n = 4; different letters represent a significant difference within the same column ($p < 0.05$).

significantly higher apparent viscosity than treated at 0.2 μs-PRT (Bi et al., 2013). Figure 17.5 shows the effect of PEF on the rheological behavior of apple juice. In Figure 17.5, the apparent viscosity in PEF-treated apple juice as a function of shear rate was shown (Figure 17.5). In the study reported by Bi et al. (2013), the apparent viscosity of all samples decreased as shear rate increased from 10 to 100 s^{-1}, suggesting that the juices followed the pseudoplasticity flow behavior. It was found that the apparent viscosity of apple juice treated at 0.25 kV/cm with 2 μs-PRT or 0 kV/cm with 0.2 μs-PRT was higher than that of the control juice (Table 17.5). It was stated that the apparent viscosity of the juices decreased with increasing the electric field strength and samples treated at 2 μs-PRT had significantly higher apparent viscosity than samples treated at 0.2 μs-PRT at the same electric field strength (Table 17.5) (Bi et al., 2013).

Turk et al. (2012) investigated the effects of PEF ($E = 650$ V/cm, $t_{PEF} = 23.2$ ms, and $q = 32$ kJ/kg) on apple mash on an industrial scale; it was determined the quantitative, qualitative, and sensory attributes of apple mash after PEF treatment. In the study stated by Turk et al. (2012), juice yield significantly increased from 71.1% to 76.3% by PEF treatment.

The significant increase (8.8%) was observed for the native polyphenol concentration and juice color significantly altered by PEF treatment. The juice from treated mash was less turbid and the typical apple odor was significantly higher than the untreated sample. It was found that the overall taste of apple juice was significantly more intense (Turk et al., 2012). In the study described by Turk et al. (2012), the fructose, glucose, and malic acid profiles of apple juice did not vary despite a significant increase of the dry matter for treated juices (Turk et al., 2012).

17.6 RECENT STUDIES ON PEF-TREATED CITRUS JUICES

It was stated that the greater the electric field strength, higher the temperature, or longer the treatment time, the greater the microbial inactivation (Wouters et al., 2001). It is accepted that the pertinent pathogen in citrus juices is generally regarded as *Salmonella* while critical and relatively PEF-resistant microorganisms in orange juice are lactic acid bacteria and pathogenic *E. coli* (Buckow et al., 2013; Parish, 1998).

Buckow et al. (2013) stated that PEF acceptance as an alternative to conventional thermal pasteurization of foodstuffs, regulators often demand the successful destruction of >5 $\log_{10}$ of pathogenic microorganisms in fruit juices. It is revealed that yeasts and molds are relatively sensitive to PEF processing and are effectively inactivated in orange juice by mild PEF conditions (<10 kV/cm for <20 μs) near room temperature (Buckow et al., 2013). It is reported that PEF inactivation of microorganisms is enhanced by the addition of antimicrobials such as nisin, benzoate, sorbate, and citric acid (CA), allowing the energy-efficient use of PEF for fruit juice preservation (Buckow et al., 2013). Table 17.6 shows the compiled data on the microbial inactivation by PEF technology in orange juices (Table 17.6) (Buckow et al., 2013).

Hartyáni et al. (2011) stated the physical-chemical and sensory properties of PEF and HHP-treated citrus juices. In the study described by Hartyáni et al. (2011), the physicochemical quality properties (pH, Brix°, electric conductivity, and color), the aroma content of most consumed citrus juices (100% orange, grapefruit, and tangerine juice) were examined (Hartyáni et al., 2011).

The applied technology was pulsed electric field (PEF) treatment with the parameters of 28 kV/cm with 50 pulses; respectively high hydrostatic pressure (HHP) technology with the parameter of 600 MPa pressure for 10 min treatment time. Table 17.7 shows the physical–chemical properties and total color difference of fruit juices in the case of control, PEF-treated and HHP-treated samples (Hartyáni et al., 2011). Table 17.8 shows the organic acid content of the fruit juices in the case of control, PEF-treated and HHP-treated juice. In the study reported by Hartyáni et al. (2011), malic and citric acid content did not decrease significantly after the treatments (Table 17.8). Respectively, in ascorbic acid content, there was a slight difference, but as an advantage of the treatment the vitamin C content was still quite stable (Hartyáni et al., 2011). It was established that the electronic nose and tongue were able to differentiate each treatment type from the control samples.

Sentandreu et al. (2006) reported more than 90% inactivation of orange juice pectin methylesterase (PME) after PEF processing at 25 kV/cm using 2-μs, square-wave bipolar pulses for up to 330 μs and at 72°C outlet temperature. Yeom et al. (2000) revealed that orange PME was partially inactivated by PEF processing at elevated temperatures (Table 17.2). Approximately 90% inactivation of PME in orange juice was reported after PEF application (35 kV/cm/59 μs) with an outlet temperature of approximately 60°C. Besides, in a subsequent study, 90% PME inactivation in

TABLE 17.6

Compiled Data on the Microbial Inactivation by PEF in Orange Juices

Microorganism	Juice pH	PEF Conditions (E, t, T_{max})[a]	Log_{10} Reduction	Reference
Staphylococcus aureus	3.7	40 kV/cm, 150 μs, 56°C	5.5	Walkling-Riberio et al., 2009b
Listeria innocua	n.d.[c]	30 kV/cm, 12 μs, 50°C	6.0	McDonald et al., 2000
Listeria innocua	3.5	40 kV/cm, 100 μs, 56°C	3.8	McNamee et al., 2010
Escherichia coli	n.d.[c]	30 kV/cm, 12 μs, 50°C	6.0	McDonald et al., 2000
Escherichia coli k12	3.5	40 kV/cm, 100 μs, 56°C	6.3	McNamee et al., 2010
Escherichia coli O157:H7 (EHEC)	3.4	22 kV/cm, 59 μs, 45°C	1.59	Gurtler et al., 2010
		20 kV/cm, 75 μs, 55°C	2.02	
		28 kV/cm, 75 μs, 55°C	3.79	
Salmonella typhimurium	3.4	22 kV/cm, 59 μs, 45°C	2.05	Gurtler et al., 2010
		20 kV/cm, 70 μs, 55°C	2.81–3.54	
Salmonella typhimurium	n.d.[c]	90 kV/cm, 50 μs, 50°C	5.9	Liang et al., 2002
Total aerobic count	4	40 kV/cm, 97 μs, ~60°C	6.2	Min et al., 2003
Yeasts and molds			5.8	
Aerobic microorganisms, yeasts and molds[b]	3.85[b]	25 kV/cm, 280 μs, T not reported	>3	Rivas et al., 2006
Aerobic plate count	n.d.[c]	29.5 kV/cm, 60 μs, T not reported	4.2	Qiu et al., 1998
Aerobic plate count	3.78	35 kV/cm, 59 μs, 60°C	4.0	Yeom et al., 2000
Yeasts and molds			4.0	
Zygosaccharomyces bailii	2.9	34.3 kV/cm, 4 μs, 20°C	3.5 (spores) 5 (veg. cells)	Raso et al., 1998
Saccharomyces cerevisiae (ascospores)	n.d.[c]	50 kV/cm, ~16 μs, 50°C	2.5	McDonald et al., 2000
Saccharomyces cerevisiae	3.4	12.5 kV/cm, 800 μs, ~10°C	5.8	Molinari et al., 2004
Saccharomyces cerevisiae	3.6	35 kV/cm, 1000 μs, 39°C	5.1	Elez-Martinez et al., 2004
Pichia fermentans	3.5	40 kV/cm, 100 μs, 56°C	4.7	McNamee et al., 2010

Byssochlamys fulva (conidiospores)	3.9	34.3 kV/cm, 30 μs, 20°C	5	Raso et al., 1998
Neosartoria fischeri (ascospores)	3.9	42.6 kV/cm, 20 μs, 34°C	<0.1	Raso et al., 1998
Lactobacillus plantarum[b]	4.19	35.8 kV/cm, 46.3μs, T not reported	2.5	Rodrigo et al., 2001
Lactobacillus plantarum	3.4	22 kV/cm, 59 μs, 45°C	2.57	Gurtler et al., 2010
		20 kV/cm, 70 μs, 55°C	3.07	
Lactobacillus lactis	3.4	22 kV/cm, 59 μs, 45°C	4.15	Gurtler et al., 2010
		20 kV/cm, 70 μs, 55°C	4.53	
Lactobacillus fermentum	3.4	22 kV/cm, 59 μs, 45°C	2.11	Gurtler et al., 2010
		20 kV/cm, 70 μs, 55°C	3.22	
Lactobacillus casei	3.4	22 kV/cm, 59 μs, 45°C	0.43	Gurtler et al., 2010
		20 kV/cm, 70 μs, 55°C	0.60	
Lactobacillus brevis	3.6	25 kV/cm, 150 μs, 32°C	1.4	Elez-Martinez et al., 2005
		35 kV/cm, 1000 μs, 32°C	5.8	
Leuconostoc mesenteroides	n.d.[c]	30 kV/cm, 15 μs, 60°C	5.1	McDonald et al., 2000

Source: Adapted from Buckow R., Ng S., and Toepfl S. 2013. Comprehensive Reviews in Food Science and Food Safety, 12, 455–467.

[a] E, electrical field strength; t, total treatment time; T_{max} = maximum (outlet) temperature.

[b] Suspended in an orange-carrot blend.

[c] n.d. = not determined.

TABLE 17.7
Physical–Chemical Properties and Total Colour Difference of Fruit Juices in Case of Control, PEF Treated, and HHP Treated Samples

Sample	Treatment	Brix (%)	pH	Conductivity (mS)	Total Color Difference (ΔE)	
Orange	Control	10.60 ± 0.01	3.65 ± 0.01	3.33 ± 0.01	Reference	
	PEF-treated	10.60 ± 0.05	3.65 ± 0.01	3.33 ± 0.01	4.8 ± 0.05	Visible
	HHP-treated	10.50 ± 0.05	3.63 ± 0.01	Not measured	9.3 ± 0.01	Big differences
Grapefruit	Control	8.90 ± 0.05	2.96 ± 0.01	3.72 ± 0.01	Reference	
	PEF-treated	8.90 ± 0.01	2.96 ± 0.01	3.78 ± 0.01	2.8 ± 0.06	Noticeable
	HHP-treated	8.70 ± 0.01	2.92 ± 0.03	Not measured	2.1 ± 0.05	Noticeable
Tangerine	Control	10.20 ± 0.05	2.95 ± 0.01	3.50 ± 0.01	Reference	
	PEF-treated	10.20 ± 0.01	3.00 ± 0.01	3.60 ± 0.01	3.9 ± 0.05	Visible
	HHP-treated	10.20 ± 0.01	2.90 ± 0.01	Not measured	2.6 ± 0.05	Noticeable

Source: Adapted from Hartyáni P. et al. 2011. *Innovative Food Science and Emerging Technologies*, 12, 255–260.

Note: Values were mean ± SD of three measurements, n = 4; different letters represent a significant difference within the same column ($p < 0.05$).

TABLE 17.8
Organic Acid Content of the Fruit Juices in Case of Control, PEF Treated, and HHP Treated Samples

Sample	Treatment	Malic Acid (mg/l)	Citric Acid (mg/l)	Ascorbic Acid (mg/l)
Orange	Control	847.50 ± 70.06[a]	5290.73 ± 207.48[a]	511.59 ± 2.04[a]
	PEF-treated	826.24 ± 0.09[a]	5222.57 ± 63.59[a]	520.64 ± 12.93[a]
	HHP-treated	755.77 ± 53.83[a]	5207.17 ± 254.89[a]	526.29 ± 17.64[a]
Grapefruit	Control	537.42 ± 49.00[a]	9923.92 ± 80.86[a]	421.18 ± 0.79[a]
	PEF-treated	494.71 ± 5.09[ab]	9933.22 ± 59.90[a]	411.38 ± 6.96[ab]
	HHP-treated	452.75 ± 49.87[b]	9751.39 ± 111.82[a]	405.42 ± 5.56[b]
Tangerine	Control	903.08 ± 112.20[a]	341.41 ± 1.41[a]	6318.03 ± 175.56[a]
	PEF-treated	944.02 ± 3.55[a]	346.45 ± 1.15[a]	6557.13 ± 7.03[a]
	HHP-treated	1191.20 ± 105.92[b]	386.49 ± 12.33[b]	7596.88 ± 171.62[b]

Source: Adapted from Hartyáni P. et al. 2011. *Innovative Food Science and Emerging Technologies*, 12, 255–260.

Note: Values were mean ± SD of three measurements, n = 4; different letters represent a significant difference within the same column ($p < 0.05$).

orange juice was achieved by application of 25 kV/cm for 2-μs pulses for 250 μs at a maximum outlet temperature of 64°C (Yeom et al., 2002).

Timmermans et al. (2011) reported that the mild heat pasteurization, HP processing, and PEF processing of freshly squeezed orange juice were comparatively evaluated, examining their impact on microbial load and quality parameters immediately after processing and during 2 months of storage. It was found that microbial counts for treated juices were reduced beyond detectable levels immediately after processing and up to 2 months of refrigerated storage. Quality parameters such as pH, dry-matter content, and Brix were not significantly different when comparing orange juices immediately after treatment and were, for all treatments, constant during storage time (Timmermans et al., 2011). It was stated that the quality parameters related to PME inactivation, such as cloud stability and viscosity, were dependent on the specific treatments that were applied. It was found that mild heat pasteurization was effective and was obtained as the most stable orange juices (Timmermans et al., 2011). On the basis of the data obtained by Timmermans et al. (2011), residual enzyme activity was clearly responsible for changes in viscosity and cloud stability during storage for PEF. Figure 17.6 shows the overview of the production, handling, and analysis of orange juice samples (Figure 17.6).

It was found that mild heat-pasteurized orange juice was significantly lighter than HP and untreated orange juice, having less red color and more yellowness. It was reported that PEF-treated samples showed the opposite: significantly darker in color than untreated, HP and heat, with significantly more red and less yellow tints (Timmermans et al., 2011). It was shown that DE values, indicated in Figure 17.7, are the sum of L*, a*, and b* values, which are more closely associated to consumer perception than singular L* a*, or b* values, since consumers do not judge each particular attribute, but the combination of them (Cserhalmi et al., 2006). According to DE data obtained by Timmermans et al. (2011), it was found that all types of orange juice showed a noticeable (DE 0.5–1.5) difference in color compared to its color on day 1 and there were no noticeable color differences between the different treatments (Figure 17.7) (Timmermans et al., 2011).

In the study described by Timmermans et al. (2011), PEF-treated orange juice gave a slightly lower Brix° after processing. It was reported that the effect of mild heat treatments on the pH of orange juice was determined and no significant differences were found between untreated and different types of treated juices. It was found that there was no variation during storage time, except for the untreated orange juice, in which the pH decreases significantly over the first 9 days (Timmermans et al., 2011).

As it is known, cloud loss is considered as a quality defect in shelf-stable citrus juices derived from the concentrate and it is one of the main reasons for the level of heating in heat pasteurization, where 90–100% of PME is inactivated (Goodner et al., 1998, 1999). Figure 17.8 shows observed sedimentation and cloud loss of untreated, mild heat- pasteurized, high-pressure-pasteurized (HPP), and PEF-processed orange juice bottles during the first 115 days of storage at 4°C (Timmermans et al., 2011).

It was stated that cloud stability was measured evaluating the degree and rate of sedimentation, by recording the height of the interface of the sediment and cloud

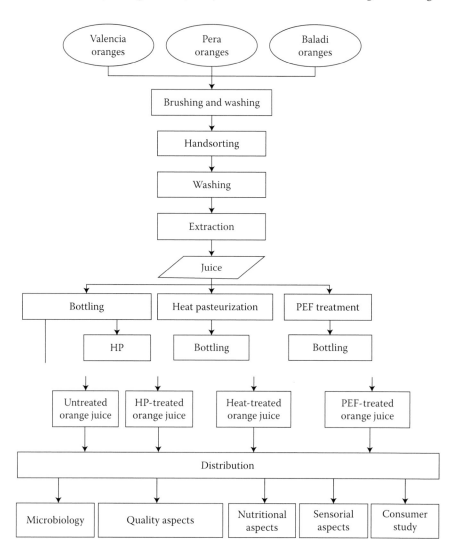

FIGURE 17.6 The overview of the production, handling, and analysis of orange juice samples. (Adapted from Timmermans R.A.H., et al. 2011. *Innovative Food Science and Emerging Technologies*, 12, 235–243.)

from bottles (Figure 17.8). The degree of sedimentation was calculated as percentage, where the height of the sediment was expressed as a function of the total height. On the basis of the data obtained by Timmermans et al. (2011), sedimentation of 0% corresponded to a completely stable orange juice, having no cloud loss (Figure 17.8). It was determined that the orange juices were sedimented according to a nonlinear response during storage time (Timmermans et al., 2011). The viscosity of the serum of centrifuged orange juice was measured during storage time and the

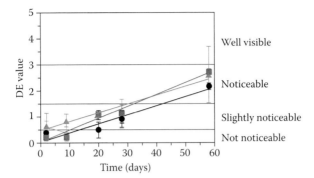

FIGURE 17.7 The total color difference (DE value) for untreated (), mild heat-pasteurized (▲), HP-processed (●), and PEF- (■) processed orange juice. (Adapted from Timmermans R.A.H. et al. 2011. *Innovative Food Science and Emerging Technologies*, 12, 235–243.)

observed alterations in viscosity were small. During storage, viscosity of mild heat-pasteurized juice remained stable, whereas the viscosity of HP- and PEF-treated juice increased (Figure 17.9). It is stated that the increase of viscosity during storage is likely induced by activity of PME (Timmermans et al., 2011).

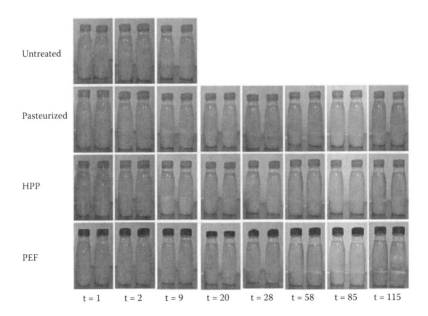

FIGURE 17.8 See color insert Observed sedimentation and cloud loss of untreated, mild heat-pasteurized, HPP-, and PEF-processed orange juice bottles during the first 115 days of storage at 4°C. (Adapted from Timmermans R.A.H. et al. 2011. *Innovative Food Science and Emerging Technologies*, 12, 235–243.)

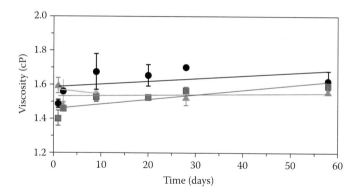

FIGURE 17.9   The serum viscosity of centrifuged (10 min at $360 \times$ g), untreated (♦), mild heat-pasteurized (▲), HP-processed (●), and PEF- (■) processed orange juice during storage at 4°C.

17.7   RECENT STUDIES ON PEF-TREATED BLUEBERRY JUICES

Berry juices are favorable due to their nutritional and healthy effects and berry juice products contain phenolic substances including flavonols, tannins, and anthocyanins with high antioxidant capacity (Altuner and Tokuşoğlu, 2013; Nindo et al., 2005; Tokuşoğlu and Stoner, 2011).

Barba et al. (2012) reported the evaluation of quality alterations of blueberry juice during refrigerated storage after HP and PEF. It was found that the physicochemical properties and levels of bioactive compounds of HHP- and PEF-treated blueberry juices were similar. During refrigerated storage at 4°C, HHP juice showed higher ascorbic acid retention, phenolic content, and antioxidant capacity than PEF juices and untreated samples. It was stated that in blueberry juice, the total color change (ΔE) in all the processed juices immediately after treatment or during storage was significantly different from the unprocessed samples ($p < 0.05$) (Table 17.9) (Barba et al., 2012). The effects of HHP and PEF treatments on color parameters [a* (greenness and redness), b*(blueness and yellowness), and L* (lightness or darkness)] of blueberry juice are shown in Table 17.9. As it is shown in Table 17.9, the application of HHP or PEF had a small effect on color changes. Morover, a significant negative correlation was observed between ΔE* and ascorbic acid ($p = 0.0016$).

In the study reported by Barba et al. (2012), the anthocyanin levels in fresh blueberry juice were 2.52 ± 0.07 mg/g. It was stated that a statistically significant increase of total anthocyanin (TA) occurred immediately after HP (as 109% retention) and PEF treatments (as 105% retention) (Barba et al., 2012).

Figure 17.10 shows the remaining ascorbic acid concentration in untreated, HP-treated and PEF-treated blueberry juices stored in refrigeration at 4°C (Barba et al., 2012). Plotting the concentration of ascorbic acid against the refrigerated storage time of the blueberry juice gave the ascorbic acid degradation rate and a 1.4th-order reaction model was found the best for explaining ascorbic acid degradation (Figure 17.10) (Barba et al., 2012). For the first week, ascorbic acid retention

TABLE 17.9

Blueberry Juice Color Evaluation during 56 Days of Refrigerated (4°C) Storage of Untreated, PEF-, and HP-Treated Samples

Color Parameters	a*	b*	L*	h⁰	ΔE
Day 0					
Untreated	$0.25 \pm 0.04^{a,b}$	$-6.83 \pm 0.02^{a,b}$	31.54 ± 0.04^a	$87.90 \pm 0.36^{a,b}$	0^a
PEF	$0.31 \pm 0.07^{b,c,d}$	$-6.81 \pm 0.04^{a,b}$	31.44 ± 0.04^a	87.09 ± 0.21^a	0.11 ± 0.03^e
HP	$0.34 \pm 0.04^{b,c,d}$	$-6.89 \pm 0.01^{b,c}$	31.52 ± 0.01^a	$87.20 \pm 0.31^{b,c,d}$	0.11 ± 0.03^e
Day 7					
Untreated	$0.29 \pm 0.13^{a,b,c}$	$-6.82 \pm 0.03^{a,b}$	31.35 ± 0.03^b	$87.59 \pm 1.06^{a,b,c}$	0.22 ± 0.02^b
PEF	$0.42 \pm 0.07^{d,e}$	$-6.76 \pm 0.02^{a,c}$	$31.24 \pm 0.02^{c,d}$	$86.59 \pm 0.72^{d,e}$	0.35 ± 0.04^d
HP	$0.37 \pm 0.06^{b,c,d,e}$	-6.81 ± 0.03^a	31.25 ± 0.03^c	$86.89 \pm 0.46^{c,d}$	$0.31 \pm 0.03^{b,c}$
Day 14					
Untreated	$0.37 \pm 0.01^{b,c,d,e}$	-6.98 ± 0.05^d	$31.30 \pm 0.02^{b,c}$	$86.91 \pm 0.03^{c,d}$	$0.31 \pm 0.01^{b,c}$
PEF	$0.35 \pm 0.01^{b,c,d}$	-6.79 ± 0.02^a	$31.23 \pm 0.01^{c,d}$	$86.97 \pm 0.17^{b,c,d}$	0.33 ± 0.01^d
HP	$0.25 \pm 0.04^{a,b}$	$-6.82 \pm 0.05^{a,b}$	31.15 ± 0.03^e	$87.87 \pm 0.35^{a,b}$	0.39 ± 0.03^d
Day 21					
Untreated	$0.38 \pm 0.01^{c,d,e}$	-6.98 ± 0.02^d	31.25 ± 0.04^c	$86.91 \pm 0.10^{c,d}$	0.35 ± 0.04^d
PEF	1.14 ± 0.17^g	-6.59 ± 0.13^f	$31.16 \pm 0.03^{d,e}$	$79.29 \pm 0.14^{h,i}$	1.00 ± 0.11^f
HP	$1.28 \pm 0.09^{h,i}$	-6.79 ± 0.05^a	$30.71 \pm 0.09^{f,g}$	$79.36 \pm 0.63^{a,h,i}$	$1.32 \pm 0.11^{g,h}$
Day 28					
Untreated	0.48 ± 0.05^e	$-6.93 \pm 0.05^{c,d}$	31.27 ± 0.05^c	85.76 ± 0.27^e	0.38 ± 0.04^d
PEF	$1.19 \pm 0.07^{g,h}$	-6.50 ± 0.03^g	30.76 ± 0.03^g	$80.18 \pm 1.40^{g,h}$	1.26 ± 0.05^g
HP	1.15 ± 0.04^g	$-6.75 \pm 0.04^{a,e}$	30.68 ± 0.05^h	80.31 ± 0.24^g	1.25 ± 0.04^g
Day 56					
Untreated	0.99 ± 0.06^f	-7.01 ± 0.06^d	30.97 ± 0.11^f	81.96 ± 0.36^f	0.95 ± 0.12^f
PEF	1.33 ± 0.03^i	$-6.68 \pm 0.01^{e,f}$	30.62 ± 0.09^h	78.73 ± 0.25^j	1.43 ± 0.04^h
HP	$1.11 \pm 0.12^{f,g}$	$-6.76 \pm 0.00^{a,e}$	30.48 ± 0.03^i	80.72 ± 0.99^g	$1.36 \pm 0.05^{g,h}$

Source: Adapted from Barba F.J. et al. 2012. *Innovative Food Science and Emerging Technologies*, 14, 18–24.

[a-j] Different letters indicate significant statistical differences in function of the applied treatment.

of untreated, HP- and PEF-treated blueberry juice was around 80% compared with the first day. It obtained higher vitamin retention in HHP blueberry juice after day 7. It was reported that at the end of the storage (day 56), lower ascorbic acid loss rates were found in the case of HHP blueberry juice (73%) compared with the respective values of PEF-treated (50%) and unprocessed (52%) juices at 4°C storage (Barba et al., 2012).

As a conclusion, nonthermal technologies including PEF may be applied to obtain safe and stable liquid foods without significant depletion of their fresh bioactive

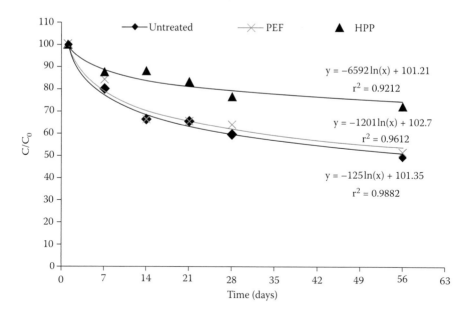

FIGURE 17.10 The remaining ascorbic acid concentration in untreated, HP-, and PEF-treated samples stored in refrigeration at 4°C. The lines interpolating the experimental data points show the fit of a 1.4th-order reaction model. (Adapted from Barba F.J. et al. 2012. *Innovative Food Science and Emerging Technologies*, 14, 18–24.)

potential. In general, temperature during processing and storage is a determinant factor affecting the health-related compounds of the processed juices. However, other process variables must be considered for each nonthermal process when aiming at understanding the nature of the changes. In-depth research is needed to study the factors involved in the generation/destruction of these compounds, as well as to elucidate the mechanistic insight of the alterations.

REFERENCES

Aguilar-Rosas S.F., Ballinas-Casarrubias M.L., Nevarez-Moorillon G.V., Martin-Belloso O., and Ortega-Rivas E. 2007. Thermal and pulsed electric fields pasteurization of apple juice: Effects on physicochemical properties and flavour compounds. *Journal of Food Engineering*, 83, 41–46.

Akdemir Evrendilek G., Jin Z.T., Ruhlman K.T., Qiu X., Zhang Q.H., and Richter E.R. 2000. Microbial safety and shelf-life of apple juice and cider processed by bench and pilot scale PEF systems. *Innovative Food Science and Emerging Technologies*, 1, 77–86.

Alkhafaji S.R., and Farid M.M. 2007. An investigation on pulsed electric fields technology using new treatment chamber design. *Innovative Food Science and Emerging Technology*, 8, 205–212.

Altuner E.M., and Tokuşoğlu Ö. 2013. The effect of high hydrostatic pressure processing on the extraction, retention and stability of anthocyanins and flavonols contents of berry fruits and berry juices. *International Journal of Food Science and Technology (IJFST)*, 48(10), 1991–1997.

Barba F.J., Jäger H., Meneses N., Esteve M.J., Frígola A., and Knorr D. 2012. Evaluation of quality changes of blueberry juice during refrigerated storage after high-pressure and pulsed electric fields processing. *Innovative Food Science and Emerging Technologies*, 14, 18–24.

Bi X., Liu F., Rao L., Li J., Liu B., Liao X., and Wu J. 2013. Effects of electric field strength and pulse rise time on physicochemical and sensory properties of apple juice by pulsed electric field. *Innovative Food Science and Emerging Technologies*, 17, 85–92.

Buckow R., Ng S., and Toepfl S. 2013. Pulsed electric field processing of orange juice: A review on microbial, enzymatic, nutritional, and sensory quality and stability. *Comprehensive Reviews in Food Science and Food Safety*, 12, 455–467.

Caminiti I.M., Noci F.F., Mūnoz A.A., Whyte P.P., Morgan D.J., Cronin D.A., Lyng J.G. 2011. Impact of selected combinations of non-thermal processing technologies on the quality of an apple and cranberry juice blend. *Food Chemistry*, 124, 1387–1392.

CDC. 1995. Center for Disease Control and Prevention. Outbreak of acute gastroenteritis attributable to *Escherichia coli* serotype O104:H21-Helena, Montana. *Morbidity Mortality Weekly Report*, 44, 501–503.

Charles-Rodríguez A.V., Nevárez-Moorillón G.V., Zhang Q.H., and Ortega-Rivas E.E. 2007. Comparison of thermal processing and pulsed electric fields treatment in pasteurization of apple juice. *Food Bioproduction Process*, 85, 93–97.

Chen Y., Yu L.J., and Rupasinghe H.P.V. 2013. Effect of thermal and non-thermal pasteurisation on the microbial inactivation and phenolic degradation in fruit juice: A mini-review. *Journal of Scientific Food Agriculture*, 93, 981–986.

Cserhalmi Z.S., Sass-Kiss A., Tóth-Markus M., and Lechner N. 2006. Study of pulsed electric field citrus juices. *Innovative Food Science and Emerging Technologies*, 7(1–2), 49–54.

Elez-Martinez P., Escola-Hernandez J., Soliva-Fortuny R.C., and Martin-Belloso O. 2004. Inactivation of Saccharomyces cerevisiae suspended in orange juice using high-intensity pulsed electric fields. *Journal of Food Protection*, 67, 2596–2602.

Elez-Martinez P., Escola-Hernandez J., Soliva-Fortuny R.C., and Martin-Belloso O. 2005. Inactivation of Lactobacillus brevis in orange juice by high-intensity pulsed electric fields. *Food Microbiology*, 22, 311–319.

Elez-Martínez P., and Martín-Belloso O. 2007. Effects of high intensity pulsed electric field processing conditions on vitamin C and antioxidant capacity of orange juice and "Gazpacho", a could vegetable soup. *Food Chemistry*, 102, 201–209.

Elez-Martínez P.P., Soliva-Fortuny R.C., and Martín-Belloso O.O. 2006. Comparative study on shelf life of orange juice processed by high intensity pulsed electric fields or heat treatment. *European Food Research Technology*, 222, 321–329.

FDA. 2001. U.S. Food and Drug Administration. Hazard analysis and critical control point (HACCP); procedures for the safe and sanitary processing and importing of juices; final rule. *Federal Register*, 66, 6138–6202.

Fincan M., DeVito F., and Dejmek P. 2004. Pulsed electric field treatment for solid–liquid extraction of red beetroot pigment. *Journal of Food Engineering*, 64, 381–388.

Fraser P.D., Truesdale M.R., Bird C.R., Schuch W., and Bramley P.M. 1994. Carotenoid biosynthesis during tomato fruit development (evidence for tissue-specific gene expression). *Plant Physiology*, 105(1), 405–413.

Goodner J.K., Braddock R.J., and Parish M.E. 1998. Inactivation of pectinesterase in orange and grapefruit juices by high pressure. *Journal of Agricultural and Food Chemistry*, 46(5), 1997–2000.

Goodner J.K., Braddock R.J., Parish M.E., and Sims C.A. 1999. Cloud stabilization of orange juice by high pressure processing. *Journal of Food Science*, 64(4), 699–700.

Grimi N., Lebovka N., Vorobiev E., and Vaxelaire J. 2009. Effect of a pulsed electric field treatment on expression behavior and juice quality of Chardonnay grape. *Food Biophysics*, 4, 191–198.

Grimi N., Mamouni F., Lebovka N., Vorobiev E., and Vaxelaire J., 2010. Acoustic impulse response in apple tissues treated by pulsed electric field. *Biosystems Engineering*, 105 (2), 266–272.

Gurtler J.B., Bailey R.B., Geveke D.J., and Zhang H.Q. 2011. Pulsed electric field inactivation of *E. coli* O157:H7 and non-pathogenic surrogate *E. coli* in strawberry juice as influenced by sodium benzoate, potassium sorbate, and citric acid. *Food Control*, 22, 1689–1694.

Hamilton W.A., and Sale A.J.H. 1967. Effects of high electric fields on microorganisms. II. Mechanism of action of the lethal effect. *Biochimica et Biophysica Acta*, 148 (3), 789–800.

Hartyáni P., Dalmadi I., Cserhalmi Z., Kántor D.B., Tóth-Markus M., and Sass-Kiss A. 2011. Physical–chemical and sensory properties of pulsed electric field and high hydrostatic pressure treated citrus juices. *Innovative Food Science and Emerging Technologies*, 12, 255–260.

Heinz V., Alvarez I., Angersbach A., and Knorr D. 2002. Preservation of liquid foods by high intensity pulsed electric field—Basic concept for process design. *Trends in Food Science Technology*, 12, 103–111.

Huang K., and Wang J. 2009. Designs of pulsed electric fields treatment chambers for liquid foods pasteurization process: A review. *Journal of Food Engineering*, 95, 227–239.

Hülsheger H., Potel J., and Niemann E.G. 1983. Electric field effects on bacteria and yeast cells. *Radiation of Environmental Biophysics*, 22,149–162.

Knorr D., Angersbach A., Eshtiaghi M., Heinz V., and Lee D.U. 2001. Processing concepts based on high intensity electric field pulses. *Trends in Food Science and Technology*, 12, 129–135.

Lado B., and Yousef A. 2002. Alternative food preservation technologies: Efficacy and mechanisms. *Microbes Infection*, 4, 433–440.

Lebovka N.I., Praporscic I., and Vorobiev E. 2003. Enhanced expression of juice from soft vegetable tissues by pulsed electric fields: Consolidation stages analysis. *Journal of Food Engineering*, 59, 309–317.

Liang Z., Mittal G.S., and Griffiths M.W. 2002. Inactivation of Salmonella typhimurium in orange juice containing antimicrobial agents by pulsed electric field. *Journal Food Protection*, 65, 1081–1087.

McDonald C.J., Lloyd S.W., Vitale M.A., Petersson K., and Innings F. 2000. Effect of pulsed electric fields on microorganisms in orange juice using electric field strengths of 30 and 50 kV/cm. *Journal of Food Science*, 65, 984–989.

McNamee C., Noci F., Cronin D.A., Lyng J.G., Morgan D.J., and Scannell A.G.M. 2010. PEF based hurdle strategy to control Pichia fermentans, Listeria innocua and *Escherichia coli* k12 in orange juice. *International Journal of Food Microbiololgy*, 138, 13–18.

Min S., Jin Z.T., Min S.K., Yeom H., and Zhang H.Q. 2003. Commercial-scale pulsed electric field processing of orange juice. *Journal Food Science*, 68, 1265–1271.

Molinari P., Pilosof A.M.R., and Jagus R.J. 2004. Effect of growth phase and inoculum size on the inactivation of Saccharomyces cerevisiae in fruit juices, by pulsed electric fields. *Food Reviews International*, 37, 793–798.

Morales-de la Peña M., Salvia-Trujillo L., Rojas-Graü M.A., and Martín-Belloso O. 2010a. Impact of high intensity pulsed electric field on antioxidant properties and quality parameters of a fruit juice–soymilk beverage in chilled storage. *LWT—Food Science and Technology*, 43, 872–881.

Morales-de la Peña M., Salvia-Trujillo L., Rojas-Graü M.A., and Martín-Belloso O. 2010b. Isoflavone profile of a high intensity pulsed electric field or thermally treated fruit juice–soymilk beverage stored under refrigetarion. *Innovative Food Science and Emerging Technologies*, 11, 604–610.

Nagy S., Rouseff R., and Lee H. 1989. Thermally degraded flavors in citrus juice products. *ACS Symposium Series*, 409, 331–345.

Ngadi M.O., Yu L.J., and Ortega-Rivas E.A.M. 2009. Food quality and safety issues during pulsed electric field processing, in *Processing Effects on Safety and Quality of Foods*, ed., Ortega-Rivas E. CRC Press, Boca Raton, FL, pp. 433–471.

Nguyen M.L., and Schwartz S.J. 1999. Lycopene: Chemical and biological properties. *Food Technology*, 53(2), 38–45.

Nindo C.I., Tang J., Powers J.R., and Singh P. 2005. Viscosity of blueberry and raspberry juices for processing applications. *Journal of Food Engineering*, 69, 343–350.

Noci F., Riener J., Walkling-Ribeiro M., Cronin D., Morgan D., and Lyng J. 2008. Ultraviolet irradiation and pulsed electric fields (PEF) in a hurdle strategy for the preservation of fresh apple juice. *Journal of Food Engineering*, 85, 141–146.

Odriozola-Serrano I., Aguiló-Aguayo I., Soliva-Fortuny R., Gimeno-Añó V., and Martín-Belloso O. 2007. Lycopene, vitamin C, and antioxidant capacity of tomato juice as affected by high-intensity pulsed electric fields critical parameters. *Journal of Agricultural and Food Chemistry*, 55, 9036–9042.

Odriozola-Serrano I., Soliva-Fortuny R., and Martín-Belloso O. 2008. Phenolic acids, flavonoids, vitamin C and antioxidant capacity of strawberry juices processed by high-intensity pulsed electric fields or heat treatments. *European Food Reserach Technology*, 228, 239–248.

Odriozola-Serrano I., Soliva-Fortuny R., and Martín-Belloso O. 2008a. Changes of health-related compounds throughout cold storage of tomato juice stabilized by thermal or high intensity pulsed electric field treatments. *Innovative Food Science and Emerging Technologies*, 9, 272–279.

Odriozola-Serrano I., Soliva-Fortuny R., Gimeno-Añó V., and Martín-Belloso O. 2008b. Modeling changes in health-related compounds of tomato juice treated by high-intensity pulsed electric fields. *Journal of Food Engineering*, 89, 210–216.

Odriozola-Serrano I., Soliva-Fortuny R., Hernández-Jover T., and Martín-Belloso O. 2009. Carotenoid and phenolic profile of tomato juices processed by high intensity pulsed electric fields compared with conventional thermal treatments. *Food Chemistry*, 112, 258–266.

Odriozola-Serrano I., Soliva-Fortuny R., and Martin-Belloso O. 2009a. Impact of high-intensity pulsed electric fields variables on vitamin C, anthocyanins and antioxidant capacity of strawberry juice. *LWT—Food Science and Technology*, 42, 93–100.

Oms-Oliu G., Odriozola-Serrano I., Soliva-Fortuny R., and Martin-Belloso O. 2009. Effects of high-intensity pulsed electric field processing conditions on lycopene, vitamin C and antioxidant capacity of watermelon juice. *Food Chemistry*, 115, 1312–1319.

Ortega-Rivas E., Zárate-Rodríguez E., and Barbosa-Cánovas G.V. 1998. Apple juice pasteurization using ultrafiltration and pulsed electric fields. *Food and Bioproducts Processing*, 76(4), 193–198.

Parish M.E. 1998. *Coliforms, Escherichia coli* and *Salmonella* serovars associated with a citrus-processing facility implicated in a *Salmonellosis* outbreak. *Journal of Food Protection*, 61, 280–844.

Praporscic I., Lebovka N., Vorobiev E., and Mietton-Peuchot M. 2007. Pulsed electric field enhanced expression and juice quality of white grapes. *Separation and Purification Technology*, 52, 520–526.

Pérez-Conesa D., García-Alonso J., García-Valverde V., Iniesta M.D., Jacob K., Sánchez-Siles L.M., Ros G., and Periago M.J. 2009. Changes in bioactive compounds and antioxidant activity during homogenization and thermal processing of tomato puree. *Innovative Food Science Emerging Technology*, 10, 179–188.

Qiu X., Sharma L., Tuhela L., Jia M., and Zhang Q.H. 1998. An integrated PEF pilot plant for continuous nonthermal pasteurization of fresh orange juice. *Trans ASAE*, 41, 1069–1074.

Ramaswamy H.S., Chen C., and Marcotte M. 2005. Novel processing technologies for food preservation, in *Processing Fruits: Science and Technology*, ed.,Barrett D.M. CRC Press, Boca Raton, FL, pp. 201–204.

Raso J., Calderon M.L., Gongora M., Barbosa-Canovas C., and Swanson B.G. 1998a. Inactivation of mold ascospores and conidiospores suspended in fruit juices by pulsed electric fields. *Lebensm-Wiss Technology*, 31, 668–672.

Raso J., Calderon M.L., Gongora M., and Barbosa-Canovas G. 1998b. Inactivation of Zygosaccharomyces bailii in fruit juices by heat, high hydrostatic pressure and pulsed electric fields. *Journal of Food Science*, 63, 1042–1044.

Riener J., Noci F., Cronin D.A., Morgan D.J., Lyng J.G. 2008. Combined effect of temperature and pulsed electric fields on apple juice peroxidase and polyphenoloxidase inactivation. *Food Chemistry*, 109, 402–407.

Rivas A., Rodrigo D., Mart'ınez A., Barbosa-Canovas G.V., and Rodrigo M. 2006. Effect of PEF and heat pasteurization on the physical-chemical characteristics of blended orange and carrot juice. *LWT-Food Science and Technology*, 39, 1163–1170.

Rodrigo D., Barbosa-Cánovas G.V., Martinez A., and Rodrigo M. 2003. Weibull distribution function based on an empirical mathematical model for inactivation of *Escherichia coli* by pulsed electric fields. *Journal of Food Protection*, 66, 1007–1012.

Rodrigo D., Martinez A., Harte F., Barbosa-Canovas G.V., and Rodrigo M. 2001. Study of inactivation of Lactobacillus plantarum in orange-carrot juice by means of pulsed electric fields: comparison of inactivation kinetics models. *Journal of Food Protection*, 64, 259–263.

Rodríguez-Amaya D.B. 1997. *Carotenoids and Food Preparation: The Retention of Provitamin A Carotenoids in Prepared, Processed and Stored Foods*, OMNI, Arlington.

Ross A.I.V., Griffiths M.W., Mittal G.S., and Deeth H.C. 2003. Combining nonthermal technologies to control foodborne microorganisms. *International Journal of Food Microbiology*, 89, 125–138.

Saint-Cricq de Gaulejac N., Vivas N., de Freitas V., Bourgeois G. 1999. The influence of various phenolic compounds on scavenging activity assessed by and enzymatic method. *Journal of Scientific Food Agriculture*, 79, 1081–1090.

Sánchez-Moreno C., Cano M.P., De Ancos B., Plaza L., Olmedilla B., Granado F., Elez-Martínez P., Martín-Belloso O., and Martín A. 2004. Pulsed electric fields-processed orange juice consumption increases plasma vitamin C and decreases F2-isoprostanes in healthy humans. *Journal of Nutritional Biochemistry*, 15, 601–607.

Sánchez-Moreno C., Cano M.P., De Ancos B., Plaza L., Olmedilla B., Granado F., Martín A. 2003. High-pressurized orange juice consumption affects plasma vitamin C, antioxidative status and inflammatory markers in healthy humans. *Journal of Nutrition*, 133, 2204–2209.

Sánchez-Moreno C., De Ancos B., Plaza L., Elez-Martínez P., and Cano M.P. 2009. Nutritional approaches and health-related properties of plant foods processed by high pressure and pulsed electric fields. *Critical Reviews in Food Science and Nutrition*, 49, 552–576.

Sánchez-Moreno C., Plaza L., Elez-Martínez P., de Ancos B., Martín-Belloso O., and Cano P. 2005. Impact of high pressure and pulsed electric fields on bioactive compounds and antioxidant capacity of orange juice in comparison with traditional thermal processing. *Journal of Agricultural and Food Chemistry*, 53(11), 4403–4409.

Schilling S., Alber T., Toepfl S., Neidhart S., Knorr D., Schieber A., and Carle R. 2007. Effects of pulsed electric field treatment of apple mash on juice yield and quality attributes of apple juices. *Innovative Food Science and Emerging Technologies*, 8, 127–134.

Schilling S., Schmid S., Jäger H., Ludwig M., Dietrich H., Toepfl S., Knorr D., Neidhart S., Schieber A., and Carle R. 2008. Comparative study of pulsed electric field and thermal processing of apple juice with particular consideration of juice quality and enzyme deactivation. *Journal of Agricultural and Food Chemistry*, 56, 4545–4554.

Schilling S., Toepfl S., Ludwig M., Dietrich H., Knorr D., Neidhart S. Schieber A., and Carle R. 2008. Comparative study of juice production by pulsed electric field treatment and enzymatic maceration of apple mash. *European Food Research and Technology*, 226, 1389–1398.

Sentandreu E., Carbonell D., Rodrigo D., and Carbonell J.V. 2006. Pulsed electric fields versus thermal treatment: Equivalent processes to obtain equally acceptable citrus juices. *Journal of Food Protection*, 69, 2016–2018.

Song H.P., Byun M.W., Jo C., Lee C.H., Kim K.S., and Kim D.H. 2007. Effects of gamma irradiation on the microbiological, nutritional, and sensory properties of fresh vegetable juice. *Food Control*, 18, 5–10.

Timmermans R.A.H., Mastwijk H.C., Knol J.J., Quataert M.C.J., Vervoort L., Van der Plancken I., Hendrickx M.E., and Matser A.M. 2011. Comparing equivalent thermal, high pressure and pulsed electric field processes for mild pasteurization of orange juice. Part I: Impact on overall quality attributes. *Innovative Food Science and Emerging Technologies*, 12, 235–243.

Toepfl S.S., Heinz V.V., and Knorr D.D. 2007. High intensity pulsed electric fields applied for food preservation. *Chemical Engineering Process*, 46, 537–546.

Toepfl S., Mathys A., Heinz V., and Knorr D. 2006. Review: Potential of emerging technologies for energy efficient and environmentally friendly food processing. *Food Reviews International*, 22, 405–423.

Tokuşoğlu Ö., and Stoner G. 2011. Phytochemical bioactives in berry fruits (Chapter 7—Part II. *Chemistry and Mechanisms of Beneficial Bioactives in Fruits and Cereals*), in *Fruit and Cereal Bioactives: Sources, Chemistry and Applications*, eds., Tokuşoğlu Ö. and Hall C. CRC Press, Taylor & Francis Group, Boca Raton, Florida, USA, 459pp. ISBN: 9781439806654; ISBN-10:1439806659.

Tournas V.H., Heeres J., and Burgess L. 2006. Moulds and yeasts in fruit salads and fruit juices. *Food Microbiology*, 23, 684–688.

Turk M.F., Vorobiev E., and Baron A. 2012. Improving apple juice expression and quality by pulsed electric field on an industrial scale. *LWT—Food Science and Technology*, 49, 245–250.

Vorobiev E., and Lebovka N. (Eds.). 2008. *Electrotechnologies for Extraction from Plant Foods and Biomaterials*, Springer, New York.

Yeom H.W., Streaker C.B., Zhang Q.H., and Min D.B. 2000. Effects of pulsed electric fields on the activities of microorganisms and pectin methyl esterase in orange juice. *Journal of Food Science*, 65, 1359–1363.

Yeom H.W., Zhang Q.H., and Chism G.W. 2002. Inactivation of pectin methyl esterase inorange juice by pulsed electric fields. *Journal of Food Science*, 67, 2154–2159.

Walkling-Ribeiro M., Noci F., Cronin D.A., Lyng J.G., and Morgan D.J. 2008a. Inactivation of *Escherichia coli* in a tropical fruit smoothie by a combination of heat and pulsed electric fields. *Journal of Food Science*, 73, 395–399.

Walkling-Ribeiro M.M., Noci F.F., Cronin D.A., Riener J.J., Lyng J.G., and Morgan D.J. 2008b. Reduction of *Staphylococcus aureus* and quality changes in apple juice processed by ultraviolet irradiation, pre-heating and pulsed electric fields. *Journal of Food Engineering*, 89, 267–273.

Walkling-Ribeiro M., Noci F., Riener J., Cronin D.A., Lyng J.G., and Morgan D.J. 2009b. The impact of thermosonication and pulsed electric fields on *Staphylococcus aureus* inactivation and selected quality parameters in orange juice. *Food Bioprocess Technology*, 2, 422–430.

Wouters P.C., Alvarez I., and Raso J. 2001. Critical factors determining inactivation kinetics by pulsed electric field food processing. *Trends in Food Science Technology*, 12, 112–121.

Zárate-Rodríguez E., Ortega-Rivas E., and Barbosa-Cánovas G.V. 2001. Effect of membrane pore size on quality of ultrafiltered apple juice. *International Journal of Food Science Technology*, 36, 663–667.

Zhang Y., Gao B., Zhang M., Shi J., and Xu Y. 2010. Pulsed electric field processing effects on physicochemical properties, flavor compounds and microorganisms of longan juice. *Journal of Food Process Preservation*, 34, 1121–1138.

18 Improving Liquid Egg Quality by Pulsed Electrical Field Processing

Özlem Tokuşoğlu, Gustavo V. Barbosa-Cánovas, and Howard Q. Zhang

CONTENTS

18.1 INTRODUCTION

The eggs of birds, primarily domestic, are used as human food. Edible eggs consist of the yolk, white, membranes, and shell (AEB, 2013). Figure 18.1 shows the main parts of whole egg (Figure 18.1) (AEB, 2013). Mostly consumed are chicken egg weight 55–65 g, goose eggs 110–180 g, turkey eggs about 110 g, guinea fowl eggs 45 g, and quail eggs 8–10 g. The eggs most often used fresh are from chickens and these eggs consist of approximately 9.5% eggshell including shell membrane, 63% albumen, and 27.5% yolk (Cotterill and Geiger, 1977).

The main components are water (75%), proteins (12%), and lipids (12%), as well as carbohydrates and minerals (Burley and Vadehra, 1989; Li-Chan et al., 1995). The proteins are distributed throughout the egg, with the majority found in the egg yolk and egg white, and a small proportion in the eggshell and shell membrane (Sugino et al., 1997).

18.2 LIQUID EGG PRODUCTS AND THERMAL PASTEURIZATION

Chicken eggs are divided into table eggs, dietetic eggs, and reject eggs based on weight, storage length, and quality. Table eggs and dietetic eggs are classified by weight and quality into two grades. For convenient transport and storage, fresh eggs are mixed, pasteurized, and frozen or powder dried. More than 30% of the total number of eggs are in the form of egg-derivative products (Amiali et al., 2005;

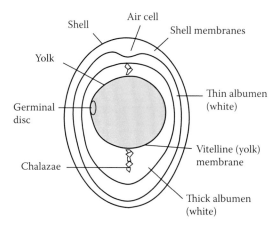

FIGURE 18.1   See color insert Main parts of egg. (Adapted from AEB. 2013. American Egg Board. http://www.aeb.org/foodservice-professionals.)

Tokuşoğlu, 2013). The use of liquid whole egg (LWE) and egg products (egg powder, etc.) is widely extended in a large number of food industries and establishments. Figure 18.2 schematically shows the industrial egg handling, grading, and processing (Sanovo, 2013).

Especially, liquid egg products (LEPs) including LWE, egg whites, liquid egg yolk, and scrambled egg mix are mostly utilized as ingredients in the bakery and catering industries. LEP delivers the same taste and performance as shell eggs, while saving time and effort, and can be used in place of shell egg for baking, scrambled eggs, omelettes, desserts, quiche, Yorkshire puddings, and meringues. Liquid egg mixes are primarily functional food ingredients.

Currently, the egg industry's primary method of improving the microbiological safety of LWE is thermal treatment, varying from minimum treatments of

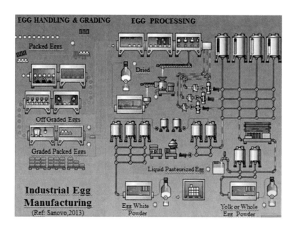

FIGURE 18.2   Schematized industrial egg-manufacturing process. (Adapted from Sanovo 2013. *Egg Production. Sanovo Egg Group.* SANOVO International A/S, Denmark.)

60°C/3.5 min up to high-intensity treatments of 70°C/1.5 min. Currently, the preservation method of LEP is thermal pasteurization (60°C/3.5 min), but undesirable effects containing viscosity decreasing, alterations and damaging in functional properties, and reduction of foaming capability can occur (Calderón-Miranda et al., 1999; Tokuşoğlu, 2013). Conalbumin is the most thermosensible protein of eggs and it is stated that slight overheating can cause conalbumin coagulation and also the film formation on the heat-exchange surface (Martín-Belloso et al., 1997; Tokuşoğlu, 2013).

Recently, consumers are demanding more minimally processed foods and that the preserving of the food quality and functional properties of foods meets the needs of customers. The demand toward fresh-like products with little or no degradation of nutritional and organoleptic properties has led to the study of new technologies in food preservation. Pulsed electric fields (PEFs) as a new nonthermal technology used to inactivate microorganisms mainly in liquid foods allows obtaining products with similar properties to fresh food and food products (Bayraktaroğlu et al., 2012; Mañas and Pagán, 2005; Tokuşoğlu, 2013).

PEF, a novel, nonthermal food preservation technology, has been shown to inactivate both spoilage and pathogenic microorganisms, while minimizing alterations in the physical, chemical, and organoleptic qualities of the food, such as those observed under conventional thermal processing. PEF involves the high-voltage short pulses (pulse durations of microseconds) application to foods placed between two electrodes and the pulsed energy destroys the bacterial cell membrane by mechanical effects with minimal heating of the food. Electrical conductivity can be used to monitor important alterations in a food product during PEF processing (Barbosa-Cánovas et al., 1999).

PEF technology is considered a very promising alternative to pasteurization processes when processing high thermal-sensitive liquid foods such as whole liquid egg (Amiali et al., 2005; Sampedro et al., 2006). Electrical conductivities of selected LEPs (whole egg, egg yolk, egg white, and products made with egg) have been determined online during PEF treatments (Amiali, 2005; Amiali et al., 2004). PEF technology offers to obtain microbiologically safe LEPs with high quality and long shelf life in refrigeration, compared with the traditional heat treatments. It is stated that PEF offers the microorganism inactivation, overall quality including yolk color and physicochemical stability, and food functionality including gel formation, structural changes, and properties in egg products (Bayraktaroğlu et al., 2012; Sampedro et al., 2006; Tokuşoğlu, 2013).

18.3 MICROORGANISM INACTIVATION BY PEF IN EGG TECHNOLOGY

Eggs are a nutritious part of our daily diet and a useful ingredient in foods owing to their functional properties. Egg and egg-based products and egg-originated foods may be responsible for a large number of food-borne illnesses each year, constituting an obstacle to the well-being of populations and a source of high economic losses (Bufano, 2000).

It is stated that there are various studies focusing on PEF inactivation of different target microorganisms in egg such as *S. enteritidis, L. innocua, E. coli* and *P. Fluorescens* (Sampedro et al., 2006). Previously, *E. coli* was used to study the effectiveness of two types of PEF treatment including continuous or batch, frequency and pulse width and it was found no differences in the inactivation achieved between the different types of treatments and frequencies (Martín-Belloso et al., 1997). The inactivation was a function of pulse number and the pulse width, and it was greatest with a pulse duration of 4 μs, which may be explained by the higher amount of energy applied (Sampedro et al., 2006).

Salmonella, and mainly serovars *Salmonella enteritidis* and *Salmonella typhimurium*, are the microorganisms responsible for most food-borne illnesses associated with egg and egg-based foods consumption (EFSA and ECDC, 2009). PEF has been demonstrated to be effective in inactivating different target microorganisms in LWE and egg products such as distinct *Salmonella* serotypes, *Escherichia coli, Bacillus cereus, Staphylococcus aureus, Listeria innocua,* and *Pseudomonas fluorescens* (Amiali, 2005, 2007; Bazhal et al., 2006; Calderón-Miranda et al., 1999; Hermawan et al., 2004; Huang et al., 2006; Jin et al., 2009; Martín-Belloso et al., 1997; Miranda et al., 1999; Monfort et al., 2010a,b, 2011, 2012a,b; Pina-Perez et al., 2009).

In the beginning, some problems concerning the performance of PEF in LWE were stated. The electrical resistivity of liquid egg is low, and therefore, a long treatment time is required to accomplish the needed inactivation. Besides, the presence of fat in liquid egg performs a protective role for microorganisms and proteins, diminishing the effect of PEF, because they may absorb the active radicals and ions resulting from the discharges (Sampedro et al., 2006; Tokuşoğlu, 2013). Another problem was the high electrical conductivity (0.37–0.17 S/m), which reduces applying possibilities of high-intensity treatments. A solution was proposed that the ultrafiltration or dialyzation treatment has been applied to reduce conductivity without altering the protein content, and pH of the original product (Sampedro et al., 2006; Tokuşoğlu, 2013).

PEF could be a possible technology to hygienize LWE since it has been observed in other products that PEF can inactivate vegetative cells of spoiling and pathogenic bacteria at ambient temperature, consequently diminishing the negative impact of heat on the quality properties (Barbosa-Cánovas et al., 2001). It is stated that PEF comprises the high-power electric field pulses application for short duration (in the order of 1 μs), using electric field strength in the range from 20 to 80 kV/cm (Barbosa-Cánovas and Rodríguez, 2002; Rastogi, 2003). It is also stated that the microbial lethality of PEF normally rises with PEF intensity (Bazhal et al., 2006).

Bazhal et al. (2006) reported an additional 2-log reduction for *E. coli* O157:H7 in LWE treated with PEF at low field intensities ranging from 9 to 15 kV/cm in conjuction with thermal treatment at 60°C compared to thermal treatment alone. Bazhal et al. (2006) reported a synergistic effect of temperature with electric field on the inactivation of *E. coli* O157:H7 within a given temperature range (Bazhal et al., 2006). The exponential decay model was used to predict the survival fraction rate of *E. coli* O157:H7 during PEF application with different pulses, different field intensities, and different treatment temperatures as 50°C, 55°C, and 60°C, as 0–138 pulses, and as 9–15 kV/cm, respectively, by Bazhal et al. (2006).

Martín-Belloso et al. (1997) reported the effectiveness of two types of PEF treatment including continuous and batch, as frequency, and pulse width for *E. coli* inactivation. It was found that there were no differences in the inactivation achieved between the different types of treatments and frequencies, whereas 5–6 log of *E. coli* reduction was achieved in 26 kV/cm, 2–4 μs at 37.2°C.

Calderón-Miranda et al. (1999) stated *L. innocua* inactivation in LWE by PESs and nisin additive. The highest extent of microbial inactivation with PEF was 3.5 log cycles (U) for an electric field intensity of 50 kV/cm and 32 pulses. Treatment of LWE by PEF was conducted at low temperatures, 36°C being the highest.

L. innocua exposed to 10 IU nisin/mL following PEF exhibited a decrease in population of 4.1 U for an electric field intensity of 50 kV/cm and 32 pulses. Exposure of *L. innocua* to 100 IU nisin/mL following PEF resulted in 5.5 U for an intensity of 50 kV/cm and 32 pulses (Calderón-Miranda et al., 1999). Góngora-Nieto et al. (1999) indicated the inactivation of *P. fluorescens* in LWE by hurdle technology including PEF. Using the pulsed electrical field/high hydrostatic pressure (PEF/HHP) alone or combined with antimicrobials in these studies, PEF was applied as 38 kV/cm and 130 pulses and was achieved as 90% inactivation of *P. fluorescens* (Góngora-Nieto et al., 1999).

It was reported that a maximum of 4.3-log reduction in *S. enteritidis* counts after subjecting inoculated prewarmed (55°C) LWE to a 25 kV/cm treatment at 200 Hz with 2.12-μs pulses totaling 250 μs (Hermawan et al., 2004).

Amiali et al. (2007) reported that increasing the applied electric field intensity, treatment time, and process temperature resulted in increased bacterial inactivation. At 30 kV/cm and 40°C, the populations of *E. coli* O157:H7 and *S. enteritidis* were reduced by ~5 logs for egg yolk. The inactivation rate constants increased from 0.004 to 0.098 μs^{-1} for *S. enteritidis* whereas for *E. coli* O157:H7, the constants increased from 0.009 to 0.039 μs^{-1} as processing temperature increased from 20°C to 40°C. Amiali et al. (2007) also revealed that *S. enteritidis* was more resistant to PEF inactivation than *E. coli* O157:H7 at lower processing temperatures for egg yolk (Amiali et al., 2007).

Zhao et al. (2007) determined the combined effects of heat and PEF treatment (20°C and 40°C, 0–800 μs at 30 kV/cm) on microbial inactivation inoculated in liquid egg whites. The effective inactivation conditions were achieved by using *S. enteritidis, E. coli* and *S. aureus*, common spoilage, and pathogenic microorganisms in egg products (Zhao et al., 2007).

Monfort et al. (2010a) described the inactivation of *S. enteritidis* and the heat-resistant *Salmonella senftenberg* to be 775 W in terms of treatment time and specific energy at electric field strengths ranging from 20 to 45 kV/cm. It was put forward that the target microorganism for PEF technology in LWE varied with PEF treatment intensity. On the basis of data given by Monfort et al. (2010a), *S. enteritidis* was found to be the most PEF-resistant strain in the conditions of the electric field strengths >25 kV/cm. They also reported that inactivation level depended only on the specific energy fort of this *Salmonella* serovar. 1-, 2-, and 3-log$_{10}$ reductions were found with 106, 272, and 472 kJ/kg of treatments, respectively (Monfort et al., 2010a).

Monfort et al. (2010b) reported *S. typhimurium* and *S. aureus* inactivation strategies by PEF in LWE. It was reported that maximum inactivation levels of 4 and 3 log$_{10}$ cycles of the population of *S. typhimurium* and *S. aureus* were achieved with treatments of 45 kV/cm, 30 μs and 419 kJ/kg, and 40 kV/cm for 15 μs and 166 kJ/kg, respectively.

Monfort et al. (2011) indicated that a PEF treatment of 25 kV/cm and 100 kJ/kg followed by a heat treatment of 55°C and 2 min reduced more than 8 $\log_{10}$ cycles of the population of *S. enteritidis* in LWE combined with 2% triethyl citrate (TC) (E1505), with a minimal impact on its protein-soluble content. The presence of additives, such as 10 mM EDTA (ethylenediaminetetraacetic acid) or 2% TC (E1505), increased the PEF lethality of 1 $\log_{10}$ cycle and generated around 1.5 $\log_{10}$ cycles of cell damage, resulting in the reduction of undamaged cells of 4.4 and 3.1 $\log_{10}$ cycles, respectively (Monfort et al., 2011).

Monfort et al. (2011) stated that the combined treatments are effective against different *Salmonella* serovars (Table 18.1). The inactivation level achieved is similar or even higher than traditional low-intensity heat-pasteurization treatments of 60°C/3.5′. For the high-intensity ultrapasteurization process at 70–71°C/1.5′, an alternative treatment has been proposed with similar *Salmonella* lethality (Table 18.1). These results indicate that proposed treatments by Monfort et al. (2011) could be alternatives to the current heat pasteurization treatments of LWE at both levels, low- and high-intensity pasteurization (Table 18.1).

Salmonella spp inactivation in LWE using PEF, heat, and additives was also performed by Monfort et al. (2012a). It was reported that the lethal effectiveness on seven different *Salmonella* serovars of the application, in static and continuous conditions, of PEF was followed by heat treatments in LWE with additives such as EDTA or TC by Monfort et al. (2012a). In that study decribed by Monfort et al. (2012a,b), the compared study to heat treatments was also carried out. The PEF (25 kV/cm and 75–100 kJ/kg) followed by heat (52°C/3.5 min, 55°C/2 min, or 60°C/1 min) in LWE with 2% TC permitted the reduction of heat-treatment time from 92-fold at 52°C to 3.4-fold at 60°C, and 4.8-fold at 52°C in LWE with EDTA for a 9-$\log_{10}$ reduction of *S. enteritidis* population (Monfort et al., 2012a).

PEF technology has been an effective new inactivation and novel pasteurization technique for LWE (Figure 18.3). PEF lethality hardly increased, but markedly

---

TABLE 18.1

$\log_{10}$ Reductions of Various *Salmonella* Serovars in LWEs by Heat Treated or PEF Treated Followed by Heat Treatments in Presence of TC Preservative

Salmonella Serovar	Low Pasteurization		High Pasteurization	
	PEF + 52°C/3.5′ + 2% TC	60°C/3.5′	PEF + 60°C/3.5′ + 1% TC	71°C/1.5′
Dublin	>8.0	4.3	>8.0	>8.0
Enteritidis	7.7	7.4	>8.0	>8.0
Senftenberg	5.5	2.0	>8.0	>8.0
Typhi	>8.0	>8.0	>8.0	>8.0
Typhimurium	6.9	>8.0	>8.0	>8.0
Virchow	>8.0	>8.0	>8.0	>8.0

Source: Adapted from Monfort S. et al. 2011. *International Journal of Food Microbiology* 145, 476–482.

FIGURE 18.3 PEF systems in OSU, USA. (From Zhang H.Q. 2012. *Advanced Nonthermal Processing in Food Technology: Effects of Quality and Shelf Life of Food and Beverages.* May 07–10, 2012, Kuşadasi-Turkey.)

augmented the energetic costs. On the basis of the above-mentioned results, it was recommended to apply treatments of energy levels higher than 250 kJ/kg. PEF applications followed by heat treatments in the presence of the additive TC could be an alternative to the current heat pasteurization treatments of LWE. These novel treatments showed higher lethal effectiveness against the above-mentioned food-borne microorganisms than the current LWE pasteurization treatments with a similar or even lower decrement of the soluble protein content. PEF could be a possible technology for hygienize LWE; it can inactivate vegetative cells of spoiling and pathogenic bacteria at ambient temperature, consequently diminishing the negative impact of heat on the quality properties.

18.4 FOOD QUALITY CONTROL BY PULSE ELECTRICAL FIELDS (PEF) IN EGG TECHNOLOGY

Some studies have focused on alterations in the physicochemical properties of egg and egg-derived products following the treatment with pulses.

Qin et al. (1995) reported the color, viscosity, and sensory properties of the liquid eggs with the addition of citric acid (0.15% w/v) after PEF treatments in continuous condition. The PEF treatment (35 kV/cm, W = 2 μs) included three steps as one step of four pulses and two steps of three pulses at 45°C. After the treatment, liquid eggs were packaged aseptically and stored at refrigeration temperature for up to 4 weeks. Qin et al. (1995) determined the decrease in viscosity and the increase in color as β-carotene. In sensory acceptance tests, no sensory differences were determined between scrambled eggs prepared from fresh egg and a PEF-treated egg sample (Qin et al., 1995).

Physical attributes and sensory evaluation of sponge cake made with PEF-treated eggs by Ma et al. (1997) and PEF conditions were 48 kV/cm, 20 pulses, and 2 μs for laboratory-size prototype system. Ma et al. (1997) stated that no differences were found between density and whiteness for sponge cake made with PEF-treated egg and PEF-untreated egg. The strength of the sponge cake made with PEF-treated egg was greater than that of PEF-untreated egg.

Góngora-Nieto et al. (1999) used the pulsed electrical field/high hydrostatic pressure (PEF/HHP) alone or combined with antimicrobials in LWE by hurdle technology. With PEF conditions (38 kV/cm,130 pulses) that could inactivate the 90% inactivation of *P. fluorescens,* there was no color difference between PEF-treated samples and the control eggs while a slight difference was found in color after HHP treatment (20–40 Kpsi, 5 min) (Góngora-Nieto et al., 1999).

Fernández-Díaz et al. (2000) performed the effects of PEFs on ovalbumin (2%) and dialyzed egg white. With the analysis of reactive sulfhydryl (SH) groups, the investigation of UV (ultraviolet) absorbance of aromatic amino acids, the determination of thermal gelation, texture of gels, and the determination of water-holding capacity in PEF-treated eggs were analyzed. In the PEF treatment of eggs, a discontinuous, exponential method was used and condition parameters are as follows: 0.7–2.3 J/pulse∗mL, 100 Hz, T < 29°C, E = 27–33 kV/cm, and W = 0.3 μs (20 nF) and 0.9 μs (80 nF). Fernández-Díaz et al. (2000) stated that PEF treatment did not cause notable changes in the egg proteins. Sensory panel results found that there were no differences between PEF-treated eggs and PEF-untreated eggs. It was reported that the reactivity of SH content of egg white proteins (EWPs) increased. No changes in polarity and conformation of tyrosine and tryptophan were found while a slight reduction was determined in gelling properties (Fernández-Díaz et al., 2000).

Liquid egg fractions including egg white and egg yolk are an adequate source of all essential amino acids, many vitamins, and important minerals, (Souci et al., 2008). Liquid egg white and/or liquid egg yolk possess highly desirable functional and nutritional properties. Morover, EWP has been extremely good for gelling, foaming, and emulsifying properties (Yang et al., 2009).

Heat induces the egg protein oxidation leading to alterations in the SH content, and specifically the unfolding of ovalbumin and other EWPs, resulting in the modification of the functional properties of egg and then resulting in the coagulation (Van Der Plancken et al., 2005). Therefore, the ovalbumin and EWP quality is so important.

Hermawan et al. (2004) reported no changes in viscosity, electrical conductivity, Brix, pH, and color (*L*, *a*, and *b* values) of the combined PEF at 55°C treated LWE sample compared with the nontreated samples. The applicated continuous PEF conditions were 200 pps, 1.2 mL/s, E = 25 kV/cm, and W = 2.12 μs in 250 μs.

Zhao et al. (2007) stated the combined effects of heat and PEF treatment (20°C and 40°C, 0–800 μs at 30 kV/cm) on liquid egg white quality. An increasing in emulsifying capacity and stability was found whereas the nonsignificant increasing was found in foaming capacity ($p < 0.05$). Surface-free SHs and hydrophobicity of EWPs increased with the increment of the PEF treatment time owing to the partial unfolding of EWP. About 50% of the trypsin inhibitory activity of ovomucoid in liquid egg white decreased when the treatment time extended to 800 μs (Zhao et al., 2007). The combined treatment containing heat and PEF could be applied to process liquid egg whites to get the best quality product.

The study described by Monfort et al. (2011) stated that, the PEF treatment application (25 kV/cm and 100 kJ/kg) in the presence of a TC additive (E1505), which can be added to LEPs to improve their whipping properties, reduced the heat resistance to a posterior thermal treatment of *Salmonella* serovars up to 100-fold, permitting

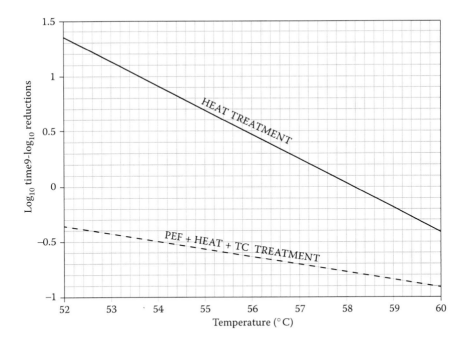

FIGURE 18.4 Thermal death time (TDT) lines for 9-$\log_{10}$ S. *enteritidis* reductions treated by heat treatment (continuous line) or PEF (25 kV/cm; 100 kJ/kg) followed by heat in the presence of 2% TC preservative (dotted line) in LWE. (Adapted from Monfort S. et al. 2011. *International Journal of Food Microbiology* 145, 476–482.)

more than 8-$\log_{10}$ reductions with heat treatments of 52°C/3.5′, 55°C/2.5′, or 60°C/1.5′ (Figure 18.4).

Marco-Moĺes et al. (2011) stated that the effect of high-intensity pulsed electric fields (HIPEF) processing (field strength: 19, 32, and 37 kV/cm) was compared to the traditional heat pasteurization (66°C for 4.5 min) (Marco-Moĺes et al., 2011). In the study described by Marco-Moĺes et al. (2011), thermal treatment caused an increasing in the viscosity of LWE, especially in nonhomogenized samples and HIPEF treatments did not modify the original LWE color, whereas thermally treated samples developed an opaque appearance. Marco-Moĺes et al. (2011) reported that heat pasteurization had a significant impact on both the water-soluble protein content of the LWE (19.5% to 23.6% decreasing) and the mechanical properties of the egg gels (up to 21.3% and 14.5% increasing in hardness and cohesiveness, respectively). It was expressed that high-intensity pulsed electric fields (HIPEF) treatments better preserved the food matrix structure.

Zhao et al. (2011) demonstrated the changes in the functional properties and structure of EWPs induced by PEF. It was shown that PEF treatment, including 25–35 kV/cm for 100–800 μs could change the structure of EWPs (Zhao et al., 2011). With these PEF conditions, the emulsifying and foaming capabilities were enhanced. It was reported that unfolding protein molecules associated with the exposure of SH and hydrophobic groups was induced by PEF (Zhao et al., 2011).

Zhao et al. (2011) stated that the solubility, emulsifying, and foaming capability decreased. In this study, researchers suggested that when electric field intensity and treatment time were enhanced, protein aggregates formed, owing to the increased exposure of SH and hydrophobic groups (Zhao et al., 2011).

In another study described by Monfort et al. (2012b), physicochemical and functional properties of LWE treated by PEF application were followed by heat in the presence of the additive TC. The PEF treatment (25 kV/cm, 75 kJ/kg) was followed by heat (52°C/3.5 min, 55°C/2 min, or 60°C/1 min) in the presence of 2% triethyl citrate (LWE–TC–PEF-HT), and by current heat pasteurization treatments of 60°C/3.5 min or 64°C/2.5 min (LWE–HT). LWE–TC–PEFHT were applied (Monfort et al., 2012b). It was shown that a more similar color to the raw LWE (R-LWE) was found, the foaming and emulsifying capacities were higher than raw LWE (R-LWE) as 48% and 26% levels, respectively. In that study, designed treatments improved the foaming and emulsifying capacities of R-LWE as 17% and 9%, respectively and kept similar texture-gelling properties in treated LWE (Monfort et al., 2012b). Monfort et al. (2012b) revealed that designed PEF treatment is lethally equivalent to heat pasteurization treatments, but keeps better liquid egg functional properties. Also, the successive application of PEF and heat in the presence of TC additive could be very promising alternative techniques to the recommended heat pasteurization treatments (Monfort et al., 2012b).

It is demonstrated that PEF could affect the antioxidant activity of hydrolyzates derived from egg white and change the molecular distribution of peptide into a small molecular weight (Lin et al., 2012,2013). The optimal parameters of PEF for processing antioxidant hydrolyzates are the pulse number at 300, electric field strength at 10 kV/cm, and concentration of antioxidant hydrolyzates at 10 mg/mL (Lin et al., 2013). On the basis of data given by Lin et al. (2013), strong PEF treatment could significantly improve antioxidant activity of hydrolyzates derived from EWP in 2–4 h and the PEF could contribute some structural alterations of antioxidant peptide originated from EWP and improve its antioxidant activities.

18.5   SHELF-LIFE STUDIES OF PEF-TREATED LIQUID EGGS

Dunn et al. (1989) stated that the shelf life of pasteurized LWE samples treated with PEF and additives potassium sorbate and citric acid and heating at 60°C was determined at 28 days/at 4°C and 10 days/at 10°C. Using PEF conditions were 36 kV/cm and 25 pulses. The product was microbiologically stable (no detectable coliforms) for 31 days, using the microbial limits for LEPs recommended by Vanderzant and Splittstoesser (1992).

28 days/at 4–6°C of shelf life for PEF-treated fresh liquid egg with citric acid (0.15% w/v) was reported by Qin et al. (1995). In this study, continuous system PEF with three steps at 45°C was used while 35 kV/cm and 2 µs of conditions were applied for one step of four pulses and two steps of three pulses (Qin et al., 1995). Ma et al. (1997) performed the inactivation of *E. coli* in fresh LWE using PEF and the shelf life of the product was found as 4 weeks at refrigeration temperature using the continuous system with five steps when the conditions were 48 kV/cm, 2 µs, and 20 pulses (Ma et al., 1997).

Góngora-Nieto et al. (1999) stated both types of PEF treatments for LWE with citric acid (0.15% and 0.5%). Stepwise and continuous recirculation modes were used to process the LWE. For stepwise conditions including 3–4 steps of 10 pulses, 30 L/h (0.15%), four steps of 30 pulses, 39 L/h (0.5%), and for continuous conditions including 30–266 pulses, 30 L/h (0.15%) and 30 pulses, 39 L/h (0.5%); the shelf-life studies were carried out and the shelf life of the product was found as 20 days at 4°C (0.15%) and 26 days at 4°C (5%) for applied conditions (Góngora-Nieto et al., 1999). For the same treatments, the stepwise mode was more effective than the continuous system, and also more effective in the sample containing 0.5% of citric acid than in the sample with 0.15% citric acid (Góngora-Nieto et al., 1999).

In the above-mentioned shelf-life studies, as in the microorganism inactivation studies, antimicrobial additive compounds and moderate temperature were used in combination with PEF to extend the shelf life of the LEPs (Dunn et al., 1989; Góngora-Nieto et al., 1999; Ma et al., 1997; Qin et al., 1995; Vanderzant and Splittstoesser, 1992).

REFERENCES

AEB. 2013. American Egg Board. http://www.aeb.org/foodservice-professionals.

Amiali M. 2005. *Inactivation of Escherichia coli O157:H7 and Salmonella enteritidis in Liquid Egg Products Using Pulsed Electric Field*. Department of Bioresource Engineering, Macdonald Campus, McGill University, Montreal, Quebec.

Amiali M., Ngadi M.O., Smith J.P. and Raghavan V.G.S. 2004. Inactivation of *Escherichia coli* O157:H7 in liquid dialyzed egg using pulsed electric fields. *Transactions of IChemE, Part C. Food and Bioproducts Processing* 82(C2), 151–156.

Amiali M., Ngadi M.O., Smith J.P. and Raghavan G.S.V. 2007. Synergistic effect of temperature and pulsed electric field on inactivation of *Escherichia coli* O157:H7 and *Salmonella enteritidis* in liquid egg yolk. *Journal of Food Engineering* 79, 689–694.

Barbosa-Cánovas G.V., Góngora-Nieto M.M., Pothakamury U.R. and Swanson B.G. 1999. *Preservation of Foods with Pulsed Electric Fields*. Academic Press, San Diego, California, USA.

Barbosa-Cánovas G.V., Pierson M.D., Zhang Q.H. and Schaffner D.W. 2001. Pulsed electric fields. *Journal of Food Science* 65, 65–79. Supplement.

Barbosa-Cánovas G.V. and Rodríguez J.J. 2002. Update on nonthermal food processing technologies: Pulsed electric field, high hydrostatic pressure, irradiation and ultrasound. *Food Australia* 54(11), 513–520.

Bayraktaroğlu N., Tokuşoğlu Ö. and Barbosa-Cánovas G.V. 2012. Effects of pulsed electric fields (PEF) on shelf life of industrial liquid egg and egg based products. *Advanced Nonthermal Processing in Food Technology: Effects of Quality and Shelf Life of Food and Beverages*. May 07–10, 2012, Kuşadasi-Turkey. Oral presentation. *In ANPFT2012 Proceeding Book*, p. 227. ISBN: 978-975-8628-33-9, Celal Bayar University Publishing, Turkey.

Bazhal M.I., Ngadi M.O., Raghavan G.S.V. and Smith J.P. 2006. Inactivation of *Escherichia coli* O157:H7 in liquid whole egg using combined pulsed electric field and thermal treatments. *LWT Food Science and Technology* 39, 419–425.

Bufano N.S. 2000. Keeping eggs safe from farm to table. *Food Technology* 54(8), 192.

Burley R.W. and Vadehra D.V. 1989. *The Avian Egg, Chemistry and Biology*. Wiley, New York.

Calderón-Miranda M.L., Barbosa-Cánovas G.V. and Swanson B.G. 1999. Inactivation of *Listeria innocua* in liquid whole egg by pulsed electric fields and nisin. *International Journal of Food Microbiology* 51, 7–17.

Cotterill O.J. and Geiger G.S. 1977. Egg product yield trends from shell eggs. *Poultry Science* 56, 1027–1031.

Dunn J.E., Pearlman J.S. and La Costa R. 1989. *Methods and Apparatus for Extending the Shelf Life of Fluid Food Products.* U.S. Patent 4,838,154.

EFSA & ECDC. 2009. European Food Safety Authority & European Centre for Diseases and Control. The community summary report on trends and sources of zoonoses and zoonotic agents in the European Union in 2007. http://www.efsa.europa.eu/EFSA/efsa_loc ale-1178620753812_1211902269834.htm

Fernández-Díaz M.D., Barsotti L., Dumay E. and Cheftel J.C. 2000. Effects of pulsed electric fields on ovalbumin and dialyzed egg white. *Journal of Agricultural and Food Chemistry* 48, 2332–2339.

Góngora-Nieto M.M., Seignour L., Riquet P., Davidson P.M., Barbosa-Cánovas G.V. and Swanson B.G. 1999. Hurdle approach for the inactivation of *Pseudomonas fluorescens* in liquid whole egg. Food engineering: Nonthermal processing. IFT Annual Meeting, Chicago, IL, USA, p. 83A-2.

Hermawan N., Evrendilek G.A., Dantzer W.R., Zhang Q.H. and Richter E.R. 2004. Pulsed electric field treatment of liquid whole egg inoculated with *Salmonella enteritidis*. *Journal of Food Safety* 24, 71–85.

Huang E., Mittal G.S. and Griffiths M.W. 2006. Inactivation of *Salmonella enteritidis* in liquid whole egg using combination treatments of pulsed electric field, high pressure and ultrasound. *Biosystems Engineering* 94, 403–413.

Jin T., Zhang H., Hermawan N. and Dantzer W. 2009. Effects of pH and temperature on inactivation of *Salmonella typhimurium* DT104 in liquid whole egg by pulsed electric fields. *International Journal of Food Science and Technology* 44, 367–372.

Li-Chan E.C.Y., Powrie W.D. and Nakai S. 1995. The chemistry of eggs and egg products. In *Egg Science and Technology*, 4th ed., Stadelman, W. J. and Cotterill, O. J., eds., Haworth Press, New York, pp. 105–175.

Lin S., Guo Y., You Q., Yinc Y. and Liua J. 2012. Preparation of antioxidant peptide from egg white protein and improvement of its activities assisted by high-intensity pulsed electric field. *Journal of Science Food Agriculture* 92, 1554–1561.

Lin S., Guo Y., You Q., Yin Y., Liu J. and Luo P.G. 2013. Effects of high intensity pulsed electric field on antioxidant attributes of hydrolysates derived from egg white protein. *Journal of Food Biochemistry* 37, 45–52.

Ma L., Chang F.J. and Barbosa-Cánovas G.V. 1997. Inactivation of *E. coli* in liquid whole egg using pulsed electric fields technology. New frontiers in food engineering. *Proceedings of the Fifth Conference of Food Engineering*, New York, USA, pp. 216–221.

Mañas P. and Pagán R. 2005. Microbial inactivation by new technologies of food preservation. *Journal of Applied Microbiology* 98, 1387–1399.

Marco-Molés R., Rojas-Grau M.A., Hernando I., Pérez-Munuera I., Soliva-Fortuny R. and Martín-Belloso O. 2011. Physical and structural changes in liquid whole egg treated with high-intensity pulsed electric fields. *Journal of Food Science* 76(2), 257–264.

Martín-Belloso O., Vega-Mercado H., Qin B.L., Chang F.J., Barbosa-Cánovas G.V. and Swanson B.G. 1997. Inactivation of *Escherichia coli* suspended in liquid egg using pulsed electric fields. *Journal of Food Processing and Preservation* 21, 193–208.

Monfort S., Gayán E., Raso J., Condón S. and Álvarez I. 2010a. Evaluation of pulsed electric fields technology for liquid whole egg pasteurization. *Food Microbiology* 27, 845–852.

Monfort S., Gayán E., Saldaña G., Puértolas E., Condón S., Raso J. and Álvarez I. 2010b. Inactivation of *Salmonella typhimurium* and *Staphylococcus aureus* by pulsed electric fields in liquid whole egg. *Innovative Food Science and Emerging Technologies* 11, 306–313.

Monfort S., Gayán E., Condón S., Raso J. and Álvarez I. 2011. Design of a combined process for the inactivation of *Salmonella enteritidis* in liquid whole egg at 55°C. *International Journal of Food Microbiology* 145, 476–482.

Monfort S., Saldaña G., Condón S., Raso J. and Álvarez I. 2012a. Inactivation of *Salmonella* spp. in liquid whole egg using pulsed electric fields, heat, and additives. *Food Microbiology* 30, 393–399.

Monfort S., Mañas P., Condón S., Raso J. and Álvarez I. 2012b. Physicochemical and functional properties of liquid whole egg treated by the application of pulsed electric fields followed by heat in the presence of triethyl citrate. *Food Research International* 48, 484–490.

Pina-Perez M.C., Silva-Angulo A.B., Rodrigo D. and Martinez-Lopez A. 2009. Synergistic effect of pulsed electric fields and CocoanOX 12% on the inactivation kinetics of *Bacillus cereus* in a mixed beverage of liquid whole egg and skim milk. *International Journal of Food Microbiology* 130, 196–204.

Qin B.L., Pothakamury U.R., Vega-Mercado H., Martín-Belloso O.M., Barbosa-Cánovas G.V. and Swanson B.G. 1995. Food pasteurization using high-intensity pulsed electric fields. *Food Technology* 12, 55–60.

Rastogi N.K. 2003. Application of high-intensity pulsed electrical fields in food processing. *Food Reviews International* 19(3), 229–251.

Sampedro F., Rodrigo D., Martínez A., Barbosa-Cánovas G.V. and Rodrigo M. 2006. Review: Application of pulsed electric fields in egg and egg derivatives. *Food Science and Technology International* 12, 397–406.

Sanovo 2013. *Egg Production. Sanovo Egg Group*. SANOVO International A/S, Denmark.

Souci S.W., Fachmann W. and Kraut H. 2008. *Food Composition and Nutrition Tables*. 7th ed., MedPharm Scientific Publishers, Stuttgart, Germany, 1364p.

Sugino H., Nitoda T. and Juneja L.R. 1997. General chemical composition of hen eggs. In *Hen Eggs, Their Basic and Applied Science*, Yamamoto, T., Juneja, L. R., Hatta, H. and Kim, M., eds., CRC Press, New York, pp. 13–24.

Tokuşoğlu Ö. 2013. *Table Egg Quality and Technology (Yemeklik Yumurta Kalite ve Teknolojisi) Graduate Course Notes*. Celal Bayar University, Manisa, Turkey [in Turkish].

Vanderzant C. and Splittstoesser D.F. 1992. *Compendium of Methods for the Microbiological Examination of Foods*. 3rd ed, American Public Health Association, Washington, DC, USA.

Van Der Plancken I., Van Loey A. and Hendrickx M.E.G. 2005. Changes in sulfhydryl content of egg white proteins due to heat and pressure treatment. *Journal of Agricultural and Food Chemistry* 53, 5726–5733.

Yang R.X., Li W.Z., Zhu C.Q., Zhang Q. 2009. Effects of ultra-high hydrostatic pressure on foaming and physical-chemistry properties of egg white. *J. Biomedical Science and Engineering*, 2, 617–620.

Zhang H.Q. 2012. PEF technology—A potential to be realized in food preservation. *Advanced Nonthermal Processing in Food Technology: Effects of Quality and Shelf Life of Food and Beverages*. May 07–10, 2012, Kuşadasi-Turkey.

Zhao W., Yang R.J., Tang Y.L. and Lu R.R. 2007. Combined effects of heat and PEF on microbial inactivation and quality of liquid egg whites. *International Journal of Food Engineering* 3(4), ISSN (Online), 1556–3758, DOI: 10.2202/1556-3758.1256.

Zhao W., Yang R.J. and Zhang W.B. 2011. Changes in functional properties and structure of egg white proteins induced by pulsed electric fields. *Journal of Food Science* 32(9), 91–96.

19 PEF Systems for Industrial Food Processing and Related Applications

Michael A. Kempkes and Özlem Tokuşoğlu

CONTENTS

19.1 INTRODUCTION

Pulsed electric field (PEF) processing involves the application of short-duration, very-high-voltage pulses to organic material, electroporating bacterial and other cells, and perforating the cell membranes. Electroporation has been investigated for a range of applications, from nonthermal pasteurization of juices, enhanced extraction from vegetative cells and algae, and enhanced drying, to wastewater treatment. This chapter describes the key PEF treatment parameters, and the specialized equipment required to implement this process in industrial processes. It also describes the interactions between the process parameters and the electrical design and performance of PEF systems, as a guide for potential adopters of this technology. Finally, this chapter presents initial commercial applications of PEF processing, and guidelines for its future adoption.

PEF processing is a low-temperature, nonthermal, nonchemical, low-impact process that induces electroporation of cell membranes in plant and animal cells. (FDA, Center for Food Safety and Applied Nutrition, 2000) PEF processing involves the application of short-duration (1–20 μs), very-high-voltage pulses that create a high-voltage field (from 1 to 50 kV/cm) across a liquid (Barbosa-Canovas et al., 1999). Depending on the desired effect, this field can be adjusted to open the cell membranes in plant cells (1–10 kV/cm) or kill resident bacteria, molds, and other microorganisms at higher field strengths, via electroporation of the cell membrane (Figure 19.1). By using multiple treatment chambers to apply pulses to a stream of fluid in a continuous-flow process, kill ratios of 5–9-log reductions against bacteria in juices and other liquid foods have been demonstrated, similar to those resulting from pasteurization (Jeyamkondan et al., 1999). Unlike pasteurization, however, the food is not heated during PEF processing; so its taste and nutritional value remain essentially indistinguishable from fresh, untreated product—while maintaining the level of food safety associated with pasteurization. PEF-processed products simply taste fresher than pasteurized products, yet have equivalent safety and shelf life.

When used for nonthermal pasteurization, PEF achieves very high kill rates, making it a practical alternative to thermal pasteurization (Clark, 2006). When used for extraction, drying, or wastewater processing, this same electroporation process releases contents from internal to the cells, making them readily available to downstream processing (digestion, separation, etc.).

Multiple experiments have demonstrated that the shelf life of PEF-processed food is comparable to that yielded by pasteurization, with little or no impact on the taste, color, or nutritional value of the food (Min, 2003). PEF processing is particularly beneficial to fresh juices, beer, and other foods that are susceptible to changes in their flavor caused by the heat of pasteurization. The commercial debut of PEF processing occurred in 2005, with Genesis Juice's introduction of PEF-processed juices to the consumer market (Figure 19.2), followed by the first large-scale PEF installation for wastewater treatment in 2006. These represent the first known commercial introductions of PEF-processed foods. After a several-year hiatus, there has been a rapid adoption of PEF processing in applications since 2010. PEF processing is being used to improve the performance of industrial processes such as the pretreatment of

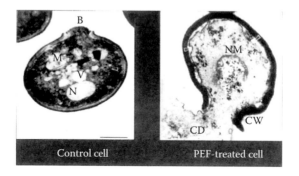

FIGURE 19.1 Untreated and PEF treated yeast cell, showing dramatic effects of electroporation.

FIGURE 19.2   Genesis Juice on sale in USA, 2005.

wastewater sludge prior to anaerobic digestion, the extraction of sugars and starches from plants, the release of lipids from algae for biofuel production, and enhanced drying of vegetable materials—all due to the ruptured cells releasing their intracellular liquids more easily into their surroundings (Gaudreau et al., 2008). PEF processing has also been reintroduced for nonthermal pasteurization of juices and other liquid foods, particularly in Europe and Asia since 2010 (Buckow, 2013). This chapter will describe the key PEF parameters, and the specialized equipment required to implement this process in liquid food processing, as a guide for adopters of this technology.

19.2   PEF UTILITY

The origins of PEF processing are deep and varied. As early as 1960, researchers demonstrated that voltage fields could disrupt biological cells, with microbial inactivation initially demonstrated in 1967 (Toepfl, 2006). Only since the mid-1990s, however, has there been the necessary confluence of applied research developments in high-voltage equipment, and commercial interest in nonthermal processes to move PEF technology from the laboratory and into commercial operation. Three key steps in this development were the Dunn and Pearlman patent on the PEF process for disinfection (4,695,472) in 1987, the development of the cofield flow treatment chambers at the Ohio State University (OSU) in 1997, and the initial use of solid-state, high-voltage pulse modulators for PEF in 2000, with the introduction of the first commercial-scale PEF system for OSU built by Diversified Technologies, Inc. (DTI, Figure 19.3). This combination of advances laid the groundwork for the movement of PEF processing from laboratory to commercialization in the last decade (Casey et al., 2002).

In a parallel and more recent set of efforts, PEF is also being commercialized as a technique for permeabilization of plant and animal tissues for a variety of purposes—typically focused on extraction, drying, or preprocessing of the tissue for subsequent chemical or microbial processing. In all cases, the electroporation-induced rupture of the plant or animal tissue cells opens these cells to allow the

FIGURE 19.3   First large-scale PEF system, built by DTI for Ohio State University in 2000.

exchange of materials between the internal and external environment, such as the simplified release of water (enhanced drying) or internal materials, such as extraction of sugar from sugar beets and accelerated fermentation (Toepfl, 2006).

Increasing commercial adoption of PEF processes has been reported in the last 3 years, including the use of PEF for wastewater treatment (prior to anaerobic digestion of wastewater sludge), extraction, and drying. In general, the focus of these activities is process acceleration, reduction of energy costs, or both. More recently, a number of researchers and firms have investigated the use of PEF processing as an aid to the extraction of lipids from algae for biofuels (Figure 19.4), with initial successes leading to the first commercial installations of PEF equipment in this area in 2013 (Frey et al., 2012; Kempkes, 2012).

Disinfection (also referred to as nonthermal pasteurization) has historically been the primary focus for PEF related to food products—using PEF to induce electroporation

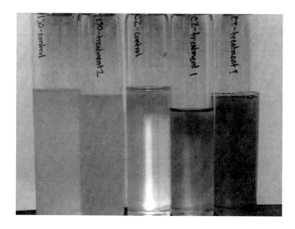

FIGURE 19.4   PEF treated algae. The release of chlorophyll is a visible marker for electroporation of the algae cells.

and kill microbes in liquid or pumpable foods. Generally, the desire has been to achieve disinfection levels similar to thermal pasteurization, but without the damage to taste, nutrition, and other characteristics of unprocessed liquids that can be significantly degraded by heating the food (i.e., thermal processing). Yeom substantial research over the last 20 years has clearly demonstrated that PEF can achieve disinfection of liquids across a wide range of microbes and products, at relatively low temperatures compared to pasteurization (e.g., 20–50°C). Literally thousands of studies, and a number of books, have been published showing the impact of PEF on microbes and foods in various combinations, under a wide range of treatment protocols. In general, these studies have focused on two related areas: showing microbial kill, and assessing the impact of PEF on food quality (taste, nutrition, etc.) across a wide range of pumpable products ranging from juices to semisolids, such as sausages.

A primary result of these studies is that it is possible to achieve high microbial kill levels (as high as 9 logs, or only one survivor out of 1 billion initial microbes), with proper treatment conditions and careful processing (Barbosa-Canovas, 2000). Given that the standard for pasteurization is typically a 5-log reduction in microbial survival, PEF is clearly capable of replacing pasteurization as a disinfection step (FDA, Center for Food Safety and Applied Nutrition, 2000). As PEF moves into commercial applications, a key area of interest has been balancing the competing factors of high levels of disinfection, minimal impact on the taste of the food, and cost. Unfortunately, the perceived risk of using a new technology, such as PEF, to achieve food safety requirements, has significantly slowed the adoption of PEF systems in this area. There are commercial PEF applications for nonthermal pasteurization in Europe and Asia (Figure 19.5), but few of these are publicized—the companies currently utilizing this technology are typically not drawing attention to PEF.

One key finding from these studies is that, to the best of our knowledge, PEF is only effective in killing vegetative microbes, yeasts, and molds—PEF appears ineffective against spores or viruses (Barbosa-Canovas et al., 1999). This is not surprising, given that PEF is believed to work through the mechanism of electroporation, and spores and viruses do not have active membranes that the electric field can impact. For this reason, the preponderance of PEF studies focused on food disinfection have

FIGURE 19.5   PEF-treated juices in the Netherlands, 2013.

targeted acidic products, such as citrus juices and tomato sauces, where spore regeneration is not an issue. Researchers are still investigating the use of PEF, either by itself or in combination with other treatments, to kill resistant bacterial spores, but without notable success to date. This work, if ultimately successful, would significantly expand the applicability of PEF disinfection to a larger range of liquid foods.

The commercialization of PEF processing in areas that do not impact food safety—where PEF is used as a pretreatment prior to extraction, digestion, or drying—has been more rapid. These processes are typically justified solely by their economic impact—the PEF-enabled process is more effective and less costly than the alternatives. Wastewater treatment and algal lipid extraction are two of these early processes transitioning into commercial operation, with juice extraction also being reported (Gaudreau et al., 2008). Again, the commercial interests of the companies adopting PEF processing have limited the publicity surrounding these applications, but the demand for larger PEF systems from manufacturers attests to their reality—the market for PEF systems is no longer limited to small laboratory/R&D (research and development) units.

As a result of these developments, PEF processing has made the leap from laboratory exploration to a growing commercial processing market in several different industries. This commercial demand, in turn, is leading to the development of less-expensive, integrated PEF systems for the industry.

19.3   KEY PROCESS PARAMETERS

Any discussion of PEF system design must be based on the development of an effective PEF treatment protocol (i.e., field strength and treatment time) for the desired application. In designing a PEF system, this protocol (or at least its outer boundaries) must be known. The PEF treatment protocol can vary significantly across different applications and products. The published literature provides numerous examples of protocol that may be used as starting points in this development, but rarely can they be used without optimization for the specific product and desired PEF effect. The specific protocol is developed from experiments involving multiple trials at different electrical field strengths and durations, followed by the assessment of the product after treatment to determine where the desired effect is achieved (typically with the lowest energy use). Development of the protocol may also include varying the properties of the liquid being treated, through concentration or dilution, grinding or slicing, or the addition of other chemicals to enhance the PEF process. Development of the optimal treatment protocol is a critical step to commercial PEF processing.

In general, the smaller the cell being treated, the higher the field strength required. These field strengths can be as low as a few hundred volts for large plant cells, to 40 kV/cm for some strains of bacteria. A typical nonthermal pasteurization treatment protocol might require the application of a 35 kV/cm field for a minimum of 50 μs to yield a target bacterial reduction (typically, a 5-log reduction) of a target organism (such as *E. coli*) in a liquid food substrate (Min, 2003). Extraction, on the other hand, typically requires a much lower field strength to achieve the desired results. Even when the required field strength is known, treatment times required may vary significantly, depending on the structure of those cells (Toepfl, 2006). Finally, there is weak inverse correlation between field strength and treatment time that has been observed (lower fields at longer treatment times can achieve similar results to higher fields/shorter treatments). For commercial applications, the desire is to minimize the amount of energy required to achieve the desired result, which minimizes the cost of the PEF process itself.

For the purposes of designing a PEF system, we assume that the boundaries of the protocol for the desired application are known. From that point, the characteristics of the product, and the desired processing capacity, are the two major remaining considerations to be accounted for in the system design. These are typically expressed as the liquid conductivity and flow rate. These two factors, when combined with the desired protocol, form the basis for designing the PEF system.

19.3.1 Conductivity/Flow Rate

The power required to apply a given protocol is determined by the conductivity of the fluid, and the desired flow rate. The electric field (in V/m or more commonly, kV/cm) is set by the treatment protocol; so, the energy required to deliver this field to a volume of liquid is a direct function of the fluid conductivity. Conductivity (σ) is a measure of the electron mobility within a volume of material—a measure of how easily it passes electrical current. Conductivity is expressed as siemens/meter (S/m), or, to get to whole numbers for typical fluids, it is expressed as mS/cm. It is the reciprocal of resistivity, which is the electrical resistance of a volume of material. In most PEF research, conductivity is used as a measure because it is readily measured with inexpensive meters, applicable to the liquid itself, and allows direct calculation of the current that will pass through a volume of fluid through application of Ohm's law.

Since power is energy per unit time, the energy required per liter, and the number of liters to be processed per unit time, give the total average power required in the PEF system. The typical calculation for average power in a PEF system is pulse voltage (volts, from the modulator) × peak current (amps, into the treatment chambers) × pulsewidth (seconds) × the pulse frequency (Hz). The power required increases linearly with flow rate and conductivity for a given protocol. A 100-kW average power system, therefore, can by definition process 5 × the volume of the product per hour as a 20-kW system, for the same protocol. More critically, power increases by the square of the required field strength—making field strength the most critical parameter to establish correctly, to minimize electrical power requirements in the PEF process. Since electroporation is a threshold effect, small changes in the field strength can result in significant changes in the electroporation levels achieved, creating a very high need for balance between effectiveness (at higher than threshold field strengths) and cost (Gaudreau et al., 2004).

A laboratory PEF system typically processes a few liters per hour (Figure 19.6), and throughput is not a critical factor. Scaling to a pilot plant typically requires operation at tens to hundreds of liters per hour. Commercial systems, however, must be capable of processing thousands to tens of thousands of liters per hour—representing 2–3 orders of magnitude higher throughput (and average power) than a pilot system. Fortunately, research at OSU and other organizations have demonstrated that the PEF process itself operates independent from flow rate—so long as the field strength

FIGURE 19.6   Lab-scale PEF system for process development of low volume operations (DTI).

and dose (total treatment time) are maintained. This allows treatment protocols to be developed in the laboratory and pilot-scale systems, and these protocols can directly transition to commercial-scale PEF operations (Gaudreau et al., 2001).

Flow rate impacts several other major PEF system characteristics. For the most common treatment chamber design, the cofield flow chamber (discussed below) the diameter of the treatment chamber must be sized to pass the desired flow at reasonable pressure drops. The presence of particulates and "chunks" in the flow can also drive the diameter of the chamber, to prevent clogging. At the same time, to achieve uniform field strength within the treatment chamber, the gap across which the voltage is applied must increase with pipe diameter (maintaining the gap at more than 1.2× the diameter provides reasonable uniform fields). Larger chambers require lower fluid pressure in the treatment chambers, but require higher absolute voltages to maintain a given field strength. For example, doubling the treatment chamber diameter allows (nominally) 4× the flow at a given pressure, but requires twice the peak voltage (since the electrode spacing must also be doubled) and twice the average current, to maintain the same field strength and treatment time. In this example, the peak power increases by a factor of 4, but the average power remains the same. Unfortunately, the cost of the PEF system is highly dependent on the peak power required; so, there are practical limits to the size of the treatment chambers that are feasible. In general, smaller chambers are more cost-effective than larger chambers; so, this trade-off is a critical function of PEF system design.

19.3.2 PULSE SHAPE/FREQUENCY

One critical issue in PEF system performance is that the electrical pulse shape delivered a given PEF system. PEF is generally believed to work on a voltage threshold, and only exposure time above that critical field strength is believed to have the desired electroporation result (Buckow, 2013). In an ideal PEF pulse, there would be zero rise and fall time, and the flattop of the pulse would be concentrated at a constant voltage. All of the pulse energy, therefore, would be at the desired voltage. In practice, however, this pulse shape is not realizable. All pulses have finite rise and fall times, where the full voltage is not present, and it is not possible to achieve a perfect flattop (Figure 19.7). Characterizing the actual pulse voltage and time is therefore subject to estimates (or, more commonly, ignored all together).

To illustrate this, Figure 19.8 shows three normalized voltage waveforms for typical real pulses: a rectangular, exponential, and half-sine wave. These are three simplified versions of the most common modulator pulse shapes. They all have the same peak voltage (V_{max} = 100), and total pulsewidth (with the exponential cutoff at ~25% V_{max}). In this simplified example, researchers using each of these three waveforms could report results as if these pulse characteristics (V = 100, t = 100) were the same. However, their total energy and time above any arbitrary threshold could vary by more than a factor of 6, depending on the pulse shape each researcher used, and how they measured voltage and pulsewidth. Given this disparity, it is not surprising that there are wildly different biological results reported by different researchers, even though identical pulse conditions are given for the experiments.

Local heating and electrolysis combine to limit the pulsewidths to microseconds for high-field PEF applications. For these shorter pulsewidths, the disparity between the ideal and actual pulse shapes is critical, since more of the total pulsewidth consists of the rise and fall times. This translates into wasted power, which does not contribute to the PEF treatment. Longer pulsewidths (milliseconds or seconds, rather than microseconds) are possible at lower field strengths, which can simplify their pulse requirements for applications where low field strengths are possible, such as plant tissue permeabilization.

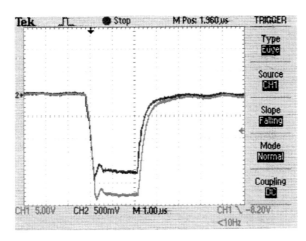

FIGURE 19.7   Sample pulse from a lab PEF system. 2HkV, 30 A, 2 µS pulse width.

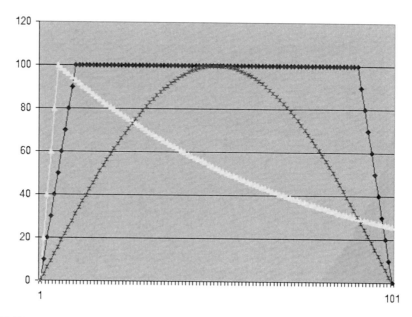

FIGURE 19.8   Representative voltage wave forms: Decaying exponential, square wave, and sinusoidal. All three have the same nominal voltage and pulse width, but widely varying PEF efficacy.

It is critical, therefore, to approximate the ideal waveform as closely as possible, and to account for the variations from the ideal pulse shape when translating treatment protocols across different PEF systems. Treatment protocols must take into account the pulse shape and measurement thresholds. As an example, Figure 19.7 shows PEF voltage and current pulses from a DTI pilot-scale PEF system processing orange juice. The DC (direct current) voltage setting is 24 kV, and this peak voltage is maintained for just fewer than 2 μs. The commanded pulsewidth (at the input of the modulator) is approximately 2.5 μs, and the difference is attributable to the pulse risetime. The total pulsewidth, taking into account both rise and fall times, could be interpreted to be as long as 3 μs. If 20 pulses are applied to each element of the juice, the reported treatment time from this pulse could be as low as 35 μs, or as high as 60 μs. This is nearly a 2:1 difference in reported treatment time, for the same pulse. Similarly (but less significantly for this case), the peak voltage could be reported as several kV above or below the nominal 24 kV, depending on where the measurement is made during the pulse. How these parameters are measured and reported significantly impacts the results when attempting to replicate protocols on different PEF systems, with different pulse shapes.

There is a strong need within the PEF research community to standardize the reporting of PEF results, allowing data from different organizations, using different PEF systems and waveforms, to be compared and assessed on a common basis. At a minimum, the voltage pulse shape should be shown, with measurement points for peak voltage and pulsewidth. This would allow other PEF researchers and process developers to make the necessary adjustment from the reported results and their own capabilities when conducting tests.

The final parameter to be assessed in the design of a PEF system is the required pulse frequency. The total energy delivered to a volume of fluid is known from the treatment protocol. As more liquid is processed, the average power (which is energy per unit time) goes up linearly. In conventional pulsed power systems, such as radar transmitters, increasing the average power is typically achieved by increasing the pulsewidth, allowing the pulse frequency to remain at reasonable levels for traditional modulator technologies (i.e., <1 kHz). It will always be more efficient to run at longer pulsewidths, since the rise and fall times of the pulse account for a lower percentage of pulse energy. For PEF systems, however, it is generally not possible to run at pulsewidths longer than approximately 10 μs (at high fields) before arcing occurs. Without the ability to increase the pulsewidth, the only remaining options are to increase the peak power in each pulse (which directly increases the cost of the PEF system), or to operate with more pulses per second (increased pulse frequency). This need for higher pulse frequencies becomes critical as PEF systems are scaled to commercial flow rates, as discussed in the next section.

19.4   PEF SYSTEM OVERVIEW

There are three unique elements to a PEF system, in comparison to a traditional pasteurizer. First, a DC power supply transitions the AC (alternating current) power available from the utility into high-voltage, DC power. The second major element of the PEF system is the pulse modulator, which transforms that DC power into short, high peak power pulses. Finally, there is the treatment chamber, where the high-voltage pulses are applied to the liquid itself. The next section describes each of these subsystems, with general assessments of the alternatives available (Gaudreau et al., 2001).

19.4.1   POWER SUPPLIES

A DC power supply converts the AC power available from the utility into high-voltage, DC power. DC power supplies are typically rated in terms of their average power (in watts). There are three basic DC power supply architectures. The simplest is a transformer rectifier, which operates at line frequency (50–60 Hz). These supplies are the least expensive, but can be very large at high power. They cannot typically provide high levels of voltage regulation, and their output voltage is subject to fluctuations in the input voltage. They are, however, less expensive on a "$ per Watt" basis, especially at high-power levels (hundreds of kWs).

The second basic power supply architecture is a switching power supply (Figure 19.9). In this design, the input power is rectified and "chopped" at high frequency (10–50 kHz) into a transformer rectifier. Since the size of the transformer decreases as the chopping frequency increases, it is possible to build very compact, high-power DC supplies in this way. Switching supplies provide highly regulated and rapidly adjustable output voltage, which supports tight control of the PEF process parameters, independent of the modulator architecture. Switching power supplies are typically used in applications requiring up to ~500 kW, which is sufficient for most anticipated commercial applications. The drawback of switching power supplies is their cost, which can be 2–5 times higher than simple transformer rectifiers of similar power.

FIGURE 19.9   DTI 200 kW switching power supply.

At very-high-power levels (above ~500 kW), the optimal solution may be the combination of a transformer rectifier (for unregulated power) with a high-frequency voltage regulator, to provide the final voltage control. This class of power supplies appears to be most applicable to very-high-throughput PEF processing.

The trade-off between these power supply designs is typically made by assessing the overall cost of the PEF process. A linear supply has lower capital costs, but may require operating above the threshold field strength to ensure complete electroporation as the voltage from the supply varies—that increases the cost of the electricity required. A switching power supply, on the other hand, will be more expensive at the outset, but may recoup these additional costs over time through lower energy use, since the threshold field strength can be maintained very accurately.

19.4.2   MODULATORS

The design and construction of PEF pulse modulators builds on the extensive development of pulse modulators for other applications, including radar and particle accelerators, since World War II. Ideally, the modulator for a PEF system will provide pulses of very consistent voltage, with fast rise and fall times, at the range of pulsewidths and pulse frequencies needed for the desired PEF process parameters. The desirability of a modulator design is typically based on how well it meets these criteria.

The function of pulse modulators is to switch electricity on and off very rapidly, at high voltage and current. This can be done mechanically, although this is slow and difficult to operate at any frequency, and not compatible with PEF processing. It can also be done by creating a controlled short circuit, in a device such as a spark gap, thyratron, or using a solid-state device such as a silicon-controlled rectifier (SCR). This is referred to as a closing switch. Closing switches must be able to remain open, holding off the full input voltage, until they are commanded to close. When they are closed, current will continue to flow through the switch until the input power is dissipated. Typically, closing switches must wait until there is zero current (or a reverse voltage) to open again, and prepare for the next pulse.

Switching can also be accomplished by allowing current to alternately flow and be interrupted. Traditionally, vacuum tubes (such as tetrodes) were used for this purpose, but modern pulse modulators rely nearly exclusively on solid-state devices, primarily insulated-gate bipolar transistors (IGBTs) connected in series/parallel combinations. In these opening and closing switches, current can be turned both on and off at any time, but the switch must be able to withstand the stresses of opening under full current, as well as holding off the full voltage when open.

Along with the two classes of switch (closing, and opening and closing), there are three fundamental modulator designs available—pulse-forming networks (PFNs, Figure 19.10), "hard switches" (Figure 19.11), and transformer-coupled systems (Figure 19.12). All these designs have their origins in the days of vacuum tube-based modulators, but they have transitioned to solid-state switches over the past decade.

The PFN shown in Figure 19.10 holds a predetermined amount of energy in its capacitors, and creates a shaped pulse through the combination of capacitors and inductors in the network. This allows modification of the pulse parameters, such as risetime and pulsewidth, but only by manually changing the values of the capacitors and inductors themselves. For best performance, PFNs and transformer-coupled systems must be impedance matched to the load—meaning that the voltage and current

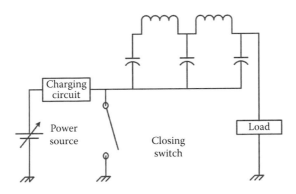

FIGURE 19.10 PFN modulator, typically using a thyratron or thyristors as a closing switch. When the switch is closed, all of stored energy in the PFN will be delivered to the load.

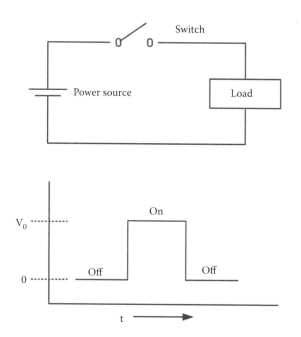

FIGURE 19.11 Hard switch modulator, using IGBTs or FETs as a switch, when the switch closer, voltage is applied to the load unit the switch is opened.

must have a single optimal relationship. This means that the pulse shape will change if the conductivity of the load changes. Finally, after a pulse, the PFN must be completely recharged from a power supply, which can limit the frequency of pulses that can be generated with this design.

The alternative is to use a "hard switch", capable of switching the full voltage on and off directly (Figure 19.11). This has a number of benefits to PEF processing—it allows flexibility in both pulsewidth and pulse frequency. Since hard switches are typically of low impedance, the modulator can provide a consistent, repeatable output voltage over a range of conductivity. Solid-state switches are ideally suited to both of these requirements. In this configuration, the power supply must provide the full voltage delivered by the modulator. The modulator simply stores this energy between pulses, and delivers it in a short time.

The alternative to hard switching at the output voltage is a pulse transformer-coupled design (Figure 19.12). In this design, the switching required to create a pulse occurs at relatively low voltage (typically <20% of the desired output voltage), which can simplify the power supply design. The resulting low-voltage pulse is passed

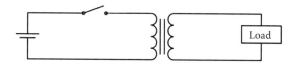

FIGURE 19.12 Hybrid modulator, combining a Lord switch with a pulse transformer.

through a pulse transformer, which increases the pulse voltage by a factor related to the ratio of the primary to secondary turns in the transformer itself. This is a critical simplification when it is difficult or impossible to switch the high voltage directly—to create a 20-kV pulse, it is possible to use only a 2-kV switch and power supply, and a 10:1 turns ratio pulse transformer. The trade-off, however, is that the same energy must be present at both the primary and the secondary of the pulse transformer. This means that to create a 20-kV, 100-A pulse (2 MW peak power), the 2-kV switch must handle 1000 A to provide the same 2 MW of peak power.

Transformer-coupled modulators have traditionally been built using a closing switch and a PFN on the primary of the pulse transformer (the combination of Figures 19.10 19.12). Modern modulators, however, can also use a solid-state switch rather than the PFN. This design is also referred to as a "hybrid modulator," since it combines elements of a hard switch and a PFN/transformer design. The primary switch can consist of a single switch driving a single transformer, or multiple switches operating in parallel. These designs eliminate several drawbacks of the PFN (e.g., fixed pulsewidth, limited frequency), and allow scaling to very high peak power output.

The pulse transformer, however, has some key drawbacks for PEF processing. First, the transformer core must be "reset" between pulses to avoid core saturation (in a monopolar system). Second, the pulse transformer typically wastes approximately 10% of the total power in the PEF system—turning it into unusable heat. The most critical factor, however, is similar to PFN systems—the pulse transformer pulse can only be optimized for a single-field strength and conductivity, and the pulse shape will vary as these parameters change. Even a single product, such as orange juice, can vary considerably in conductivity due to both changes in the raw material, and its temperature changes—even within the duration of a single pulse. Changing the conductivity can significantly change the pulse waveform of a transformer-coupled system—that in turn impacts the actual electroporation effectiveness of the system.

A hard switch, on the other hand, maintains a more consistent pulse shape across a range of conductivities, even though the switch performance (risetime, etc.) may be the same in both cases. This variability limits impedance-matched modulator designs (using pulse transformers and/or PFNs) from consistent performance in a PEF system without additional compensation.

Finally, the issue of peak current from the modulator can be an important limit to practical modulator sizing. There are trade-offs between peak current capability and pulse frequency that must be made in designing the modulator. The design of a specific modulator for a commercial application, therefore, is multifaceted, and the constraints will vary according to the modulator topology and manufacturer. The goal in every case is to reliably achieve the desired PEF protocol on the targeted product(s), at the lowest cost.

Finally, there has been considerable discussion over the last decade about the need for bipolar modulators for PEF, versus the monopolar modulators described above. Bipolar modulators can produce both positive and negative polarity pulses, typically in an alternating waveform. From an equipment standpoint, a bipolar modulator is much more complex and expensive. The biggest difference is that it requires the equivalent of four monopolar solid-state switches, regardless of how they are arranged. Since this switch is typically the most expensive element in a PEF system, requiring four switches can significantly increase the cost of a PEF system. From a process standpoint, there are conflicting results in the studies that have examined mono-versus bipolar PEF performance. No compelling arguments for bipolar pulsing have emerged to date that appear to justify its increased cost for industrial PEF systems. If bipolar pulsing is truly required for the desired PEF process, however, it is certainly feasible. The first commercial-scale PEF system built for OSU in 2000 was a bipolar system (Gaudreau et al., 2001). That is, however, the first and last bipolar system DTI has built to date.

19.4.3 Treatment Chambers

The third major element of a PEF system is the treatment chamber (Figures 19.13 and 19.14) where the high-voltage pulses are applied to the liquid. The key attributes of the treatment chamber are its ability to minimally impact the fluid flow, while ensuring that a consistent electric field is applied to all elements of the flow (Salengke and Zhang, 2002). Unfortunately, these two attributes are often in conflict, making the design of the treatment chamber a critical exercise in system optimization.

It is critical that the treatment chamber design produces a consistent field within the treated volume, and be as immune as possible to arcing (electrical breakdown). The two basic factors affecting treatment chamber design and operation are the physical configuration of the chamber, and the electrode and insulator materials. There are many chamber designs that have been developed and patented over the last 20 years, but the prevailing approach for pumpable fluids is the cofield flow chamber design, developed and patented by OSU (Yin, 1997). This design has been shown to

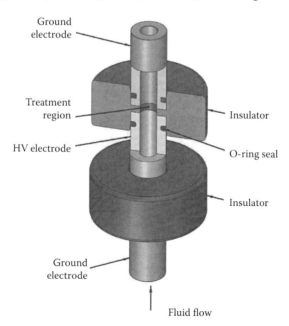

FIGURE 19.13 Co-field flow treatment chamber, developed at Ohio State University.

FIGURE 19.14 Commercial co-field flow treatment chambers built by DTI for liquid processing.

provide the optimal balance between the flow and field requirements. One critical attribute of this design, however, is that to maintain consistent field strengths, the length of the gap must be larger than the diameter by an appreciable amount (20% or more). This places a limit on the practical size of the cofield flow chamber at high fields, since commercial modulators above ~150 kV would not be practical or cost-effective today. 160 kV is required to provide 40 kV/cm across a 4-cm gap; so, the largest practical chamber diameter would be limited to ~3 cm at 40 kV/cm.

For the treatment of bulk foods (such as whole apples or potatoes), alternative treatment chamber designs are required, such as parallel plates surrounding the product being transported in a water bath (Figure 19.15 shows one commercial example from DIL/ELEA) (Mathys, 2012). These are feasible since the required field strengths are generally low for these applications (<2 kV/cm; Toepfl, 2006).

All the PEF treatment chambers have two primary elements: the electrodes, which are conductors that pass the electrical voltage into the liquid being processed,

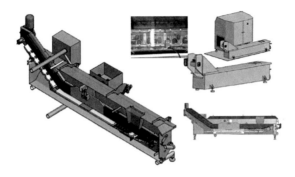

FIGURE 19.15 Commercial PEF treatment conveyor for bulk products, such as potatoes, built by ELEA.

and the insulators between these electrodes, which maintain the "gap" between electrodes, where the voltage field is applied (Figure 19.13).

Insulators are a key aspect of this design. Substantial research has been conducted in the high-voltage world, for radar, power distribution, and other applications, to characterize insulator capabilities and failure mechanisms. Multiple insulators exist from which are food-grade materials, allowing their use in PEF treatment chambers. The key is to utilize a material that can withstand both the electrical stresses within the treatment chamber and the mechanical stresses imposed by rapidly pulsing the electrodes (due to the electromagnetic force applied during each pulse). As PEF systems scale to higher peak and average powers, the design of the insulators has become its own technical challenge, both in terms of materials (e.g., ceramics, crystalline structures such as quartz, and polymers) and their shape and interface with the electrodes. The internal shape of the insulators can significantly impact the uniformity of the electric fields within the treatment chamber—as well as the cost of their manufacture. Figure 19.16 shows field plots for several simple variations of the cofield treatment chamber geometry. Researchers have developed relatively sophisticated designs to optimize the field strength uniformity (Bauman et al., 2011) and reduce areas of high fields ("hot spots," typically at the electrode–insulator junction) where arcing is likely to occur. Translating these optimized designs into cost-effective products, however, is still a challenge.

Electrodes are also an active area of research, but the range of possible electrode materials is being narrowed rapidly. There are three major factors that impact electrode life: erosion, cathodization, and deposition. Erosion relates to physical wear of the electrode as the fluid passes through it. Many foods, especially those that are acidic or highly particulate, will cause electrode erosion even in the absence of high-voltage pulses. This is not a significant problem, or it would be experienced in every liquid-processing system, including pasteurization systems. More critically, the application of high-voltage pulses leads to cathodization of the negative electrode, where electrode ions are transported out of the electrode by the voltage itself. Minimizing this effect, while maintaining food safety, is therefore a major area of research (Li, 2005).

Early PEF electrodes were built from carbon (graphite), with the intent of avoiding any contamination of the treated product by nonfood materials. Unfortunately, these early electrodes had very short lifetimes. The electrical current would erode them

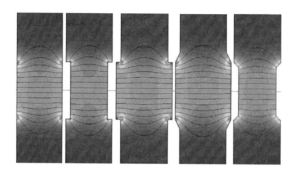

FIGURE 19.16   Field strength models of different co-field flow treatment chambers.

rapidly, and deposits would develop on the electrode surfaces, which acted as electrical insulators. These early studies reported electrode life ranging from 10s to 100s of hours. When an electrode needs to be replaced, the cost incurred is not only the cost of the electrode, but also the cost of the system downtime for the electrode replacement *and* the cost of cleaning the entire process line before operation can be resumed.

Since these early PEF, considerable research has been performed to develop treatment chambers with extended operational lifetimes, supporting extended, continuous operation of the PEF systems in both R&D and commercial applications. The predominant electrode materials, in order of increasing durability, are stainless steel, titanium, and platinum (which is orders of magnitude more durable than stainless steel). The key is that miniscule amounts of the electrode material are introduced into large volumes of fluid, and these very low concentrations are acceptable in every known country, allowing the use of these materials in PEF system (Li, 2005). DTI and others have worked with a variety of materials, including combinations of materials in electrodes, and industrially acceptable electrode life has been repeatedly demonstrated in different applications (Gaudreau et al., 2004).

Wear on the electrode is also caused by arcing and corona discharge within the electrode. This is not a normal product of PEF processing. It occurs when the voltage is not well controlled, or when dissimilar particulates within the fluid cause areas of field stress beyond the breakdown voltage of the liquid. Solid-state PEF systems are specifically designed to identify and respond to arcing very quickly by removing voltage from the chambers, limiting arc energy, and the pitting at the arc site on the electrode (Figure 19.17). This capability alone appears to minimize the effect of arcing on the electrode (as well as the product quality).

The final factor related to treatment chamber design and lifetime is anodization/deposition of material on the positive electrode. This is the major argument for bipolar pulsing: Alternating the polarity of the pulses applied in the treatment chamber may prevent molecules from depositing on the positive electrode (since it becomes the negative electrode in the next pulse). There are two sources of potential deposition on the electrode, the cathode and the liquid itself. Cathode material anodization

FIGURE 19.17   PEF electrode after use, showing erosion of the electrode and deposition of processed material on the electrical.

is typically associated with DC systems (this is called electroplating), but is unlikely to be a major issue in a pulsed system with fluid flow through the electrodes. Proteins and other ionized molecules within the liquid can deposit on the electrodes, however, especially in areas of high field strength. The anecdotal evidence for this is widespread, but is primarily associated with electrodes with very rough surfaces (such as the graphite electrodes used in early PEF systems). Operating with short pulses at higher pulse frequencies minimizes this effect. There are, however, still issues to be addressed with specific high-protein products (such as milk), which seem to create deposits on the electrodes relatively rapidly.

In most PEF systems, the voltage is applied to the fluid across multiple treatment chambers, which are used in (fluid flow) series as the fluid passes through the PEF system (Figure 19.14). The typical range is one to six treatment chambers (providing two to 12 treatment zones). This allows the desired treatment time to be applied at lower pulse frequencies (at higher total current), and helps ensure that every element of the flow receives the desired field strength and duration, for more consistent treatment. It can also provide the opportunity to cool the product between treatment chambers, to keep the temperature within a narrow range.

19.5 PEF SYSTEM TRADE-OFFS AND OPTIMIZATION

As the previous sections have demonstrated, there are many factors that impact the design of a PEF system. There are technical constraints on all levels of the equipment, operational requirements that must be met, and financial criteria that must be achieved to make PEF processing a profitable endeavor. Balancing these competing criteria is the key to the optimization of PEF systems for commercial processes. Our approach is to begin with the fixed parameters facing every PEF system. All the key PEF system design parameters trace back to the following three elements:

- Required process protocol (field strength, treatment time)
- Product characteristics (conductivity, viscosity)
- Desired throughput/flow rate

The critical, and often least-defined, parameters come from the treatment protocol itself. A starting point for these conditions can be typically found in the PEF research literature. Experimentation is required around this starting point to define the optimal process conditions, balancing the needs for effectiveness, product quality, and power utilization. As discussed earlier, if a laboratory or pilot PEF system is used for this experimentation, it is critical that the pulse shapes in the pilot system are as similar as possible to those of the commercial system to allow the results to be readily scaled.

These PEF protocols (field strength and treatment time), combined with the fluid conductivity and desired flow rate, allow calculation of the average power required in the PEF system, which determines the overall system size. The system voltage and treatment chamber size (which are directly related) are the first parameters to be selected. A 35 kV/cm field strength requirement, for example, can be satisfied by a 35 kV pulse power system, with a 1 cm gap distance, or a 17.5-kV system with 0.5-cm gaps, or any other combination that results in the desired field strength. The treatment chamber size, in turn, dictates both the current required in each chamber for each pulse (due to the chamber volume and fluid conductivity), and the chamber diameter (which, with viscosity, determines the pressure drop in each treatment chamber). Finally, the number of chambers, pulsewidth, and pulse frequency can be adjusted to ensure that the fluid receives the desired total treatment time (Gaudreau et al., 2004).

The peak power required from the modulator decreases considerably as the gap distance is reduced, but this increases the speed of the material passing through the chambers, resulting in the need for higher pulse frequencies, as well as operating at higher pressure. The selection of a specific design for a given set of conditions, therefore, may hinge on two major factors: the ability of the modulator to operate at the higher pulse frequency, and the pump pressure required to maintain the fluid velocity through the chambers. Operating at high pulse frequencies increases the switching losses in the modulator proportionately, and can create thermal problems in the switches if not carefully designed and cooled. Pump pressure (calculable from the fluid velocity and viscosity, number of chambers, and chamber diameter) is typically limited for a commercial system, as it affects not just the pumps, but fittings, pipes, and other components, including the treatment chambers themselves.

All these parameters are interactive, and must be simultaneously assessed against the cost and complexity (and often, even the feasibility) of the required pulse modulator. At DTI, we perform these optimizations regularly, and have developed software tools to support these interactive design decisions. Figure 19.18 shows one

	Value
*Flow rate (cm^3/s)	3.0
Process time (min/L)	5.6
*Electric field (kV/cm)	30
*Conductivity (mS/cm)	5
*t dose (us)	50
*t pulse (us)	5
Cell diameter (cm)	0.50
Number of cells	2
Cell length (cm)	0.64
Cell area (cm^2)	0.196
Velocity(cm/s)	15
Transit time (ms)	41.6
Pulses per transit	5.0
Voltage (kV)	19.1
Peak current (A)	59
Average current (mA)	35
Power (W)	675
Frequency (Hz)	120
ΔT per chamber (°C)	54

FIGURE 19.18 Sample PEF processing conditions. * Items are input, while the remainder are calculated, and must be kept within the capabilities of the PEF system itself.

example of the interaction of these parameters for a laboratory PEF system. They are not, however, generalizable across PEF system designs—they are based on DTI's approach to pulse modulator design. Other modulator designs/manufacturers would lead to different optimizations.

Optimization of a commercial PEF system clearly requires a close interaction between the manufacturer, and the capabilities of the plant in terms of pumping, cooling, and so on.

19.6   PEF COSTS

For commercial systems, the relative cost of a PEF system, compared to traditional thermal processing or other nonthermal processes, is primarily made up of two elements: the capital cost for a level of capacity (in \$/liter/hour), and the cost of the electricity needed to provide the PEF treatment itself. The electricity cost is made up of two parts: the protocol (which is typically not subject to change), and the efficiency of the PEF system itself. As described earlier, the closer the pulse shape of the PEF system is to the ideal pulse (zero rise and fall time, perfect stationary flattop), the higher the system efficiency. For a commercial system, operating nearly continuously, improvements in system efficiency (even at the expense of higher initial system costs) are required to yield the lowest overall processing cost over the lifetime of the plant.

PEF system capacity directly affects the capital cost (on a per-liter basis). Moving to higher capacity means increasing the peak and average power in the system, resulting in higher PEF hardware costs. These costs do not increase linearly with the added capacity, however. The cost per liter of capacity goes down in larger systems, making them much more economical than pilot systems on that basis. A (very) general rule from assessing the size and cost of commercial PEF systems is that the cost increases by the square root of the capacity increase. A 10× higher capacity system will be approximately 3–3.5× the capital cost. This relationship holds for the range of pilot and commercial systems possible (~20 kW–1 MW). Low-power, lab-scale systems are dominated by the minimum infrastructure required for a PEF system, such as controls and basic functionality; so, this rule does not apply as power decreases below ~20 kW.

Beyond a certain power and voltage, the sheer scale of the equipment (in terms of voltage, peak power, and average power) exceeds commercial pulse power system capabilities; so, cost increases faster than the additional capacity, due to the engineering required. When that size is reached, it makes more sense to increase plant capacity by replicating the largest feasible system, rather than designing and building ever-larger systems. DTI's current assessment is that this crossover point is reached around 150 kV and 1 MW of average power, but these parameters are increasing over time, as pulsed power systems for other applications are developed.

There have been several recent studies comparing the cost of PEF to thermal pasteurization and alternative nonthermal processes for liquid foods. In general, these have reported that the cost of PEF treatment is at 1.5–3× the cost of thermal pasteurization, with significant variations in the assumptions used in these calculations (Sampedro et al., 2013). The major contributors to this difference are the equipment costs, and the fact that PEF systems require electricity rather than steam for their input energy. The input energy costs are not likely to narrow over time, but more efficient PEF system designs and processes can minimize the amount of energy required, just as pasteurization systems have become more efficient over time. PEF is, at the same time, significantly cheaper than UV (ultraviolet) light or high-pressure processing. In the future, we expect that the cost of PEF systems will continue to decline, as more companies introduce commercial PEF systems, and the overall costs of pulse power systems and components decrease. PEF may never be less expensive than thermal pasteurization, but it will closely approach cost competitiveness in the future.

19.7   PEF PROCESSING AND COMMERCIALIZATION STATUS

PEF processing has been the subject of focused R&D for over 20 years. In its earliest days, the emphasis was on validating and defining the ability of high-voltage pulses to kill bacteria, molds, and other potential pathogens. In the early 2000s, programs in both the United States and Europe focused on scaling the results discovered in the laboratory to commercially viable processes and systems (Buckow, 2013). Since that time, several key developments have brought PEF processing to the brink of wide-scale commercial adoption. Genesis Juice Corporation introduced the first FDA (Food and Drug Administration)-approved, PEF-processed juices to the U.S. market in August, 2005 (Figure 19.2). Unfortunately, the financial impact of no sales for over 18 months while they instituted PEF processing, combined with the capital requirements of rapid growth, proved financially fatal to Genesis. The company was forced to cease operation before selling its brand name in mid-2007. At the time, however, market acceptance of PEF-processed juice was strong and growing. At approximately the same time, DTI introduced a standard pilot-scale PEF system, capable of treating 100–500 L/h of juices or other products, allowing researchers to process a variety of products for both R&D and preproduction process definition (Gaudreau, 2006). As a result, Genesis and DTI were awarded the Institute of Food Technologists' Food Technology Industrial Achievement Award in 2007 (Zhang, 2008).

The world's first known commercial deployment of large-scale PEF technology occurred in 2006 in Mesa, Arizona (Figure 19.19). This system, rated at 10,000 L/h capacity, was deployed for wastewater treatment, rather than food processing. In over 6 years of operation to date, this system has demonstrated the ability to meet stringent operational requirements for efficiency, reliability, electrode life, and unattended operation (Gaudreau, 2006). Since that time, DTI has developed a lower-cost, industrialized version of that system, shown in Figure 19.20. This system has the same basic performance as the system deployed in 2006, at approximately one-third of its cost. As of 2013, over one dozen of these systems have been deployed, representing, to our knowledge, the largest commercial PEF success to date (Kempkes, 2013).

In parallel, a number of lab- and pilot-scale systems have been deployed to research facilities around the world, including the United States, Europe, and Australia. The availability of these systems has increased the amount of PEF process development, in addition to research on the basic PEF effectiveness. Several companies have

FIGURE 19.19   Commercial PEF system from DTI, capable of delivering up to 300 kW at 40 kV (200b).

entered the PEF systems market, either from the pulsed power industry or from the food industry. Five companies (DTI, ELEA (Figures 19.15 and 19.21), PurePulse, Scandinova, and Steribeam) are known to be producing commercial PEF systems at this time, of varying sizes, and focused on a range of applications (Kempkes, 2013). Most of these commercial systems are being introduced for nonpasteurization

FIGURE 19.20   Industrial PEF system developed by DTI from the commercial system in Figure 19.19 (2012), but at 1/3 of the cost.

FIGURE 19.21   Pilot PEF system from DIL/Elea, delivering up to 25 kW.

applications (extraction, etc.), but there are several food products reported to have been introduced in Europe and Asia (Buckow, 2013). As these commercial systems are deployed, manufacturers are also simplifying their setup and operation, through the use of programmable logic controller (PLC) control systems and monitoring (Figure 19.22), which greatly reduce the expertise required by the operator.

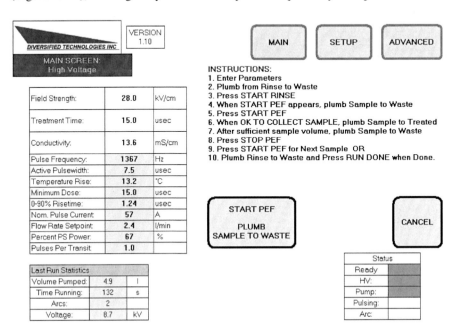

FIGURE 19.22   Sample PLC screen for operation of the DTI Lab PEF system shown in Figure 19.6.

19.8   SUMMARY/CONCLUSIONS

Multiple researchers have shown PEF processing to be equivalent to pasteurization in terms of pathogen reduction for a wide range of liquid foods. For foods that are heat sensitive, there are considerable benefits in taste, color, and nutritional value from the nonthermal PEF process. Despite some early hiccups, nonthermal pasteurization using PEF is beginning to emerge (again) as a commercial process after all these years in the laboratory. There have been several reasons given for this delay in adoption, including costs and regulatory issues. With the reintroduction of PEF-treated juices into the market in the past 2 years, it appears that these issues have been addressed (Buckow, 2013). These PEF-treated products into the market are likely to attract others into this field, as the perceived risk of PEF-treated food products diminishes, accelerating this growth.

The application of PEF to other industrial processes directly builds on the research in food processing, and new applications of PEF are emerging at a significant pace. Since the early research on PEF as a nonthermal pasteurization process, the applications of PEF have extended into a range of other processes dependent on electroporation of cells, including drying, extraction, and preprocessing of both foods and other streams (e.g., algae and wastewater). Commercial markets for PEF systems are now growing at a rapid pace, for the first time.

There are now several dozen commercial PEF systems in operation around the world. R&D into PEF processing continues at a number of institutions around the world (Sperber, 2011), with a similar number of R&D PEF systems deployed in research organizations. Commercial PEF systems are now readily available to food processors and industrial users, from a number of suppliers, for the full range of PEF applications.

Finally, the emergence of a diverse set of companies capable of building PEF systems, and assisting processors in their selection and operation, points to a strong and vibrant outlook for PEF processing in the future. The threshold between R&D and commercialization of PEFs has been crossed.

REFERENCES

Barbosa-Canovas, G., Gongora-Nieto, M., Pothakamury, U., and Swanson, B. 1999. *Preservation of Foods with Pulsed Electric Fields.* Elsevier Inc., New York, NY. ISBN: 978-0-12-078149-2.

Barbosa-Canovas, G. and Gould, G. 2000. *Innovations in Food Processing.* CRC Press LLC., Boca Raton, Florida. ISBN: 1-56676-782-2.

Barbosa-Canovas, G. and Zhang, Q. 2001. *Pulsed Electric Fields in Food Processing.* Technomic.

Barbosa-Canovas, G. and Zhang, Q. 2001. *Pulsed Electric Fields in Food Processing*: Fundamental Aspects and Applications. Technomic Publication Company, Inc., Lancaster, Pennsylvania. ISBN: 1-56676-783-0.

Barbosa-Canovas, G., Tapia, M., and Pilar Cano, M. 2004. *Novel Food Processing Technologies.* C.H.I.P.S. Publishing, Weimar, Texas, 720 p.

Baumann, P., Buckow R., Knoerzer, K., and Schroeder, S. 2011. Effect of dimensions and geometry of co-field and co-linear pulsed electric field treatment chambers on electric field strength and energy utilization. *Journal of Food Engineering*, 105: 64–70.

Buckow, R., Ng, S., and Toepfl, S. 2013. Pulsed electric field processing of orange juice: A review on microbial, enzymatic, nutritional, and sensory quality and stability. *Comprehensive Reviews in Food Science and Food Safety*, 12, 455–467.

Casey, J., Gaudreau, M., Hawkey, T., Kempkes, M., and Roth, I. 2002. *Solid-State Modulators for Commercial Pulsed Power Systems. Proceedings of the Pulsed Power Conference*, Las Vegas, NV.

Clark, J. 2006. Pulsed electric field processing. *Food Technology*, 60: 66–67.

Dunn, J.E. and Pearlman, J.S. 1987. *Methods and Apparatus for Extending the Shelf Life of Fluid Food Products*. U.S. Patent 4,695,472.

Frey, W., Eing, C., Goettel, M., Gusbeth, C., Posten, C., and Straessner, R. 2012. *Application of Pulsed Electric Field Treatment for Microalgae Processing. Proceedings of the Euro-Asian Pulse Power Conference*, Karlsruhe, Germany.

Gaudreau, M., Hawkey, T., Kempkes, M., and Petry, J. 2004. *Design Considerations for Pulsed Electric Field Processing. Proceedings of the European Pulsed Power Conference*, Hamburg, Germany.

Gaudreau, M., Hawkey, T., Kempkes, M., and Petry, J. 2006. *Scaleup of PEF systems for food and Waste Streams. Proceedings of the 3rd Innovative Food Centre Conference*, Bedford, MA.

Gaudreau, M., Hawkey, T., Kempkes, M., and Petry, J. 2008. *PEF Systems for Food and Waste Streams. Proceedings of the Power Modulator Conference*, Las Vegas, NM.

Gaudreau, M., Hawkey, T., Kempkes, M., Petry, J., and Zhang, Q.H. 2001. *A Solid-State Pulsed Power System for Food Processing. Proceedings of the International Food Technology Conference*, New Orleans, LA.

Jeyamkondan, S., Jayas, D.S., and Holley, R.A. 1999. Pulsed electric field processing of foods. *Journal of Food Protection*, 62:1088–1096.

Kempkes, M. 2013. *PEF Systems for Industrial Applications. Proceedings of the Institute of Food Technologists*, Chicago, IL.

Kempkes, M., Reinhardt, N., and Roth, I. 2012. *Enhancing Industrial Processes by Pulsed Electric Fields. Proceedings of the Bio and Food Electrotechnologies Conference*, Salerno, Italy.

Li, S.Q. and Zhang, Q.H. 2005. *Electrode Erosion under High Intensity Pulsed Electric Fields. Proceedings of the International Food Technology Conference*, New Orleans, LA.

Mathys, A. 2012. *Alternative Processing Technologies for Natural and Fresh Food Products. European Symposium on Food Safety*, Warsaw, Poland.

Min, S., Jin, T., and Zhang, Q. 2003. Commercial scale pulsed electric field processing of tomato juice. *Journal of Agricultural and Food Chemistry*, 51: 3338–3344.

Raso J., Heinz V. 2006. Pulsed Electric Fields Technology for the Food Industry C.H.I.P.S. Publishing, Weimar, Texas, p. 245.

Salengke, S.K. and Zhang, Q.H. 2002. *Modeling of Pulsed Electric Field (PEF) Processes. Proceedings of the International Food Technology Conference*, Anaheim, CA.

Sampedro, F., McAloon, A., Yee, W., Fan, X., Zhang, Q.H., and Geveke, D.J. 2013. Cost analysis of commercial pasteurization of orange juice by pulsed electric fields. *Innovative Food Science and Emerging Technologies*, 17: 72–78.

Sperber, B. 2011. Making pasteurization cool. *Plant Operations Magazine*, May 2011.

Toepfl, S. 2006. *Pulsed Electric Field (PEF) for Permeabilization of Cell Membranes in Food and Bioprocessing—Applications, Process and Equipment Design and Cost Analysis*. PhD thesis. Technical University of Berlin.

FDA, Center for Food Safety and Applied Nutrition. 2000. *Kinetics of Microbial Inactivation for Alternative Food Processing Technologies. Pulsed Electric Fields*, College Park, MD.

Yeom, H.W., Streaker, C.B., Zhang, Q.H., and Min, D.B. 2000. Effects of pulsed electric fields on the activities of microorganisms and pectin methyl esterase in orange juice. *Journal of Food Science*, 65(8):1359–1363.

Yin, Y., Zhang, Q.H., and Sastry, S.K. 1997. *High Voltage Pulsed Electric Field Treatment Chambers for the Preservation of Liquid food Products*. U.S. Patent 5,690,978.

Zhang, Q.H. 2008. *A Pulsed Electric Field Case Study*. Presented at the *Non-Thermal Processing Symposium*, Madrid, Spain.

Index

Printed and bound by CPI Group (UK) Ltd, Croydon, CR0 4YY

21/10/2024

01777112-0009